Interdisciplinary Applied Mathematics

Volume 3

Interdisciplinary Applied Mathematics

Daniel D. Joseph   Yuriko Y. Renardy

Fundamentals of Two-Fluid Dynamics

Part I: Mathematical Theory and Applications

With 205 illustrations, 69 in color

Springer Science+Business Media, LLC

Daniel D. Joseph
Department of Aerospace
    Engineering and Mechanics
University of Minnesota
Minneapolis, MN 55455 USA

Yuriko Y. Renardy
Department of Mathematics
Virginia Polytechnic Institute
    and State University
Blacksburg, VA 24061 USA

Editors

F. John
Courant Institute of
    Mathematical Sciences
New York University
New York, NY 10012
USA

L. Kadanoff
Department of Physics
James Franck Institute
University of Chicago
Chicago, IL 60637
USA

J.E. Marsden
Department of
    Mathematics
University of California
Berkeley, CA 94720
USA

L. Sirovich
Division of
    Applied Mathematics
Brown University
Providence, RI 02912
USA

S. Wiggins
Applied Mechanics Department
Mail Code 104-44
California Institute of Technology
Pasadena, CA 91125
USA

Cover illustration: Cylinder coated with 12500 cs silicone oil cusps in air. (See color plate II.6.5 (a-b).)

Mathematics Subject Classifications (1991): 76E05, 76E15, 76E30, 35B20

Library of Congress Cataloging-in-Publication Data
Joseph, Daniel D.
    Fundamentals of two-fluid dynamics / Daniel D. Joseph, Yuriko Y.
Renardy.
        p.   cm. — (Interdisciplinary applied mathematics : v. 3/4)
    Includes bibliographical references and index.
    Contents: part 1. Mathematical theory and applications — part
2. Lubricated transport, drops and miscible liquids.
    ISBN 978-1-4613-9295-8

    1. Fluid dynamics.   I. Renardy, Yuriko Y.   II. Title.
III. Series.
QC151.J67   1992
620.1'064 — dc20   92-34044

Printed on acid-free paper.
© 1993 Springer Science+Business Media New York
Originally published by Springer-Verlag New York, Inc. in 1993
Softcover reprint of the hardcover 1st edition 1993

Production managed by Henry Krell; manufacturing supervised by Vincent Scelta.
Photocomposed copy prepared from the authors' TeX file.

9 8 7 6 5 4 3 2 1

ISBN 978-1-4613-9295-8   ISBN 978-1-4613-9293-4 (eBook)
DOI 10.1007/978-1-4613-9293-4

Preface

Two-fluid dynamics is a challenging subject rich in physics and practical applications. Many of the most interesting problems are tied to the loss of stability which is realized in preferential positioning and shaping of the interface, so that interfacial stability is a major player in this drama. Typically, solutions of equations governing the dynamics of two fluids are not uniquely determined by the boundary data and different configurations of flow are compatible with the same data. This is one reason why stability studies are important; we need to know which of the possible solutions are stable to predict what might be observed. When we started our studies in the early 1980's, it was not at all evident that stability theory could actually work in the hostile environment of pervasive nonuniqueness. We were pleasantly surprised, even astounded, by the extent to which it does work. There are many simple solutions, called basic flows, which are never stable, but we may always compute growth rates and determine the wavelength and frequency of the unstable mode which grows the fastest. This procedure appears to work well even in deeply nonlinear regimes where linear theory is not strictly valid, just as Lord Rayleigh showed long ago in his calculation of the size of drops resulting from capillary-induced pinch-off of an inviscid jet. In two-fluid problems, there are many sources for instability and the active ones may be determined to a degree by analysis of different terms which arise in the energy budget of the most dangerous disturbance.

Though we have presented many results from nonlinear analysis of two-fluid problems, this side of the subject is not yet well-developed. We are also certain that the direct simulations of two-fluid problems which have commenced only in the years just passed have a potentially huge domain for increased understanding.

Applications of two-fluid dynamics range from manufacturing to lubricated transport. Different mechanisms which are unique to the flow of two fluids can be exploited for this purpose. Density-matching can be used to depress the effect of gravity, or of centripetal acceleration in rotating systems, allowing one to manipulate the places occupied and the shapes of the interfaces between fluids. Viscosity segregation can be used to promote mixing and demixing, to promote say the displacement of one fluid by another as in the problem of oil recovery, or to segregate one molten plastic from another by encapsulation. The lubrication of one fluid by another is a particularly important branch of two-fluid dynamics; if one fluid has a surpassingly large viscosity, it may be lubricated by a less viscous fluid. The most beautiful of the lubricated flows are the rollers discussed in chapter II of the first part, *Mathematical Theory and Applications*. The most useful of the lubricated flows are the water-lubricated pipelines, which are discussed in chapters V through VIII of the second part, *Lubricated Transport, Drops and Miscible Liquids*. It is our hope that this book will lead to deeper understanding of the principles and applications of two-fluid dynamics.

The topics treated in this book are displayed in the table of contents. It has been divided into two self-contained parts. Most of the techniques of analysis used in the study of two-fluid problems are developed in the first part, *Mathematical Theory and Applications*. The analyses are fully worked there, with enough details to teach students. Four of the six chapters of the second part, *Lubricated Transport, Drops and Miscible Liquids*, treat problems which arise in the study of water-lubricated pipelining. We have done a much more complete comparison of theory and observations than was previously available with the serious intent of advancing this energy-efficient technology. Chapter IX is about immiscible vortex rings and is a report and explanation of experimental results. Chapter X develops a new theory of binary mixtures of miscible incompressible liquids in which the density changes and the stresses induced by diffusion are considered. As always, it is certain that a number of excellent studies of two-fluid dynamics which deserve mention have not been mentioned.

Our research for this project could not have been done without the help of certain persons: Mike Arney, Runyan Bai, Nick Baumann, Gordon Beavers, Kangping Chen, Howard Hu, Paul Mohr, John Nelson, Ky Nguyen, Luigi Preziosi and Michael Renardy. We are especially indebted to Chen, Hu and Preziosi for their excellent analytical and numerical studies of lubricated pipelining and to Bai for the design and execution of very elegant experiments. We thank Michael Renardy for reading through the manuscript.

The work of Joseph was supported mainly by the Department of Energy, Office of Basic Energy Sciences and also by the fluid mechanics branch of the National Science Foundation, the mathematics division of the Army Research Office, by the Army High Performance Computing Center, and the Minnesota Supercomputer Institute. Joseph's research on water-lubricated pipelining was funded initially under a special small NSF grant for innovative research involving the lubricated transport of coal-oil dispersions. Joseph is grateful to Steve Traugott for this initial grant which was later picked up by Oscar Manley at the DOE.

Renardy's research was funded by the National Science Foundation under Grant No. DMS-8902166. This project was begun during the Winter Quarter of 1989 at the Institute for Mathematics and Its Applications at the University of Minnesota.

Yuriko dedicates this book with love to her father Sadayuki Yamamuro ("Papa, arigato"), and to her mother Akiko ("osewani narimashita"). Dan dedicates this book to Adam, Bai, Chris, Claude, Dave, Geraldo, Harry, Howard, John, Kangping, Luigi, Mike, Paul, Pushpendra and Terrence.

February 1992   Minneapolis, Minnesota
                                                 Blacksburg, Virginia

Contents

Contents of *Part II: Lubricated Transport, Drops and Miscible Liquids*

Chapter I
Introduction

Bicomponent flows are coupled flows of a fluid and another constituent, either a solid or a fluid. An example is the flow of two immiscible fluids. A second example is a liquid and small solid particles such as fluidized beds and suspensions. An important example, not treated here, is the flow of liquid and a gas, called two-phase flow. We will concentrate on the flow of two immiscible liquids.

Many configurations are possible for the flow of two immiscible fluids: layers, slugs, rollers, sheets, bubbles, drops and dynamic emulsions and foams. These structures are often topologically different from the initial rest configurations from which they arise: the places which are occupied by each of the fluids often have no continuous correspondence to the places which are occupied by them at rest. We will see that there is a strong tendency for two fluids to arrange themselves so that the low viscosity constituent is placed in the region of high shear. This gives rise to a kind of gift of nature, in which the lubricated flows are stable, and it opens up very interesting possibilities for technological applications, in which one fluid is used to lubricate another.

In order to control and use these flows, we must determine where the two fluids will be. In terms of physical reasoning, the configurations that are ultimately achieved in practice are controlled by the problem of placements and the problem of shapes. In the problem of placements, we must describe where the constituents are to go, and the evolution processes involved in the transport of the two fluids to the places they ultimately occupy. This problem is concerned with the fingering, breakup and fracture of the liquids; processes which are not included in the usual statements, such as the Navier-Stokes equations, governing the flow dynamics. An attempt at solving the problem of placements is the idea of the ad-hoc "viscous dissipation principle" described in chapter II. In the problem of shapes, we must describe the geometric form of the interfaces between flowing fluids, once their relative positions have been established. A part of this problem is concerned with the study of stability and bifurcation of various interface shapes, and the numerical simulation of initial-value problems.

I.1 Examples

The following are several applications associated with the flow of two immiscible liquids, as well as related applications of liquid-solid systems. The goal here is to describe some of the problems of nascent technological interest that need to be investigated for flows of two liquids.

I.1(a) Fingering

Oil Recovery. One well-known problem of two fluids arises in the problem of oil recovery. The idea is to push out the oil which is trapped in the ground by flooding with water. One difficulty with this technology is that the water fingers through the oil, so that more water than oil is recovered. Saffman and Taylor [1958] showed, using Darcy's law, that a flat interface of water pushing oil would be unstable to short waves, but a flat interface of oil pushing water is perfectly stable.

The technological problem for oil recovery is to find methods to frustrate the undesirable effects of fingering instability. Different techniques like the addition of polymers and surfactants have been tried with partial success.

Air Entrainment. The tendency of thin fluids to finger through thick fluid is a general one, not restricted to flow through porous media. There are many situations in which air entrainment is a problem. This kind of entrainment can occur, for example, under certain conditions in the lubrication of bodies in contact; consider a shallow bath of oil and a cylinder dipped partly in it. As the cylinder is rotated, the oil adheres to it. This film of oil becomes thinner as it is drawn up and around. When the cylinder has completed a rotation, the position where the oil initially adhered to it becomes immersed in the bath again, but under certain conditions, it emerges dragging much less oil than at the first rotation. This may be due to the presence of a thin air film which covers the original thin layer of oil. If the original thin layer of oil can be destroyed before re-entering the bath, a very thick oil film is formed on emergence from the bath.

Formation of Emulsions. Fingering instability of two fluids leads to emulsification. The emulsification of two flowing liquids under shear was studied by Joseph, Nguyen and Beavers [1984]. They created fingers by shearing, in a manner related to the fingering in a Hele-Shaw cell, but with moving boundaries. Emulsification under shear in two fluids flowing down pipes is described by Russell, Hodgson and Govier [1959] and by Charles, Govier and Hodgson [1961]. Joseph, Nguyen and Beavers [1984] noted that the fingers undergo capillary instabilities in which the tips of the low viscosity finger are pinched off. These tips form drops of low viscosity liquid, leading to emulsions of low viscosity liquid in a continuous high viscosity phase. The dynamic emulsions separate when the shearing stops.

Fingering instabilities are not at all well understood. They are of fundamental importance in determining the configuration of two fluids in motion.

I.1(b) Lubricated Pipelining

There are significant reserves of heavy viscous crude oils in the United States, Canada, Venezuela and Europe which are increasingly important. Heavy crudes may have viscosities of 1000 poise at room temperature. These viscous crudes cannot be transported by the usual pipeline methods: the resistance to the flow arises mainly from friction at the pipe wall. It is customary to reduce the viscosity of the oil either through the addition of a hydrocarbon diluent or through the installation of heating equipment at short intervals along the pipeline. The former method can be used in only the unusual case in which there is an abundant supply of light oil in the same region as the heavy oil; heating is inconvenient and costly. However, if one could replace the viscous fluid just along the wall by a much less viscous immiscible one, such as water, then the work required to transport the oil would be significantly lowered. For horizontal pipelining,

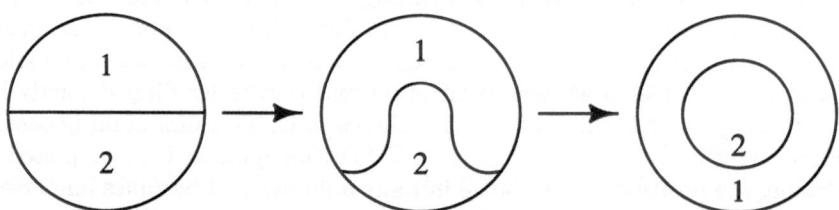

Fig. 1.1. [Renardy, 1988, Rocky Mountain Mathematics Consortium] Evolution of flat interface into the encapsulated arrangement. Fluid 2 is the more viscous.

such an arrangement is plausible if the densities of the fluids are similar. In fact, experiments on the addition of a small amount of a less viscous fluid to the pipeline transport of a viscous fluid showed that the less viscous fluid has a strong tendency to migrate to the wall, exactly where it is needed! (see figure 1.1) [Charles and Redberger 1962; Gemmell and Epstein 1962; Russell and Charles 1959; Russell, Hodgson and Govier 1959; Charles, Govier and Hodgson 1961]. Moreover, by using additives in the less viscous liquid, the pipe wall could be protected from corroding. The measured pressure drops indicated that the addition of water can greatly reduce the pressure gradient. Oliemans and Ooms [1988] have written a comprehensive review of pipe flows of oil and water, theory and experiment, prior to 1984. The related pressure-driven flow of oil and water in a rectangular channel has been studied experimentally by Charles and Lilleleht [1965] and Kao and Park [1972].

Various arrangements of the oil and water occur in the aforementioned experiments. This type of nonuniqueness is typical of flows of immiscible liquids. Some of the arrangements which appear in horizontal pipes are:

1.   stratified flow with heavy fluid below,
2.   concentric oil in water (core-annular) flow,
3.   water drops in oil, and
4.   oil drops in water; these include large drops and slugs of oil, lubricated by water.

It was found that the core-annular flow had the greatest volume flux for a given pressure drop, among all of the realized arrangements. The pressure drop over the pipe could be even smaller than the pressure drop in water alone at the same value of the volume flux. The lubricating water layer was 10 % to 40 % of the pipe diameter. The power requirements for moving the small amount of water in core-annular flow is negligible.

Many of the properties of the observed flows are not understood or are understood only vaguely from a theoretical point of view. The following three items are examples of such properties. First, the interface between the oil and water in core-annular flow is rippled, whereas the unrippled cylindrical interface ought to be stable in some regimes, according to linearized stability theory. Secondly, there is a minimum speed, observed in experiments, below which the core-annular flow is unstable and gives way to oil slugs and drops in water. This effect of surface tension needs investigation. Thirdly, there is a maximum speed, observed in experiments, above which core-annular flow is replaced by emulsion of water in oil. These water drops may arise from fingering, followed by a capillary instability, and the water drops may then congregate at the pipe walls, lubricating the flow. This needs deeper study.

The oil and water used in the experiments of Charles, Govier and Hodgson [1961] were matched for density by adding the appropriate amount of carbon tetrachloride to the oil. Density matching would be inconvenient and expensive in the actual applications of pipeline technology. In the applications for flow in horizontal pipes, it is necessary to deal with effects of gravity, which causes the oil-core to be eccentric (see Oliemans and Ooms [1988] for a photograph and discussion of the effects of gravity). Fortunately, the oil core need not touch the upper wall. A lubricating layer can persist. The exact hydrodynamics which maintains the lubrication layer at the top of the pipe is not understood. A lubricating film model for core-annular flow has been developed [Oliemans 1986; Ooms, Segal, Cheung and Oliemans 1985]: this agrees with experiments in some details and not in others.

From the point of view of applications, it is important that core-annular flow can be established from a stratified fluid at rest, with oil floating on water. The technology which is required to develop core-annular flow from stratified flows uses specially designed nozzles as well as the addition of surfactants to increase the attractivity of water to the pipe walls [Oliemans and Ooms 1988]. It is of interest to map out the parameter ranges where complete encapsulation of oil by water is possible. Yu and Sparrow [1969]'s observations of interface distortion increasing with flow rate suggest that in low-viscosity Newtonian liquids, forces other than interfacial tension play a key role. It would be of interest to understand the factors which promote the rate of encapsulation of high by low viscosity liquids.

I.1(c) Segregation and Lubrication of Solids in Liquids

We could imagine that nature's gift which places the lubricating fluid in the regions of high shear where it will reduce drag [Merkle and Deutsch 1990] could lead to segregation of different constituents in a composite fluid or mixture. This type of flow differentiation was achieved in experiments of Bhattacharji [1967] modeling different types of magma flows which occur in nature. Highly compacted central zones of crystals together with crystal-free lubrication layers at the walls developed. The zoning of magmas by viscosity in volcanic conduits was studied by Carrigan and Eichelberger [1990]. Molecular dynamic simulations [Rothman and Keller 1988; Stockman, Stockman and Carrigan 1990; Joseph 1990c] modeling viscous segregation in immiscible fluids via lattice gas automata do lead to such segregations in which the low viscosity constituent migrates into the region of high shear. The method of cellular automata is not, and is not meant to be, an exact description of the flow's dynamics: it models the Navier-Stokes equation only in an average sense. However, it is encouraging that several variants of this method lead to identical results and that they all give rise to lubrication.

Another manifestation of the segregation of mixtures to promote lubrication occurs in fluid-solid mixtures. In these mixtures, there are regimes of flow in which the solids migrate into the regions of low shear where they compact, forming a composite fluid of huge viscosity. Solids-poor regions of relatively pure liquid are left in the high-shear regions near the walls, lubricating the flow. Lubricated flows of particulates described below are very similar to lubricated flows of two liquids, one of which is much more viscous. Lubricated flows of particulates in pipes strongly resemble the water-lubricated flows of viscous oil cores down pipes. The basic problem we need to solve in the case of lubricated flows of particulates is to determine the mechanism, the hydrodynamic glue which forces the particles to cluster. The underlying cause may be related to the Segré-Silberberg effect [Segré and Silberberg 1962; Russell and Charles 1959]; the equilibrium position of a small solid sphere in a Poiseuille flow in a round pipe is at a radius of approximately 0.6 of the pipe radius. We can consider the possibility that a suspension of neutrally buoyant solid particles could be modeled by an effective continuum of two interpenetrating fluids. In such a model, the fraction of solids per unit volume would depend on the flow. Drew [1986] has shown that a model of this type leads to a lubrication type of solution with clear water on the wall of the pipe.

Lubricated regimes of neutrally buoyant suspensions of small spheres, disks or rods in Poiseuille flow were observed in the experiments of Karnis, Goldsmith and Mason [1966]. The working fluid was a Newtonian oil of about 25P. For small spheres in low concentrations, the velocity profile of the mixture was parabolic. For larger values of the concentration C and the radius ratio b/R_0 of particle to pipe radius, a partial plug-flow developed, with particles depleted in the annulus $r_c < r < R_0$ and concentrated where $r < r_c$. The particles in the annulus rotate and the particles in the core do not. As b/R_0 was increased in the most concentrated suspensions, the apparent viscosity decreased. This is a lubrication effect due to the formation of a particle-depleted water annulus near the wall. The partial plug-flow developed at about a value of $Cb/R_0 = 5$. Partial plug-flow also developed in concentrated suspensions of rods and disks.

Recent experiments using laser-induced photochemical anemometry [Falco, Klewicki and Nocera 1990] on concentrated mixtures of silica gel in water show that '...even at concentrations as high as 52%, a film of liquid builds up around the walls of the pipe. The film thickness varies considerably but can be instantaneously as large as a tenth of the pipe radius. The wavy core which is of even higher solid-loading than the initial mixture appears to snake down the pipe'. This wavy core flow looks very much like the corkscrew waves shown in the photographs of chapters VII and VIII.

Results similar to those reported by Karnis *et al.* and Falco *et al.* have been established by an elegant use of nuclear magnetic resonance (NMR) imaging by Altobelli, Givler and Fukushima [1990] who studied steady flow in a horizontal tube of heavy particles in SAE-80 gear oil. For slow flow, the suspensions are vertically stratified; faster flows undergo a dilation not unlike that described by Leighton and Acrivos [1986]. For increasing flow rates, the dilated particle bed gives way to the formation of a plume-like structure. At high velocities, the particles form a highly concentrated core in the tube center and the velocity profile flattens as shown in plate I.1.2.

Graham, Altobelli, Fukushima, Mondy and Stephens [1991] did NMR imaging of suspensions of spheres undergoing flow between rotating cylinders. They find that particles migrate from the higher shear rate regions near the rotating inner cylinder to the lower shear rate regions near the stationary outer wall, establishing large concentration gradients after only a short time. They studied unimodal and bimodal suspensions and found that the bimodal suspension forms concentric cylindrical sheets parallel to the axis of the Couette device, which rotate relative to each other. Radial (but no significant axial) migration of particles was observed. They say that 'This particle migration and structure formation is believed responsible for torque reductions and other anomalous behavior witnessed during rheological testing of concentrated suspensions reported by Leighton and Acrivos [1986]'. Their observation about the torque reduction could be viewed in the light of the computation given in section I.3 (f) in which it is shown that the configuration of two liquids which minimize the torque in the flow between cylinders which rotate at prescribed angular velocity with gravity and axial variation neglected, is the one in which the low-viscosity constituent migrates to the inner wall where the shear is greatest.

I.1(d) Lubricated Pipelining of Solid Particulates

The tendency for solids in solid-liquid mixtures to segregate into solid-rich core regions surrounded by solid-poor liquid regions can be used to reduce the cost of transporting particulates in pipes (see, for example, Kennedy [1966]. The core flow of suspendable solids has already been discussed in I.1 (c). The lubricated pipe flow of suspended solids has been put into commercial practice for the transport of coal and gelsonite [Dauber 1957; Bond 1957], and would find a rich application in the pipeline transport of fluidized solids.

Several major limitations exist for this mode of pipelining: energy is required to maintain the solids in suspension, and this becomes particularly important when the specific gravity of the solids is high; there are some solid commodities which would be damaged through direct contact with the supporting fluid; and further, considerable difficulty may be associated with the separation of the suspended solids from the supporting fluid at the point of delivery.

Various strategies may be employed to improve the performance of lubricated transport of particulates. Obviously, it is necessary to character- ize the conditions, volume fractions, flow rates, density difference, particle shapes, and so on, which promote lubrication. In difficult cases, it may be possible to enhance lubrication by the addition of a third carrier phase to the system. One strategy is to add fibers to the suspension. Another is to use a second carrier fluid as in the case of water-lubricated coal-oil mixtures.

There is a well-documented regime of lubricated flow in the pipelining of pulp: papermaking fibers in water (see, for example, Norman, Moller, Ek and Duffy [1977]). Moller and Duffy [1978] describe three regimes of flow of such suspensions as follows:

'At low flow rates, the suspension flows as a plug of fibers and water, and all the shear occurs in the thin annulus adjacent to the pipe wall. At high flow rates, water, fibers, and fiber aggregates are in complex turbulent motion. At intermediate flow rates, there is a transition regime between plug flow and turbulent flow where a central, in-tact plug is surrounded by a turbulent fiber-water annulus. This is termed the transition flow regime, and it starts at the onset of drag reduction and extends through the region of developing drag reduction to the maximum level and beyond that to the point of fully developed turbulent flow.'

The core mode of transport of fibers evidently involves the entangle- ment of fibers suggesting, on a macroscopic scale, mechanisms which are believed responsible for non-Newtonian behavior of rubber-like liquids. The hydrodynamics involved in the various modes of fiber transport should be better understood.

By using specially prepared suspensions of elastic, flexible fiber at low suspension concentrations, particles can be trapped in the fiber structure in a stable condition and can be prevented from colliding or dragging on the pipe wall. In this way, one can achieve hydraulic transport of solid particles and capsulized materials in pipelines with friction losses lower than water [Walmsley and Duffy 1987]. This technique appears also to be useful for non-settling slurries like coal-water mixtures, as well as to settling slurries. According to Heywood [1986], '...with fiber concentrations of up to 5%, fibers form a central plug with water annulus.'

Another strategy for transporting coal is to mix it with oil and lubricate it with water.

During the last decade (especially since the oil crisis of 1973), considerable effort has been directed towards the incorporation of coal particles in fuel oil to form liquid fuels. These mixtures offer the potential for conserving fuel oil, and also for using coal in existing equipment that was originally designed to handle heavy fuel oil.

Since 1977, British Petroleum has been developing mixtures containing up to 50% by weight of coal ground to a particle size smaller than the conventional pulverized coal as a potential replacement for fuel oil in utilities and industrial steam production. These mixtures are known as coal-oil dispersions (CODs). These have been transported [Veal, Wall and Groszek 1979], handled and burnt successfully in equipment designed for heavy fuel oil [Anderson, Veal and Withers 1982; Denham, Wall and Whitehead 1982].

The particles in CODs are assumed to be weakly aggregated, forming a three-dimensional lattice structure, which extends throughout the suspension and prevents sedimentation. In this respect, CODs are very similar to coal-oil mixtures which contain larger coal particles and are stabilized by means of added surfactants or polymers [Rowell, Vasconcellos, Sala and Farinato 1981].

The idea is to investigate a lubricated pipeline technology for the transportation of CODs. This is a tricomponent problem, with coal, oil and water. We have found that the density of a COD can be matched with that of water and therefore will have a specific gravity not too different from water. It is also found that drops of CODs in water are stable in the sense that the coal does not come out of the oil. This is an important property: we say that the coal particles bond to oil. The bonding property may be related to wetting properties of oil and water relative to coal.

It is well known that both oil and coal are hydrophobic. The tenacity of the bonding between coal and oil in CODs is such that the COD may be regarded as a single fluid and this composite fluid is hydrophobic. In some preliminary tests of agitation of COD drops in water, Joseph [1988] has found that the coal did not come out of the oil. It would be interesting to see if the integrity of a COD can be upset under the shearing conditions which prevail in lubricated pipe flow. It is even possible that shearing may increase the stability of the dispersion. Chung and Hogg [1985] conclude, after a theoretical study of the stability of fine-particle dispersions, that "suspensions which are stable with respect to Brownian coagulation can rarely be destabilized by shear except at very high shear rates".

The lubrication of heavy crudes with water is enhanced when the pipe walls are preferentially wet by water rather than by oil. We expect to confront the same problem in transporting CODs. In the crude oil case, the wall may be treated by surfactants to produce desired effects. Even in the worst case, when a film of oil wets the pipe wall, lubricating water layers are possible. This is a promising method for the cheap transport of CODs which has the additional advantage of being easily separated from the water at the point of delivery.

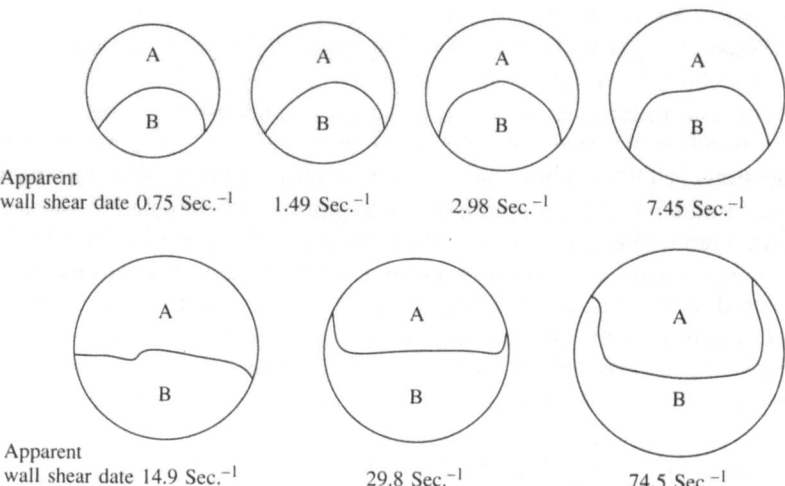

Apparent
wall shear date 0.75 Sec.$^{-1}$ 1.49 Sec.$^{-1}$ 2.98 Sec.$^{-1}$ 7.45 Sec.$^{-1}$

Apparent
wall shear date 14.9 Sec.$^{-1}$ 29.8 Sec.$^{-1}$ 74.5 Sec.$^{-1}$

Fig. 1.3. [Figure 2 of Southern and Ballman, 1973, John Wiley & Sons] Fluids A and B are viscoelastic fluids, and their viscosities vary depending on the apparent wall shear rate. In the upper set of arrangements, fluid B is the more viscous, and fluid A tends to wrap around fluid B. In the lower left, the viscosities of the fluids are about equal and the flat interface is retained. In the lower middle, fluid A is the more viscous and fluid B tends to wrap around fluid A. In the lower right arrangement, fluid A is the more viscous.

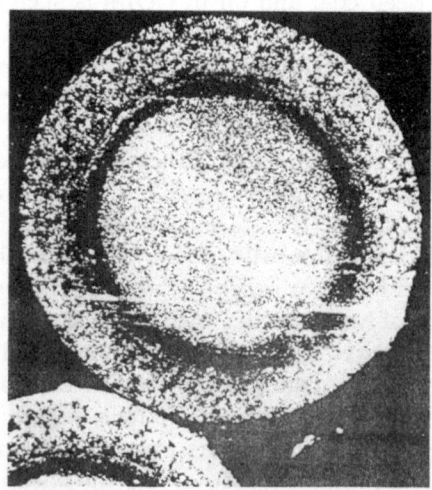

Fig. 1.4. [Figure 5 of Everage, 1973, John Wiley & Sons] Typical cross section of nylon-nylon bicomponent filament illustrating complete encapsulation of higher viscosity material (nylon-6 II) by lower viscosity material (nylon-6 I).

I.1(e) Manufacturing

There are many manufacturing applications involving co-extrusion and co-drawing of different materials. Basic understanding of the science and mechanics behind the applications listed below is meager.

Co-extrusion of plastics and other composite polymeric materials. In the extrusion of two molten polymers through circular and rectangular pipes [Southern and Ballman 1973; MacLean 1973; Everage 1973; Han 1973, 1975; Yu and Han 1973; Lee and White 1974; White and Lee 1975; Khan and Han 1976; Schrenk and Alfrey 1978; Jones and Thomas 1989], experimental data show the tendency for the lower-viscosity component to wrap around the higher-viscosity component (see figure 1.3). This tendency appears to be influenced more by a difference in the viscosity of the two components than by a difference in other properties, such as elasticity. In some instances, the less viscous fluid eventually encapsulates the more viscous fluid as in figure 1.1, the final configuration appearing to be independent of the initial arrangement of the fluids. This encapsulation phenomenon is related to the stability of the lubricated flow discussed above in (b).

In the experiments of Southern and Ballman [1973] and Han [1975], encapsulation of the type exhibited in figure 1.1 is documented, and the effects of viscosity and elasticity stratifications are examined. Everage [1973] shows a photograph of complete encapsulation with a centrally located high viscosity nylon completely encapsulated by an annular ring of low viscosity liquid (see figure 1.4).

The co-extrusion of two components side-by-side has been used in the fiber industry [Sisson and Morehead 1953; Brand and Backer 1962; Hicks, Ryan, Taylor and Tichenor 1960; Hicks, Tippets, Hewett and Brand 1967; Buckley and Phillips 1969; Blais, Carlsson, Suprunchuk and Wiles 1971], in the spinning of bicomponent fibers with self-crimping characteristics: for example, bicomponent acrylics that resemble natural wools, bicomponent nylons with improved stretch.

Another related example in polymer processing is the extrusion of multi-layer polymer melts through a slit die to form a film having the desired optical and mechanical properties [Minagawa and White 1975].

The Compound Jet. In the commercial area of ink-jet printing, one of the conventional methods is to force an electrically charged fluid out of a small nozzle under high pressure and to direct the flow with an electric field. Among some restrictions of this method is the clogging of the nozzle by pigments in the ink. To overcome this, a compound jet has been suggested [Hermanrud 1981; Hertz and Hermanrud 1983]: a primary fluid jet emerges from a nozzle below the surface of a stationary secondary fluid (see figure 1.5). The jet in the air then consists of a core of the primary jet surrounded by a layer of the secondary fluid, having a parabolic velocity profile on leaving the nozzle. Experiments by Hertz and Hermanrud [1983], detailing

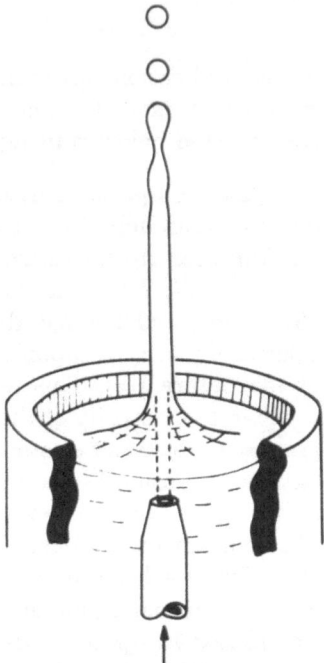

Fig. 1.5. [Figure 1 of Hertz and Hermanrud, 1983] Generation of a compound jet. A high-speed liquid jet emerges from a nozzle a short distance below the surface of a stationary fluid. After breaking the surface, the primary jet carries a concentric layer of the stationary (secondary) fluid, which, owing to viscous forces, has been entrained by the primary jet during its passage through the secondary fluid.

the instabilities of the compound jet, show that it tends to be more stable if the outer thin layer is the less viscous of the fluids. The nature of these instabilities needs more theoretical investigation.

There are many potential technological applications for compound jets other than ink jets. Two of these applications are the production of polymeric composites discussed above, and the production of filaments, discussed next. It is desirable to study compound jets for a wide range of fluids including non-Newtonian fluids.

High Quality Conducting and Superconducting Filaments Using the Method of Glass-coated Melt Spinning (Co-drawing). Glass-coated melt spinning is a co-extrusion or co-drawing procedure for obtaining high quality metal fibers of small diameter for various applications. Manfré, Servi and Ruffino [1974] have described a co-drawing method to obtain copper microwires for electrical and electronic applications by co-drawing. They write that co-drawing is the best method for manufacturing such fibers.

Goto and Waku [1985] have used the same method to produce long superconducting filaments with high superconducting transition temperatures (of greater than 10 K). They note that superconductors with high superconducting transition temperatures are usually brittle and difficult to manufacture into tapes or wires; this would be useful in large-scale engineering applications such as the production of superconducting magnets and power transmission lines. They use the following co-drawing technique for fabricating fine filaments directly from the metal in one state:

'... A 1 g mixture of lead, bismuth and germanium of appropriate composition was placed in a Pyrex glass tube and melted by induction heating in an argon atmosphere. When the glass tube containing the molten alloy was drawn the alloy was stretched to form a glass-coated metallic filament and was coiled on a winding drum. The glass coating was removed in a 45% hydrogen flouride aqueous solution.'

The method of co-drawing is based on the following simple idea. The softened glass is wound upon the spinneret. The filament is encapsulated in the glass and is drawn by the glass and not directly. Filaments obtained in this way may have desirable properties which cannot be obtained by other methods of manufacturing.

Fiber Industry. There are many instances in the fiber industry, where materials are manufactured from several components through processes mentioned in this chapter. For specifics, see ISF-85 [1985]; examples include the manufacture of the fiber Belltron, which prevents the accumulation of static electricity on people walking on carpets made from it, the production of ultrafine fibers for synthetic suede, and the spinning of polymer blends that are aesthetically pleasant (e.g., to touch).

I.1(f) Lubricated Extensional Flows: A Rheological Application

Viscoelastic fluids often exhibit elongation while flowing, and it is of interest to measure their ability to stretch in extensional flows. In rheological experiments on extensional flow, a lubricated planar-die rheometer has been used [Macosko, Ocansey and Winter 1982; Walters 1984,1985; Williams and Williams 1985] to measure planar extensional viscosity. The boundaries of the rheometer are shaped so that a viscoelastic fluid can be forced into it from two opposing directions in the plane, meet in the center and leave through two exits perpendicular to the entrances (see figure 1.6). Hyperbolic streamlines are desired, and the shape of the boundaries reflects this, but the no-slip condition at the walls prevents the desired velocity field. A small amount of a second less viscous fluid is introduced to the flow and this migrates to the walls, where most of the shearing takes place, and the more viscous test fluid then produces the desired flow field (a planar stagnation flow).

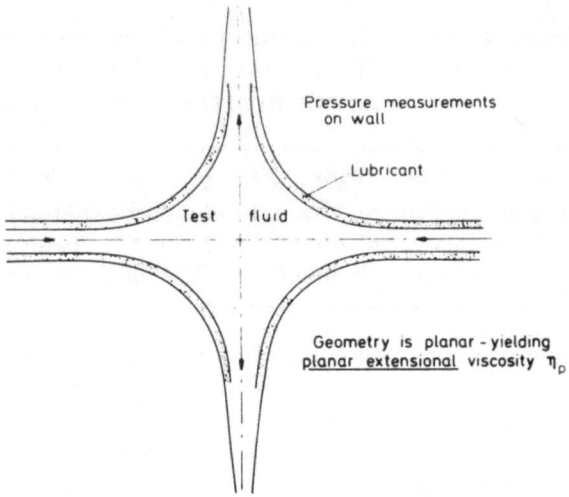

Fig. 1.6. [Figure 3 of Williams and Williams, 1985, J. Non-Newt. Fluid Mech. 19, 53, Elsevier] Lubricated die.

I.1(g) Microgravity Through Density Matching

Artificial microgravity environments have been simulated by matching densities. The idea is that gravity enters into the interface dynamics through the relative weight per unit volume which is equal to the density difference times gravity. However, it is not possible to suppress gravitational effects in bodies with internal density gradients, like those which arise from temperature gradients.

Density matching is presently applied in engineering studies of the flow of two fluids when interest is focused on some other effect, say on the effect of viscosity. For example, in the experiments on core-annular flow in horizontal pipes by Charles, Govier and Hodgson [1961], density matching is used to suppress the tendency of the oil core to float upward into an eccentric position.

There are many flows in which the hydrodynamic forces which distort the interface are independent of gravity. The shape of the interface is determined by a balance of these forces and the forces associated with interfacial tension and gravity. In all such cases, we may amplify the deformation by using density matching to suppress gravity. The deformations produced in this way can be enormous. For example, the shape of the interface between two viscoelastic liquids outside a rotating rod is determined by a balance of surface tension and gravity against normal stresses and centrifugal acceleration. If the diameter of the rod is smaller than a certain critical value, normal stresses will dominate and the fluid will climb the rod against gravity [Joseph and Beavers 1977]. The interfacial tension between liquids is usually relatively small. When interfacial tension is neglected, the height of rise of the climbing interface is proportional to the reciprocal of the density difference. Thus, as the density difference tends to zero, there is a singularity: the rise in height of the interface tends to infinity. This prediction was verified in experiments [Beavers and Joseph 1979].

The shape of free surfaces on liquids in motion is a sensitive indicator of the state of stress in the liquids. It is possible and convenient to determine parameters in the constitutive equations for viscoelastic fluids by measuring the shape of the free surfaces. Measurements of a rod climbing constant have been made in many different fluids by Joseph, Beavers, Cers, Dewald, Hoger and Than [1984] (see Joseph [1990a]). This constant gives a certain linear combination of the first and second normal stresses at low rates of shearing. The distortion of the free surface on a liquid flowing down a tilted trough gives the second normal stress at low rates of shearing [Pipkin and Tanner 1972]. We can say that the rod and the tilted trough are free surface rheometers for normal stresses at low rates of shearing. These normal stress effects can be enormously amplified by density matching [Sturges and Joseph 1980].

Bisch, Lasek and Rodot [1982] have used the principles of density matching to simulate the effects of weightlessness on the vibrations of constrained liquid spheres. They were interested in problems of crystal growth in a weightless environment.

I.1(h) Geophysical Applications

Underlying many geophysical processes are coupled motions of many components, motivating the use of models composed of coupled motions of two immiscible fluids.

Modeling the Earth's Interior. The Earth's crust and mantle have been modeled as a two-layer fluid system [Busse 1978, 1989]: heat sources in the crust induce convection in the entire mantle which is coupled back to the motion in the crust. Models of the Earth's mantle (see also Richter and Johnson [1974])are sometimes based on the assumption that convection takes place in two chemically uniform layers [Busse 1981, 1982], the upper and the lower mantles. Earthquakes originate in the upper mantle: the scale of convection cells in the lower mantle may determine the scale of the flow in the upper mantle via viscous coupling. This type of modeling allows one to think about laboratory experiments to test competing explanations.

Magma Deposits. In the study of minerals present in volcanic activity, Carrigan and Eichelberger [1990] obtain data from a conduit (vertical pipe) that leads up to a past position of volcanic eruption, which is now covered by a lava dome. Drilling through the conduit gives the chemical data on its makeup as a function of distance from its wall. The results indicate that the lower viscosity magma is situated in the zone near the wall of the conduit and higher viscosity magma occupies the low shear zone at the center, displaying the encapsulation property discussed in the items (b) - (c) above. They note that this encapsulation process may contribute significantly to the chemical stratigraphy of the products of eruption, and influence zonations that are commonly attributed solely to preeruption layering of the source magma chamber.

Modelling of Rock Formations. The geological phenomenon of rock folding is a process in which stratified layers of rock subjected to lateral compression buckle to form wave-like patterns. Wollkind and Alexander [1982] have modeled a rock folding situation by means of a two-dimensional layered Newtonian fluid which consists of a single layer embedded between two less viscous half-spaces. The single layer represents a rock stratum which is more resistant to deformation than its embedding medium. They examined the occurrence of plane folding by means of a linear stability analysis of an exact solution, which corresponds to both the layer and the surrounding medium undergoing a uniform parallel-layer shortening at a constant rate of compressive strain, such that the interfaces separating the two are parallel planes moving apart symmetrically. Their results are relevant to the explanation of small scale folds, as well as accounting for some of the observed features in the Castile Formation of southern New Mexico.

The deformation of rock glaciers has been studied by Loewenherz and Lawrence [1989] with a two-layer model flowing down an incline at zero Reynolds number. Rock glaciers are composed of debris and ice and display wavelike ridges.

I.1(i) Transient Flow of Two Immiscible Liquids in a Rotating Container

There is a well-developed branch of fluid dynamics associated with rotating flows (*cf.* Greenspan [1969]). One of many outstanding results in this field is that the time it takes to spin up or spin down a rotating fluid after a sudden change of angular velocity is greatly altered by the presence of Ekman suction at the end plates of a container. A classic paper on this was written by Wedemeyer [1969]. In the case of spin-up, the fluid is sucked to the end plates where it is brought quickly up to the new rate of rotation and then is returned to the fluid through side wall boundary layers attached to Ekman layers at their outer edges. The convective transport of angular momentum from the end plates is much more efficient than diffusion and can reduce the spin-up time by orders of magnitude.

There appears to be no literature on the spin-up of two immiscible liquids. This is a potentially important problem in the control of stability of liquid-carrying spinning artillery shells which are stabilized by the addition of small amounts of water [Miller 1990]. In this application, the light liquid fill can be viscoelastic so that the two-fluid problem can involve at least one viscoelastic liquid. The problem of two-fluid spin-up is also important for understanding the different types of relaxation curves which can develop in spinning drop tensiometers.

The dynamics associated with the spin-up of two immiscible liquids is readily visualized by watching the highly visible motion of the interface. Some still photographs from the video sequence showing how spin-up and spin-down affects the interface in different cases are exhibited and described below.

Plate I.1.7 shows a typical sequence of spin-up events of SAE 30 motor oil in water, with some tentative explanations of the rotating flow mechanisms underway. The equilibrium bubble (a) at angular velocity Ω_1 is attached to the end plate. The angular velocity is increased from Ω_1 to Ω_2 (b). The bubble is sucked to the wall by suction into the Ekman layer. The strong suck tears the oil apart and the bubbly mixture of oil and water spreads in the Ekman layer (b). The small oil bubbles in the bubbly mixture cannot enter into the return flow in the Stewartson layers on the cylinder walls because they will be pushed inward by the centrifuging of the heavier water in any region not participating in the Ekman pumping. At the end plates, more water and oil bubbles are pulled through the Ekman layer where they acquire the new angular velocity Ω_2 rapidly, as in (b) and (c). After a time, the spin-up to $\Omega_2 > \Omega_1$ of the two fluids away from the end plates has begun to take effect and the intensity of the pumping in the Ekman layer decreases. In this regime, the bubbly mixture at the end cap collapses, as in (d), (e) and (f), and the oil is jetted out in a column of smaller diameter appropriate to equilibrium rotation at the new angular velocity Ω_2.

In plate I.1.8, we see different dynamics associated with increasing the volume to 60% of SAE 30 motor oil in an 80% glycerin plus water solution. In any equilibrium flow in which the rate of rotation is such as to make the parameter J, defined in section II.2, larger than 4, the oil will rotate as a perfect cylindrical column of diameter D, except for capillary regions near the end caps. This is the case when $\Omega = \Omega_1$ in plate I.1.8 (a). The spin-up to $\Omega_2 > \Omega_1$ of the oil away from the end cap is shown in plates I.1.8 (b) - (d). The diameter D is determined by the volume of oil for Ω's larger than the critical one for each $J = 4$. Plates I.1.8 (e) - (f) show what happens in spin-down when the angular velocity of the container is reduced. In this case, it seems likely that the liquids are sucked to the end cap through the Stewartson layers on the cylinder wall and on the oil-water interface where they are spun down and ejected, giving rise to an effect opposite to spin-up. The effects of Ekman suction can be suppressed by increasing the viscosity of either fluid.

I.2 Formulation of Equations

In this section, we collect the equations that determine the motion of the fluids (cf. chapter 3 of Joseph [1990 a]).

We use the following notation: $\mathbf{X}$ denotes the Lagrangian coordinate for a point in a continuum in some reference configuration $V(\mathbf{X})$;

$$\xi(\mathbf{X}, \tau), \quad \tau \leq t,$$

denotes the Eulerian coordinate for that particle at time τ;

$$\xi(\mathbf{X}, t) = \mathbf{x} \tag{2.1}$$

denotes the position of the particle at time t;

$$\mathbf{u}(\xi, \tau) = \frac{\partial \xi}{\partial \tau} = \left(\frac{\partial \xi}{\partial \tau} \right)_{\mathbf{X}} \tag{2.2}$$

denotes the velocity of the point $\mathbf{X}$;

$$\mathbf{a}(\xi, \tau) = \frac{\partial^2 \xi}{\partial \tau^2} \tag{2.3}$$

denotes the acceleration;

$$\mathbf{F} = \nabla \xi(\mathbf{X}, \tau) \tag{2.4}$$

denotes the deformation gradient or the Jacobian matrix;

$$J = \det \mathbf{F} \tag{2.5}$$

denotes the Jacobian. The transformation of volumes between Eulerian and Lagrangian coordinates is given by

$$dV(\xi) = J dV(\mathbf{X}). \tag{2.6}$$

Conservation of Mass. Let $\rho[\xi(\mathbf{X}, \tau), \tau]$ be the density at the point ξ, where $\xi(\mathbf{X}, \tau)$ is the position at time τ of the particle which was at $\mathbf{X}$ at time τ_0. We can choose the Lagrangian coordinate $\mathbf{X}$ to be the position the particle occupies at time τ_0 so that $\xi(\mathbf{X}, \tau_0) = \mathbf{X}$. Let V be any material volume; that is, no mass crosses the boundary ∂V of V. Then the equation expressing the conservation of mass is

$$\int_{V(\tau)} \rho[\xi, \tau] dV = M = \int_{V_0} \rho_0 dV \tag{2.7}$$

where $\rho > 0$, $\rho_0 = \rho[\mathbf{X}, \tau_0]$, $M > 0$ is a constant (independent of τ). Transforming $V(\tau)$, we have

$$\int_{V_0} \left(\rho[\xi(X, \tau), \tau] J - \rho_0 \right) dV = 0. \tag{2.8}$$

Since V_0 is arbitrary, the integrand vanishes, and

$$\rho J = \rho_0. \tag{2.9}$$

This expresses the notion that mass is conserved.

We will be dealing with incompressible materials in this book, and for them, $\rho[\mathbf{X}, \tau]$ is independent of τ. Then $\rho = \rho_0$ and

$$J = 1. \tag{2.10}$$

This means that the volume is conserved for an incompressible material.

I.2(a) Transport Identities

The evolution of the Jacobian for a material volume $V(t)$ is governed by

$$\frac{dJ}{dt} = J \operatorname{div} \mathbf{u}(\mathbf{x}, t), \tag{2a.1}$$

where J is the Jacobian of the transformation from $\xi(\mathbf{X}, t) = \mathbf{x}$ to $\mathbf{X}$, and d/dt is the material derivative $\partial/\partial t + \mathbf{u} \cdot \nabla$. Equations (2.9) and (2a.1) imply that

$$\frac{d\rho}{dt} + \rho \operatorname{div} \mathbf{u} = 0. \tag{2a.2}$$

This or (2a.1), together with (2.10) imply that for an incompressible fluid, we have

$$\operatorname{div} \mathbf{u} = 0. \tag{2a.3}$$

Let $V(t)$ be a material volume across which no mass passes and let $V(t_0)$ be the location of this volume in a reference configuration with coordinate $\mathbf{X}$. Let $f(\mathbf{x}, t)$ be a smooth field in $V(t)$; then

$$\frac{d}{dt} \int_{V(t)} f dV = \int_{V(t_0)} \frac{d}{dt}(fJ) dV_0 = \int_{V(t)} \left(\frac{df}{dt} + f \operatorname{div} \mathbf{u}\right) dV \tag{2a.4}$$

and since (2a.2) holds, we have

$$\frac{d}{dt} \int_{V(t)} \rho f dV = \int_{V(t)} \rho \frac{df}{dt} dV. \tag{2a.5}$$

Now let $f(\mathbf{x}, t)$ be any field in $V(t)$ which has a simple discontinuity (a finite jump in value, with limits existing on both sides of the discontinuity) across the surface Σ shown in figure 2.1.

The surface Σ represents an interface between two fluids and travels at the same velocity as the fluid particles there. We consider limits in which $V(t)$ tends to zero, while Σ is held fixed; that is, $V(t)$ is collapsed onto Σ. Define

$$[\![f]\!] = f_1(\mathbf{x}, t) - f_2(\mathbf{x}, t) \tag{2a.6}$$

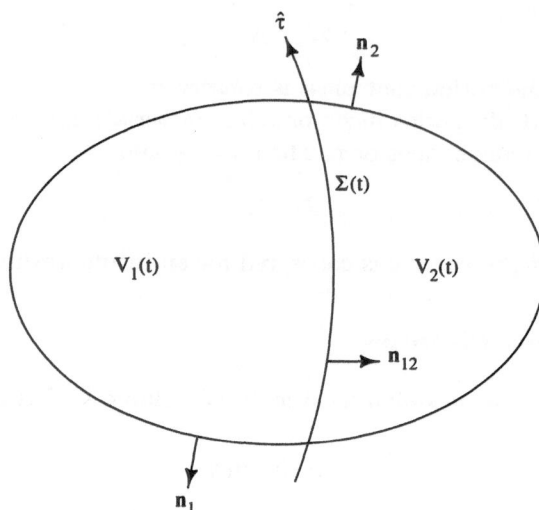

Fig. 2.1. [Figure 3.1. of Joseph, 1990, *Fluid Dynamics of Viscoelastic Liquids*, Springer-Verlag] $V(t) = V_1(t) \cup V_2(t)$ is a material volume and $\Sigma(t)$ is a surface across which the otherwise smooth field $f(\mathbf{x}, t)$ has a simple discontinuity, with smooth derivatives on either side of Σ. $\mathbf{n_1}$ is the outward normal on ∂V_1 and $\mathbf{n_2}$ on ∂V_2. $\mathbf{n_{12}}$ is from V_1 to V_2 on Σ and $\mathbf{n_{12}} = \mathbf{n_1}$ there.

as the jump in f at the point $\mathbf{x} \in \Sigma$. There are two jump identities. The first one is

$$\lim_{V \to \Sigma} \int_{\partial V} \mathbf{n} f dS = \int_{\Sigma} -\mathbf{n_{12}} [\![f]\!] d\Sigma \qquad (2a.7)$$

where dS is an element on the surface ∂V. The proof of (2a.7) is obvious from figure 2.1.

We next assume that Σ is not necessarily an interface between two immiscible liquids, but that it may move with a speed different from the surrounding fluid. This type of assumption is used in section IV.10 where we deal with change of phase.

The second jump identity is

$$\lim_{V \to \Sigma} \frac{d}{dt} \int_V f dV = \int_{\Sigma} [\![fv]\!] d\Sigma \qquad (2a.8)$$

where

$$v = \mathbf{n_{12}} \cdot (\mathbf{u}_\Sigma - \mathbf{u}). \qquad (2a.9)$$

$\mathbf{u}(\mathbf{x}, t)$ is the velocity at a point $\mathbf{x}$ on this or that side of Σ and $\mathbf{u}_\Sigma$ is the velocity of Σ at that point. In general, $\mathbf{u}_\Sigma \neq \mathbf{u}_l(\mathbf{x}, t)$, $\mathbf{x} \in \Sigma$, where $\mathbf{u}_1$ is the particle velocity on the left side at Σ and $\mathbf{u}_2$ the particle velocity on the right. If Σ is an interface between two fluids, the right hand side of (2a.8) is zero. To prove (2a.8), we note that

$$\frac{d}{dt} \int_{V_l} f(\mathbf{x}, t) \, dV_l = \int_{V_l} \frac{\partial f}{\partial t} \, dV_l + \int_{\partial V_l} f \mathbf{n}_l \cdot \mathbf{v}_l \, dS_l \qquad (2a.10)$$

holds for $l = 1$ and $l = 2$, where $\mathbf{v}_l(\mathbf{x}, t)$ is the velocity of points $\mathbf{x} \in \partial V_l$ and $\mathbf{v}_l(\mathbf{x}, t) = \mathbf{u}(\mathbf{x}, t)$ for $\mathbf{x} \in \partial V$ and $\mathbf{u}$ is the particle velocity and $\mathbf{v}_l(\mathbf{x}, t) = \mathbf{u}_\Sigma(\mathbf{x}, t)$ for $\mathbf{x} \in \Sigma$. Then

$$\int_{\partial V_1} f \mathbf{n}_1 \cdot \mathbf{v}_1 \, dS_1 + \int_\Sigma f \mathbf{u}_1 \cdot \mathbf{n}_1 \, d\Sigma - \int_\Sigma f \mathbf{u}_1 \cdot \mathbf{n}_1 \, d\Sigma$$

$$= \int_{\partial V_1} f \mathbf{u} \cdot \mathbf{n}_1 dS_1 + \int_\Sigma f(\mathbf{u}_\Sigma - \mathbf{u}_1) \cdot \mathbf{n}_1 \, d\Sigma$$

$$= \int_{V_1} \text{div}(f\mathbf{u}) \, dV_1 + \int_\Sigma f(\mathbf{u}_\Sigma - \mathbf{u}_1) \cdot \mathbf{n}_{12} \, d\Sigma. \qquad (2a.11)$$

We get the same equation (2a.11) with subscript 2; then, we add the two equations (2a.10) noting that $\mathbf{n}_1 = \mathbf{n}_{12} = -\mathbf{n}_{21}$ on Σ and get

$$\frac{d}{dt} \int_V f(\mathbf{x}, t) \, dV = \int_V \left(\frac{\partial f}{\partial t} + \text{div} f \mathbf{u} \right) dV + \int_\Sigma [\![fv]\!] \, d\Sigma \qquad (2a.12)$$

leading to (2a.8). We remind the reader that

$$\int_V (\cdot) \, dV = \int_{V_1} (\cdot) \, dV + \int_{V_2} (\cdot) \, dV.$$

Since

$$\frac{d}{dt} \int_V \rho dV = 0$$

for every material volume V, we find from (2a.8) that

$$\rho_1 v_1 = \rho_2 v_2 \overset{\text{def}}{=} m. \qquad (2a.13)$$

When Σ is simply an interface between two different fluids, $m = 0$. Replacing f by ρf in (2a.8), we get

$$\lim_{V \to \Sigma} \frac{d}{dt} \int_V \rho f dV = \int_\Sigma m[\![f]\!] d\Sigma. \qquad (2a.14)$$

From (2a.12) and (2a.2), we get

$$\frac{d}{dt} \int_V \rho f dV = \int_V \rho \frac{df}{dt} dV + \int_\Sigma m[\![f]\!] d\Sigma. \qquad (2a.15)$$

I.2(b) Balance of Momentum

We balance the momentum of a material volume $V(t)$:

$$\frac{d}{dt}\int_V \rho\mathbf{u}\, dV = \int_{\partial V} \mathbf{t}\, dS + \int_V \rho\mathbf{g}\, dV + \int_{\partial\Sigma} \sigma\hat{\tau}dl, \qquad (2b.1)$$

change of   tractions on   body   force due to

momentum   the boundary   forces   membrane stress

where the interface Σ has a surface tension $\sigma(\mathbf{x})$. Here, $\hat{\tau}$ lies in Σ and is the normal to the curve $\partial\Sigma = \Sigma \cap \partial V$: this curve denotes the intersection of the surface Σ with the surface of the volume. $\hat{\tau}$ points out of $\partial\Sigma$. The last term in (2b.1) is a line integral and can be written using the analogue of the divergence theorem for surfaces as

$$\int_\Sigma \nabla_\| \sigma\, d\Sigma - \int_\Sigma (\nabla_\| \cdot \mathbf{n_{12}})\sigma\mathbf{n_{12}}\, d\Sigma, \qquad (2b.2)$$

where $\nabla_\|$ is the surface gradient $\mathbf{t_1}(\mathbf{t_1}\cdot\nabla) + \mathbf{t_2}(\mathbf{t_2}\cdot\nabla)$ with $\mathbf{t_1}$ and $\mathbf{t_2}$ being the two tangents to the interface. That is,

$$\nabla_\| = \nabla - \mathbf{n_{12}}(\mathbf{n_{12}}\cdot\nabla) \qquad (2b.3)$$

and

$$\nabla_\| \cdot \mathbf{n_{12}} = -2H = -(\frac{1}{R_1} + \frac{1}{R_2}) \qquad (2b.4)$$

where H is the mean curvature and R_1 and R_2 are principal radii of curvature.

Now we replace the traction vector $\mathbf{t}$ (force/area) with a stress tensor $\mathbf{T}$ related to $\mathbf{t}$ by the Cauchy-Fourier formula

$$\mathbf{t} = \mathbf{T}^T\mathbf{n} \qquad (2b.5)$$

where $\mathbf{n}$ is the outward normal on ∂V. In the absence of couple stresses, the stress tensor is symmetric:

$$\mathbf{T}^T = \mathbf{T}; \qquad (2b.6)$$

this is the case for materials we will be dealing with in this book, so we will drop the superscript T in what follows. In particular, the stress in an incompressible Newtonian fluid is

$$\mathbf{T} = -p\mathbf{I} + \mu(\nabla\mathbf{u} + (\nabla\mathbf{u})^T) \qquad (2b.7)$$

After applying (2a.15) to (2b.1), we get

$$\int_V \rho\frac{d\mathbf{u}}{dt}\, dV + \int_\Sigma m[\![\mathbf{u}]\!]d\Sigma = \int_{\partial V} \mathbf{T}\mathbf{n}\, dS + \int_V \rho\mathbf{g}\, dV$$
$$+ \int_\Sigma \nabla_\| \sigma\, d\Sigma + \int_\Sigma 2H\sigma\mathbf{n_{12}}\, d\Sigma. \qquad (2b.8)$$

It is easy to verify that

$$\int_{\partial V} \mathbf{Tn} \, dS = \int_{\partial V_1} \mathbf{Tn_1} \, dS + \int_{\partial V_2} \mathbf{Tn_2} \, dS - \int_{\Sigma} [\![\mathbf{T}]\!] \mathbf{n_{12}} \, d\Sigma. \qquad (2b.9)$$

After using the divergence theorem, we get

$$0 = \int_V \left(\rho \frac{d\mathbf{u}}{dt} - \text{div } \mathbf{T} - \rho \mathbf{g} \right) dV$$

$$+ \int_{\Sigma} m[\![\mathbf{u}]\!] + [\![\mathbf{T}]\!] \mathbf{n_{12}} - \nabla_{\|} \sigma - 2H\sigma \mathbf{n_{12}} \, d\Sigma. \qquad (2b.10)$$

Standard arguments based on the fact that (2b.10) holds for any material volume and surface of discontinuity lead then to

$$\rho \frac{d\mathbf{u}}{dt} = \text{div } \mathbf{T} + \rho \mathbf{g}, \quad \mathbf{x} \in V_1 \text{ or } V_2, \qquad (2b.11)$$

$$m[\![\mathbf{u}]\!] + [\![\mathbf{T}]\!] \mathbf{n_{12}} = \nabla_{\|} \sigma + 2H\sigma \mathbf{n_{12}}, \quad \mathbf{x} \in \Sigma. \qquad (2b.12)$$

I.2(c) Balance of Energy

If we neglect radiative transfer and other body sources of energy, we have

$$\frac{d}{dt} \int_V \rho \left(e + \frac{|\mathbf{u}|^2}{2} \right) dV + \frac{d}{dt} \int_{\Sigma} \sigma \, d\Sigma$$

$$= - \int_{\partial V} \mathbf{q} \cdot \mathbf{n} \, dS + \int_{\partial V} \mathbf{u} \cdot \mathbf{t} \, dS + \int_V \rho \mathbf{u} \cdot \mathbf{g} \, dV + \int_{\partial \Sigma} \sigma \hat{\underline{\tau}} \cdot \mathbf{u} \, dl, \quad (2c.1)$$

where $\hat{\underline{\tau}}$ is the vector defined after equation (2b.1) and $\mathbf{t}$ is defined in (2b.5). On the left hand side, the first two terms express the rate of change of the specific internal plus kinetic energies, and the third denotes the rate of change of the stored energy of the surface. On the right hand side, the four terms express, respectively, the heat in, the power of the traction vector $\mathbf{Tn}$ at the boundary surface, the power of the body force, and the power of the surface tension. The vector $\mathbf{q}$ denotes the heat flux.

The interface is known to satisfy (cf. Joseph [1976])

$$\frac{d}{dt} \int_{\Sigma} \sigma \, d\Sigma - \int_{\partial \Sigma} \sigma \hat{\underline{\tau}} \cdot \mathbf{u}_{\Sigma} \, dl = \int_{\Sigma} \frac{\partial \sigma}{\partial t} - \mathbf{u}_{\Sigma} \cdot \nabla_{\|} \sigma - 2H\sigma \mathbf{u}_{\Sigma} \cdot \mathbf{n_{12}} \, d\Sigma,$$

$$(2c.2)$$

where H is defined in (2b.4).

Arguing, as in (2b.9) and (2b.10), we find, using (2a.5), that

$$\int_V \left(\rho \frac{d}{dt} (e + \frac{|\mathbf{u}|^2}{2}) + \text{div} \mathbf{q} - \text{div}(\mathbf{Tu}) - \rho \mathbf{u} \cdot \mathbf{g} \right) dV$$

$$= \int_{\Sigma} \left(-m[e + \frac{|\mathbf{u}|^2}{2}] + [\![\mathbf{q}]\!] \cdot \mathbf{n}_{12} - [\![\mathbf{u} \cdot \mathbf{Tn}_{12}]\!] \right.$$
$$\left. - \frac{\partial \sigma}{\partial t} + \mathbf{u}_{\Sigma} \cdot \nabla_{\|} \sigma + 2H\sigma \mathbf{u}_{\Sigma} \cdot \mathbf{n}_{12} \right) d\Sigma. \tag{2c.3}$$

Standard arguments then lead us to

$$\rho \frac{d}{dt} \left(e + \frac{|\mathbf{u}|^2}{2} \right) + \operatorname{div} \mathbf{q} - \operatorname{div}(\mathbf{Tu}) - \rho \mathbf{u} \cdot \mathbf{g} = 0 \tag{2c.4}$$

for $\mathbf{x} \in V_1$ or $\mathbf{x} \in V_2$. We differentiate the term $\rho\frac{d}{dt}(\frac{\mathbf{u} \cdot \mathbf{u}}{2})$ and use the expression for $\rho d\mathbf{u}/dt$ from (2b.11). The term $\rho \mathbf{u} \cdot \mathbf{g}$ cancels out and

$$\rho \frac{de}{dt} + \operatorname{div} \mathbf{q} - \operatorname{tr}(\mathbf{LT}) = 0 \tag{2c.5}$$

where the last term comes from $\mathbf{u} \cdot \operatorname{div}\mathbf{T} - \operatorname{div}(\mathbf{Tu})$. Also, (2c.3) yields

$$-m[e + \frac{|\mathbf{u}|^2}{2}] + [\![\mathbf{q}]\!] \cdot \mathbf{n}_{12} - [\![\mathbf{u} \cdot \mathbf{Tn}_{12}]\!] = \frac{\partial \sigma}{\partial t} - \mathbf{u}_{\Sigma} \cdot \nabla_{\|} \sigma - 2H\sigma \mathbf{u}_{\Sigma} \cdot \mathbf{n}_{12} \tag{2c.6a}$$

for $\mathbf{x} \in \Sigma$. The second term represents work due to traction forces, and the terms on the right hand side represent work due to surface tension forces. The energy balance (2c.6a) may be written relative to an observer moving with the interface as

$$[\![\mathbf{q}]\!] \cdot \mathbf{n}_{12} = \frac{\partial \sigma}{\partial t} + [\![(\mathbf{u} - \mathbf{u}_{\Sigma}) \cdot \mathbf{Tn}_{12}]\!] + m[e + \frac{|\mathbf{u} - \mathbf{u}_{\Sigma}|^2}{2}]. \tag{2c.6b}$$

To derive this, add the projection of (2b.12) with $\mathbf{u}_{\Sigma}$ to (2c.6a). Equation (2c.5) is useful for the calculation of temperature in non-isothermal flow: for example, the Bénard problem (chapter III).

The quantity $\operatorname{tr}(\mathbf{LT})$ is the stress power or dissipation, heating due to friction, and this is defined as

$$\operatorname{tr}(\mathbf{LT}) = \frac{1}{2}\operatorname{tr}(\mathbf{DT}) \qquad \text{when} \quad \mathbf{T} = \mathbf{T}^T. \tag{2c.7}$$

For incompressible fluids,

$$\mathbf{D} = \frac{1}{2}(\nabla\mathbf{u} + (\nabla\mathbf{u})^T), \qquad \mathbf{L} = \nabla\mathbf{u}. \tag{2c.8}$$

The stability analysis of convection problems in this book utilizes the Oberbeck-Boussinesq approximation (for this we refer to the works of Joseph [II, 1976] and Drazin and Reid [1982]) which neglects the viscous heating in (2c.5).

For incompressible Newtonian fluids, let us look at the dissipation term in (2c.7). In indicial notation, we have $\mathbf{D}^2 = D_{ij}D_{jk}$ and therefore, the trace of $\mathbf{D}^2$ is $D_{ij}D_{ji}$. Since $\mathbf{D}$ is symmetric (by (2c.8)), $\operatorname{tr} \mathbf{D}^2$ is $D_{ij}D_{ij}$. We define the notation

$$\mathbf{A} : \mathbf{B} = \sum_{i,j} A_{ij} B_{ij}, \tag{2c.9}$$

where $\mathbf{A}$ and $\mathbf{B}$ are square matrices. Thus,

$$\mathbf{D} : \mathbf{D} = \operatorname{tr} \mathbf{D}^2. \tag{2c.10}$$

In addition,

$$\mathbf{TD} = \mathbf{DT} = 2\mu \mathbf{D}^2. \tag{2c.11}$$

Therefore,

$$\operatorname{tr} \mathbf{TD} = 2\mu \mathbf{D} : \mathbf{D}. \tag{2c.12}$$

We will refer to this in section I.3 (f).

I.2(d) Boundary Conditions

Consider the boundary between a solid and fluid or between immiscible liquids across which no mass flows. Then setting $m = 0$ in (2b.12) yields

$$[\![\mathbf{T}]\!]\mathbf{n}_{12} = [\![\mathbf{t}]\!] = \nabla_\| \sigma + 2H\sigma\mathbf{n}_{12} \tag{2d.1}$$

shows that the tangential tractions and the normal component of the jump $[\![\mathbf{t}]\!]$ are balanced by surface tension. In particular, if surface tension is a constant, the tangential tractions are continuous.

Using (2d.1), (2c.6) reduces to

$$[\![\mathbf{q}]\!] \cdot \mathbf{n}_{12} = \frac{\partial \sigma}{\partial t}. \tag{2d.2}$$

For example, if surface tension is constant, this says that the heat flux across the interface is continuous.

The kinematic free-surface condition holds at the interface, i.e., if the interface is described by $h(\mathbf{x}(t), t) = 0$, then

$$\frac{\partial h}{\partial t} + \mathbf{u} \cdot \nabla h = 0. \tag{2d.3}$$

If we do not assume the no-slip condition at a solid boundary (or an interface), then a term accounting for frictional heating must be added to the heat flux balance as follows:

$$[\![\mathbf{u}]\!] \cdot \mathbf{t} = [\![\mathbf{q}]\!] \cdot \mathbf{n}_{12}. \tag{2d.4}$$

This is equation (30) of section 3.4 of the book of Joseph [1990a]. If the stress and heat flux are continuous across a solid-liquid boundary for all tractions $\mathbf{t}$, then the velocity must be continuous there:

$$[\![\mathbf{u}]\!] = 0. \tag{2d.5}$$

I.2(e) Summary

The stability analyses in this book will focus on isothermal shearing flows and convection, for which we summarize the governing equations below.

(i) **Isothermal Shearing Flows.** Shearing flows composed of two immiscible fluids with different viscosities and densities will be considered in chapters IV and later. The fluids are separated by an interface and are driven by prescribed forces of the usual type. Each fluid is governed by the Navier-Stokes equations (2b.7) and (2b.11), and assumed to be incompressible (2a.3).

The interface between the two fluids is an unknown, at which the normal n_{12} is denoted by n and the orthonormal tangential vectors are t_1 and t_2. There is a constant surface tension $\sigma(x) = S^*$. Let μ denote the viscosity, D the dimensional extra stress tensor of (2c.8), p the pressure and H the sum of principal curvatures. Across the interface, the velocity is continuous:

$$[\![u]\!] = 0. \tag{2e.1}$$

From (2d.1), the shear stress is continuous:

$$t_l \cdot [\![2\mu D]\!] \cdot n = 0, \ l = 1, 2, \tag{2e.2}$$

and the jump in the normal stress is balanced by surface tension :

$$n \cdot [\![2\mu D]\!] \cdot n - [\![p]\!] + 2HS^* = 0. \tag{2e.3}$$

The kinematic free-surface condition (2d.3) holds at the interface.

The volumes of the two fluids are given. Appropriate boundary conditions are imposed to complete the formulation, e.g., the no-slip condition at solid walls, periodicity in the unbounded direction. There are at least six dimensionless parameters: a Reynolds number, a Froude number, a surface tension parameter, the volume ratio, the ratio of viscosities and the ratio of densities. The problem is to describe the motion and the spatial arrangement of each fluid.

(ii) **Convection.** The effect of heating will be considered in chapter III: the two fluids have different viscosities, densities, coefficients of cubical expansion, thermal diffusivities κ, and thermal conductivities k.

Temperature differences will be assumed to be small and the Oberbeck-Boussinesq approximation is applied. Let Θ denote the temperature. In addition to the formulation of (i) above, the linear heat equation,

$$\frac{D\Theta}{Dt} = \kappa \, \triangle \, \Theta, \tag{2e.4}$$

is applied. This is equivalent to (2c.5) where the div q gives rise to $k\triangle\Theta$, the third term is neglected; the internal energy e is the product of the specific heat and the temperature, and $\kappa = k$/specific heat. Across the interface, additional jump conditions are the continuity of heat flux (2d.2):

$$[\![k\mathbf{n} \cdot \nabla\Theta]\!] = 0, \qquad (2e.5)$$

and the continuity of temperature:

$$[\![\Theta]\!] = 0. \qquad (2e.6)$$

On top of the dimensionless parameters in (i) above, there are at least five more: a Rayleigh number which is a measure of the temperature difference in the system, a Prandtl number which measures the relative effect of momentum diffusivity versus thermal diffusivity for one of the fluids, and ratios of the thermal properties of the fluids.

I.3 Nonuniqueness of Steady Solutions

A major problem in the theory of flows involving more than one fluid lies in their nonuniqueness: the position of the interface is one of the unknowns, and the equations may permit an infinite number of different interface configurations. The theoretical problem is plagued by a degree of nonuniqueness without parallel in the theory of flow of one fluid. Of these theoretically possible flows, only some, but still many, may be realized in experiments. This nonuniqueness for two immiscible liquids was documented and discussed by Joseph, Nguyen and Beavers [1984]. They note that the realized flows are such as to position the low viscosity constituent in regions of high shear. The variety of ways that nature achieves these placements is amazing. The mechanisms at work in arranging the flowing components are only vaguely understood but they appear to follow along lines which were expressed in anthropomorphic terms by Joseph, Nguyen and Beavers:

High viscosity liquids are lazy. Low viscosity liquids are the victims of the laziness of high viscosity liquids because they are easy to push around.
Some examples of nonuniqueness are described below.

I.3(a) Bubbles

Immiscible liquids of the same density will form spheres of one liquid in another. The steady distribution of such spheres, their sizes and their placement, seem not to be unique. In fact, Plateau [1873] suspended masses of olive oil in a mixture of alcohol (lighter than oil) and water (heavier than oil) of the same density. Spheres of many centimeters were obtained. It is possible to have big spheres of oil in alcohol-water and big spheres of alcohol-water in oil. For such spheres, the conditions at the interface (2e.1)-(2e.3) are satisfied with the steady solution

$$\llbracket \mathbf{u} \rrbracket = 0, \quad \mathbf{D} = 0, \quad 2HS^* = \llbracket p \rrbracket, \quad H = 1/R, \qquad (3a.1)$$

where R is the radius of each sphere. There is nothing here to determine the size and placement of spheres from the given data.

The type of stationary configuration with bubbles of different sizes in immiscible fluids of matched density which was studied by Plateau is shown in plate I.3.1. This plate shows the result of matching the density ($\rho = 1.04$ g/cm^3) of dibutyl phthalate with a glycerol (heavier, $\rho = 1.25$ g/cm^3) and water (lighter, $\rho = 1.00$ g/cm^3) solution.

The nonunique stationary configurations that are achieved by matching density appear to be stable to small disturbances. However, there is a tendency for bubbles that touch to collapse into one bubble. Perhaps there is a selection mechanism based on the stability to large disturbances, in which the stable configuration is the one that minimizes the surface area. This type of criterion would lead to large bubbles, even one large one, rather than many small ones. There appears to be no simple criterion. In order to determine the sizes of the bubbles, what is needed here is a study of the dynamics, including the various types of instabilities that arise; for example, fingering. The physics of these bubbles is dominated by surface tension, and is related to our study of the rollers in chapter II.

I.3(b) Parallel Shear Flows

Consider plane Couette flow in layers. An example consisting of four layers of two different fluids is shown in figure 3.2. This flow, like other steady parallel shear flows, is characterized by alternate layers of fluids, where the speeds and stresses are matched at each interface. The two fluids may have any number of contiguous layers: this problem has a continuum of solutions in which the fluid with viscosity μ_1 is in N layers of total height l_1, separated by layers of fluid with viscosity μ_2 whose total height is l_2. The heights of the constituent layers and their number are otherwise arbitrary. Here, as in other steady parallel shear flows, a possible interface position is a streamtube, and this results in an infinite number of possible arrangements.

Figure 3.3 shows three possible ways of arranging two different fluids in three layers for plane Poiseuille flow. The fluid of viscosity μ_2 is the more viscous. Another case of nonuniqueness of steady flow of two immiscible fluids is Couette flow in circular rings between rotating cylinders. Figure 3.4 illustrates two possibilities for the case of equal density but different viscosity.

Two-fluid flow through a vertical pipe has been modelled as a Hagen-Poiseuille flow with fluids of equal density. Mathematically, anti-plane shear flow (exclusively axial flow with only one non-zero component of velocity which depends on the coordinates perpendicular to the axial coordinate) of two fluids in a cylindrical pipe of arbitrary cross-section has a continuum of solutions. Here, there is no flow within the cross-section perpendicular

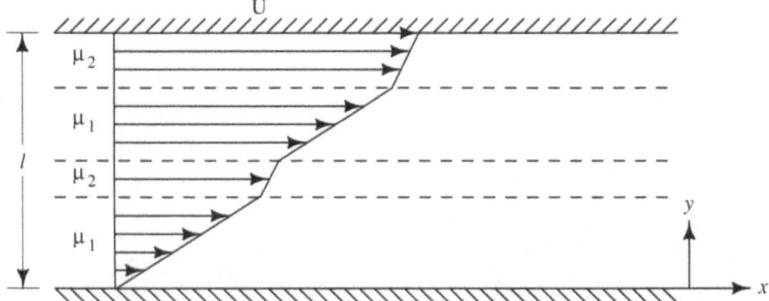

Fig. 3.2. Layered Couette flow. The high viscosity fluid is the one with viscosity $\mu_2 > \mu_1$. The two layers consisting of Fluid 2 have a total height equal to l_2 and the two layers consisting of Fluid 1 have a total height of l_1, where $l_1 + l_2 = l$. The horizontal velocity (or velocity profile) is a linear function of y, and increases continuously from $u(0) = 0$ to $u(l) = U$.

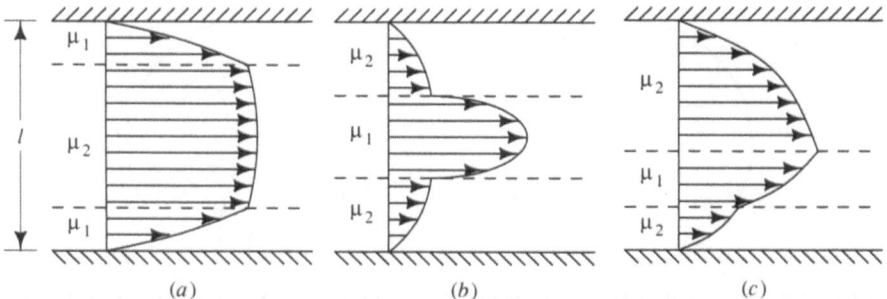

Fig. 3.3(a–c). [Figure 4 of Joseph, Nguyen and Beavers, 1984] Layered plane Poiseuille flow in three layers. The configurations in (a) and (b) are the centrally located ones. The configuration in (a) maximizes the mass flux for a given pressure gradient, among the continuum of contiguous layers of high viscosity (μ_2) layers of total height l_2 and low viscosity (μ_1) layers of total height l_1, such that $l_1 + l_2 = l$.

to the axis of the pipe, so the pressure must be a constant in each of the fluids within a cross-section (but depends on the axial coordinate). At the interface, because of the parallel nature of the flow, the jump in the normal stress is the jump in the pressures, and hence is a constant. By equation (2e.3), this constant must equal the product of the surface tension with the principle radius of curvature. Therefore, if the surface tension is zero, any interface curvature is allowed. If the surface tension is not zero, the curvature must be a constant, so the interface is a circle or a circular arc terminating at the pipe wall (see figure 3.5, drawn for a pipe of circular

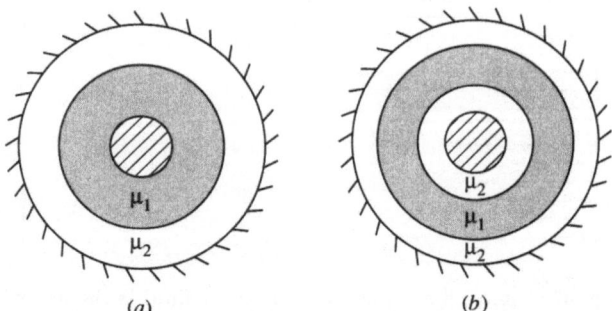

Fig. 3.4(a–b). Couette flow in circles of immiscible liquids of the same density but different viscosities μ_1 and μ_2: (a) two layers and (b) three layers, with the same total volume of fluids. The number and thickness of these layers are arbitrary.

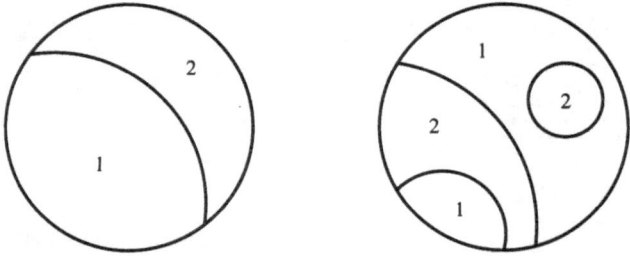

Fig. 3.5. [Figure 2 of Renardy, 1988, Rocky Mountain Math Consortium] Some possible arrangements for anti-plane shear flow in a pipe of circular cross-section, in the presence of surface tension.

cross-section). In either case, there is an infinite number of possible interface positions.

Nonunique arrangements of components in bicomponent flows appear to be a general property going far beyond the examples exhibited here. For example, we could generate nonunique two-dimensional 'Poiseuille' flows by, say, a wavy perturbation of the solid boundary. As another example, consider how the circular flows in figure 3.4 perturb when there is a small density difference. We would expect to see slightly displaced circles.

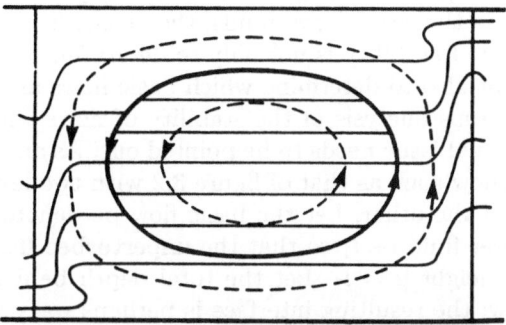

Fig. 3.6. [Figure 2 of Busse, Geophysical Research Letters, 9, 519, 1982, copyright by the American Geophysical Union.] Sketch of boundary layer model of high Rayleigh number convection. The horizontal lines in the interior of the convection cell indicate isotherms. The solid streamline separates fluids A and B.

I.3(c) Two-Fluid Convection

Busse [1982] describes possible solutions for two-fluid convection (see figure 3.6) as follows:

"Convection in a fluid layer heated from below with two immiscible components A and B of the same average density exhibits different states of motion. Besides a solution describing convection in sublayers, other solutions in which fluid B is surrounded by streamlines of fluid A or vice versa are possible."

I.3(d) Rotating Couette Flow

Many possible flows of two liquids between rotating cylinders are possible (see chapter II). For example, we may imagine *layered Couette flow* between rotating cylinders in which one fluid is on the inner rotating cylinder and the other one is on the outside cylinder [Renardy and Joseph 1985 a]. This type of flow is illustrated in figure 3.4. A completely different possibility is the family of *banded Couette flows* with the two liquids arranged in bands as in figure 5.10 in section II.5 (see also Joseph, Nguyen and Beavers [1984]).

I.3(e) Nonuniqueness and Stability

The natural thing to do when faced with nonuniqueness may be to study the stability of each possible arrangement. This is frustrated by the fact that not all the solutions may be known. For example, experiments [Moffatt 1977] show that a thin film of constant radius on a rotating rod is unstable, but there are stable rigid motions, that may not be known a priori, which have rings and bumps for the same data. There may also be more than one stable solution. In the case of two fluids, the "basic flows" are not unique, so that the study of stability is not only to determine when the basic flow loses stability but also to determine which basic flows are stable.

For the rigorous analysis of the stability of flows with interfaces, the following conceptual issue needs to be pointed out. As an example, imagine a parallel shear flow such as that of figure 3.2 with two layers of fluids, one flowing on top of the other. Let the basic flow be denoted by $\mathbf{U}$, and the depth of the lower fluid be l_1 so that the unperturbed flat interface in the basic flow is at height $y = l_1$. Let the total depth be denoted by l. Now perturb this flow: the resulting interface is perhaps wavy and is located at height $y = l_1 + h(x)$ where $h(x)$ must be found as part of the solution. Across this interface, we have to deal with continuity conditions (see section I.2 (e)). Take, for example, the continuity of velocity. Let the total perturbed velocity be denoted by $\mathbf{U} + \hat{\mathbf{u}}$ where $\hat{\mathbf{u}}$ is the perturbation. The continuity of velocity is then expressed as $\mathbf{U_1} + \hat{\mathbf{u}}_1 = \mathbf{U_2} + \hat{\mathbf{u}}_2$ at $y = l_1 + h(x)$, where subscipts 1 and 2 refer to the lower and upper fluids, respectively. But now note that the basic flows $\mathbf{U_1}$ and $\mathbf{U_2}$ are defined for the domains $0 \leq y \leq l_1$ and $l_1 \leq y \leq l$, respectively, so that if the perturbed interface height $l_1 + h(x)$ lies in $0 \leq y \leq l_1$, then there is the complication of having to make sense out of what is meant by $\mathbf{U_2}$ there. Similarly, if the perturbed interface lies in $l_1 \leq y \leq l$, then we have to define what is meant by $\mathbf{U_1}$ there.

In general, the position of the interface changes when it is perturbed so that the basic flow, defined for the domain with the unperturbed interface, is not fully defined when the interface is perturbed. To get around this, the usual procedure is to define $\mathbf{U}$ at the perturbed interface by its Taylor expansion about the unperturbed position $y = l_1$: $\mathbf{U}(l_1 + h) = \mathbf{U}(l_1) + h\mathbf{U}'(l_1) + \dots$. This procedure "extends" the definition of the basic flow $\mathbf{U}$ into domains in which it was not defined to begin with.

It is thus useful to extend the domain of the basic flow into neighboring domains in which it was not originally defined. When the extended basic flow is stable, the perturbations, including the perturbed interface, ultimately decay to zero. This method of extension is always used but almost never acknowledged. The method of extension is especially useful for nonlinear problems like bifurcation problems in which the final interface differs from the starting one.

I.3(f) Nonuniqueness and Variational Principles

A short-cut to doing stability analyses is to obtain the configuration of the two fluids from a variational principle. The reader should not be confused by the word "principle", which we take here to mean some statement of physics in which some people believe but cannot prove.

It has been suggested by Southern and Ballman [1973], Everage [1973] and MacLean [1973] that the configuration which is achieved by two liquids in rectilinear motion down a round pipe is the one which minimizes the viscous dissipation, say the total mass flux is maximized for a given pressure gradient. If this variational "principle" were used for the pipeline transport of oil with the addition of water, we would maximize the mass flux of oil plus water. Another "principle" worked on already by Russell and Charles [1959] is to maximize the flux of oil alone for a given pressure gradient. Both of these variational problems give rise to unique solutions, if the class of admissible variations is appropriately prescribed.

There are various ways to generalize variational principles expressing the notion that the dissipation is extremalized. A more precise way to formulate this extremal problem is in terms of fluxes and forces as follows (using the notation of section I.2).

We assume that the flow is steady: $\partial \mathbf{u}/\partial t = 0$, and that gravity is absent. We also assume creeping flow: $\mathbf{u} \cdot \nabla \mathbf{u}$ is neglected. The equation of motion (2b.11) is then div $\mathbf{T} = 0$ where $\mathbf{T}$ is given in (2b.7). Multiply by $\mathbf{u}$ and integrate over the volume V. In indicial notation, this is

$$0 = \int_V u_i \partial_j T_{ij} = -\int_V \partial_j u_i T_{ij} + \int_{\partial V} u_i T_{ij} n_j + \int_\Sigma [\![\mathbf{u} \cdot \mathbf{t}]\!] dS, \qquad (3f.1)$$

where $T_{ij} n_j = t_i$, $\mathbf{t}$ is defined in (2b.5), Σ is the interface, and ∂V is the exterior boundary. We assume the continuity of velocity $[\![\mathbf{u}]\!] = 0$, and the continuity of tractions $[\![\mathbf{T}]\!] \cdot \mathbf{n_{12}} = 0$ on each and every interface: hence, the last term in (3f.1) vanishes. Also, since the product of two matrices $\mathbf{AB}$ is represented in indicial notation by $A_{ij} B_{jk}$, the trace of $\mathbf{AB}$ is $A_{ij} B_{ji}$: thus, $\partial_j u_i T_{ij}$ is the trace of $\mathbf{TL}$ where $\mathbf{L}$ is defined in (2c.8). We note that tr $(\mathbf{TL}) = \mathrm{tr}(\mathbf{TD}[\mathbf{u}])$ where $\mathbf{D}$ is given in (2c.8). Therefore, (3f.1) yields

$$\int_{\partial V} \mathbf{u} \cdot (\mathbf{Tn}) dS = \int_V \mathrm{tr}\ (\mathbf{TD}[\mathbf{u}]) dV. \qquad (3f.2)$$

For the case of incompressible Newtonian fluids, using (2c.12), this equation is

$$\int_{\partial V} \mathbf{u} \cdot (\mathbf{Tn}) dS = \int_V 2\mu \mathbf{D} : \mathbf{D} dV. \qquad (3f.3)$$

This equation expresses a balance between the work done by tractions on the left hand side and viscous dissipation of energy. If the flow were non-isothermal, non-steady, includes gravity, or the effect of surface tension, then this equation would be modified as in (2c.3).

In (3f.2), we could identify the flux with $\mathbf{u}$ on ∂V and the force with $\mathbf{Tn}$ there, whilst tr $(\mathbf{TD}[\mathbf{u}])$ gives the dissipation.

We could try to maximize the flux for a given force, or to minimize the force for a given flux, over a set of arrangements of the two fluids in a set of nonunique steady solutions.

Rotating Couette Flow. A very simple example of such a calculation is presented below for the case of two-fluid flow between concentric cylinders (see figure 3.4 and section I.3 (d)). The fluids are assumed to have the same density: the variational approach just outlined is not useful for the case of different densities.

We use cylindrical coordinates (r, θ, x). If the velocity is (u, v, w), then the stress tensor (2b.7) is (see, for example, Appendix 2 of Batchelor [1970])

$$T^{<xx>} = -p + 2\mu\frac{\partial w}{\partial x}, \quad T^{<rr>} = -p + 2\mu\frac{\partial u}{\partial r},$$

$$T^{<\theta\theta>} = -p + \frac{2\mu}{r}\left(\frac{\partial v}{\partial \theta} + u\right), T^{<r\theta>} = \mu\left(r\frac{\partial}{\partial r}\left(\frac{v}{r}\right) + \frac{1}{r}\frac{\partial u}{\partial \theta}\right),$$

$$T^{<\theta x>} = \mu\left(\frac{\partial v}{\partial x} + \frac{1}{r}\frac{\partial w}{\partial \theta}\right), \quad T^{<xr>} = \mu\left(\frac{\partial u}{\partial x} + \frac{\partial w}{\partial r}\right). \tag{3f.4}$$

The angular velocity of each cylinder is prescribed, Ω_2 at the outer cylinder located at radius $r = b$, Ω_1 at the inner cylinder located at $r = a$, $b > a$. The two fluids have prescribed volumes. We choose the set of arrangements to be the placement of each fluid in annular layers, with the velocity

$$\mathbf{u} = \mathbf{e}_\theta v(r). \tag{3f.5}$$

The conditions at the boundaries of the flow are that

$$v(r) = \begin{cases} a\Omega_1 & \text{at } r = a, \\ b\Omega_2 & \text{at } r = b. \end{cases} \tag{3f.6}$$

The stress tensor (3f.4) simplifies to

$$T^{<xx>} = T^{<rr>} = T^{<\theta\theta>} = -p, \quad T^{<\theta x>} = T^{<xr>} = 0,$$

$$T^{<r\theta>} = 2\mu D^{<r\theta>} = \mu r\frac{\partial}{\partial r}\left(\frac{v}{r}\right). \tag{3f.7}$$

The extremal problem is to minimize the torque between the rotating cylinders. Suppose that for this Couette flow between cylinders, we allow the two fluids to be in N annular rings or layers. The normal to the interface $\mathbf{n}$ is the radial vector $\mathbf{e_r}$. In equation (3f.3), the V is the cross-sectional area at a fixed value of the axial coordinate x, and the exterior boundary ∂V consists of the circles at $r = a$ and b. The left hand side of (3f.3) is (see equation (4.9) of Joseph, Nguyen and Beavers [1984])

$$\int_{\partial V} \mathbf{u}\cdot(\mathbf{Tn}) = \int_{\partial V} v(r)\,\mathbf{e}_\theta\cdot(\mathbf{Tn}), \tag{3f.8}$$

where $\mathbf{Tn} = T^{<rr>}\mathbf{e_r} + T^{<r\theta>}\mathbf{e}_\theta$ and $\mathbf{e}_\theta\cdot\mathbf{Te_r} = T^{<r\theta>}$,

$$= \int_{\partial V} v(r)T^{<r\theta>}rd\theta = \int_0^{2\pi} v(b)T^{<r\theta>}(b)bd\theta - \int_0^{2\pi} v(a)T^{<r\theta>}(a)ad\theta$$

and using the boundary conditions (3f.6),

$$= 2\pi(b^2\Omega_2 T^{<r\theta>}(b) - a^2\Omega_1 T^{<r\theta>}(a)). \qquad (3f.9)$$

The torque density is defined to be the product of the radius with the force density, or $r\mathbf{e_r} \times \mathbf{Tn}$; the total torque at the radius $r = r_0$ is

$$\int_0^{2\pi} \left[r\mathbf{e_r} \times \mathbf{Tn}r\right]_{r=r_0} d\theta = r_0^2 \mathbf{e_x} T^{<r\theta>}(r_0)2\pi.$$

This will be denoted by $M2\pi\mathbf{e_x}$: we can verify that M is a constant for $a \leq r \leq b$ by substituting for $v(r)$, for example, from equations (3f.15) - (3f.16). Since M is a constant, we can write (3f.9) as

$$= (\Omega_2 - \Omega_1)2\pi M. \qquad (3f.10)$$

It is clear from the right hand side of (3f.3) that $(\Omega_2 - \Omega_1)M$ is positive.

After integrating the right hand side of (3f.3) over θ, substituting for $D^{<r\theta>}$ from (3f.7), and dividing through by 2π, we have

$$M(\Omega_2 - \Omega_1) = \frac{\mu}{2}\int_a^b r^3\left[\frac{d(v/r)}{dr}\right]^2 dr$$

$$= \frac{\mu_1}{2}\int_a^{r_{(1)}} r^3\left[\left(\frac{v}{r}\right)'\right]^2 dr + \frac{\mu_2}{2}\int_{r_{(1)}}^{r_{(2)}} r^3\left[\left(\frac{v}{r}\right)'\right]^2 dr + \frac{\mu_1}{2}\int_{r_{(2)}}^{r_{(3)}} r^3\left[\left(\frac{v}{r}\right)'\right]^2 dr + \dots,$$

$$(3f.11)$$

and the continuity of velocity and shear stress (2e.1) - (2e.2) yields

$$\left[\!\left[\mu\left(\frac{v}{r}\right)'\right]\!\right] = \left[\!\left[v\right]\!\right] = 0 \quad \text{at } r_{(1)}, r_{(2)}, \dots \qquad (3f.12)$$

The normal stress condition (2e.3) yields the continuity of the pressure.

We want to choose the arrangements of layers and the placement of viscosities so as to minimize $|M|$ when Ω_2 and Ω_1 are prescribed. It will suffice to solve this problem for a two-layer configuration.

We may always consider the problem posed for two adjacent layers. we find that the minimizing solution has the lower-viscosity fluid on the inside. We conclude that we can take $N = 2$, with the more viscous fluid on the outside and the less viscous fluid inside.

Suppose $r_{(1)} = d$ is the interface between two layers lying between $r = a$ and $r = b$. The fluid with viscosity μ_2 occupies the region $a \leq r \leq d$, and the fluid with viscosity μ_1 occupies the region $d \leq r \leq b$. The equations of motion are

$$\rho\frac{v^2}{r} = \frac{\partial p}{\partial r}, \quad \frac{\partial p}{\partial x} = 0, \qquad (3f.13)$$

$$\frac{1}{r}\frac{\partial p}{\partial \theta} = \mu(\nabla^2 v - \frac{v}{r^2}),$$

and for Couette flow between rotating cylinders,

$$\frac{\partial p}{\partial \theta} = 0.$$

Therefore,

$$\frac{1}{r}\frac{d}{dr}\left(r\frac{dv}{dr}\right) = \frac{v}{r^2}. \tag{3f.14}$$

Solving this together with the interface conditions (3f.12) and the boundary conditions (3f.6), we find that $v(r)$ is given by:

$$v_1 = A_1 r + B_1/r,$$

$$A_1 = \left(\mu_2\left(\frac{\Omega_1}{b^2} - \frac{\Omega_2}{d^2}\right) + \mu_1\Omega_2\left(\frac{1}{d^2} - \frac{1}{a^2}\right)\right)/q,$$

$$q = \mu_2\left(\frac{1}{b^2} - \frac{1}{d^2}\right) + \mu_1\left(\frac{1}{d^2} - \frac{1}{a^2}\right),$$

$$B_1 = (\Omega_2 - \Omega_1)\mu_2/q \tag{3f.15}$$

in $d \leq r \leq b$; and in $a \leq r \leq d$ by

$$v_2 = A_2 r + B_2/r.$$

$$A_2 = \left(\mu_2\Omega_1\left(\frac{1}{b^2} - \frac{1}{d^2}\right) + \mu_1\left(\frac{\Omega_1}{d^2} - \frac{\Omega_2}{a^2}\right)\right)/q,$$

$$B_2 = \mu_1 B_1/\mu_2. \tag{3f.16}$$

Since

$$M(\Omega_2 - \Omega_1) = \frac{\mu_2}{2}\int_a^d r^3\left[\left(\frac{v_2}{r}\right)'\right]^2 dr + \frac{\mu_1}{2}\int_d^b r^3\left[\left(\frac{v_1}{r}\right)'\right]^2 dr$$

$$= \frac{a^2 b^2(\Omega_2 - \Omega_1)^2(ka^2 + b^2)\mu_1\mu_2}{(b^2 - a^2)(b^2\mu_1 + ka^2\mu_2)}, \tag{3f.17}$$

where $k = (b^2 - d^2)/(d^2 - a^2)$, is positive, we may, without losing generality, consider the case for which M and $\Omega_2 - \Omega_1$ are positive. It is immediate that $M(k, \mu_1, \mu_2)$ is a monotonically increasing function of μ_2, from zero at $\mu_2 = 0$ to

$$\frac{b^2(\Omega_2 - \Omega_1)(ka^2 + b^2)\mu_1}{(b^2 - a^2)k}. \tag{3f.18}$$

It follows from monotonicity that M is larger when $\mu_2 > \mu_1$ than when $\mu_2 < \mu_1$. So we minimize the torque by putting the lower-viscosity fluid $\mu_- = \mu_2$ on the inner cylinder up to the radius $r = d$.

The situation when one of the liquids occupies an infinite region has to be treated separately. For example, when $b \to \infty$ and $\Omega_2 \to 0$ then we solve (3f.14) with $v(\infty) = 0$ to find $v_1(r) = D_1/r$ for $d \leq r < \infty$, $v_2(r) = C_2 r + D_2/r$ for $a \leq r \leq d$, where C_2, D_1 and D_2 are determined by the conditions (3f.12) at $r = d$ and $v(a) = \Omega_1 a$. This yields

$$|M| = \frac{|\Omega_1|a^2 d^2 \mu_1 \mu_2}{(\mu_2 - \mu_1)a^2 + \mu_1 d^2}, \qquad (3f.19)$$

which, for a fixed d is smaller when $\mu_2 > \mu_1$.

So, in every pair of layers, the arrangement that minimizes the torque has the low-viscosity liquid on the inside. It follows that the optimal arrangement of the layers for minimum torque is the one with two layers and the less viscous fluid on the inner cylinder.

Plane Couette Flow. Consider Couette flow in the $x - y$ plane with the two fluids lying between the planes $y = 0$ and l as in figure 3.2. The velocity $\mathbf{u}$ is $(u(y), 0)$ with $u(0) = 0$ and $u(l) = U$. The tensor $\mathbf{D}$ in (2c.8) is

$$\mathbf{D} = \begin{pmatrix} 0 & \frac{1}{2}u'(y) \\ \frac{1}{2}u'(y) & 0 \end{pmatrix}. \qquad (3f.20)$$

The preferred arrangement for this problem may be studied using (3f.3) in a reduced form, where V denotes the line integral from $y = 0$ to l at a fixed value of x , and ∂V represents the endpoints. The left hand side of (3f.3) is $[u(y)T^{<xy>}]_{y=0}^{y=l}$. We have

$$UT^{<xy>}(l) = \int_0^l \mu u'^2(y)dy. \qquad (3f.21)$$

The value of $T^{<xy>}(l)$ depends only on $\mu u'(l)$, which is independent of the number and size of the layers for the following reason. Suppose $u_1(y)$ and $u_2(y)$ are the velocities in fluids 1 and 2 respectively. Continuity of shear stress yields

$$\mu_1 u_1' = \mu_2 u_2' \qquad (3f.22)$$

and in fact for Couette flow, $\mu u'$ is a constant throughout the flow because the velocity field is linear in y. We can calculate what this constant is provided that the total volumes l_1 and l_2 of the fluids 1 and 2 are prescribed. Since $\mu u'$ is a constant, we have that u' is a constant in each fluid, no matter how many layers there are. We also have

$$u_1' l_1 + u_2' l_2 = U. \qquad (3f.23)$$

These are two equations for the two unknowns u_1' and u_2' and so the left hand side of (3f.21) is a constant, independent of the number and size of layers.

It follows that the variational problem for the preferred arrangement of layers in plane Couette flow has no solution.

In pipe flow of two liquids with different viscosities under an applied pressure drop, experiments show that the low viscosity liquid will tend to encapsulate the high viscosity liquid. If the effects of gravity are negligible, the phases will arange themselves so that whatever may have been the initial configuration, the high viscosity phase will ultimately be centrally located (see figure 1.1). This property has been convincingly demonstrated in experiments with very viscous viscoelastic liquids (polymer melts), as well as in the flow of oil and water in which the water migrates to the pipe wall, forming a lubrication layer (see section I.1 (b)).

Theoretical explanations for the slow envelopment phenomenon were based upon extremalizing energy dissipation [Everage 1973; Southern and Ballman 1973; MacLean 1973]. Maclean [1973], who considered planar layered flow, and Everage [1973], who studied a cylindrical geometry, both invoked a variational principle to show that the phase configuration with the high-viscosity component centrally located is favored over several other configurations.

We will state an ad-hoc "viscous dissipation principle" which reduces to the previously mentioned ones [Everage 1973; Maclean 1973].

Plane Poiseuille Flow. For simplicity, we first consider the case of layered plane Poiseuille flows of the type shown in figure 3.3 in which the liquids are arranged in parallel layers. The velocity is $\mathbf{u} = (u(y), 0)$ where u may be a quadratic function of y, and the tensor $\mathbf{D}$ is again of the form (3f.20). There is a pressure gradient in the x-direction.

The total number N of the layers and the size of the layers, subject to total height constraints, are left undetermined. Now we shall write the energy balance (3f.3). We choose the volume here to be a plane area of channel height l and length L (along the axis x). Since $\mathbf{u}$ vanishes on the solid walls ($y = 0, l$), the boundary integral on the left of (3f.3) is over planes perpendicular to x at x and $x + L$:

$$\int_{\partial V} \mathbf{u} \cdot (\mathbf{T} \cdot \mathbf{n}) = \int_0^l \left(-uT^{<xx>}|_x + uT^{<xx>}|_{x+L} \right) dy$$

$$= (-p_{x+L} + p_x) \int_0^l u \, dy = -\frac{\Delta p}{L} LQ = GLQ, \qquad (3f.24)$$

where $Q = \int_0^l u \, dy$ is the volume flux and $G = -\Delta p/L$ ($\Delta p < 0$ so $G > 0$) is the pressure gradient.

On the other hand,

$$2 \int \mu \mathbf{D} : \mathbf{D} = \int \mu u'^2(y) dV$$

$$= L \int_0^l \mu u'^2(y) dy. \qquad (3f.25)$$

Hence

$$GQ = \int_0^l \mu u'^2(y) dy. \qquad (3f.26)$$

We next recall that between 0 and l are N contiguous layers with fluids of different viscosities. We can suppose that the layer nearest the bottom is occupied by a fluid of viscosity μ_1, the next layer has μ_2, then μ_1 again, and so on. So besides the total number N of layers and their sizes, we need to know if μ_1 is the larger viscosity. We suppose that the total volume (height) of high-viscosity (μ_+) fluid is given as l_+ and l_- is the volume of low-viscosity fluid and $l_+ + l_- = l$. The N layers are divided into intervals

$$[0, y_{(1)}], [y_{(1)}, y_{(2)}], [y_{(2)}, y_{(3)}], \ldots, [y_{(N-1)}, l]. \qquad (3f.27)$$

With G given, we maximize

$$Q = \int_0^{y_{(1)}} u_{(1)}(y)dy + \int_{y_{(1)}}^{y_{(2)}} u_{(2)}(y)dy + \ldots + \int_{y_{(N-1)}}^{l} u_{(N)}(y)dy, \quad (3f.28)$$

where

$$QG = \mu_1 \int_0^{y_{(1)}} u_{(1)}'^2(y)dy + \mu_2 \int_{y_{(1)}}^{y_{(2)}} u_{(2)}'^2(y)dy + \mu_1 \int_{y_{(2)}}^{y_{(3)}} u_{(3)}'^2(y)dy + \ldots$$
$$(3f.29)$$

Now we change variables:

$$u_{(1)} = \frac{G}{\mu_1} v_{(1)}, \quad u_{(2)} = \frac{G}{\mu_2} v_{(2)}, \quad u_{(3)} = \frac{G}{\mu_1} v_{(3)}, \ldots \qquad (3f.30)$$

Since the equation for plane Poiseuille flow is

$$\mu u'' = -G, \qquad (3f.31)$$

the equation satisfied by $v_{(n)}$ is

$$v_{(n)}'' = -1. \qquad (3f.32)$$

Thus, $v_{(n)}' = -y + Y_n$ where Y_n is a constant. This constant is found from the continuity of shear stress at each interface

$$[\![\mu u']\!] = 0 \qquad (3f.33)$$

to be independent of n; let Y_n be denoted by Y. Then

$$v_{(n)}' = -y + Y. \quad y \in [0, l] \qquad (3f.34)$$

Because of the sign of u'' in (3f.31), v' must change sign between $y = 0$ and $y = l$ and the value of Y is found from that position. For example, in (a) and (b) of figure 3.3, this change of sign occurs at the mid-plane where $v_{(n)}'(Y) = 0$ so that $Y = l/2$, while in (c), this occurs at an interface. Inserting the change of variables (3f.30) into (3f.29), we get

$$\frac{Q}{G} = \frac{1}{\mu_1} \int_0^{y_{(1)}} (y - Y)^2 dy + \frac{1}{\mu_2} \int_{y_{(1)}}^{y_{(2)}} (y - Y)^2 dy + \frac{1}{\mu_1} \int_{y_{(2)}}^{y_{(3)}} (y - Y)^2 dy + \ldots$$
$$(3f.35)$$

$$= \int_0^l \frac{1}{\mu}(y - Y)^2 dy, \qquad (3f.36)$$

We will use the following equation. Since u vanishes at $y = 0$ and l, we have $\int_0^l u' dy = 0$. From (3f.34), $u' = -G(y - Y)/\mu$ so that

$$\int_0^l \frac{1}{\mu}(y - Y) dy = 0$$

or,

$$\int_0^l \frac{1}{\mu}(y - \frac{l}{2}) dy = \int_0^l \frac{1}{\mu}(Y - \frac{l}{2}) dy. \qquad (3f.37)$$

Equation (3f.36) is

$$\int_0^l \frac{1}{\mu}\left((y - \frac{l}{2}) - (Y - \frac{l}{2})\right)^2 dy$$

$$= \int_0^l \frac{1}{\mu}(y - \frac{l}{2})^2 dy - 2 \int_0^l \frac{1}{\mu}(y - \frac{l}{2})(Y - \frac{l}{2}) dy + \int_0^l \frac{1}{\mu}(Y - \frac{l}{2})^2 dy. \qquad (3f.38)$$

Using (3f.37) on the second term, (3f.38) becomes

$$\int_0^l \frac{1}{\mu}(y - \frac{l}{2})^2 dy - \int_0^l \frac{1}{\mu}(Y - \frac{l}{2})^2 dy. \qquad (3f.39)$$

We maximize this by maximizing each term separately. The second term is non-positive, so it is maximized by setting

$$Y = \frac{l}{2}. \qquad (3f.40)$$

The first term is maximized if we place the fluid with the larger $1/\mu$ at the place where $(y - \frac{l}{2})^2$ is the largest. Both this condition and (3f.40) can be simultaneously satisfied if the arrangement of the fluids is symmetric about the mid-plane with the less viscous fluid outside, just as in part (a) of figure 3.3. Thus, the solution to maximizing Q is $N = 3$ and μ_1 is the low-viscosity μ_-.

The same considerations, but in more complicated form, enter into the rigorous solution of the extremal problem for pipe flow [Joseph, Renardy and Renardy 1984] and reproduced below.

Pipe Flow. We consider flow in a cylindrical pipe, whose cross-section Ω is a bounded domain with a smooth boundary. The pipe is occupied by two fluids of equal densities, but different viscosities. Fluids 1 and 2 occupy regions Ω_1 and Ω_2, respectively, so that $\Omega = \Omega_1 \cup \Omega_2$. An example is the right-most arrangement in figure 1.1. The sizes of Ω_1 and Ω_2 are given, but their locations are unknown. The viscosity μ is a step function $\mu = \mu_1$ in Ω_1, μ_2 in Ω_2. We study stationary antiplane shear flow, i.e., if x and y denote transverse coordinates, and z the axial coordinate, then the velocity field has only one component, namely $u(x, y)$ in the z-direction. The flow is forced by a given pressure gradient $\partial p/\partial z = -G$. For this situation, the equation of equilibrium reads

$$div(\mu \nabla u) = -G, \qquad (3f.41)$$

with the boundary condition

$$u = 0, \qquad (3f.42)$$

on the boundary of Ω. On the interface between the two fluids, the velocity and shear stress have to be continuous: these two conditions are automatically satisfied if equation (3f.41) is taken in the distributional sense. Surface tension is ignored. Equations (3f.41)-(3f.42) are the Euler equations for the variational problem

$$min \int_\Omega F_\mu(u), \quad u \in H_0^1(\Omega), \qquad (3f.43)$$

where

$$F_\mu(u) = \frac{1}{2}\mu(\nabla u)^2 - Gu. \qquad (3f.44)$$

$2G$ is the Lagrange multiplier. Here, the integral expression is to be minimized for a prescribed viscosity step function μ.

Since $F_\mu(u)$ is a strictly convex functional of u, the following result on nonuniqueness is immediate: for any $\mu \in L^\infty(\Omega)$ such that $\mu \geq \epsilon > 0$, there exists one and only one $u \in H_0^1(\Omega)$, which solves the problem (3f.43), and hence problem (3f.41)-(3f.42). Thus, for any given arrangement of the fluids, there is a corresponding flow field. However, only certain interface positions are observed in experiments, others are not. The question thus arises: which interface positions are stable?

The viscous dissipation principle states that the stable solution is the one that minimizes (3f.43) with respect to both u and μ. Thus, μ is allowed to vary in a set of functions taking constant values in the two regions, whose measures are prescribed. The idea is that the integral should be minimized not only over u but also over the choice of possible interfaces, subject to the constraint that the volume ratio of the two fluids is prescribed.

If we denote by D the rate of viscous dissipation,

$$D = \int_\Omega \mu |\nabla u|^2$$

and by W the mechanical work done by the fluid, then

$$F_\mu(u) = \frac{1}{2}D - W. \qquad (3f.45)$$

We have seen that minimizing (3f.43) over u yields the equation of motion (3f.41) and the conditions mentioned thereafter. Using these equations, we have, for any steady flow,

$$D = W.$$

Hence, the flow that minimizes F_μ is the one that maximizes D or W. In addition, since

$$W = G \int_\Omega u, \qquad (3f.46)$$

it maximizes $\int_\Omega u$, the volume flux of the fluid. Thus, the principle states that the most favored configuration is the one which minimizes viscous energy dissipation for given volume flux, or, equivalently, maximizes the flow rate for a given pressure gradient. Alternatively, we may say that the pressure gradient is minimized for a given flow rate, or that the flow rate is maximized for a given pressure gradient.

We comment on the analogy with the preceding calculation for plane Poiseuille flow. From (3f.45) - (3f.46), we have that (3f.44) becomes $-\frac{1}{2} \int Gu$, which is $-\frac{1}{2}GQ$ where Q is given after (3f.24). We maximized Q with given G in the plane Poiseuille flow calculation, so this is analogous to the problem (3f.43).

The flow of a single fluid at zero Reynolds number is indeed described by the variational formulation [Temam 1979] in (3f.43) but the above variational formulation for bicomponent flows has no firm mathematical basis.

For a pipe with non-circular cross-section, the viscous dissipation principle does not have a solution with a smooth interface [Lurie, Cherkaev and Fedorov 1982]. Rather, minimizing sequences lead to patterns involving layered structures with thinner and thinner alternating layers of the two fluids. In the limit, this leads to a region that is not filled by either fluid, but by an anisotropic mixture (the viscosity in one direction is different from the viscosity in another direction). Results available in the area of structural optimization [Lurie, Cherkaev and Fedorov 1982; Tartar 1975; Raitum 1978,79; Kohn and Lipton 1986] show that a modified formulation of the problem allowing such anisotropic mixtures does lead to the existence of minimizers, predicting flows with three regions: regions occupied by each of the fluids, and, instead of a smooth interface, a third region occupied by an anisotropic mixture, consisting of layered composites of the two fluids. In reality, surface tension would not permit the formation of layered composites, so that if one were to believe in the viscous dissipation principle, one can only guess as to how the third region ought to be interpreted.

For a pipe with circular cross-section, however, the symmetry leads to a simplification, and a classical solution to problem (3f.43) can be constructed explicitly by elementary means [Joseph, Renardy and Renardy 1984]. Following an idea of Everage [1973], let Ω be a circular disk, and let u_0 be the solution of problem (3f.41)-(3f.42): $\triangle u_0 = -G$, $u_0 = 0$ on the boundary of Ω. We put

$$u = \frac{u_0}{\mu} + \bar{u}. \qquad (3f.47)$$

Then the integral in (3f.43) is equal to

$$\int_{\Omega} \nabla u_0 \nabla u - Gu + \int_{\Omega_1} \frac{\mu_1}{2}(\nabla \bar{u})^2 - \frac{1}{2\mu_1}(\nabla u_0)^2 + \int_{\Omega_2} \frac{\mu_2}{2}(\nabla \bar{u})^2 - \frac{1}{2\mu_2}(\nabla u_0)^2,$$

$$(3f.48)$$

where Ω_2 denotes the region occupied by the more viscous fluid ($\mu_2 > \mu_1$). The first of the integrals is zero. The term

$$\int_{\Omega_1} \frac{1}{2\mu_1}(\nabla u_0)^2 + \int_{\Omega_2} \frac{1}{2\mu_2}(\nabla u_0)^2 \qquad (3f.49)$$

is maximal if and only if $(\nabla u_0)^2$ takes its smallest values in Ω_2; i.e., if Ω_2 is a disk at the center of the pipe. In this case, the boundary of Ω_1 is a line on which u_0 is constant. If we then choose $\bar{u} = 0$ in the outer region, and $\bar{u}$=constant in the inner region, such that u is continuous, then the continuity of velocity and shear stress across the interface is satisfied. The expression

$$\int_{\Omega_1} \frac{\mu_1}{2}(\nabla \bar{u})^2 + \int_{\Omega_2} \frac{\mu_2}{2}(\nabla \bar{u})^2 \qquad (3f.50)$$

becomes zero, which is clearly it minimal value. Therefore, there is a unique minimizer with a smooth interface and it is the concentric arrangement with the more viscous fluid located at the core; this agrees with experimental observation.

Does the "viscous dissipation principle" really explain encapsulation? The validity of the principle can be assessed by a study of the linear stability of the concentric arrangement. The results show that the viscous dissipation principle does not hold because stability depends crucially on the volume ratio, surface tension effects, the Reynolds stresses in the fluids, and other things. In addition, the principle has other weaknesses. For example, there are flows for which it gives no prediction: we have seen that plane Couette flow is one of them. Nevertheless, experiments do suggest that the realized flows with boundary velocities prescribed do "minimize" the dissipation in the sense that the lower-viscosity liquid tends to migrate into the regions of greatest shear, thus minimizing the shearing of the high viscosity liquid. There are evidently too many constraints posed by the dynamics of the flow to give this observation a precise mathematical meaning.

Chapter II
Rotating Flows of Two Liquids

II.1 Rigid Motions of Two Liquids Rotating in a Cylindrical Container

The problem of shapes of interfaces between fluids which rotate rigidly without shear, when gravity is neglected, is determined by a balance of the capillary force against pressure forces associated with centripetal accelerations. When there is no rotation, and no other constraints, surface tension will pull the interface into a sphere.

The results which are discussed in this chapter are taken from papers by Joseph, Nguyen and Beavers [1984, 1986], called JNB; Joseph, Renardy, Renardy and Nguyen [1985], called JRRN; Guillopé, Joseph, Nguyen and Rosso [1985], called GJNR; Joseph and Preziosi [1987], called JP, and other authors for which we do not use acronyms.

II.1(a) Steady Rigid Rotation of Two Fluids

Rigid motions of a fluid are possible provided that the fluid rotates steadily about a fixed axis. Drops, bubbles, different types of fluids in all types of containers may rotate rigidly. Various kinds of perturbations of rigid motion are also of interest.

A single liquid which fills a container rotating steadily around some fixed axis will eventually rotate with the container. But in the case of two fluids, it is necessary to determine the places occupied by the two fluids and the shape of the interfaces between the two fluids.

Exercise II.1(a).1. A container is filled with an incompressible fluid and is rotating with angular velocity $\Omega(t)$. Show that rigid motions of the fluid in the container are possible only when Ω is independent of t.■

We consider the motion of two liquids that occupy the region between two coaxial cylinders of radii R_1 and R_2 which rotate with a common constant angular velocity Ω (see figure 1.1). In cylindrical coordinates, this region is denoted by

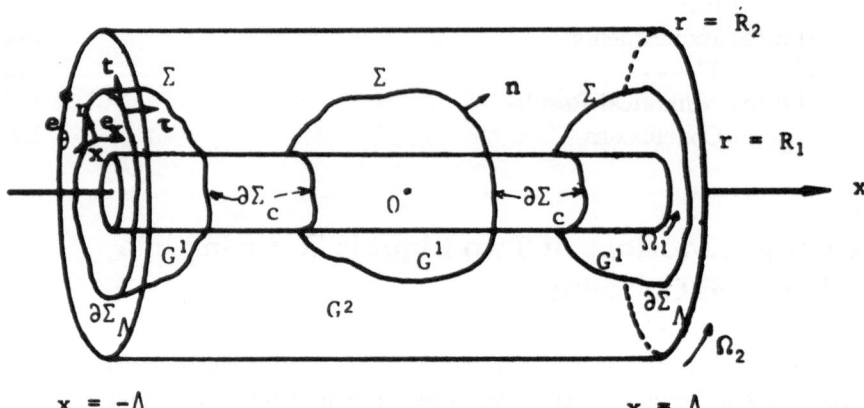

Fig. 1.1. [Figure 1 of Guillopé, Joseph, Nguyen and Rosso, 1987, Gauthier-Villars] $G = G^1 \cup G^2 \cup \Sigma$.

$$G = \{\mathbf{x} = (r, \theta, x), \; R_1 < r < R_2, \; 0 \le \theta < 2\pi, \; -\Lambda < x < \Lambda\}. \quad (1a.1)$$

Liquid one is in G^1 and two is in G^2, $G^1 \cup G^2 \cup \Sigma = G$. The interface between G^1 and G^2 is called Σ and is located at

$$r = R(\theta, x, t). \quad (1a.2)$$

This interface may consist of disjoint parts. The stress is given by $\mathbf{T} = -p\mathbf{I} + 2\mu\mathbf{D}$ (see (2b.7) and (2c.8) of chapter I) where, for $\mathbf{u} = (u, v, w)$ in cylindrical coordinates,

$$D_{rr} = \frac{\partial u}{\partial r}, \quad D_{\theta\theta} = \frac{1}{r}\frac{\partial v}{\partial \theta} + \frac{u}{r}, \quad D_{xx} = \frac{\partial w}{\partial x},$$

$$D_{r\theta} = \frac{r}{2}\frac{\partial}{\partial r}\left(\frac{v}{r}\right) + \frac{1}{2r}\frac{\partial u}{\partial \theta}, \quad D_{\theta x} = \frac{1}{2r}\frac{\partial w}{\partial \theta} + \frac{1}{2}\frac{\partial v}{\partial x}, \quad D_{xr} = \frac{1}{2}\frac{\partial u}{\partial x} + \frac{1}{2}\frac{\partial w}{\partial r}.$$

$$\quad (1a.3)$$

The incompressibility condition $div\ \mathbf{u} = 0$ is

$$\frac{1}{r}\frac{\partial}{\partial r}(ru) + \frac{1}{r}\frac{\partial v}{\partial \theta} + \frac{\partial w}{\partial x} = 0. \quad (1a.4)$$

The equations of motion are

$$\rho\frac{d\mathbf{u}}{dt} = -\nabla\phi + 2\mu div\ \mathbf{D}, \quad (1a.5)$$

where

$$\phi = p + \rho g r \sin \theta, \tag{1a.6}$$

for the case of horizontal cylinders.

We neglect gravity in the analysis to follow, and these equations are then

$$\frac{\partial u}{\partial t} + \mathbf{u} \cdot \nabla u - \frac{v^2}{r} = -\frac{1}{\rho}\frac{\partial p}{\partial r} + \frac{\mu}{\rho}(\nabla^2 u - \frac{u}{r^2} - \frac{2}{r^2}\frac{\partial v}{\partial \theta}),$$

$$\frac{\partial v}{\partial t} + \mathbf{u} \cdot \nabla v + \frac{uv}{r} = -\frac{1}{\rho r}\frac{\partial p}{\partial \theta} + \frac{\mu}{\rho}(\nabla^2 v + \frac{2}{r^2}\frac{\partial u}{\partial \theta} - \frac{v}{r^2}),$$

$$\frac{\partial w}{\partial t} + \mathbf{u} \cdot \nabla w = -\frac{1}{\rho}\frac{\partial p}{\partial x} + \frac{\mu}{\rho}\nabla^2 w, \tag{1a.7}$$

where

$$\nabla^2 = \frac{\partial^2}{\partial x^2} + \frac{1}{r}\frac{\partial}{\partial r}(r\frac{\partial}{\partial r}) + \frac{1}{r^2}\frac{\partial^2}{\partial \theta^2}, \quad \nabla = (\frac{\partial}{\partial r}, \frac{1}{r}\frac{\partial}{\partial \theta}, \frac{\partial}{\partial x}). \tag{1a.8}$$

At the interface, the kinematic free surface condition is

$$u = \frac{\partial R}{\partial t} + \frac{v}{R}\frac{\partial R}{\partial \theta} + w\frac{\partial R}{\partial x}. \tag{1a.9}$$

Jumps across the interface are designated by $[\![\cdot]\!] = (\cdot)_1 - (\cdot)_2$. We require that the jump in the velocity and shear stress, and the difference between the jump in the normal stress and the surface tension force vanish (see section I.2(d)):

$$[\![\mathbf{u}]\!] = 0, \quad -([\![p]\!] + 2H\sigma)\mathbf{n} + [\![2\mu\mathbf{D}]\!] \cdot \mathbf{n} = 0, \tag{1a.10}$$

where

$$2H =$$
$$\frac{RR_{\theta\theta}(1 + R_x^2) + RR_{xx}(R^2 + R_\theta^2) - R^2(1 + R_\theta^2) - 2R_\theta^2 - 2RR_\theta R_x R_{x\theta}}{(R^2 + R_\theta^2 + R^2 R_x^2)^{3/2}}, \tag{1a.11}$$

$$\mathbf{n} = \frac{\nabla F}{|\nabla F|}, \quad F(r, \theta, x, t) = r - R(\theta, x, t) = 0, \quad \nabla F = \mathbf{e_r} - \mathbf{e_\theta}\frac{R_\theta}{R} - \mathbf{e_x}R_x. \tag{1a.12}$$

The jump in the shear stress will be found to be automatically zero in the types of motions we will discuss.

The boundary conditions are that at $r = R_1$, $\mathbf{u} = \Omega R_1 \mathbf{e_\theta}$ and at $r = R_2$, $\mathbf{u} = \Omega R_2 \mathbf{e_\theta}$.

We neglect gravity and look at a class of solutions of the equations of motion that satisfy the boundary conditions and all of the interface conditions except possibly the normal stress condition. We call these *candidates* for rigid motions. A subclass of these solutions satisfies the normal stress condition as well, and we call these *equilibrium* solutions. Candidates for the velocity and pressure fields for rigid motions, with gravity neglected, are

$$(\mathbf{u_0}, p_0) = (\Omega r \mathbf{e}_\theta, \rho \Omega^2 \frac{r^2}{2} + c). \qquad (1a.13)$$

The velocity is continuous across the interface, no matter where it is, and $\mathbf{D}$ vanishes, so that the shear stress is automatically zero. We call (1a.13) a candidate because it need not satisfy the normal stress condition

$$-[\![p_0]\!] = 2H\sigma \quad \text{on} \quad \Sigma, \qquad (1a.14)$$

where

$$[\![p_0]\!] = [\![\rho]\!] \frac{1}{2} \Omega^2 R^2 + [\![c]\!], \qquad (1a.15)$$

if $R(\theta, x, t)$ were an arbitrary function of θ, x and t. For example, if R were a constant, then the normal stress condition could be satisfied, but for an arbitrary $R(\theta, x, t)$, (1a.14) together with (1a.11) is a partial differential equation that would have to be solved and this may not always be possible. Candidates such as the R =constant case which fit the normal stress condition (1a.14) are equilibrium solutions.

We next comment on gravity, and the conditions under which it is negligible. In order for the effect of gravity to be negligible, we require that in the normal stress condition at the interface,

$$-[\![\phi]\!] + [\![\rho]\!]gR \sin \theta + [\![2\mu D_{nn}]\!] - 2H\sigma = 0, \qquad (1a.16)$$

the pressure term in ϕ dominates $\rho g R \sin \theta$; then, the secondary motions induced by gravity are expected to be small. In the case of the candidates (1a.13), if we let d denote the mean value of $R(\theta, x) = R(x)$, then we require $[\![p_0]\!] >> [\![\rho]\!]gd$ or, from (1a.15),

$$\frac{[\![p_0]\!]}{[\![\rho]\!]gd} \approx \frac{\Omega^2 d}{2g} = F^2 >> 1, \qquad (1a.17)$$

where F is a Froude number. This condition means that gravity is dominated by centrifugal force when the Froude number is large.

There is another situation in which the dynamical effects of gravity are negligible, which is easiest to understand as the rigid motion of thin coating films of very viscous liquids. In this case, it is p which is independent of θ and secondary motions are suppressed by the fact that the force of gravity is not sufficient to make a thin viscous liquid flow. The analysis of Moffatt [1977] and the analysis and experiments of Preziosi and Joseph [1988] show that the effect of gravity on a thin film rotating in air on a cylinder of radius a, is small when

$$Sh_0^2 << 1, \qquad S = \frac{ga}{\nu\Omega} \qquad (1a.18)$$

where h_0 is the maximum film thickness $h(\theta, x)$, and ν is the kinematic viscosity of the liquid. The parameter S is essentially (R/F^2) where R is the Reynolds number Uh_0/ν, and $U = \Omega a$. When the film thickness is small and the viscosity is large, the motion approaches rigid rotation. When Sh_0^2

is at the critical value of 1, the rate of rotation of the rod is insufficient to maintain the load and some of the fluid will drop off the rod, as honey drops off a slowly rotating knife. At this critical condition, the maximum surface velocity is approximately $\Omega a/2$ (see equation 13 of Moffatt [1977]). The removal of fluid reduces h_0 so that Sh_0^2 is less than 1.

We shall now proceed with $g = 0$.

II.1(b) Disturbance Equations

Set

$$\mathbf{u} = \mathbf{u_0} + \hat{\mathbf{u}},$$

$$p = p_0 + \hat{p}, \tag{1b.1}$$

where $\mathbf{u_0}$ and p_0 are given in (1a.13), and $\hat{\mathbf{u}}$ and $\hat{p}$ represent the disturbances. Then

$$\rho\left[\frac{\partial \hat{\mathbf{u}}}{\partial t} + \hat{\mathbf{u}} \cdot \nabla \mathbf{u_0} + \mathbf{u_0} \cdot \nabla \hat{\mathbf{u}} + \hat{\mathbf{u}} \cdot \nabla \hat{\mathbf{u}}\right] = -\nabla \hat{p} + \mathrm{div}\hat{\mathbf{S}}, \tag{1b.2}$$

$$\hat{\mathbf{S}} = 2\mu\hat{\mathbf{D}}, \tag{1b.3}$$

where $\hat{\mathbf{u}}$ is solenoidal and satisfies the no slip condition on the walls of the cylinders.

At the interface Σ, the continuity of velocity and the stress conditions read

$$[\hat{\mathbf{u}}] = \mathbf{0}, \tag{1b.4}$$

$$-[\hat{p}]\mathbf{n} + [\hat{\mathbf{S}}] \cdot \mathbf{n} = [p_0]\mathbf{n} + 2H\sigma\mathbf{n}. \tag{1b.5}$$

For any integrable function f which is equal to f_1 in G^1 and f_2 in G^2, we define

$$< f >= \int_{G^1} f_1 d\mathbf{x} + \int_{G^2} f_2 d\mathbf{x}. \tag{1b.6}$$

For any g defined on Σ, we define

$$< g >_\Sigma = \int_\Sigma g d\Sigma, \tag{1b.7}$$

where, for the interface given by (1a.2) and (1a.12), we have

$$d\Sigma = (1 + \frac{R_\theta^2}{R^2} + R_x^2)^{1/2} R d\theta dx = |\nabla F| R d\theta dx. \tag{1b.8}$$

II.1(c) Energy Equation for Rigid Motions of Two Fluids

We define the kinetic energy to be

$$\mathcal{E}[\hat{\mathbf{u}}] =< \rho\frac{\hat{\mathbf{u}} \cdot \hat{\mathbf{u}}}{2} > \tag{1c.1}$$

(see section I.2(c)) and the dissipation to be

$$\mathcal{D}[\hat{\mathbf{u}}] = < 2\mu\mathbf{D}[\hat{\mathbf{u}}] : \mathbf{D}[\hat{\mathbf{u}}] > . \tag{1c.2}$$

Using Reynolds' transport theorem, we have (cf. equation I.(2a.5))

$$\frac{d}{dt}\int_G \rho f dx = \int_G \rho \frac{df}{dt} dx \tag{1c.3}$$

where

$$G = G^1 \cup G^2$$

is a material volume,

$$\mathbf{u} \cdot \mathbf{n} = 0 \text{ on the boundary of } G \tag{1c.4}$$

and

$$\frac{df}{dt} = \frac{\partial f}{\partial t} + \mathbf{u} \cdot \nabla f. \tag{1c.5}$$

Hence, setting $f = \hat{\mathbf{u}}^2/2$, we get

$$\frac{d}{dt}\int_G \frac{1}{2}\rho\hat{\mathbf{u}}^2 dx = \int_G \rho\hat{\mathbf{u}} \cdot \frac{d\hat{\mathbf{u}}}{dt} dx \tag{1c.6}$$

where, after using (1b.2) for $d\hat{\mathbf{u}}/dt$, we get

$$= \int_G \rho\hat{\mathbf{u}} \cdot (-\hat{\mathbf{u}} \cdot \nabla\mathbf{u}_0) + \hat{\mathbf{u}} \cdot (-\nabla\hat{p} + div \ \hat{\mathbf{S}}) dx. \tag{1c.7}$$

We consider the latter term in (1c.7):

$$\int_G -\hat{\mathbf{u}} \cdot \nabla\hat{p} \ dx = \int_\Sigma [\![-\hat{p}\mathbf{n}]\!] \cdot \hat{\mathbf{u}} d\Sigma, \tag{1c.8}$$

where, on the right hand side, the contribution from the volume integral of $\hat{p} div \ \hat{\mathbf{u}}$ vanishes.

Also,

$$\int_G div \ \hat{\mathbf{S}} \cdot \hat{\mathbf{u}} dx = -\int_G \hat{\mathbf{S}} : \nabla\hat{\mathbf{u}} dx + \int_\Sigma [\![\hat{\mathbf{S}}]\!] \cdot \mathbf{n} \cdot \hat{\mathbf{u}} d\Sigma \tag{1c.9}$$

Equations (1c.8) and (1c.9) are added, and we use (1b.5) for the jump in the normal stress to find

$$\int_G -\hat{\mathbf{u}} \cdot \nabla\hat{p} \ dx + \int_G div \ \hat{\mathbf{S}} \cdot \hat{\mathbf{u}} dx = \int_\Sigma [\![p_0]\!]\mathbf{n} \cdot \hat{\mathbf{u}} d\Sigma - \mathcal{D}[\hat{\mathbf{u}}] + \int_\Sigma 2H\sigma\mathbf{n} \cdot \hat{\mathbf{u}} d\Sigma. \tag{1c.10}$$

This completes the reduction of the last term in (1c.7).

We next show that the first term on the right hand side of (1c.7) vanishes. This is easier to see in Cartesian coordinates, in which $\mathbf{u}_0 = (-\Omega y, \Omega x, 0)$, so that

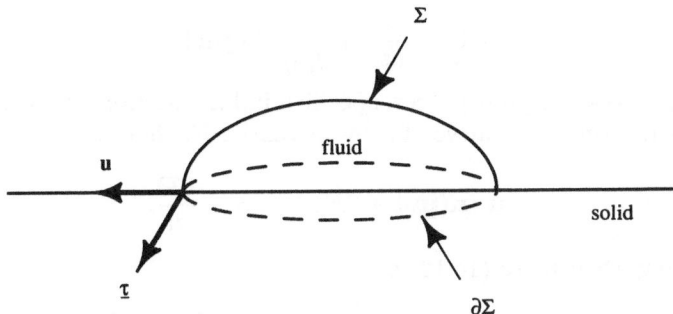

Fig. 1.2. Sketch of a fluid with surface Σ lying on a solid, with which the contact line is the boundary $\partial\Sigma$. The contact line moves with velocity $\mathbf{U}$, the vector $\underline{\tau}$ is perpendicular to $\partial\Sigma$ in the plane tangent to Σ.

$$\nabla\mathbf{u_0} = \begin{pmatrix} 0 & -\Omega & 0 \\ \Omega & 0 & 0 \\ 0 & 0 & 0 \end{pmatrix} \tag{1c.11}$$

which is antisymmetric. Thus, $(\hat{\mathbf{u}} \cdot \nabla)\mathbf{u_0} \cdot \hat{\mathbf{u}} = \hat{\mathbf{u}} \cdot (\nabla\mathbf{u_0}) \cdot \hat{\mathbf{u}} = 0$. Thus, the integrand for the first term on the right of (1c.7) vanishes.

Putting together (1c.7) and (1c.10), we have

$$\frac{d\mathcal{E}}{dt} + \mathcal{D}[\hat{\mathbf{u}}] = <\hat{\mathbf{u}} \cdot \mathbf{n}(\llbracket p_0 \rrbracket + 2H\sigma) >_\Sigma. \tag{1c.12}$$

We make use of the transport theorem for surface areas (see, for example, equation (96.11) in volume II of Joseph [1976]):

$$\frac{d}{dt}\int_\Sigma \sigma d\Sigma = \int_\Sigma -2\sigma H\mathbf{u} \cdot \mathbf{n}d\Sigma + \int_{\partial\Sigma} \sigma\underline{\tau} \cdot \mathbf{U}dl, \tag{1c.13}$$

when σ is a constant. The left hand side is then written as

$$\sigma\frac{d|\Sigma|}{dt}, \tag{1c.14}$$

where $|\Sigma|$ is the area of Σ, $\underline{\tau}$ is the outward normal to $\partial\Sigma$, lying on Σ, and $\mathbf{U}$ is the velocity of a point of the contact line $\partial\Sigma$. Figure 1.2 illustrates this for a blob of fluid lying on a solid, with a circular area in contact with the solid. In the figure, the surface of the blob is Σ and the boundary of the area in contact with the solid is $\partial\Sigma$. In other words, $\partial\Sigma$ is a contact line. The vector $\underline{\tau}$ is perpendicular to the boundary $\partial\Sigma$ and lies in the tangent plane to Σ. The vector $\mathbf{U}$ is the velocity with which the contact point shown in the figure on $\partial\Sigma$ is moving. The terms in (1c.13) are relevant only if the interface touches a solid boundary. For example, if Σ were a closed surface, not touching any solids, then there would be no $\partial\Sigma$.

Since $\hat{\mathbf{u}} = \mathbf{u} - \mathbf{u_0}$, we may write

$$< \hat{\mathbf{u}} \cdot \mathbf{n}([\![p_0]\!] + 2H\sigma) >_\Sigma = < \mathbf{u} \cdot \mathbf{n}[\![p_0]\!] >_\Sigma - < \mathbf{u_0} \cdot \mathbf{n}([\![p_0]\!] + 2H\sigma) >_\Sigma$$

$$+ \sigma\left(-\frac{d|\Sigma|}{dt} + \int_{\partial\Sigma} \mathbf{U} \cdot \underline{\tau} dl \right). \tag{1c.15}$$

Under certain circumstances, specified in later sections, we can express (1c.15) as the time derivative of some potential $\mathcal{P}$, that is,

$$< \hat{\mathbf{u}} \cdot \mathbf{n}([\![p_0]\!] + 2H\sigma) >_\Sigma = -\frac{d\mathcal{P}}{dt}. \tag{1c.16}$$

We may then write (1c.12) as

$$\frac{d}{dt}(\mathcal{E} + \mathcal{P}) = -\mathcal{D}. \tag{1c.17}$$

We will look for a velocity field and interface position which minimize $\mathcal{E} + \mathcal{P}$ for the following reasons. If the solution is at a minimum of $\mathcal{E} + \mathcal{P}$ then (1c.17) says that $\mathcal{E} + \mathcal{P}$ can only decrease. On the other hand, it cannot decrease any further if the solution is already at a minimum (if it is at a minimum, then $\mathcal{D}$ must be 0. Otherwise, (1c.17) would be impossible), so the solution is trapped at its minimum. In addition, if a solution is close to a minimum of $\mathcal{E} + \mathcal{P}$, then (1c.17) says that it will approach the minimum. Therefore, the minima of $\mathcal{E} + \mathcal{P}$ are the stable solutions. Thus, the problem of finding stable solutions can be framed as a minimization problem.

II.1(d) The Interface Potential

We proceed now to the terms on the right of (1c.15). These were computed by JRRN for the periodic problem and by Guillopé and Joseph in JP for the results reproduced below.

This calculation allows for a contact line to be on either of the cylinders or on the end walls. The boundary $\partial\Sigma$ of the interface Σ lying on any solid is assumed to have a finite number of components, say

$$\partial\Sigma = \partial\Sigma_\Lambda \cup \partial\Sigma_c. \tag{1d.1}$$

Here, $\partial\Sigma_\Lambda$ is a contact line on an end wall at $x = \pm\Lambda$, and is represented by a graph

$$r = R(\theta, \pm\Lambda, t) \tag{1d.2}$$

and $\partial\Sigma_c$ is a contact line on one of the cylinders, say the inner one. The interface there is locally represented by

$$x = \chi(r, \theta, t), \tag{1d.3}$$

so that the contact line is given by

$$x = \chi(R_1, \theta, t) \tag{1d.4}$$

$$= x_1(\theta, t). \tag{1d.5}$$

We assume that χ is continuously differentiable (derivatives of χ appear in the calculations) and periodic in θ.

We will take the following representation for Σ:

$$r = R(\theta, x, t) \quad \text{for} \quad x_1(\theta, t) < x < \Lambda, \ 0 \leq \theta < 2\pi. \tag{1d.6}$$

Thus, $\partial \Sigma_\Lambda$ is given by the curve

$$r = R(\theta, \Lambda, t). \tag{1d.7}$$

The function R is continuously differentiable (derivatives of R appear in the calculations) and periodic in θ.

Reduction of $< \mathbf{u_0} \cdot \mathbf{n} >_\Sigma$. We begin with the second term on the right of (1c.15). Since the total volume of each incompressible fluid is conserved, we deduce that

$$< \mathbf{u} \cdot \mathbf{n} >_\Sigma = 0, \tag{1d.8}$$

where, from (1a.12) and (1a.9), we have

$$\mathbf{u} \cdot \mathbf{n} = \mathbf{u} \cdot \frac{\nabla F}{|\nabla F|} = \frac{1}{|\nabla F|} \frac{\partial R}{\partial t}. \tag{1d.9}$$

Using (1a.15),

$$< \mathbf{u_0} \cdot \mathbf{n} [\![p_0]\!] >_\Sigma = < \mathbf{u_0} \cdot \mathbf{n} R^2 >_\Sigma \frac{1}{2} [\![\rho]\!] \Omega^2 + < \mathbf{u_0} \cdot \mathbf{n} >_\Sigma [\![c]\!],$$

and the latter term vanishes for the same reason as in (1d.8): $< \mathbf{u_0} \cdot \mathbf{n} >_\Sigma = 0$. Therefore,

$$< \mathbf{u_0} \cdot \mathbf{n} [\![p_0]\!] >_\Sigma = < \mathbf{u_0} \cdot \mathbf{n} R^2 >_\Sigma \frac{1}{2} [\![\rho]\!] \Omega^2. \tag{1d.10}$$

We show that this vanishes. We define

$$I_1 = < \mathbf{u_0} \cdot \mathbf{n} R^2 >_\Sigma = -\Omega \int_0^{2\pi} \int_{x_1(\theta)}^{\Lambda} R_\theta R^3 \, dx \, d\theta$$

where $\mathbf{u_0} = \Omega R \mathbf{e}_\theta$, $\mathbf{n}$ is given in (1a.12), and $d\Sigma$ in (1b.8). Using Leibniz's rule, we find that

$$\frac{d}{d\theta} \int_{x_1(\theta)}^{\Lambda} R^4 \, dx = 4 \int_{x_1(\theta)}^{\Lambda} R^3 R_\theta \, dx - \frac{dx_1}{d\theta} R^4(\theta, x_1(\theta)). \tag{1d.11}$$

We integrate this over θ from 0 to 2π. The first term yields

$$\left[\int_{x_1(\theta)}^{\Lambda} R^4 \, dx \right]_{\theta=0}^{2\pi} \tag{1d.12}$$

and this expression is the same at $\theta = 0$ and 2π due to periodicity. The last term in (1d.11) yields

$$\int_0^{2\pi} \frac{dx_1}{d\theta} R^4(\theta, x_1(\theta))d\theta \qquad (1d.13)$$

where

$$R(\theta, x_1(\theta)) = R_1 \qquad (1d.14)$$

because $x_1(\theta)$ is a contact line on a cylinder. Thus, (1d.13) is

$$R_1^4 \int_{\theta=0}^{2\pi} \frac{dx_1}{d\theta} d\theta = R_1^4 \Big[x_1(\theta)\Big]_0^{2\pi} \qquad (1d.15)$$

and this is 0 by periodicity. Therefore, $I_1 = 0$.

Reduction of $< \mathbf{u} \cdot \mathbf{n}[\![p_0]\!] >_\Sigma$. This calculation is similar to the above. We have, analogously to (1d.10),

$$< \mathbf{u} \cdot \mathbf{n}[\![p_0]\!] >_\Sigma = \frac{\Omega^2[\![\rho]\!]}{2} < \mathbf{u} \cdot \mathbf{n}R^2 >_\Sigma . \qquad (1d.16)$$

Then, using (1b.8) $(d\Sigma = R|\nabla F|d\theta dx)$ and (1d.9), we get

$$I_2 = < \mathbf{u} \cdot \mathbf{n}R^2 >_\Sigma = \int_0^{2\pi} \int_{x_1(\theta)}^\Lambda R_t R^3 dx d\theta. \qquad (1d.17)$$

Using Leibniz's rule,

$$\frac{d}{dt} \int_0^{2\pi} \int_{x_1(\theta,t)}^\Lambda R^4(\theta, x, t)dx d\theta = 4I_2 - \int_0^{2\pi} \frac{\partial x_1}{\partial t} R^4(\theta, x_1(\theta))d\theta$$

$$= 4I_2 - R_1^4 \frac{d}{dt} \int_0^{2\pi} x_1(\theta, t)d\theta, \qquad (1d.18)$$

using (1d.14).

It follows that I_2 is the time derivative of some function, and

$$< \mathbf{u} \cdot \mathbf{n}[\![p_0]\!] >_\Sigma = \frac{d}{dt} \Big(\frac{\Omega^2[\![\rho]\!]}{8} \big(< R^3|\nabla F|^{-1} >_\Sigma + \int_0^{2\pi} R_1^4 x_1(\theta, t)d\theta \big) \Big). \qquad (1d.19)$$

Reduction of $< \mathbf{u_0} \cdot \mathbf{n}2H\sigma >_\Sigma$. We have

$$< \mathbf{u_0} \cdot \mathbf{n}2H\sigma >_\Sigma = -\sigma\Omega \int_0^{2\pi} \int_{x_1(\theta,t)}^\Lambda 2HRR_\theta dx d\theta, \qquad (1d.20)$$

using the expression for $\mathbf{u_0}$ in (1a.13), for $\mathbf{n}$ in (1a.12) and for $d\Sigma$ in (1b.8). We use the identity

$$RR_\theta 2H = \frac{\partial}{\partial \theta}\Big(-\frac{R(1+R_x^2)}{|\nabla F|}\Big) + \frac{\partial}{\partial x}\Big(\frac{RR_x R_\theta}{|\nabla F|}\Big), \tag{1d.21}$$

which can be verified from (1a.11), to find

$$< \mathbf{u_0} \cdot \mathbf{n}2H\sigma >_\Sigma =$$

$$-\sigma\Omega \int_0^{2\pi}\int_{x_1(\theta,t)}^\Lambda \frac{\partial}{\partial\theta}\Big(-\frac{R(1+R_x^2)}{|\nabla F|}\Big) + \frac{\partial}{\partial x}\Big(\frac{RR_x R_\theta}{|\nabla F|}\Big) dx d\theta. \tag{1d.22}$$

The second term is $-\sigma\Omega$ times

$$\int_0^{2\pi}\int_{x_1(\theta,t)}^\Lambda \frac{\partial}{\partial x}\Big(\frac{RR_x R_\theta}{|\nabla F|}\Big) dx d\theta = \int_0^{2\pi}\Big(\frac{RR_x R_\theta}{|\nabla F|}\big|_\Lambda - \frac{RR_x R_\theta}{|\nabla F|}\big|_{x_1}\Big) d\theta. \tag{1d.23}$$

For the first term in (1d.22), we use Leibniz's rule

$$\frac{d}{d\theta}\int_{x_1}^\Lambda -\frac{R(1+R_x^2)}{|\nabla F|} dx = \int_{x_1}^\Lambda \frac{\partial}{\partial\theta}\Big(-\frac{R(1+R_x^2)}{|\nabla F|}\Big) dx + \frac{\partial x_1}{\partial\theta}\Big[\frac{R(1+R_x^2)}{|\nabla F|}\Big]_{x_1}. \tag{1d.24}$$

Since $R_1 = R(\theta, x_1(\theta))$, implicit differentiation of this yields $R_x x_1'(\theta) + R_\theta = 0$ or

$$\frac{\partial x_1}{\partial\theta} = -\frac{R_\theta}{R_x}. \tag{1d.25}$$

Integrating (1d.24) over θ, the left hand side vanishes by periodicity and we find

$$\int_0^{2\pi}\int_{x_1}^\Lambda \frac{\partial}{\partial\theta}\Big(-\frac{R(1+R_x^2)}{|\nabla F|}\Big) dx d\theta = R_1 \int_0^{2\pi}\Big[\frac{R_\theta}{R_x}\frac{(1+R_x^2)}{|\nabla F|}\Big]_{x_1} d\theta. \tag{1d.26}$$

Collecting results from (1d.22), (1d.23) and (1d.26), we find that

$$< \mathbf{u_0} \cdot \mathbf{n}2H\sigma >_\Sigma = -\sigma\Omega \int_0^{2\pi}\Big(\frac{R_x R_\theta R}{|\nabla F|}\big|_\Lambda + \frac{R_1 R_\theta}{R_x |\nabla F|}\big|_{x_1}\Big) d\theta. \tag{1d.27}$$

We may re-express the second term on the right in terms of χ defined in (1d.3). In order to do this, we define a new variable F_1, analogous to F of (1a.12):

$$F_1 = \chi(r, \theta, t) - x, \tag{1d.28}$$

and

$$|\nabla F_1| = (1 + \chi_\theta^2/R_1^2 + \chi_r^2)^{1/2}. \tag{1d.29}$$

From (1d.29) and (1d.25),

$$\frac{R_1 R_\theta}{R_x |\nabla F|}\big|_{x_1} = -\frac{R_1 \chi_\theta \chi_r}{|\nabla F_1|}\big|_{R_1} \tag{1d.30}$$

Reduction of $\int_{\partial\Sigma} \mathbf{U} \cdot \underline{\tau} dl$. The calculation of $\mathbf{U} \cdot \underline{\tau}$ on Σ (see exercise below) implies that

$$\int_{\partial \Sigma} \mathbf{U} \cdot \underline{\tau}\, dl = \int_0^{2\pi} \left(\frac{R R_x R_t}{|\nabla F|}\Big|_\Lambda - \frac{R_1 \chi_r \chi_t}{|\nabla F_1|}\Big|_{R_1} \right) d\theta. \qquad (1d.31)$$

Let α_Λ be the angle between the interface Σ and the end walls and α_c the angle between the interface Σ and the cylinders. Then

$$\cos \alpha_\Lambda = \mathbf{n} \cdot \mathbf{e_x} = -\frac{R_x}{|\nabla F|},$$

$$\cos \alpha_c = \mathbf{n} \cdot \mathbf{e_r} = \frac{\chi_r}{|\nabla F_1|}. \qquad (1d.32)$$

Exercise II.1(d).1. (GJNR)   Calculation of $\underline{\tau} \cdot \mathbf{U}$ on $\partial \Sigma$.
The interface Σ is described locally by the equation

$$r = R(\theta, x, t), \ x_1(\theta) \le x \le x_2(\theta), \ 0 \le \theta < 2\pi$$

or by

$$x = \chi(r, \theta, t), \ r_1(\theta) \le r \le r_2(\theta), \ 0 \le \theta < 2\pi.$$

Assume that $\partial \Sigma$ has a part on the side walls

$$\partial \Sigma_\Lambda = \{\mathbf{x} = R(\theta, \pm\Lambda, t)\mathbf{e_r} \pm \Lambda \mathbf{e_x}\}$$

and a part on the cylinders

$$\partial \Sigma_c = \{\mathbf{x} = R_i \mathbf{e_r} + \chi(R_i, \theta, t)\mathbf{e_x}\}.$$

Define
   $\mathbf{n} = \nabla F/|\nabla F|$, the unit vector normal to Σ, from fluid 1 to fluid 2, with
   $F(\mathbf{x}) = r - R(\theta, x)$ or $F(\mathbf{x}) = \chi(r, \theta) - x$;
   $\mathbf{t} = d\mathbf{x}/dl$, a unit vector tangent to $\partial \Sigma$;
   $\underline{\tau} = \mathbf{n} \times \mathbf{t}$, a unit vector normal to $\partial \Sigma$ on Σ;
   $\mathbf{U} = d\mathbf{x}/dt$, the velocity of a point on the contact line $\partial \Sigma$.
The function $\underline{\tau} \cdot \mathbf{U}$ is calculated as follows

$$\underline{\tau} \cdot \mathbf{U} = \mathbf{U} \cdot (\mathbf{n} \times \mathbf{t}) = \mathbf{n} \cdot (\mathbf{t} \times \mathbf{U}).$$

On the side walls at $x = \pm \Lambda$, we have

$$\nabla F = \mathbf{e_r} - \mathbf{e_\theta} \frac{R_\theta}{R} - R_x \mathbf{e_x},$$

$$\mathbf{t} = (\mathbf{e_r} R_\theta + \mathbf{e_\theta} R) \frac{d\theta}{dl},$$

$$\mathbf{U} = \mathbf{e_r} R_t + (\mathbf{e_r} R_\theta + \mathbf{e_\theta} R) \frac{d\theta}{dt},$$

$$\underline{\tau} \cdot \mathbf{U} = R R_x R_t |\nabla F|^{-1} \frac{d\theta}{dl}.$$

On the cylinders at $r = R_i$, we have

$$\nabla F = \mathbf{e_r} \chi_r + \mathbf{e_\theta} \chi_\theta/R_i - \mathbf{e_x},$$

$$\mathbf{t} = (\mathbf{e_\theta} R_i + \mathbf{e_x} \chi_\theta) \frac{d\theta}{dl},$$

$$\mathbf{U} = (\mathbf{e_\theta} R_i + \mathbf{e_x} \chi_\theta) \frac{d\theta}{dt} + \mathbf{e_x} \chi_t,$$

$$\mathbf{\tau} \cdot \mathbf{U} = -R_i \chi_r \chi_t |\nabla F|^{-1} \frac{d\theta}{dl}. \blacksquare$$

After introducing (1d.32) into (1d.27) and (1d.31) and collecting all the previous results, we find that

$$- < \hat{\mathbf{u}} \cdot \mathbf{n}([\![p_0]\!] + 2H\sigma) >_\Sigma = \frac{d}{dt}\left(\sigma |\Sigma| - \frac{\Omega^2 [\![\rho]\!]}{8}(< \frac{R^3}{|\nabla F|} > + \int_0^{2\pi} R_1^4 x_1 d\theta)\right)$$

$$+ \sigma \int_0^{2\pi} \left((R_t + \Omega R_\theta) R \cos \alpha_\Lambda|_{x=\Lambda} + (\chi_t + \Omega \chi_\theta) R_1 \cos \alpha_c|_{r=R_1}\right) d\theta.$$

(1d.33)

In order to reduce the interface terms (1d.33) to potential form we assume that

(i) The contact angle at the interface on the end walls depends only on the distance r from the contact line to the axis of the cylinder. (At two different points θ_1 and θ_2 at which $r_1 = r_2$, the contact angle will be the same.)

(ii) The contact angle at the interface on the cylinders depends only on the distance $\Lambda - x$ of the contact line to the end wall at $x = \Lambda$.

These assumptions mean that the contact angle at a certain point depends only on the position of that particular contact point, and not on the position of the entire contact line.

The assumptions (i) and (ii) imply the existence of two functions $\psi_\Lambda(R)$ and $\psi_c(\chi)$ such that

$$R \cos \alpha_\Lambda(R) = \psi'_\Lambda(R),$$

$$R_1 \cos \alpha_c(\chi) = \psi'_c(\chi).$$

(1d.34)

The reduction of (1d.33) to a time derivative of a potential $\mathcal{P}$, using (1d.34), is as follows. We write, since $\Omega \partial/\partial\theta$ gives the velocity on a cylinder,

$$\int_0^{2\pi} \left((R_t + \Omega R_\theta)\psi'_\Lambda(R) + (\chi_t + \Omega \chi_\theta)\psi'_c(\chi)\right) d\theta = \int_0^{2\pi} \frac{D}{Dt}(\psi_\Lambda + \psi_c) d\theta$$

$$= \frac{d}{dt} \int_0^{2\pi} (\psi_\Lambda(R) + \psi_c(\chi)) d\theta$$

(1d.35)

where $D/Dt = \partial/\partial t + \Omega \partial/\partial\theta$ is a derivative following rigid motion. The notations in (1d.35) are slightly misleading; the integration is to be carried out on each and every contact line.

It now follows that $\mathcal{P}$ in (1c.17) is given by

$$\mathcal{P} = \sigma\left(|\Sigma| + \int_0^{2\pi} (\psi_\Lambda(R) + \psi_c(\chi)) d\theta\right)$$

$$-\frac{[\![\rho]\!]\Omega^2}{8}\Big(<\frac{R^3}{|\nabla F|}>+\int_0^{2\pi}R_1^4 x_1 d\theta\Big). \qquad (1d.36)$$

Finally we note that the working of the contact line cannot always be represented by a potential. The relation of the assumptions (i) and (ii) which lead to a potential and the classical ones in which contact angles or contact lines are fixed is obscure. We note, however, that these assumptions hold trivially for the case of a fixed line or a constant angle independent of position and, in general, whenever implicit relations of the form $f_1(\alpha, R) = 0$ or $f_2(\alpha, \chi) = 0$ are valid.

II.1(e) Integrability of the Energy

We have now established all of the terms in (1c.17) and the next step is to look at the problem of finding the minima of $\mathcal{E} + \mathcal{P}$ which occur as $t \to \infty$.

In this section, we show (not rigorously) that the energy $\mathcal{E}$ approaches 0 as $t \to \infty$ in some sense. This is important because once this is shown, we can just concentrate on $\mathcal{P}$ for the following reasons. Because of the way $\mathcal{E}$ depends on the disturbance velocity $\hat{\mathbf{u}}$ (see (1c.1)), we have that as $t \to \infty$, $\hat{\mathbf{u}} \to 0$ so that the velocity field is that of rigid body rotation. The minimization problem is then simplified to one of looking for the minima of $\mathcal{P}$ alone, and this quantity depends only on the interface position Σ. If you don't want to see the details, skip to the next section.

Let $\mathbf{v}$ belong to a space X of square integrable solenoidal vectors defined in G which vanish on the solid parts of the boundary of G, or are periodic in the axial coordinate x, with period 2Λ, if the cylinders are infinitely long. Suppose further that the gradients of such functions are also square integrable in G where integration is in the sense of (1b.6). Such functions are said to lie in $H^1(G)$ and they satisfy

$$[\![\mathbf{v}]\!] = 0 \quad \text{on} \quad \partial\Sigma \qquad (1e.1)$$

in the sense of traces, where the trace of $\mathbf{v}$ is defined as $\mathbf{v}|_\Sigma \in H^{1/2}(\Sigma)$ [Lions and Magenes 1972]. Each such $\mathbf{v}$ satisfies Korn's inequality

$$<|\mathbf{v}|^2 >\leq 2k < |\mathbf{D}[\mathbf{v}]|^2 >, \qquad (1e.2)$$

for some positive constant k. Since

$$2 < |\mathbf{D}[\mathbf{v}]|^2 >=< |\nabla\mathbf{v}|^2 > + < |div\ \mathbf{v}|^2 >, \qquad (1e.3)$$

and $div\ \mathbf{v} = 0$, the constant k is Poincaré's constant.

Exercise II.1(e).1. Consider the functional

$$\lambda[\theta] = \frac{\int_0^1 \mu\theta'^2 dx}{\int_0^1 \theta^2 dx}$$

where μ, μ_1 and μ_2 are positive constants

$$\mu = \begin{cases} \mu_1, & 0 < x < l, \\ \mu_2, & l < x < 1, \end{cases} \quad \mu_2 \neq \mu_1,$$

$$\theta(0) = \theta(1) = 0,$$

$\theta(x)$ is continuous across $x = l$, and θ is twice differentiable above and below l. Show that $\lambda[\theta]$ is positive and bounded from below,

$$\tilde{\lambda} = \lambda(\tilde{\theta}) = min \ \lambda[\theta] > 0.$$

Show that

$$\tilde{\theta}'' + \lambda \tilde{\theta} = 0$$

holds above and below l and

$$\mu_1 \tilde{\theta}'_1 = \mu_2 \tilde{\theta}'_2.$$

Find $\tilde{\lambda}$.■

Exercise II.1(e).2. The above exercise is the one-dimensional equivalent of the following. Let

$$\tilde{\Lambda} = \Lambda[\tilde{\mathbf{v}}] = \min_{\mathbf{v} \in \mathbf{X}} \Lambda[\mathbf{v}]$$

where

$$\Lambda[\mathbf{v}] = \frac{\mathcal{D}[\mathbf{v}]}{\mathcal{E}[\mathbf{v}]}.$$

Find Euler's equation and the natural boundary conditions for $\tilde{\mathbf{v}}$. (This constrained minimization problem yields the Stokes equation, analogous to the equation for $\tilde{\theta}$ in exercise II.1(e).1 and the natural boundary condition is the continuity of the stresses.) ■

We have

$$\mathcal{E}[\mathbf{v}] = \frac{1}{2} \int \rho \mathbf{v}^2$$

$$\leq \max[\rho_1, \rho_2] \frac{1}{2} \int \mathbf{v}^2$$

and on using (1e.2), we find

$$\leq \max[\rho_1, \rho_2] \frac{1}{2} 2k \int |\mathbf{D}[\mathbf{v}]|^2$$

$$\leq k \max[\rho_1, \rho_2] \cdot \frac{1}{2 \min[\mu_1, \mu_2]} \mathcal{D}$$

because $\mathcal{D} = \int 2\mu |\mathbf{D}[\mathbf{v}]|^2$ (see (1c.2)). Therefore,

$$\mathcal{D}[\mathbf{v}] \geq 2\tilde{\lambda}\mathcal{E}[\mathbf{v}], \quad \forall \mathbf{v} \in \mathbf{X} \tag{1e.4}$$

where

$$\tilde{\lambda} = \frac{\min[\mu_1, \mu_2]}{k \max[\rho_1, \rho_2]}. \tag{1e.5}$$

The inequality (1e.4) holds for connected configurations as well as for bubbles, drops and emulsions.

It now follows from (1e.4) and (1c.17) that

$$\frac{d}{dt}(\mathcal{E} + \mathcal{P}) \leq -2\tilde{\lambda}\mathcal{E}. \tag{1e.6}$$

Integrating this from $t = 0$ to t, we find that

$$\mathcal{E}(t) + \mathcal{P}(t) = \mathcal{E}(0) + \mathcal{P}(0) - \int_0^t \mathcal{D}(\tau)d\tau \tag{1e.7}$$

$$\leq \mathcal{E}(0) + \mathcal{P}(0) - 2\tilde{\lambda}\int_0^t \mathcal{E}(\tau)d\tau.$$

It follows that

$$2\tilde{\lambda}\int_0^t \mathcal{E}(\tau)d\tau \leq \mathcal{E}(0) + \mathcal{P}(0) - \mathcal{E}(t) - \mathcal{P}(t). \tag{1e.8}$$

Looking at the right hand side of this equation, we have that $\mathcal{E}(0)$ and $\mathcal{P}(0)$ are prescribed, and $\mathcal{E}(t) \geq 0$. As for $\mathcal{P}(t)$, we will assume it is bounded below on the set of allowed interfaces. In fact, if G is a bounded region, $\mathcal{P}(.)$ is bounded from below. (In unbounded domains, $\mathcal{P}(.)$ need not be bounded below.) In a bounded domain, we could centrifuge all the heavy fluid to the outer cylinder wall. Hence, the right hand side of (1e.8) is bounded from above, and therefore the left hand side is also, and we conclude that $\mathcal{E}$ is integrable. If a quantity is integrable over the interval $(0, \infty)$, then we have that it decays to zero in some sense. Therefore $\mathcal{E}(\infty) = 0$. Since the energy involves the square of the velocity, we have that the disturbance $\hat{u}$ to rigid motion decays to 0 as $t \to \infty$; and this means that $\mathcal{D}[\hat{u}] \to 0$ as well. Therefore, (1c.17) yields

$$\frac{d\mathcal{P}}{dt} \to 0 \quad \text{as} \quad t \to \infty. \tag{1e.9}$$

II.1(f) Minimum of the Potential

Let us consider the limit configuration $(\mathcal{E}(\infty), \mathcal{P}(\infty))$. This is rigid motion, and (1c.17) yields

$$\mathcal{P}(\infty) - \mathcal{P}(0) = \mathcal{E}(0) - \int_0^\infty \mathcal{D}(\tau)d\tau. \tag{1f.1}$$

In chapter XII of Iooss and Joseph [1990], it is shown that a critical point $R = \tilde{R}(\theta, x)$ of the functional $\mathcal{P}[R]$ corresponds to an *equilibrium* solution, as defined before (1a.13). The proof is that the Euler equation for $\mathcal{P}$,

$$\frac{\partial}{\partial \epsilon}\mathcal{P}[\tilde{R} + \epsilon R]|_{\epsilon=0} \tag{1f.2}$$

for all admissible functions R and $\tilde{R}$, is the normal stress equation (1a.14). This means that we have a rigid motion, with R satisfying (1a.14). So critical points of $\mathcal{P}$ correspond to equilibria.

We next prove that a stable equilibrium gives rise to a local minimum for $\mathcal{P}$ in the class of initial disturbances in which the interface position is perturbed from the one given for the equilibrium solution, and the disturbance to the velocity field is zero. For such disturbances, $\mathcal{E}(0) = 0$ and (1f.1) shows $\mathcal{P}$ decreases if $\mathcal{D} \neq 0$. However, if this initial disturbance is such that $R(\theta, x, 0)$ is not an equilibrium, a motion must develop which decays to rigid motion with an equilibrium value of $R(\theta, x, \infty)$, satisfying the normal stress condition (1a.14). The potential of nonequilibrium states with no disturbance velocity $\mathcal{E}(0) = 0$ must decrease. It follows that $\mathcal{P}[\tilde{R}]$ is the smallest value that $\mathcal{P}$ can take in some neighborhood of this point in function space.

We have shown that every rigid motion that minimizes $\mathcal{P}$ is a stable solution. Every stable solution is a rigid motion (but not every rigid motion is a stable solution).

Metastable states are local minima of $\mathcal{P}$, as opposed to a global minimum. The possibility that there exist metastable states is examined in chapter XII of Iooss and Joseph [1990]. Their calculation requires the assumption that

$$\frac{\int \mathcal{D}(t)dt}{\mathcal{E}(0)} \tag{1f.3}$$

be as large as they wish (this term is related to the θ in their equation (XII.60)), which is, as they point out, a speculation. One can, of course, make $\mathcal{D}(0)/\mathcal{E}(0)$ as large as one wishes, by choosing a small $\hat{u}$ (small $\mathcal{E}$) with lots of wiggles, that is, with a large gradient $\nabla\hat{u}$ (large $\mathcal{D}$). The absolute minimum of $\mathcal{P}$ is the only stable state in a class of disturbances which allows one to enter into the basin of attraction of the absolute minimum from any metastable state but disallows the reverse transition. An example of a metastable state is shown in plate I.3.1. Any set of spheres at rest is a local minimum of the potential, since a sphere is a minimal surface, and the potential in this case is the surface area. The only global minimum is one large bubble, which has a smaller surface area for a given volume than many small bubbles, but there are many metastable states (local minima of $\mathcal{P}$) with any number of bubbles. In the experiments, there are also many disturbances present, like convective currents, which bring the spheres together and cause them to coalesce.

II.1(g) Spatially Periodic Connected Interfaces

The analysis so far has allowed for contact lines. In this section, there are
no contact lines, and this simplifies the calculations. It will be shown that
unless you have a cylindrical interface, you cannot have a spatially periodic
connected interface. The proof is by contradiction.

We assume that the interface is a graph

$$r = R(\theta, x), \tag{1g.1}$$

periodic, with period 2π in θ, with period $2\pi/\alpha = 2\Lambda$ in the axial direction
x. We will show that either $R = d$, where d is the mean radius of R, or
the minimizing solution touches the axis at $r = 0$. The latter solution is
impossible because it would penetrate the inner cylinder; we interpret it
to mean that we do not get a *connected* interface but that we get periodic
arrays of drops and bubbles which have contact lines on the inner rod. The
analysis neglects the effects of these lines on the potential, but is in good
agreement with experiments away from these lines.

The analysis of stable configurations starts from the expression (1d.36)
for $\mathcal{P}$. It is assumed that there are no end plates and that $R(\theta, x) \geq a$, with
possibly flat tangents at $R = a$. The contact line potentials (the second and
last terms on the right hand side of (1d.36)) are put to zero. Then,

$$\mathcal{P} = \sigma|\Sigma| - \frac{[\![\rho]\!]\Omega^2}{8} \left< \frac{R^3}{|\nabla F|} \right>$$

$$= \int_0^{2\pi/\alpha} \int_0^{2\pi} \left(\sigma(R^2 + R_\theta^2 + R^2 R_x^2)^{1/2} - \frac{1}{8}[\![\rho]\!]\Omega^2 R^4 \right) d\theta dx. \tag{1g.2}$$

We define

$$d^2 = \int_0^{2\pi/\alpha} \int_0^{2\pi} R^2 d\theta dx. \tag{1g.3}$$

and because of the volume constraint, d is a constant. In the analysis
which follows, we will work with a potential $\mathcal{M}$ defined in the next section. This differs from $\mathcal{P}$ by terms that are independent of R, namely the
terms $-\frac{[\![\rho]\!]\Omega^2}{8} \langle\!\langle -2R^2 d^2 + d^4 \rangle\!\rangle$ where $\langle\!\langle R^2 \rangle\!\rangle = \langle\!\langle d^2 \rangle\!\rangle$ (see next section).

II.2 The Minimum Problem for Rigid Rotation of Two Fluids

II.2(a) The Cylindrical Interface

JRRN showed that rigid motions of two liquids between concentric cylinders of radii R_1 and R_2 are stable to spatially periodic disturbances of arbitrary amplitude and that the stable interface $r = R(\theta, x)$ minimizes the potential

$$\mathcal{M} = \sigma \langle\!\langle (R^2 + R_\theta^2 + R^2 R_x^2)^{1/2} \rangle\!\rangle - \frac{[\![\rho]\!] \Omega^2}{8} \langle\!\langle (R^2 - d^2)^2 \rangle\!\rangle, \qquad (2a.1)$$

where σ is the interfacial tension, $[\![\rho]\!] = \rho_1 - \rho_2$, where ρ_1 is the density of the inner fluid, Ω is the angular velocity of the two fluids, d^2 is the spatial average of R^2,

$$\langle\!\langle R^2 \rangle\!\rangle = \langle\!\langle d^2 \rangle\!\rangle \qquad (2a.2)$$

where

$$\langle\!\langle \cdot \rangle\!\rangle = \int_0^{2\pi/\alpha} dx \int_0^{2\pi} d\theta, \qquad (2a.3)$$

and $2\pi/\alpha$ is the wavelength in the direction x. Note that $\mathcal{M}$ and $\mathcal{P}$ differ just by a constant.

We define

$$J = -\frac{[\![\rho]\!] \Omega^2 d^3}{\sigma}. \qquad (2a.4)$$

JRRN studied, in particular, the stability of the cylindrical interface. We mention two of their theorems here (Theorems 3 and 4, JRRN).

Theorem. *The cylindrical interface with constant radius $R = d$ is stable against small disturbances if and only if $J \geq 1$.*

Consider the case in which the heavy fluid is outside ($\rho_2 > \rho_1$): $J > 0$. Let $R = d + \delta$ in the potential $\mathcal{M}$ of (2a.1).

$$\mathcal{M} = \sigma d \langle\!\langle \left[(1 + \frac{\delta}{d})^2 + \frac{\delta_\theta^2}{d^2} + (1 + \frac{\delta}{d})^2 \delta_x^2 \right]^{1/2} \rangle\!\rangle - [\![\rho]\!] \frac{\Omega^2}{8} d^4 \langle\!\langle \left[(1 + \frac{\delta}{d})^2 - 1 \right]^2 \rangle\!\rangle. \tag{2a.5}$$

We linearize $\mathcal{M}$ for small δ/d:

$$\frac{\mathcal{M}}{\sigma d} \sim \langle\!\langle \left[1 + \frac{2\delta}{d} + \frac{\delta^2}{d^2} + \frac{\delta_\theta^2}{d^2} + \delta_x^2 \right]^{1/2} \rangle\!\rangle + \frac{J}{8} \langle\!\langle \left[\frac{2\delta}{d} \right]^2 \rangle\!\rangle,$$

and using $(1 + x)^{1/2} \sim 1 + x/2 - x^2/8 + \dots$ for $|x|$ small,

$$\frac{\mathcal{M}}{\sigma d} \sim \langle\!\langle 1 + \frac{\delta}{d} + \frac{\delta_\theta^2}{2d^2} + \frac{\delta_x^2}{2} \rangle\!\rangle + \frac{J}{2} \langle\!\langle \frac{\delta^2}{d^2} \rangle\!\rangle. \qquad (2a.6)$$

The volume constraint (2a.2) is $\langle\!\langle (d + \delta)^2 \rangle\!\rangle = \langle\!\langle d^2 \rangle\!\rangle$ or

$$\left\langle\!\left\langle 2\frac{\delta}{d} + \frac{\delta^2}{d^2}\right\rangle\!\right\rangle = 0. \tag{2a.7}$$

Thus, $\langle\!\langle \delta/d \rangle\!\rangle = -\langle\!\langle \delta^2/2d^2 \rangle\!\rangle$ and

$$\frac{\mathcal{M}}{\sigma d} \sim \left\langle\!\left\langle 1 + \frac{\delta_\theta^2}{2d^2} + \frac{\delta_x^2}{2}\right\rangle\!\right\rangle + \frac{1}{2}(J-1)\left\langle\!\left\langle\frac{\delta^2}{d^2}\right\rangle\!\right\rangle. \tag{2a.8}$$

Thus, for $J \geq 1$, the potential is minimized when $\delta = 0$. Therefore, the solution $R = d$ is stable against small disturbances if and only if $J \geq 1$.

Theorem. *The concentric interface $R = d$ is a global minimum of $\mathcal{M}$ among all interfaces $r = R(\theta, z)$ satisfying (2a.2) if and only if*

$$J \geq 4(1 + R_1/d)^{-2}, \quad 0 \leq R_1 \leq d. \tag{2a.9}$$

We start with the case of the heavy fluid being outside $(J > 0)$. First, we show that the criterion of the theorem is sufficient for stability. We define

$$\hat{\mathcal{M}} = \sigma\langle\!\langle R \rangle\!\rangle - [\![\rho]\!]\frac{\Omega^2}{8}\langle\!\langle (R^2 - d^2)^2 \rangle\!\rangle, \tag{2a.10}$$

by getting rid of R_x and R_θ in $\mathcal{M}$. The reason we do this is that R_x and R_θ only increase $\mathcal{M}$: $\mathcal{M} \geq \hat{\mathcal{M}}$. It is sufficient to show that $R = d$ minimizes $\hat{\mathcal{M}}$ because for a cylindrical interface, $\mathcal{M} = \hat{\mathcal{M}}$.

We set $R^2 = d^2(1 + \gamma)$; hence, γ is subject to the constraints $-1 + R_1^2/d^2 \leq \gamma \leq -1 + R_2^2/d^2$ and $\langle\!\langle \gamma \rangle\!\rangle = 0$. We then have

$$\frac{\hat{\mathcal{M}}}{\sigma d} = \langle\!\langle \sqrt{1+\gamma} \rangle\!\rangle + \frac{J}{8}\langle\!\langle \gamma^2 \rangle\!\rangle.$$

We define

$$f(\gamma) = \sqrt{1+\gamma} + \frac{J}{8}\gamma^2 - \frac{1}{2}\gamma \tag{2a.11}$$

and we have

$$\frac{\hat{\mathcal{M}}}{\sigma d} = \langle\!\langle f(\gamma) \rangle\!\rangle, \tag{2a.12}$$

where the last term in $f(\gamma)$ has been added to make $f'(0) = 0$. Note that the condition $f''(0) \geq 0$ for a local minimum retrieves the condition $J \geq 1$ of the previous theorem.

A minimum occurs either at points where $f'(\gamma) = 0$ or at the endpoints of the interval $[-1+R_1^2/d^2, -1+R_2^2/d^2]$. The condition that $f(0) \leq f(-1+R_1^2/d^2)$ and $f(0) \leq f(-1+R_2^2/d^2)$ yields the criterion $J \geq 4(1 + R_1/d)^{-2}$.

If $J < 4(1 + R_1/d)^{-2}$, then the minimum of f is at $-1 + R_1^2/d^2$. We want to show that in this case, the minimum of $\mathcal{M}$ is not at $R = d$. Actually, we shall show that $\hat{\mathcal{M}}$ is not minimized by $R = d$, but since we can make R_x arbitrarily small by stretching the period (and R_θ is taken to be zero because otherwise it would only increase $\mathcal{M}$), this will imply that $R = d$

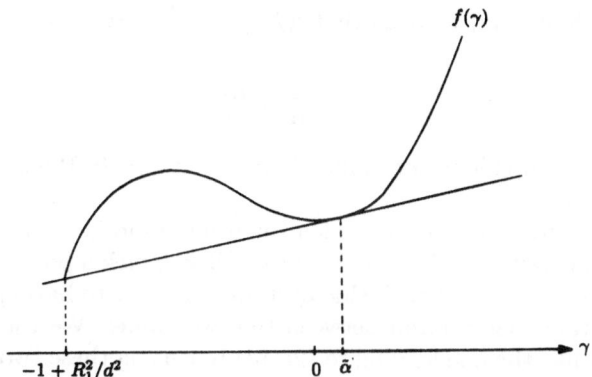

Fig. 2.1. [Figure 1 of Joseph, Renardy, Renardy and Nguyen, 1985] Sketch of $f(\gamma)$ versus γ.

does not minimize $\mathcal{M}$ either. The graph of f for $J < 4(1 + R_1/d)^{-2}$ looks like figure 2.1.

If we draw the tangent from the point $(-1 + R_1^2/d^2, f(-1 + R_1^2/d^2))$ as indicated, it will touch the graph of f at a point $(\tilde{\alpha}, f(\tilde{\alpha}))$ to the right of $\gamma = 0$. It is obvious from the graph that the straight line connecting the points $(-1 + R_1^2/d^2, f(-1 + R_1^2/d^2))$ and $(\tilde{\alpha}, f(\tilde{\alpha}))$ intersects the line $\gamma = 0$ at a value below $f(0)$. We will construct a perturbation $\gamma(x)$ of the cylindrical interface $R = d$ for which $\hat{\mathcal{M}}/\sigma d$ is less than $[\hat{\mathcal{M}}/\sigma d](\gamma = 0)$.

Consider a perturbation $\gamma(x)$ with the following properties:

(i) γ takes only the values $\tilde{\alpha}$ and $-1 + R_1^2/d^2$,
(ii) $\langle\!\langle\gamma\rangle\!\rangle = 0$.

Let p and $q = 1 - p$ be the probabilities with which γ takes the values $\tilde{\alpha}$ and $-1 + R_1^2/d^2$, respectively:

$$\gamma = \begin{cases} \tilde{\alpha} & \text{for } p \text{ values of a period } 2\pi/\alpha \\ -1 + \dfrac{R_1^2}{d^2} & \text{for } q \text{ values of the period} \end{cases}.$$

Then integrating the expression for γ over one wavelength $2\pi/\alpha$ in the x-direction and over 2π in θ and dividing through by $4\pi^2/\alpha$, we have

$$\frac{\alpha}{4\pi^2} \langle\!\langle\gamma\rangle\!\rangle = p\tilde{\alpha} + q(-1 + \frac{R_1^2}{d^2}) = 0, \quad \frac{\alpha}{4\pi^2} \langle\!\langle f(\gamma)\rangle\!\rangle = pf(\tilde{\alpha}) + qf(-1 + \frac{R_1^2}{d^2}).$$
$$(2a.13)$$

Let $(\gamma, y(\gamma))$ be a point on the tangent line of figure 2.1. After eliminating p from the two equations (2a.13), and using the equation for the line between $(\tilde{\alpha}, y(\tilde{\alpha}))$ and $(-1 + R_1^2/d^2, f(-1 + R_1^2/d^2))$, we find that

$$\frac{\alpha}{4\pi^2} \langle\!\langle f(\gamma)\rangle\!\rangle = \frac{(-1 + \frac{R_1^2}{d^2})f(\tilde{\alpha}) + \tilde{\alpha}f(-1 + \frac{R_1^2}{d^2})}{-1 + \frac{R_1^2}{d^2} - \tilde{\alpha}}$$

$$= y(0) < f(0). \tag{2a.14}$$

We have now found a $\gamma(x)$ such that $\langle\langle f(\gamma) \rangle\rangle < \frac{4\pi^2}{\alpha} f(0)$. Thus

$$\frac{\hat{\mathcal{M}}}{\sigma d} < \frac{4\pi^2}{\alpha} f(0)$$

and the right hand side is the value of $\hat{\mathcal{M}}/\sigma d$ at $\gamma = 0$. Hence $\gamma = 0$ does not minimize $\hat{\mathcal{M}}$.

Turning next to the case in which the heavy fluid is inside ($J < 0$), we find that $f(\gamma)$ is concave. Since f is concave, if you pick a point $(\gamma, f(\gamma))$ on it, you can take a point on each side of it and in order to have γ satisfy the volume constraint, let γ jump between the two values. We then carry out the argument like the one leading to (2a.14) but using these two points: the value of $\hat{\mathcal{M}}$ constructed in this way is less than its value at $(\gamma, f(\gamma))$. We conclude that the global minimum of $\mathcal{M}$ is found by constructing γ to jump between the two boundary values $-1 + R_1^2/d^2$ and $-1 + R_2^2/d^2$. Physically, this means the interface jumps between being at the inner cylinder and being at the outer cylinder. In an infinite cylinder, there will therefore be no minimizers of $\mathcal{M}$ of the form $r = R(\theta, z)$. In a finite cylinder, we cannot make R_x arbitrarily small without also making γ small, and there may be stable motions with heavy fluid inside, which have a corrugated free surface, as in the experiments of Yih [1960] and Moffatt [1977]. It would be of interest to determine these corrugated shapes as a solution for the minimum problem for the potential $\mathcal{P}$ of (1d.36) in which the effects of the inner rod are expressed in potential form.

In the light of these results, we note that Yih [1960] studied the problem of stability of a film of liquid rotating in air. He treats this problem in the linearized approximation. He studies the stability of rigid motions with a free surface of constant radius with gravity neglected. Naturally, these constant radius interfaces are unstable because J is negative. Rigid motions, with negligible gravity, are stable and can be obtained easily in experiments (see section II.3) but the free surface cannot have a constant radius.

In the experiments we will discuss later, the inner cylinder has a very small radius. In this case, we let R_1 tend to zero in the formula (2a.9) and we have the following summary: When the heavy fluid is outside, $[\![\rho]\!] < 0$, $\mathcal{M}$ is minimized by $R(\theta, x) = d$ whenever

$$J > 4. \tag{2a.15}$$

If $J < 4$, the minimizing solution is not of constant radius. The volume constraint (2a.2) eliminates solutions of constant radius other than d. The effects of the inner cylinder are neglected in the analysis of sections II.2 (c) - (d).

II.2(b) Mathematical Formulation of the Minimum Problem

In the last section, we studied conditions under which the cylindrical interface is a minimizer. We next consider what happens when the cylindrical interface is not a minimizer.

When $J = 0$, we have the classical problem of Plateau [1863], with the minimal surface equation: there is no density difference, and the problem is governed only by surface tension. The interface is then a surface of constant mean curvature.

We are going to study the θ-independent solutions. We measure all lengths in units of d, setting $r = R(x)/d$ where x and α are dimensionless. Then there is a new $\mathcal{M}$ which is the old one divided by σd and such that

$$\mathcal{M} = \langle\!\langle r(1 + r_x^2)^{1/2} + \frac{J}{8}(r^2 - 1)^2 \rangle\!\rangle \tag{2b.1}$$

where

$$\langle\!\langle r^2 - 1 \rangle\!\rangle = 0. \tag{2b.2}$$

We seek to minimize $\mathcal{M}$ among periodic functions $r(x)$, in the class of continuously differentiable functions of x, denoted by C^1, and satisfying (2b.2). To do this, we introduce a Lagrange multiplier λ and seek the minimum of $\mathcal{M} - \frac{\lambda}{2}\langle\!\langle r^2 - 1\rangle\!\rangle$ among periodic C^1 functions $r(x)$. The Euler equation for this problem is

$$\left[\frac{\partial}{\partial \epsilon}[\mathcal{M} - \frac{\lambda}{2}\langle\!\langle r^2 - 1\rangle\!\rangle]_{r=r+\epsilon\tilde{r}}\right]_{\epsilon=0} = 0$$

$$= \left[\frac{\partial}{\partial \epsilon}\left(\langle\!\langle (r+\epsilon\tilde{r})(1+(r'+\epsilon\tilde{r}')^2)^{1/2} + \frac{J}{8}((r+\epsilon\tilde{r})^2 - 1)^2 - \frac{\lambda}{2}((r+\epsilon\tilde{r})^2 - 1))\rangle\!\rangle\right)\right]_{\epsilon=0}$$

$$= \langle\!\langle \tilde{r}(1+r'^2)^{1/2} + rr'\tilde{r}'(1+r'^2)^{-1/2} + \frac{J}{2}(r^2 - 1)r\tilde{r} - \lambda r\tilde{r}\rangle\!\rangle.$$

The second term is integrated by parts, and the boundary terms vanish because both r and $\tilde{r}$ are periodic in x. We have

$$\int_0^{2\pi/\alpha} \tilde{r}\left(\sqrt{1+r'^2} - (\frac{rr'}{\sqrt{1+r'^2}})' + \frac{J}{2}(r^2 - 1)r - \lambda r\right)dx = 0$$

and since $\tilde{r}$ is arbitrary, this yields

$$\frac{1+r'^2 - r''r}{(1+r'^2)^{3/2}} + \left(\frac{J}{2}(r^2 - 1) - \lambda\right)r = 0. \tag{2b.3}$$

We may find a first integral of (2b.3) by following a change of variable first introduced by Beer [1869]. Consider the interface curve formed in the intersection of the axisymmetric interface and a plane through the axis $r = 0$ of revolution. The coordinates in this plane are (x, r) and the angle between the interface curve $r = r(x)$ and x is denoted by ψ. We define

$$v = \cos\psi \quad 0 \le v^2 \le 1 \tag{2b.4}$$

and

$$r' = \tan\psi = \frac{\sqrt{1-v^2}}{v}. \tag{2b.5}$$

Then

$$r'' = \frac{dr'}{dr}r' = \frac{d\tan\psi}{dr}\tan\psi = -\frac{1}{v^3}\frac{dv}{dr}. \tag{2b.6}$$

The Euler equation (2b.3) becomes

$$\frac{d}{dr}(rv) + \frac{J}{2}r^3 - \mu r = 0 \tag{2b.7}$$

where $\mu = \frac{J}{2} + \lambda$ is as yet undetermined and

$$v(r) = -\frac{J}{8}r^3 + \frac{\mu r}{2} - \frac{\beta}{r} \tag{2b.8}$$

where β is a constant of integration.

The solution (2b.8) is to be associated with an interface profile satisfying $r' = \sqrt{1-v^2}/v$ and the volume constraint

$$\langle\langle r^2 - 1\rangle\rangle = 4\pi \int_{r_1}^{r_2} \frac{(r^2-1)v}{\sqrt{1-v^2}}\,dr = 0. \tag{2b.9}$$

This is the equation to be satisfied by r.

We may find all the axisymmetric solutions of our problem $v(r(x))$ governed by (2b.8). There are solutions (I) of "unduloid type", $v = \cos\psi$ (see figure 2.2; the horizontal axis is the x-axis) if

$$\text{for any } r, \quad 0 \le v(r) \le 1, \tag{I}$$

and solutions (II) of "nodoid type" (see figure 2.3) if

$$\text{there exists } \hat{r} \text{ such that } v(\hat{r}) < 0. \tag{II}$$

The angle between the interface curve and the x-axis is ψ and $r' = \tan\psi$.

An unduloid is a surface of constant mean curvature, $J = 0$, which is generated by the focus of a rolling ellipse. A nodoid is a surface of constant mean curvature, $J = 0$, which is generated by the focus of a rolling hyperbola. These are representative of the two types of solutions (I) and (II) above.

II.2(c) Analysis of the Minimum Problem

It is convenient to replace the parameters (μ, β) with (r_1, r_2), the minimum and maximum values of $r(x)$ (see figures 2.2 and 2.3). Since $r'(r_1) = r'(r_2) = 0$, we have $v(r_1) = v(r_2) = 1$ and, using (2b.8), we find that

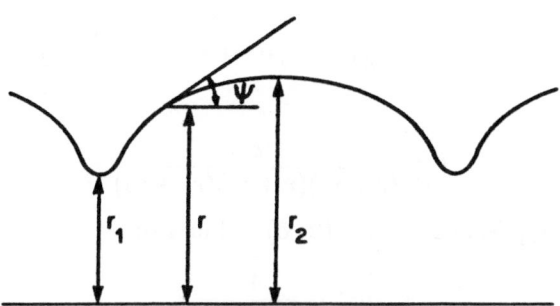

Fig. 2.2. [Figure 1 of Joseph and Preziosi, 1987] Solution (I) of unduloid type.

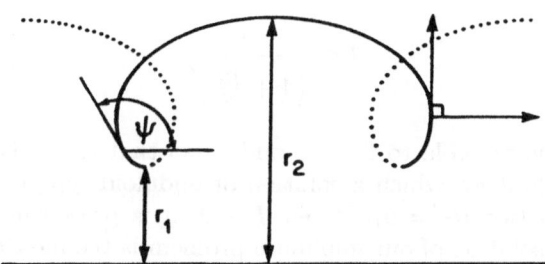

Fig. 2.3. [Figure 2 of Joseph and Preziosi, 1987] Solution (II) of nodoid type.

$$\frac{J}{8}r_i^4 - \frac{\mu}{2}r_i^2 + r_i + \beta = 0 \qquad i = 1, 2,$$

$$\beta = r_1 r_2 \left(\frac{J}{8}r_1 r_2 - \frac{1}{r_1 + r_2} \right),$$

$$\frac{\mu}{2} = \frac{J}{8}(r_1^2 + r_2^2) + \frac{1}{r_1 + r_2}, \qquad (2c.1)$$

$$rv = \frac{J}{8}(r^2 - r_1^2)(r_2^2 - r^2) + \frac{r^2 + r_1 r_2}{r_1 + r_2}.$$

Moreover, since $r' = dr/dx = \sqrt{1 - v^2}/v$,

$$x(r) = \int_{r_1}^{r} \frac{v(\zeta)}{\sqrt{1 - v^2(\zeta)}} d\zeta. \qquad (2c.2)$$

The period (i.e., the wavelength $2\pi/\alpha$) of periodic solutions is given by

$$\lambda = 2 \int_{r_1}^{r_2} \frac{v(\zeta)}{\sqrt{1 - v^2(\zeta)}} d\zeta. \qquad (2c.3)$$

Solutions (I) of the unduloid type have $0 \leq v \leq 1$. The upper bound leads us to

$$(r - r_1)(r_2 - r)\left(\frac{J}{8}(r + r_1)(r + r_2) - \frac{1}{r_1 + r_2}\right) \leq 0. \qquad (2c.4)$$

Hence,

$$J \leq \frac{8}{(r + r_1)(r + r_2)(r_1 + r_2)}, \qquad (2c.5)$$

for all $r \in [r_1, r_2]$. When $r = r_2$, (2c.5) reduces to

$$J \leq \frac{4}{r_2(r_1 + r_2)^2}. \qquad (2c.6)$$

In the problem of section II.2 (a), the fluid is confined by cylinders of radii R_1 and R_2. The condition (2c.6) is consistent with the condition (2a.9). If (2a.9) is satisfied, then since $r_1 \geq R_1/d$ and $r_2 \geq 1$, it is not possible to satisfy (2c.6), whereas if

$$J \leq \frac{4}{\left(1 + \frac{R_1}{d}\right)^2}, \qquad (2c.7)$$

then it may be possible to find r_1 and r_2 so that (2c.6) is satisfied. The largest possible J for which a solution of unduloid type is possible is obviously $J = 4$ (for $R_1 = 0$). When $J \geq 4$, it is proved in section II.2(a) that the only solution of our minimum problem is the interface of constant radius $r(x) = 1$.

The other condition $v \geq 0$ for a solution of unduloid type leads us to the inequality

$$J \geq -\frac{8(r^2 + r_1 r_2)}{(r_1 + r_2)(r_2^2 - r^2)(r^2 - r_1^2)}, \qquad (2c.8)$$

for all $r \in [r_1, r_2]$. Let us choose $r \in [r_1, r_2]$ so as to make the right side of (2c.8) as large as possible; setting

$$\frac{\partial}{\partial r}\left[\frac{(r^2 + r_1 r_2)}{(r_2^2 - r^2)(r^2 - r_1^2)}\right] = 0,$$

we find

$$r^2 = \sqrt{r_1 r_2}(r_1 + r_2 - \sqrt{r_1 r_2}),$$

$$J \geq -\frac{8}{(r_1 + r_2)^2(\sqrt{r_2} - \sqrt{r_1})^2} = -\theta. \qquad (2c.9)$$

Solutions of type (I) are possible only when J satisfies the inequalities (2c.6) and (2c.9).

If $J < -\theta$, solutions of type (I) are not possible and we get solutions (II) of nodoid type.

After introducing the new variables into (2b.1) and (2b.2), we seek the minimum of

$$M = 4\pi \int_{-\pi/2}^{\pi/2} \left(\frac{r^2 + \frac{J}{8}(r^2 - 1)^2 rv(r)}{\sqrt{g(r)}} \right) d\theta \qquad (2c.10)$$

subject to the volume constraint

$$0 = 4\pi \int_{-\pi/2}^{\pi/2} \left(\frac{(r^2 - 1)rv(r)}{\sqrt{g(r)}} \right) d\theta \qquad (2c.11)$$

where

$$g(r) = (r+r_1)(r+r_2)\left(\left[\frac{1}{r_1+r_2} - \frac{J}{8}(r^2+r_1 r_2) \right]^2 - \frac{J^2}{64}r^2(r_1+r_2)^2 \right) \qquad (2c.12)$$

is positive, and $r(\theta)$ is defined by

$$r = \frac{r_1 + r_2}{2} + \frac{r_2 - r_1}{2} \sin \theta. \qquad (2c.13)$$

We have used

$$\frac{dx}{d\theta} = \frac{1}{r'} \frac{dr}{d\theta} = \frac{v}{\sqrt{1-v^2}} \frac{r_2 - r_1}{2} \sqrt{1 - \sin^2 \theta},$$

(2c.1) is used for v, and (2c.13) gives $dr/d\theta$.

The distance $x(r)$ defined in (2c.2) between the values of x for the minimum radius r_1 and for the radius $r(x)$ is now given by

$$x(r) = \int_{-\pi/2}^{\phi} \frac{v(y)y \, d\theta}{\sqrt{g(y)}} \qquad (2c.14)$$

where $y = r_2(1 + \sin \theta)/2$ and $\phi = \arcsin[2r/r_2 - 1]$. The wavelength λ is $2x(r_2)$ where $x(r_2)$ is the distance along the x-axis between the maximum and minimum values of r. When $r = r_2$, $\phi = \pi/2$. The volume constraint gives r_2 as a function of r_1, $0 \leq r_1 \leq 1$.

The form of the solution just given was used for numerical calculations. In this unconstrained problem, we minimize M as r_1 varies (it is possible to think of constrained problems, such as the one where the contact angle at the rod at $r = R_1$ is prescribed [JP]). We have set $R_1 = 0$ and proceed as follows: for each r_1, we find r_2 from the volume constraint (2c.11). This yields $v(r)$ from (2c.1) and $g(r)$ from (2c.12). We then calculate M from (2c.10) as a function of r_1, and seek the value of r_1 which minimizes M. Numerical calculations [Preziosi 1986] show that these minimizers have vertical tangents at the axis of the cylinder: $r'(x) = \infty$ at the point where $r(x) = 0$. It is found that the minimizer of M is at $r_1 = 0$. If there were an inner cylinder of small radius R_1 in the apparatus, these solutions would cross that as well. These minimizers are drops or bubbles lined up along the axis of the cylinder. When the heavier fluid is inside ($J < 0$), we have "drops" that elongate in the radial direction as the rotation is increased. When the heavier fluid is outside ($J > 0$), then we have "bubbles" that elongate in the axial direction as the rate of rotation is increased.

Values of $r_2(J)$ and $\lambda(J)$ for unconstrained minimizers are given in tables 2.1 and 2.2. These tables indicate the shapes of the solutions by giving the maximum distance of the interface away from the axis (r_2) and the diameter (λ) in the axial direction for the drops and bubbles that are lined up along the x-axis. When $J \geq 4$ ($R_1 = 0$ here), the minimizing interface has constant radius. The period $\lambda(J)$ of the minimizing solutions with $J > 0$ varies monotonically from that of a spherical interface at $J = 0$ to that of a cylindrical interface for $J \geq 4$. When $-5.42285 \leq J < 4$, the solutions are of unduloid type. When $-8.18834 < J < -5.42285$, the minimizing solutions are of nodoid type (II). There are two values of r_2 for the range of J from -8.18839 to -7.55395. When $J < -8.18834$, there are no axisymmetric minimizers. One could, instead of J, use Chandrasekhar's drop parameter $\tilde{\Sigma} = -Jr_2^3/8$ which is also listed in table 2.1. We refer to figure 2.4 for the shapes of the solutions in this table. At the value of J for the last entry, we run into the limiting solution depicted in figure 2.4(c).

Another form of the solution just given, in which it is implicitly assumed from the start that the minimizers cross the axis, can be found in Rosenthal's [1962] study of rotating bubbles and Chandrasekhar's [1965] study of rotating drops. These authors prescribe volume rather than the mean radius. The condition on nonexistence of axisymmetric minimizing drops $J < -8.18834$ was first given by Chandrasekhar who uses a drop parameter $-Jr_2^3/8 = [\![\rho]\!]\Omega^2\alpha^3/8\sigma$ where α is the maximum radius of the drop and $r_2 = \alpha/d$. He mentions the possibility of toroidal figures of equilibrium (see figure 2.4). These figures were first discussed by Rayleigh [1914] and studied extensively by Ross [1968]. Toroidal figures of equilibrium at large negative values of J might be interpreted to mean that there are no locally stable flows with heavy fluid inside; all of the heavy liquid has been centrifuged to the outer cylinder, giving rise to robustly stable flows with $J > 4$.

Table 2.1. Parameter values for minimizing solutions for negative J's. For $J \geq -5.42285$, the solutions are of unduloid type, otherwise they are of nodoid type. $\tilde{\Sigma}$ is Chandrasekhar's drop parameter $-Jr_2^3/8$.

J	r_2	λ	λ/r_2	$\tilde{\Sigma}$
0	1.22471	2.44943	2	0
-1	1.21935	1.98166	1.62518	0.2266
-2	1.20628	1.64583	1.36438	0.4388
-2.100	1.20452	1.62048	1.34534	0.4587
-3	1.18868	1.39238	1.17137	0.6298
-4	1.16846	1.19534	1.02300	0.7976
-4.304	1.16192	1.14785	0.98789	0.8440
-5	1.14693	1.03912	0.90600	0.9429
-5.42285	1.13863	0.984440	0.86458	1.0000
-6	1.12731	0.909805	0.80706	1.0745
-7	1.12165	0.776710	0.69247	1.2348
-8	1.14944	0.580407	0.50495	1.5186
-8.18834	1.19010	0.445631	0.37449	1.7253
-8.188	1.19158	0.441463	0.37049	1.7316
-8	1.26327	0.246188	0.19488	2.0160
-7.55395	1.35115	0	0	2.3291

Table 2.2. Parameter values for minimizing solutions with positive J's.

J	r_2	λ	λ/r_2
0	1.22474	2.44949	2.0000
0.1	1.22468	2.50693	2.0470
0.2	1.22448	2.56687	2.0963
0.3	1.22413	2.62950	2.1481
0.4	1.22362	2.69503	2.2025
0.5	1.22295	2.76370	2.2599
0.6	1.22211	2.83578	2.3204
0.7	1.22107	2.91157	2.3844
0.8	1.21984	2.99141	2.4523
0.9	1.21839	3.07570	2.5244

(Continued)

Table 2.2. Continued.

J	r_2	λ	λ/r_2
1.0	1.21671	3.16487	2.6012
1.1	1.21477	3.25943	2.6832
1.2	1.21257	3.35998	2.7710
1.3	1.21009	3.46718	2.8652
1.4	1.20729	3.58183	2.9668
1.5	1.20415	3.70486	3.0767
1.6	1.20065	3.83734	3.1961
1.7	1.19676	3.98058	3.3261
1.8	1.19245	4.13611	3.4686
1.9	1.18768	4.30579	3.6254
2	1.18242	4.49184	3.7989
2.1	1.17664	4.69700	3.9926
2.2	1.17031	4.92463	4.2080
2.3	1.16338	5.17891	4.4516
2.4	1.15585	5.46506	4.7282
2.5	1.14770	5.78976	5.0447
2.6	1.13893	6.16153	5.4099
2.7	1.12957	6.59155	5.8355
2.8	1.11967	7.09467	6.3364
2.9	1.10931	7.69106	6.9332
3	1.09863	8.40886	7.6540
3.1	1.08776	9.28770	8.5393
3.2	1.07689	10.3914	9.6495
3.3	1.06614	11.8120	11.0792
3.4	1.05565	13.7087	12.9860
3.5	1.04551	16.3676	15.6551
3.6	1.03574	19.8891	19.2028
3.7	1.02630	27.0346	26.3418

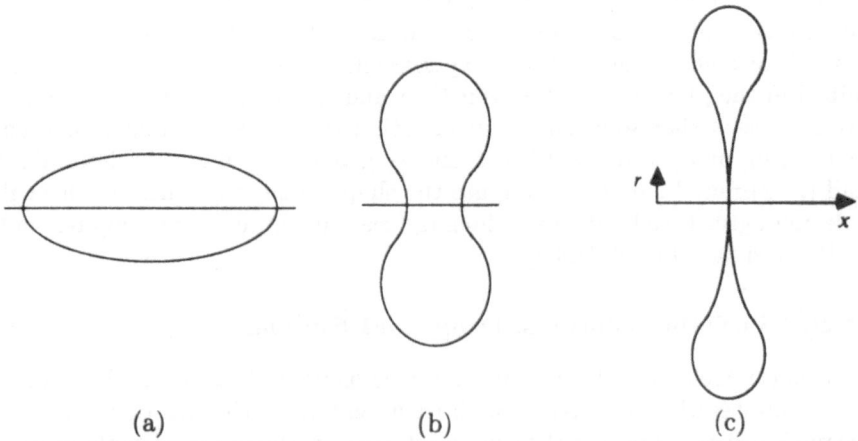

(a) (b) (c)

Fig. 2.4. [Figure 3 of Joseph and Preziosi, 1987] Schematic drawing of minimizing solutions. Minimizing solutions touch the axis with a perpendicular tangent. (a) Solutions of unduloid type are convex, $4 \geq J \geq -5.42285$. (b) Solutions of nodoid type have a point of inflection $-5.42285 \geq J \geq -8.18834$. (c) Limiting (toroidal) form of the solution for $J = -7.583908$.

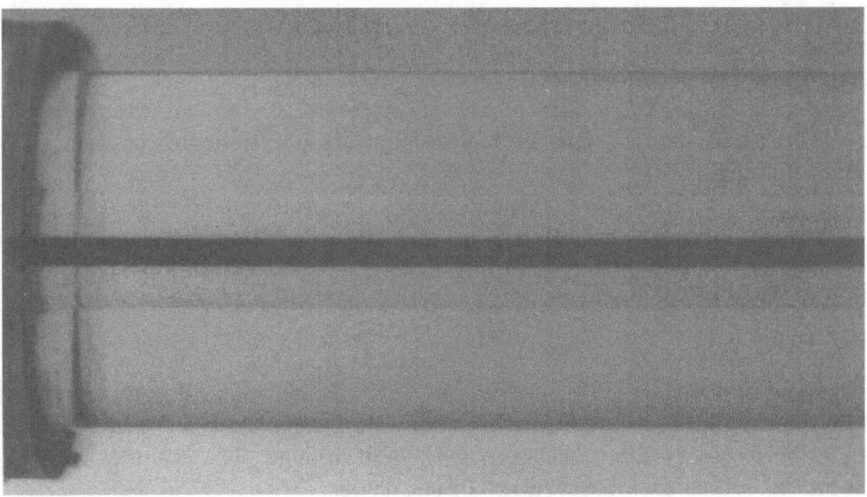

Fig. 2.5. [Figure 4 of Joseph and Preziosi, 1987] Cylindrical interface between 1000 cSt silicone oil and water when $J = 4.5$. The radius of the rod is 0.24cm.

The following procedure was used for the computation of the interface shape of rigidly rotating coating flows (see section II.3 (b)) making tangent contact at the rod. In (2b.4), $v = 1$. R_1 is assumed to be nonzero. The constrained minimization problem of tangent angle at the contact point occurs in some of the cases observed; in others, the unconstrained minimiser is used for comparison with the experiments. The value of d is first determined in the plane $\theta = \pi/2$. Given Ω, σ and $[\![\rho]\!]$, we may then compute J. We next seek that solution of our equation that makes a tangent contact at the rod, at $r_1 = R_1/d$. With d and r_1 given, r_2 is uniquely determined and (2c.2) may be used to compute the shape of the free surface. This calculation agreed well with experiments (see our figure 3.5 and figures 4a,b in Preziosi and Joseph [1988].)

II.2(d) Periodic Solutions, Drops and Bubbles

Solutions which cross the axis are regarded as periodic arrays of drops ($J < 0$) or bubbles ($J > 0$) lined up with their centers on the axis of rotation. If we put a rod of radius a at the center of the array, the resulting configuration would assume the form of an array of rollers. When the periodic solution is viewed as an array of drops or bubbles, r_2 is the "radius" of the drop in the radial direction, and $\lambda/2$ is the "radius" in the axial direction.

It is useful to distinguish bubbles and drops by a dynamic criterion. Bubbles (the heavier fluid is outside) will elongate along the x-direction which is their axis of rotation as Ω is increased. Drops (the heavier fluid is inside) contract in the axial direction and come out radially under the same conditions.

Leslie [1985] has solved the problem of rotating bubble shapes in a low-gravity environment. His drop can contact endwalls that are perpendicular to the axis of rotation. He applies the condition of constant contact angle at the endwalls.

In our experiments using end plates, we spin fast enough so that the centrifugal force governs the motion and gravity is no longer important. We then achieve a constant radius $R = d$ when $J > 4$, as in figure 2.5.

II.2(e) All the Solutions with $J < 4$ Touch the Cylinder

Minimizing solutions which cross the axis of rotation will certainly touch the inner cylinder. The prediction that solutions with $J < 4$ will touch the cylinder is completely consistent with experiments. The cylinder-touching solutions which are observed are of two types:

(i)   The interface between the two fluids intersects the cylinder at lines of contact, as shown in figures 2.6, 3.1, and plate II.4.3(c)(4), and

(ii) The interface between the two fluids makes a tangent contact with the wetted rod at $r = a$ (a=radius of inner cylinder) as shown in figures 3.5, 3.6, 3.8-10.

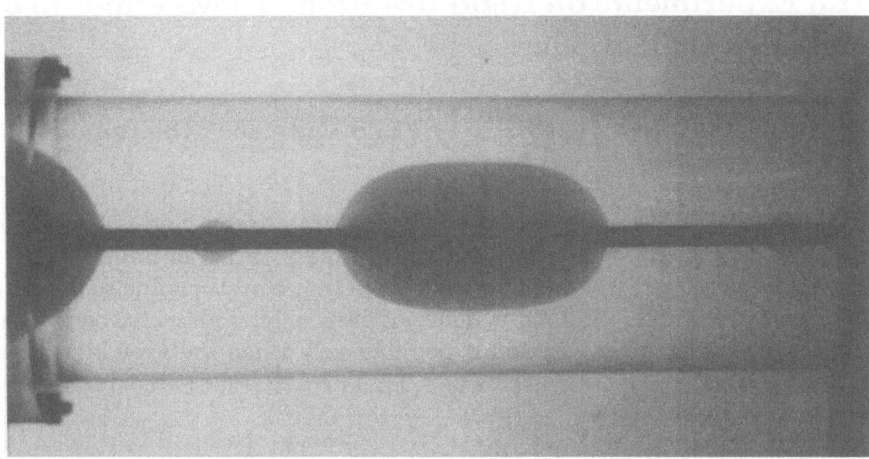

Fig. 2.6. [Figure 5 of Joseph and Preziosi, 1987] Array of bubbles of 95% silicone oil (12500 cS) dyed with 5% castor oil ($\rho = .974$, $\sigma = 20.5$) in water. The J for the central bubble is 1.87 and that for the small ones is 0.05. The radius of the aluminum rod is 0.24cm.

The physical mechanisms embodied in the difference between (i) and (ii) are associated with fundamental problems of adhesion and cohesion not considered here.

The analysis given in the previous sections II.2 (c) - (d) is not a precise realization of the cylinder-touching solutions of type (i) and (ii) which are observed in experiments. The presence of an inner rod at $r = a$ is not explicitly acknowledged in the analysis, and contact line terms in the potential have been neglected. Nevertheless, the analysis is in quantitative agreement with the observations except for some local effects at touching points on the cylinder. The effects of capillarity at lines of contact should be considered for solutions of type (i); but the interfacial tension between liquids can be small and the effects of capillarity at contact lines will then also be small. The comparison between theory and experiment shown in section II.3 shows that contact line effects are local even in cases when the interfacial tension is large; say when the radius of the cylinder a is small compared to the bubble radius. In the other case (ii), with tangent contact, and no contact line, we acknowledge a constraint on our variational problem, by looking for periodic solutions with troughs which touch the cylinder at $r = a$. This also agrees well with experiments (see figure 3.5, 3.6, 3.8-10).

II.3 Experiments on Rigid Rotation of Two Fluids in a Cylindrical Container

II.3(a) Experiments with Heavy Fluid Outside – the Spinning Rod Tensiometer

The case $J > 0$ corresponds to centrifuging, with heavy fluid outside, and $0 < J < 4$ is the domain corresponding to rigid rotation of bubbles whose long dimension increases monotonically from that corresponding to a sphere at $J = 0$ to an infinitely long cylinder of ever smaller diameter as $J \to 4$. Strictly speaking, rigid rotation is possible only when the Froude number criterion (1a.17) is satisfied, $\Omega^2 d \gg 2g$. This criterion is independent of the density difference $[\![\rho]\!]$.

If $[\![\rho]\!]$ is small, we may have

$$|J| = |\frac{[\![\rho]\!]\Omega^2 d^3}{\sigma}| < 4 \qquad (3a.1)$$

for those values of Ω such that gravity is negligible.

The spinning rod tensiometer (U. S. Patent 4,644,782) is a device for measuring interfacial tension between different liquids. This device is basically another type of spinning drop tensiometer (see Rosenthal [1962], Princen, Zia and Mason [1967]). The rod and the drop tensiometers are designed to work under conditions of negligible gravity in which (1a.17) and (3a.1) hold simultaneously.

The experimental apparatus used to obtain the results reported here is a cylindrical container of Plexiglas with inner radius 3.6cm, of length 24.5 cm, closed at each end. A rod may be inserted along the central axis. The rods are attached rigidly to the cylinder and all the parts rotate together as a rigid body. Aluminum rods of radius 0.24 cm and 0.5 cm and a Plexiglas rod of radius 1.25 cm were used. Figure 2.5 will aid the reader in visualizing the cylinder apparatus (although the condition there is $J > 4$ in contrast with the range of J of interest here).

The liquids used in these experiments were water, castor oil, soybean oil and 20, 1000 and 12500 cp silicone oils. The densities of these liquids are 1, 0.960, 0.922, 0.949, 0.967 and 0.975 respectively. The small density difference greatly reduces perturbing effects due to gravity. The effects of gravity can be reduced to negligible levels with $|J| < 4$ when the density differences are less than 0.1 and $d < 1$cm in these experiments. If we suppose that $\Omega^2 d = 2 \ gk$ for $k \gg 1$, then (3a.1) requires that $[\![\rho]\!]d^2 2gk < 4\sigma$. Under the conditions of this section, the heavy fluid was always placed outside ($[\![\rho]\!] < 0$).

Solutions of permanent form, periodic in x, with heavy fluid outside ($0 < J < 4$) were never observed. Instead of periodic solutions, there were isolated bubbles of light liquid centered on the rod. When $J > 4$, the interface is cylindrical and modified by capillarity at the end walls. The effects of capillarity are smaller when J is much larger than 4. When J is reduced from above to below 4, the interface deforms continuously until points at the interior touch the axis. At this point, there are some changes in the topology of the interface. The fluid may rupture into bubbles separated from the heavy liquid outside by well-defined contact lines. This depends on energetic considerations associated with the two fluids and the rod. The configuration of permanent form which is seen most frequently when $0 < J < 4$ is like that shown in figure 2.6 in which small bubbles and large bubbles both appear.

We define d for an isolated bubble as the radius associated with a right circular cylinder of the same length and volume. It follows that there are different J's for small and large bubbles rotating with the same Ω. This is why the small bubbles are almost spherical and the large ones are elongated. There is further discussion of the measurement of d at equation (3a.2).

Agreement between theory and experiment is demonstrated in figure 3.1 The dots represent the theory of (2c.14). The value of the interfacial tension may be selected to make theory and experiment agree for one value of Ω. The same interfacial tension then gives agreement for other values of Ω. In figures 3.1(a) and(b), we can compare the agreement between theory and experiment for two different values of Ω, with one σ. Figures 3.1(a) through (c) show that capillarity at contact points on the inner cylinder is a small effect. This was true for all the cases studied. Figures 3.1(c) and (d) compare a free bubble to a captured bubble, using the same interfacial tension. Very rapid measurements of surface tension may be obtained using elongated bubbles with aspect ratios greater than 6 by assuming that $J = 4$.

The parameter J is not convenient for the measurement of interfacial tension because the determination of the mean radius is indirect and depends on the volume of the fluid in the drop. Fortunately, the measurement of surface tension using long drops can be carried out from measurements of the maximum drop radius D. We define

$$J_D = \frac{JD^3}{d^3} = \frac{\Delta\rho\Omega^2 D^3}{\sigma}. \tag{3a.2}$$

The value of J_D is already at its asymptotic large aspect ratio $(\lambda/D \to \infty)$ value of $J_D = 4$ when $\lambda/D > 3.5$ (see table 3.1). For these long bubbles, we compute σ, after measuring D, using $J_D = 4$. A further reference is Than, Preziosi, Joseph and Arney [1988].

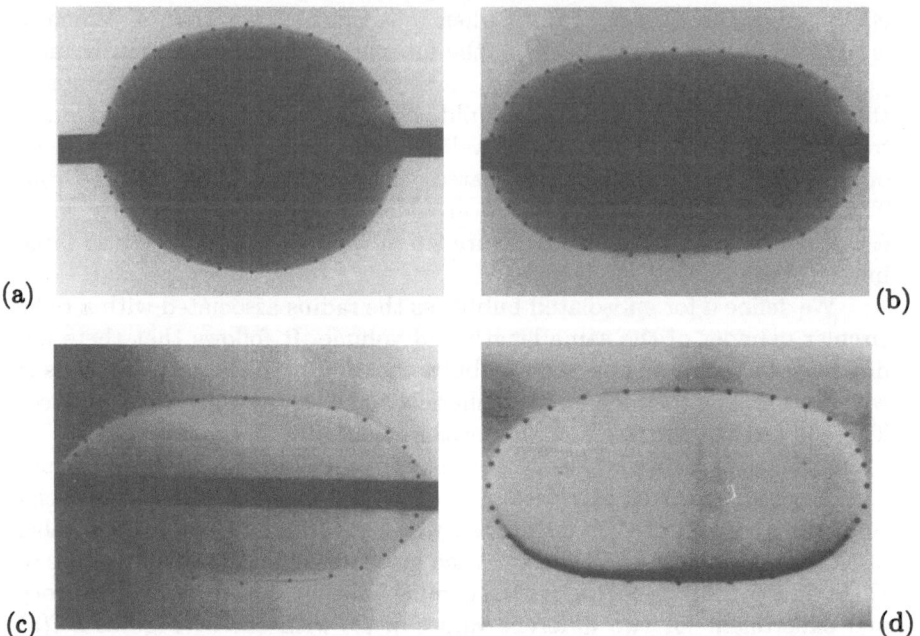

(a)   (b)

(c)   (d)

Fig. 3.1. [Figure 6 of Joseph and Preziosi, 1987] Numerical comparison. The dots are theory computed numerically (see II.(2c.14)) neglecting the contact line potential. The rod radius is 0.24 cm. (a) bubble of 95% silicone oil (12500 cs) dyed with 5% castor oil (ρ=0.974, σ=20.5) in water when J=0.72. (b) bubble of 95% silicone oil (12500 cs) dyed with 5% castor oil in water when J=1.87. (c) bubble of silicone oil (1000 cs, ρ =0.967, σ= 22.1) in water when $J = 2$. (d) bubble of silicone oil in water when $J = 2$ without inner rod.

Table 3.1. Values of J and J_D as a function of λ/D.

λ/D	J	J_D
1.000	0.100	0.184
1.048	0.300	0.550
1.101	0.500	0.915
1.226	0.900	1.628
1.385	1.300	2.304
1.598	1.700	2.914
1.899	2.050	3.365

(Continued)

Table 3.1. Continued.

λ/D	J	J_D
2.103	2.250	3.575
2.292	2.400	3.706
2.522	2.550	3.812
2.610	2.600	3.841
2.807	2.700	3.891
3.037	2.800	3.930
3.168	2.850	3.945
3.310	2.900	3.959
3.638	3.000	3.978
4.126	3.090	3.989
4.220	3.110	3.991
4.422	3.150	3.994
4.763	3.210	3.997
5.158	3.270	3.998
5.621	3.330	3.999
6.174	3.390	4.000
6.844	3.450	4.000
7.670	3.510	4.000
8.338	3.540	4.000
8.918	3.570	4.000
9.583	3.600	4.000
10.084	3.620	4.000
10.940	3.650	4.000
12.328	3.680	4.000
13.622	3.710	4.000
15.213	3.740	4.000
19.828	3.800	4.000
30.600	3.870	4.000
99.827	3.960	4.000
199.822	3.980	4.000

Spinning drop devices may be used to determine lower and upper bounds for interfacial tension between immiscible liquids. We like the idea of upper and lower bounds because the equilibrium tension is not a robust quantity and it depends strongly on impurities. The key lower and upper bounds for the equilibrium tension is the tension relaxation function [Joseph, Arney and Ma 1992].

$D(t,\Omega)$

I. [Joseph, Arney and Ma, 1992, Academic Press] Overly large initial diameter: $D(0,\Omega) > D_{eq}(\Omega)$

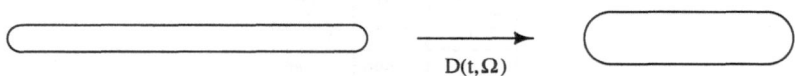

$D(t,\Omega)$

II. [Joseph, Arney and Ma, 1992, Academic Press] Overly small initial diameter: $D(0,\Omega) < D_{eq}(\Omega)$

We have seen in equation (2a.15) that when $J > 4$ (heavy fluid outside), $\mathcal{M}$ is minimized by $R(\theta, x) = d$. Actually, this solution can be realized only if there are end caps to restrain the expansion of the cylindrical-like bubble. If the bubble is free to expand as Ω increases, it will elongate with $J = 4$ in an extensional flow with length L increasing as d decreases. The bubble will be a long cylinder of nearly constant maximum radius D with rounded end caps.

A useful *relaxation function* which can easily be measured in experiments is the maximum diameter $2D(t, \Omega)$ of the evolving bubble at a fixed value of the angular velocity Ω. The value $D_{eq}(\Omega)$ is the asymptotic value of the relaxation function

$$\lim_{t \to \infty} D(t, \Omega) = D_{eq}(\Omega). \tag{3a.3}$$

This is the maximum radius of the equilibrium bubble at the given value of Ω. The two types of relaxation, I and II below, are of special interest.

To get upper and lower bounds for the equilibrium surface tension, we use equation (3a.2) and define a surface tension relaxation function

$$\sigma(t, \Omega) = \frac{(\rho_2 - \rho_1)\Omega^2 D^3(t, \Omega)}{4}. \tag{3a.4}$$

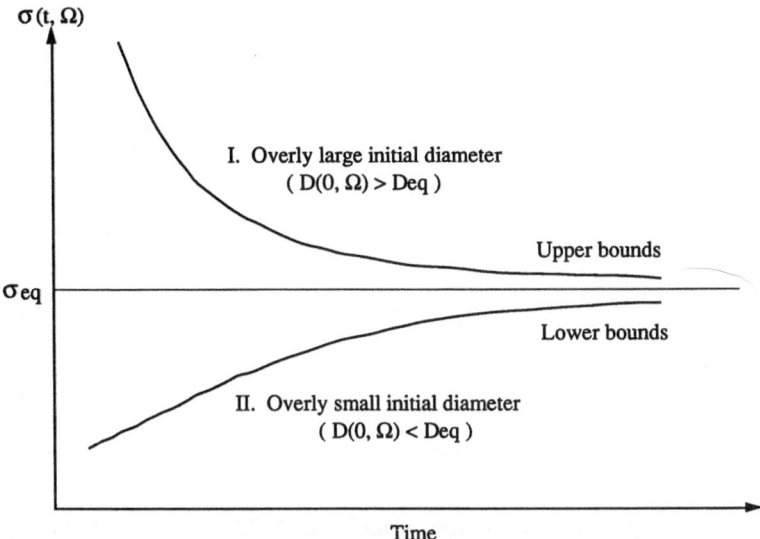

Fig. 3.2. [Figure 1 of Joseph, Arney and Ma, 1992, Academic Press] Surface tension relaxation function and upper and lower bounds for σ_{eq}.

The limiting value of this function as $t \to \infty$ is the equilibrium surface tension given by

$$\sigma_{eq} = \lim_{t \to \infty} \sigma(t, \Omega) = \frac{(\rho_2 - \rho_1)\Omega^2 D_{eq}^3(t, \Omega)}{4} \qquad (3a.5)$$

independent of Ω or $D(0)$. All the relaxation functions I lie above σ_{eq} and all those in group II lie below σ_{eq}.

You can get many different relaxation curves, depending on $D(0, \Omega)$ and Ω. They have a common asymptote σ_{eq} as in figure 3.2, where σ is denoted by T. Some measured values of relaxation curves giving upper and lower bounds are exhibited in figures 3.3-3.4. For STP in water, the value of σ_{eq} lies between the lower bound of 5.2 dyn/cm and the upper bound of 6.0 dyn/cm. For Safflower oil in glycerol, σ_{eq} lies between the lower bound of 8.9 dyn/cm and the upper bound of 12.1 dyn/cm. As evident from table 3.1, when $L/D \geq 10$, we may estimate the equilibrium surface tension from $J_D = 4$. The idea of a surface tension relaxation function which we have introduced from our discussion of lower and upper bounds needs further study from a theoretical point of view. The main idea here is the relaxation of a *system* which is related to the problem of stability. A spinning drop device equipped with an oven and desgined for measuring interfacial tension and relaxation has been described by Joseph, Arney, Gillberg, Hu, Huttman, Verdier and Vinagre [1992] and a patent application has been filed.

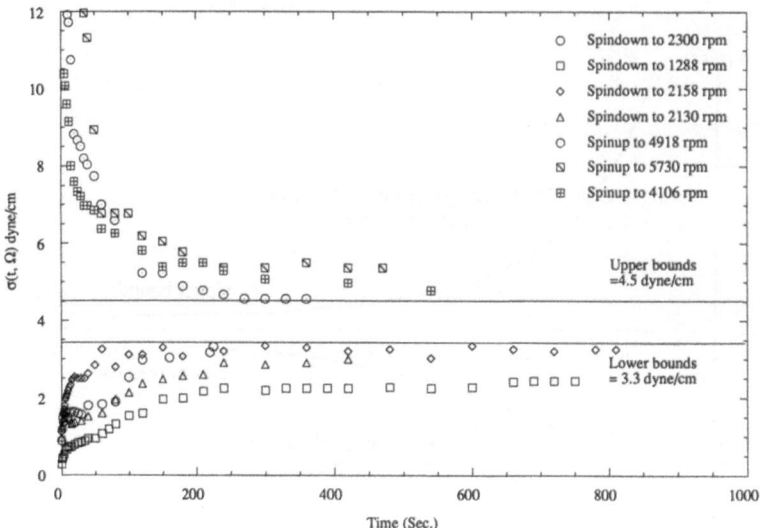

Fig. 3.3. [Figure 2 of Joseph, Arney and Ma, 1992, Academic Press] Surface tension relaxation function for STP in glycerol. The asymptotic value $\sigma(t, \Omega)$ for large time is the equilibrium tension σ_{eq}, a material parameter, independent of Ω or D, whose value lies between the lower bound of 3.3 dyn/cm and the upper bound of 4.5 dyn/cm.

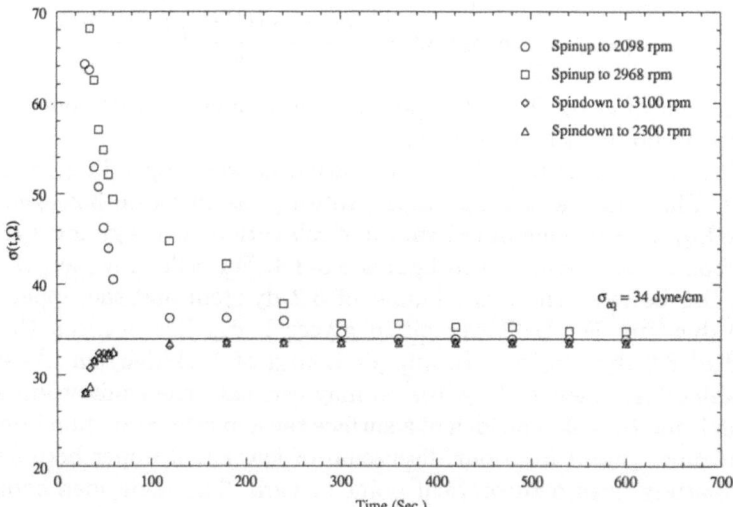

Fig. 3.4. [Figure 5 of Joseph, Arney and Ma, 1992, Academic Press] Surface tension relaxation function for 20 cs poly (dimethyl-siloxane) in water. In this case, the asymptotic value of $\sigma(t, \Omega)$ for large time was achieved. This value is the equilibrium tension σ_{eq} =34 dyn/cm.

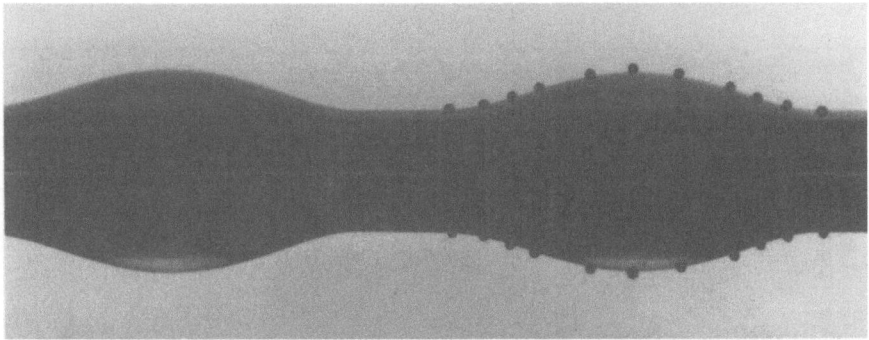

Fig. 3.5. [Figure 7 of Joseph and Preziosi, 1987] The rod is coated with STP and is rotating in air with $J = -0.95$, $a/d = 0.74$, $a=$ radius of inner cylinder, and the parameter Sh_0^2 of II.(1a.18) is 0.57. The dots compare the observed shape with the axisymmetric drop which has a tangent contact at $r = a$. The shape of this axisymmetric figure is determined by the method of Preziosi and Joseph [1988] (see end of section II.2(c)).

II.3(b) Experiments with Heavy Fluid Inside – Coating Flows

JP did experiments with heavy fluid inside ("drop" experiments), coating aluminum and Plexiglas rods rotating in air with oil. With air outside, $[\![\rho]\!] = \rho$ is not small, and gravity can be important. The coating liquids used in their studies were STP, 1000 and 6000 poise silicone oils. The viscosity of STP is about 100 poise. The dynamics which they observed are not inconsistent with the observations of Moffatt [1977] of films of golden syrup (80 poise) rotating in air. The JP study complements Moffatt's in carefully examining the axial structure of the rotating films as well as the azimuthal variations.

The apparatus used in these experiments is nearly identical to the one used in the experiments of Moffatt [1977]. A layer of liquid was first coated on the cylinder by rotating it while partially immersed in a trough; the roller was then raised from the trough while still rotating. The coating films achieved in this way could be maintained indefinitely. The films undergo many different transitions as flow parameters are changed. Three different rods were used: two aluminum rods with radius and length (1.02, 30) cm and (1.42, 45) cm, respectively, and one Plexiglas rod with radius and length (2.04, 30) cm. The liquids coated the entire rod along its whole length.

All the fluids which were mentioned in the last paragraph are sticky. Once the rod is coated with these fluids, it stays coated; dry patches do not develop. We are then obliged to reconsider the implications of the fact that in the unconstrained problem for minimizing the potential, the minimizing solutions cross the axis when $J < 4$. In the case of sticky coats on rotating

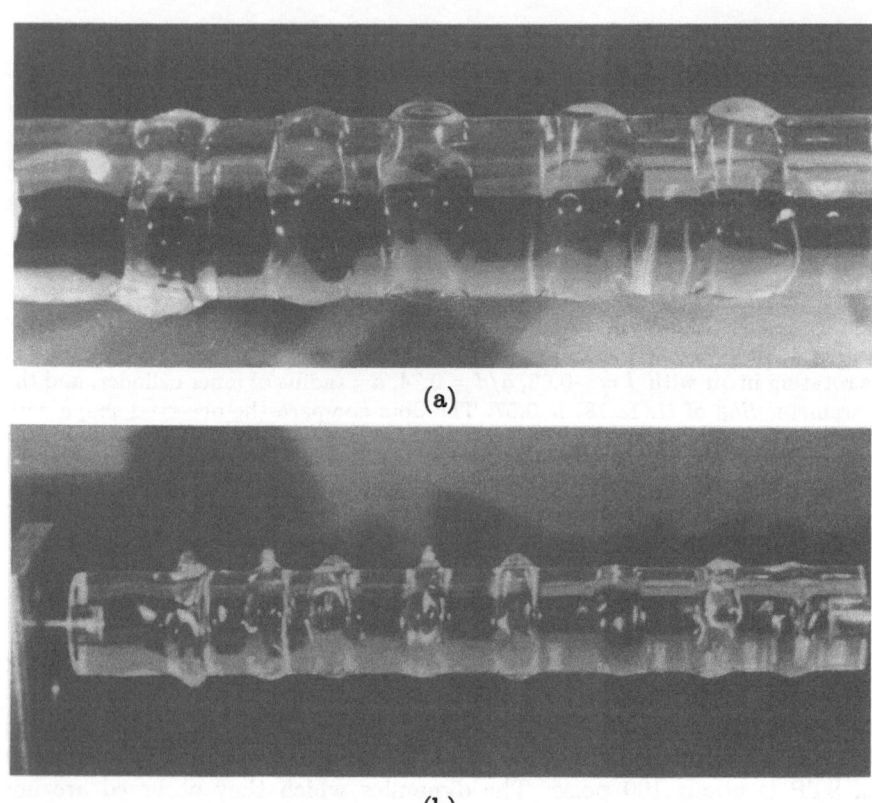

(a)

(b)

Fig. 3.6(a-b). [Figure 8 of Joseph and Preziosi, 1987] Bifurcated non-axisymmetric solutions on rings of 6000 P silicone oil in air on a 2.04 cm rotating Plexiglas rod. (b) Instability of the bifurcated rings at a higher rotation rate.

rods, we have a constrained variational problem in which we require that if the interface touches the rod, it will do so with a flat angle of contact at touching points. In fact, all the realized solutions touch the rod in just this way (see figures 3.5, 3.8-10).

The same oil which sticks on solids immersed in one fluid need not stick when immersed in another fluid. Silicone oil sticks to Plexiglas rods rotated in air, but not in water (see section II.4 on rollers) even though silicone oil preferentially wets Plexiglas.

Some gross features of rod-touching oil films rotating in air may be explained as follows. Uniform coats are unstable and undulations begin to develop along the rod. Moffatt [1977] gave a heuristic argument to explain why this instability should not equilibrate until the troughs of the wavy interface touch the cylinder. In his argument, surface tension is neglected and we suppose that a liquid is rotating in air with $p = p_a$ on $r = R(\theta, x, t)$. Then, from (1a.13),

$$p - p_a = \rho(\frac{r^2 - R^2}{2})\Omega^2, \tag{3b.1}$$

and the pressure under any bump is less than the pressure at the side of the bump. In the absence of countervailing forces, the pressure deficit would exaggerate the bump, with largest pressure gradients along lines from the point at the base of the bump where $p - p_a$ is minimum. This pressure gradient pulls in the sides of the bump, exaggerating bumpiness. This heuristic argument does not require axisymmetry; it works also for bumps and rings. Moffatt's argument was for a system with air as the outer fluid, but the same argument works without change whenever the heavy fluid is inside $[\![\rho]\!] > 0$.

When the speeds are low, axisymmetric wavy solutions with troughs that touch the cylinder minimize the interface potential (see end of section II.2(c)). It is nearly impossible to pass fluid from one wave to another. Further increases in the angular velocity lead to increases in the amplitude of the undulations and length of the troughs touching the cylinders. These features are evident in figures 3.5-6, 3.8-9, the figures on rollers in section II.4, and figures 5 and 6 of Moffatt [1977]. The undulations are then isolated from one another and the solutions may not be exactly periodic in the x-direction. Periodicity can be more closely simulated when the viscosity of the coating fluid is smaller. In this case, the transfer of fluid from one undulation to another, which is required to maintain periodicity against disturbances, is enhanced. It may be useful to think of the undulations as rotating drops constrained by tangent contact at the rod.

Bifurcation of Coating Films to Non-Axisymmetric Shapes. The argument of section II.2 applies equally to axisymmetric shapes (rings) and non-axisymmetric ones (bumps). The stability of rings for small amplitudes and their loss of stability to non-axisymmetric shapes at large amplitudes can be argued from the form of the potential $\mathcal{M}$ (see (2a.5)). We let $\Delta = \delta/d$ in equation (2a.5) and find, with lengths made dimensionless with respect to d, that

$$\frac{\mathcal{M}}{\sigma d} = \langle\!\langle \left[(1 + \Delta)^2(1 + \Delta_x^2) + \Delta_\theta^2 \right]^{1/2} \rangle\!\rangle + \frac{J}{8}\langle\!\langle (2\Delta + \Delta^2)^2 \rangle\!\rangle, \tag{3b.2}$$

with the volume constraint (2a.7). At low rates of rotation, we have axisymmetric drops of small amplitudes ($\Delta_\theta = 0$). When Ω is increased, the drops bulge outwards, the radius of each drop increases (that is, Δ increases), and the drop contracts in the axial direction, so that Δ_x increases. Note that in (3b.2), the term Δ_x^2 is multiplied by $(1 + \Delta)^2$ whereas Δ_θ^2 is not. Hence, at positions where $1 + \Delta$ is large, the motion pays a greater price in

(a)-(b)

 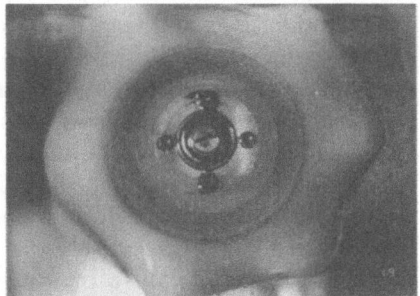

(c)-(d)

Fig. 3.7(a-d). [Figure 14 of Joseph and Preziosi, 1987] Bifurcated configurations of silicone oil in water. (a) One lobe, $\Omega = 0.47$ rad/s. (b) Two lobes, $\Omega = 0.83$ rad/s. (c) Three lobes, $\Omega = 1.49$ rad/s. (d) Six lobes, $\Omega = 3.18$ rad/s.

the potential in increasing Δ_x rather than in increasing Δ_θ. Since the idea is that the motion tries to keep the potential as low as possible, there is a value of Δ, sufficiently large, at which the motion chooses to set Δ_θ to be non-zero, rather than change the motion in the x-direction further, so that bifurcation to a non-axisymmetric interface occurs.

The argument just given shows that non-axisymmetric solutions will form on rings of relatively short length. Such bifurcation on narrow rings can be seen in figure 3.6 and in figures 6(a) and 9(b) of Moffatt [1977].

It can be seen from the way the θ variation appears in (3b.2) that the

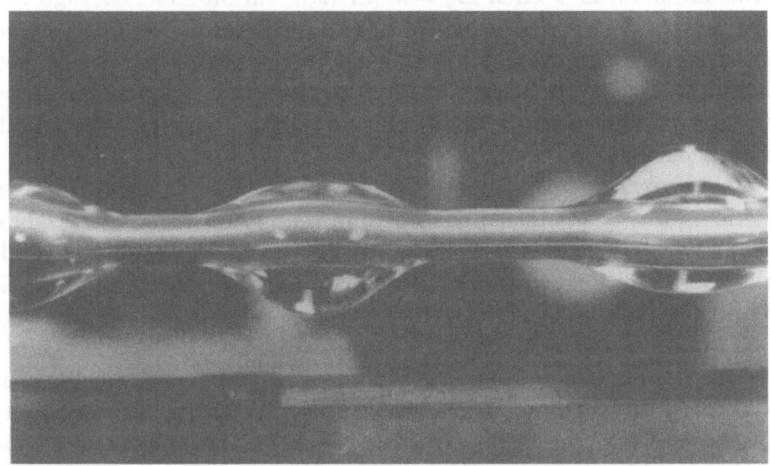

Fig. 3.8. [Figure 9 of Joseph and Preziosi, 1987] Silicone oil (6000 P) on a 1.02 cm radius aluminum rod rotating in air at 21.9 rpm. The lobes rotate much more slowly, left to right; 15, 18.8 and 13.7 rpm, respectively.

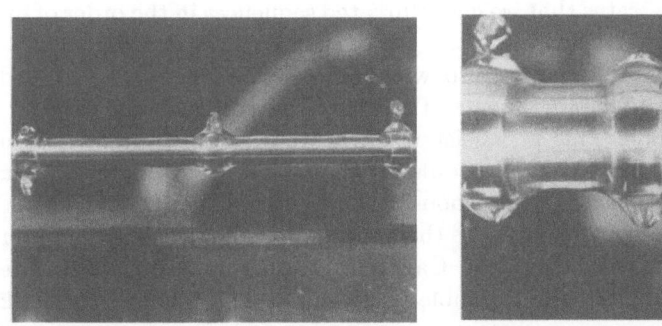
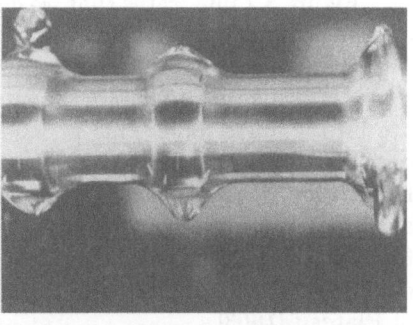

(a)   (b)

Fig. 3.9(a-b). [Figure 10 of Joseph and Preziosi, 1987] (a) The 6000 P silicone oil is thrown off radially. The effect of gravity is negligible. the configuration is nearly steady in a rotating coordinate system. $\Omega = 99.15$ rpm, cylinder radius a = 1.02 cm. (b) $\Omega = 31.76$ rpm, $a = 2.04$ cm. The surface velocities of the three rings from left to right are 31.41, 31.57 and 31.53 rpm, respectively.

first non-axisymmetric solution to bifurcate, as Ω is increased, will have a first mode azimuthal periodicity corresponding to an eccentric ring. Write $\Delta(x, \theta) = \Delta^0(x) + \Delta^1(x) \cos n\theta$, where Δ^0 is the axisymmetric solution, and linearize for small Δ^1. Then $\Delta_\theta^2 = n^2 [\Delta^1(x)]^2 \sin^2 n\theta$. The average value of $\sin^2 n\theta$ is $1/2$ and is independent of n; thus, this term is smallest for $n = 1$. The idea that the motion tries to keep the potential as low as possible implies that $n = 1$ is the favored mode for the first bifurcation. Repeated bifurcation of non-axisymmetric solutions leads to ever higher modes of azimuthal periodicity.

The bifurcation of axisymmetric figures of equilibrium to first-mode, eccentric figures is a robust phenomenon, readily observed on nearly every type of coating film (see figures 3.7-9). The same type of first-mode azimuthal periodicity was observed in the experiments of Plateau [1863](see his figure 4) as the first bifurcation of the olive-oil drop coating the rotating disk in an alcohol-water mixture.

Non-axisymmetric one-lobed shapes were analyzed by Brown and Scriven [1980] for a drop captured between two rotating disks. These calculations showed that higher-lobed shapes between disks, like those reported by Plateau and here, are indeed unstable. First-mode, eccentric figures, like the shape of drops rotating on a rod, can even be explained within the context of static figures. In his observations of static capillary bridges, Plateau observed such a bifurcation when the end plates confining the neutrally buoyant fluid were brought sufficiently close to one another. This reduces the wavelength, as in the case of rotating drops, until the price paid by further decrease of the axial wavelength in minimizing the potential is greater than that for bifurcation into the first mode. Russo and Steen [1986] have recently analyzed this bifurcation in the context of static figures.

Figure 3.7 illustrates that we get bifurcated sequences in the order of increasing azimuthal periodicity. First, $n = 1$ as in figure 3.7 (a), then $n = 2$ as in (b). Some higher values of n are shown in (c) and (d). Only the one-lobed figure is stable in these experiments. The other figures eventually degenerate into a single lobe. This is consistent with predictions given by Brown and Scriven [1980] for a related problem. The same sort of phenomenon, bifurcation of rotating drops into non-axisymmetric shapes in qualitative agreement with theory, although the theory does not acknowledge the outer fluid, was reported by Wang, Tagg, Cammack and Croonquist [1981]. The one- and two-lobed structures resemble figures 4 and 8 sketched in the work of Plateau [1863].

Intrinsically Steady and Unsteady Coating Flows. Flows which are steady in laboratory coordinates can be achieved when Ω is small. When Ω is larger, there is a range of Ω over which these flows are unsteady in every coordinate system. For sufficiently large values of Ω, thin films rotate rigidly and are steady in a rotating coordinate system.

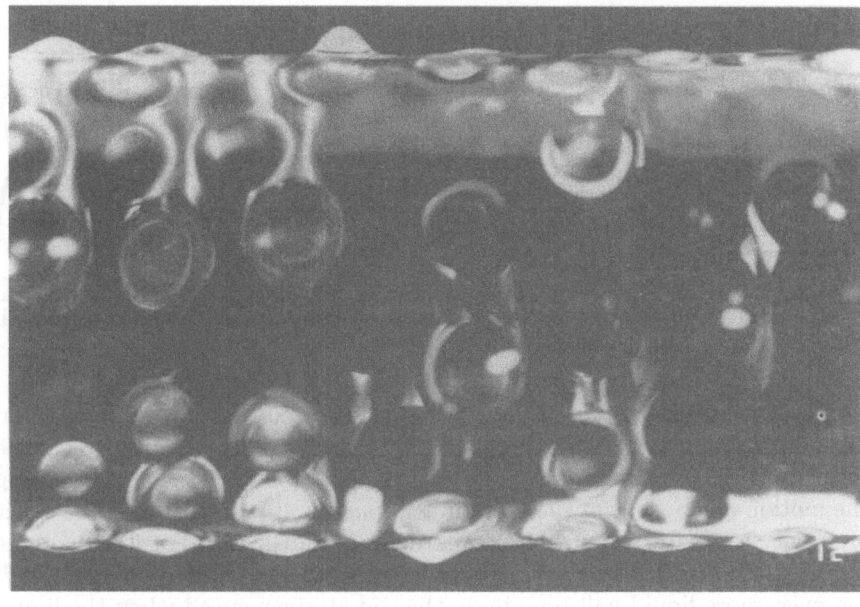

(a)

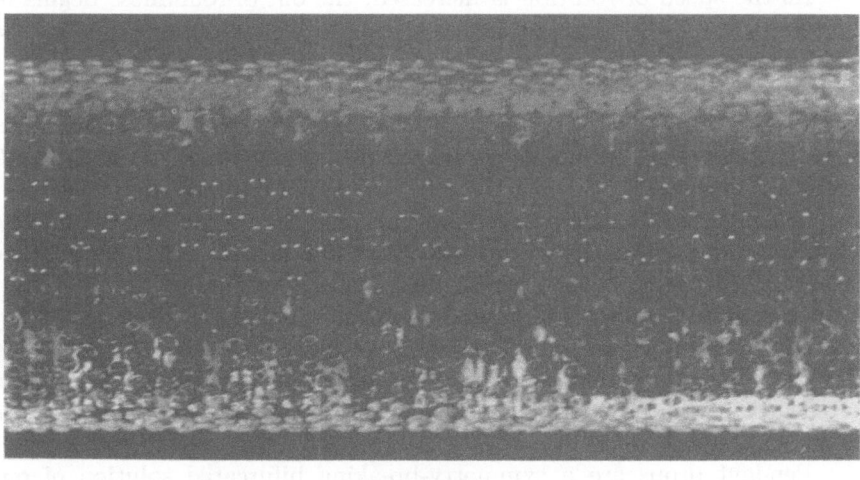

(b)

Fig. 3.10(a-b). [Figure 11 of Joseph and Preziosi, 1987] (a) Pendant drops of 6000 P silicone oil in air on a rod of radius 2.04 cm. The motion is perfectly steady in a rotating coordinate system. Gravity is negligible. Ω= 500 rpm. (b) Ω = 1000 rpm.

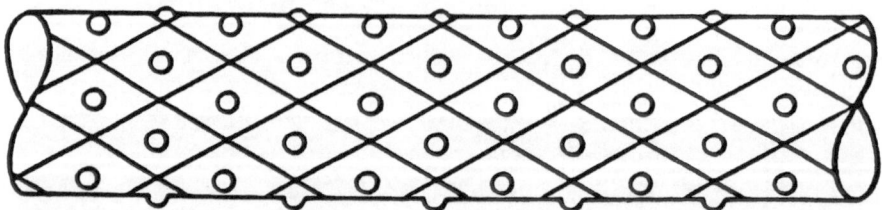

Fig. 3.11. Sketch of pendant drops on a rapidly rotating rod. This configuration rotates rigidly and is perfectly steady in a coordinate system which rotates with angular velocity Ω. The drops tend to form in a diamond-like lattice.

It is impossible to maintain an interface of constant radius on a film of liquid coating a rod rotating in air. If the coat is thick, gravity can be very effective in creating a large secondary motion with gravity opposing the motion on one side of the film and supporting it on the other. The effect of gravity is greater when there is more liquid on the rod. For fixed volume of liquid, the effect of gravity is diminished when the viscosity is increased; however, more liquid will remain on the rod at given speed when the liquid is more viscous. The two effects compete. At low speeds, an equilibrium is established with a "lop-sided" configuration as in figure 4b of Preziosi and Joseph [1988] which is steady in laboratory coordinates.

As the speed of rotation is increased, the out-of-roundness begins to increase and also to rotate relative to laboratory coordinates. This is a manifestation of bifurcation to a mode-one azimuthal variation, but it is slightly masked by out-of-roundness due to gravity (see figure 3.8). Such solutions are intrinsically unsteady. At the same time, the crest of the waves grow and rings develop, in the manner shown in figure 3.9, and in figures 7 and 8 of Moffatt [1977].

When the coating fluid is very viscous and the coating film is thin, the effect of gravity will be diminished, as shown in figure 3.9. Thin films can be created by centrifuging away excess fluid. If the speed of rotation is further increased, more liquid will be flung off the rod. At very high speeds, most of the fluid is thrown off the rod. Gravity has nothing to do with "throwing off" because ejected particles of fluid are flung out radially. An equilibrium is reached in which there are pendant drops on a rotating rod. These are shown in figure 3.10.

Pendant drops are a symmetry-breaking bifurcated solution of our coating film. They tend to form on the rings of earlier solutions with successive rows staggered so that the drops in one row lie in the interstice of the next row. This induces the diamond symmetry shown in figures 3.10 and 3.11. The pendant drops are like those which might develop under gravity on a moist ceiling with an effective gravity equal to $\Omega^2 a$ where a is the radius of the cylinder.

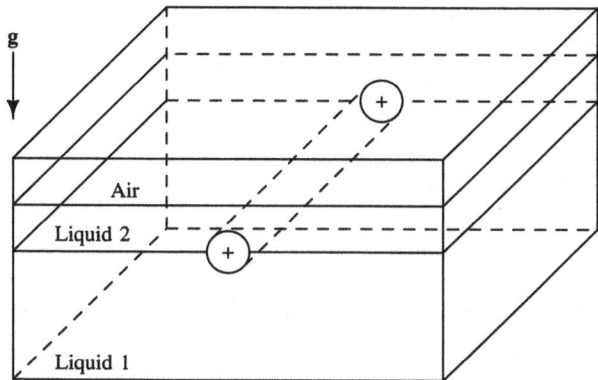

Fig. 4.1. Sketch of Plexiglas boxes used in the experiments. The [length, height, depth, rod diameter, rod material] of boxes I and II are, respectively, [20.32 cm, 22.86 cm, 10.16 cm, 5.08 cm, Plexiglas] for I, and [20.32 cm, 11.43 cm, 7.62 cm, 2.54 cm, aluminum] for II.

II.4 Experiments with Liquids on Immersed and Partially Immersed Rotating Rods. Rollers, Sheet Coatings and Emulsions

In these experiments, two fluids are put into a container, usually a rectangular box pierced by a cylinder, sometimes between coaxial cylinders, the inner one of which may rotate. Two rectangular boxes used in the experiments are described in the caption under figure 4.1

We are interested in the places to which each fluid is transported by rotating the rod, and in the shapes of the interfaces which separate the fluids. As an example of what might happen, we consider sheet coatings and use the minimum torque solution derived in section I.3 (f) to motivate the lubrication scenario shown in figure 4.2.

In all the cases, one finds a lubricated flow in which the low viscosity fluid is dragged around by the higher viscosity fluid. The type of flow which is observed seems to correlate with the viscosity and viscosity difference. There are three basic flow regimes: (i) rollers, (ii) sheets, and (iii) dynamically maintained emulsions. Mixtures of these three basic types may also occur.

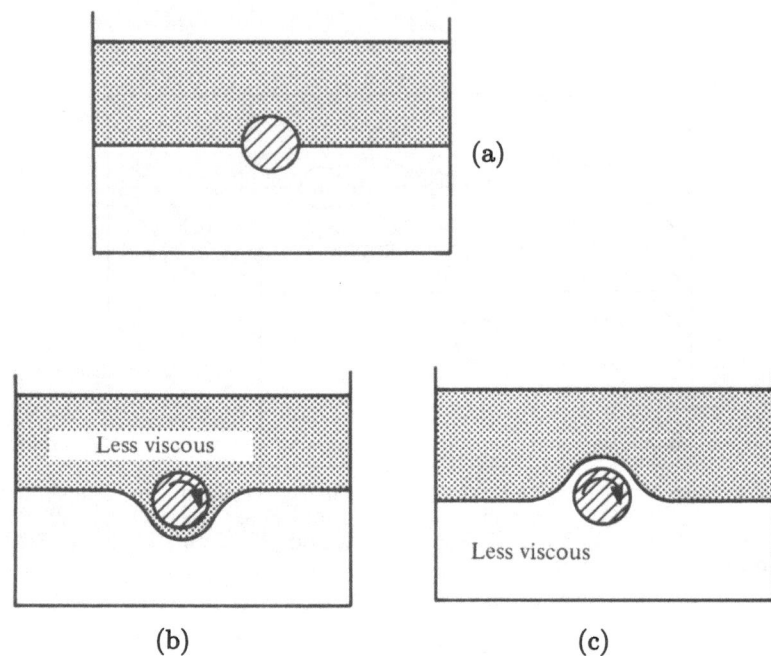

Fig. 4.2(a-c). [Figures 5 and 6 of Joseph, Nguyen and Beavers, 1984] (a) The Plexiglas box is loaded with two liquids with the undisturbed interface at the level of the horizontal diameter of the rod. The rod is then rotated in the direction of the arrow. The less viscous fluid coats the rod. This kind of coating is called sheet coating. The low-viscosity liquid shields the more viscous liquid from intense shearing. (b) Sheet coating when the light liquid is less viscous. (c) Sheet coating when the heavy liquid is less viscous.

II.4(a) Rollers

Rollers are viscous fluid bodies which rotate (nearly) as rigid solids. Rollers occur when one of the two fluids is very viscous and the viscosity ratio is large. For example, water ($\mu = 0.01$ poise) and oils ($\mu > 100$ poise) have worked well. The shape of rollers is strongly influenced by the forces which enter into the interface potential, surface tension, centripetal acceleration, and gravity, and by geometrical constraints. Rollers may rotate rigidly only if they detach from every stationary surface. The process of detachment is a remarkable event. It is an instability which is not understood. It appears as a neck-down at the sidewalls at a critical speed. The viscous liquid always separates from the wall. Sometimes, it separates cleanly, exhibiting adhesive fractures. Sometimes, a deposit of silicone oil is left on the sidewall, and the separation occurs as a cohesive fracture. What happens appears to depend on the relative strengths of cohesive and adhesive bonding.

950 Poise Silicone Oil Rollers. The photographs shown in plate II.4.3 exhibit the sidewall detachment instability leading to rollers. These figures show silicone oil (950 Poise) in water. The viscosity ratio is 95,000. The roller of silicone oil is almost perfectly round, and it has detached from the walls under hydrodynamic action. The angular velocity of the silicone roller is constant; the roller rotates as a rigid wheel. The roller is robustly stable: it withstands large perturbations and can rotate for weeks without change of form.

Since the roller is undergoing rigid-body rotation, with a much less viscous fluid surrounding it, its dynamics appears to be governed by the inviscid equations of motion. There is an unknown constant in the pressure (see equation (1a.13) in section II.1) which determines the radius of the roller, through the balance of normal stress with surface tension: this is the normal stress condition at the interface (see equations (1a.15), (1a.16)). This radius is not uniquely determined (because the $[\![c]\!]$ in (1a.15) is not uniquely determined). In fact, the roller in plate II.4.3 has captured all of the silicone oil which was originally floated on the water. Presumably, if we had floated more or less oil, we would achieve a roller with larger or smaller radius. The silicone oil is attracted to the Plexiglas rod; it is favored over water.

Gravity enters into the dynamics of the stable rollers reported in JNB and JRRN. In experiments in which the top of the roller rotates in air, the roller would centrifuge out into the air were it not for gravity, which on the small top portion of the roller exposed to air is nearly radial. A similar, but smaller effect of gravity occurs at the bottom of the roller, which is pushed up by gravity because the lighter oil is buoyant in water. Gravity tends to flatten rollers into right-circular cylinders. To a degree, the diameter of stable rollers can be controlled by gravity, with a tendency for the roller to poke its head into the air. It is possible to change the diameter of the rollers by changing the water level in the box. This effect of gravity is exhibited in figures 4.4(a)-(c). Sketches of the side view of these plates are shown in figures 4.4(d)-(f).

The principal effect of gravity may be reduced drastically by submerging the roller entirely in water, as in figures 4.5(a)-(b). When the flattening effect of gravity is absent, the shape of the interface on a stable roller is strongly influenced by interfacial tension, with bounding surfaces in nearly circular arcs, as in figure 4.5(b). Presumably, this roller is in rigid motion and the shape of the interface minimizes a potential.

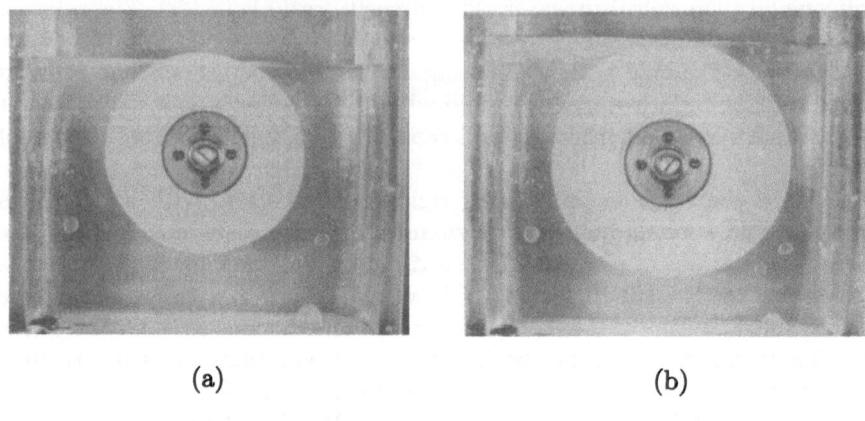

(a)   (b)

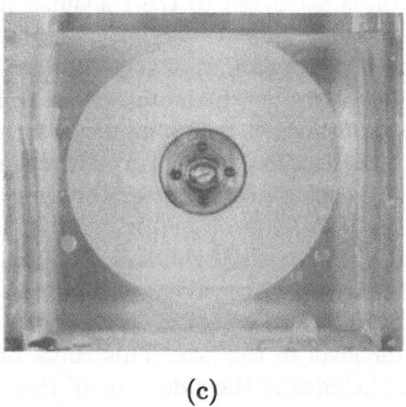

(c)

Fig. 4.4(a-c). [Figure 2 of Joseph, Renardy, Renardy and Nguyen, 1985] Roller of silicone oil ($\rho = 0.95$ g/cm^3, $\mu = 950$ P) in water at different water levels. The rod is made of Plexiglas, 2 inches in diameter, and rotates at 10 rpm. (a) The roller is very nearly in solid-body rotation with small shearing by water at the roller rim. Part of the roller is in water and the other in air. The roller is very stable, held together by hydrostatic pressure in water and gravity in air. (b) Water is added to the box. The roller becomes larger by flattening out but remains round and stable. (c) More water is added. The roller becomes even larger. The roller is now completely submerged in water and is slightly out of round due to buoyancy. (Continued)

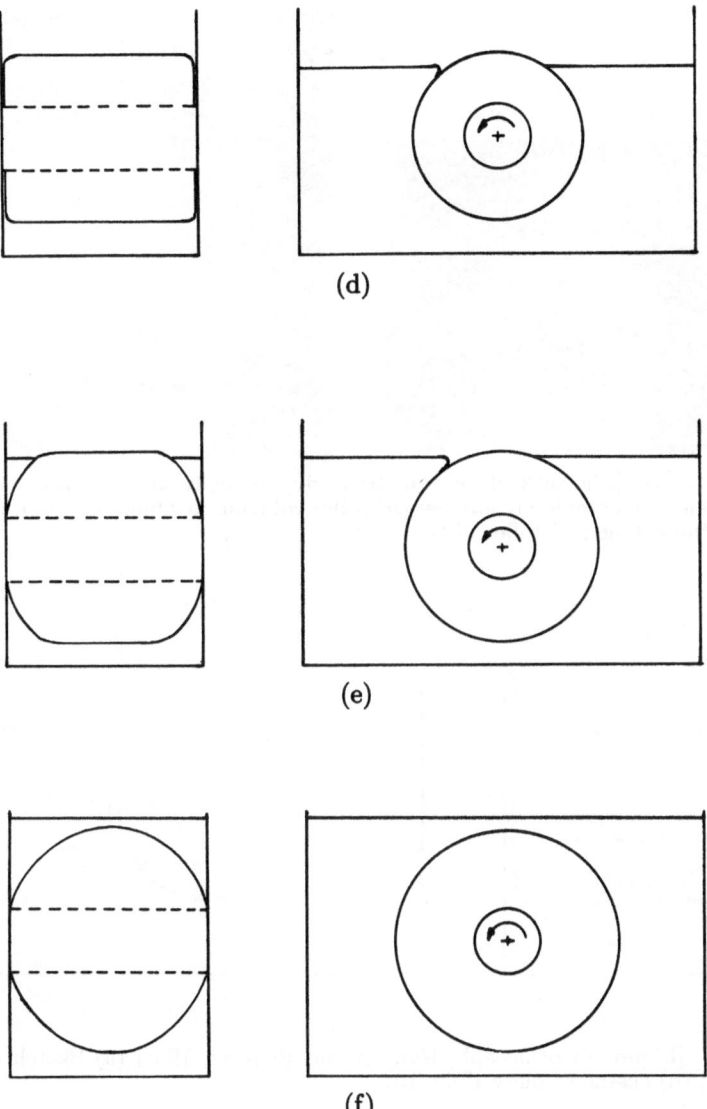

Fig. 4.4(d-f). (d) is a sketch of the side view and front view of (a). (e) is a sketch of the side view and front view of (b). Water is added to the box. The diameter of the roller becomes larger. The shape of the roller changes, conserving volume. (f) is a sketch corresponding to (c).

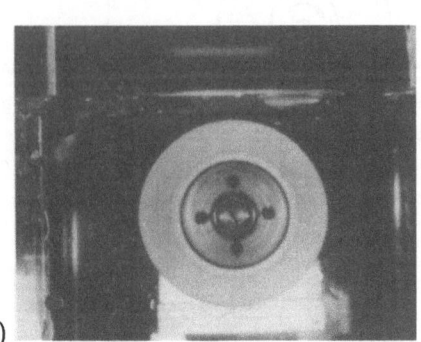
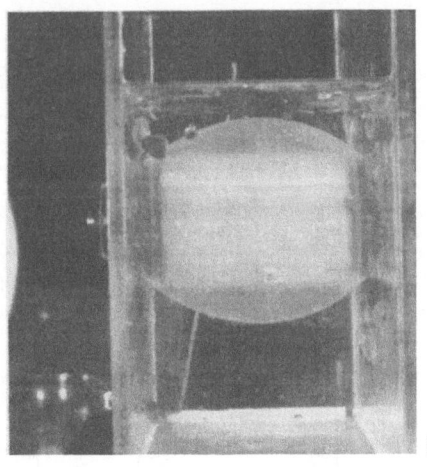

(a)   (b)

Fig. 4.5(a-b). [Figure 3 of Joseph, Renardy, Renardy and Nguyen, 1985] (a) Front view of a completely submerged roller rotating at about 1.5 rpm. (b) Side view of the submerged roller of (a).

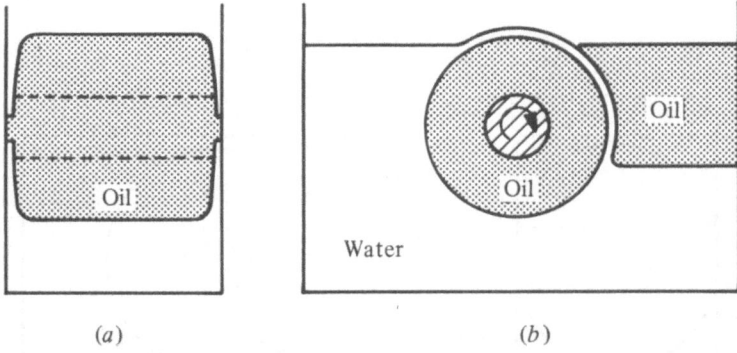

        (a)   (b)

Fig. 4.7. [Figure 13 of Joseph, Nguyen and Beavers, 1984] (a) Sketch of plate II.4.6(c); (b) sketch of plate II.4.6(b).

110 Poise STP Rollers. In plate II.4.6, we show STP lying on water. A sketch of plate II.4.6 is shown in figure 4.7. The STP wets the Plexiglas box and water wets the aluminum rod. The rod was put into steady rotation. Rotating clockwise transported all of the STP on the left side of the box (that is, the upward-motion side of the rod) to the roller of STP, or to a stagnant region of STP on the right (that is, the downward-motion side of the rod). A thin sheet of water was pulled between the STP roller and the stagnant STP (see plate II.4.6(b) and figure 4.7 (b)). Astonishingly, the roller of STP also separated from the wall of the box even though it is well known that STP is very sticky to Plexiglas. This hydrodynamically generated separation of STP from the Plexiglas side wall is shown in plate II.4.6(c).

The STP roller rotates as a rigid wheel, lubricated by water from all sides. The stagnant STP on the right barely moves. The configuration of the roller evidently reduces the total dissipation to a very small value associated mainly with shearing water in a lubrication layer.

It was not possible to maintain an STP roller without stabilizing it with the dead STP at the right of plate II.4.6(b). Some of the STP was removed from the stagnant region. But, at a critical value of the depth of the dead STP, the roller bifurcated into another roller with triangular symmetry. This triangular figure was unstable but not violently so. Some STP was added back to the stagnant region and stability was recovered. However, the newly stable roller was lopsided, so a screwdriver was used to mould it, as does a potter at his wheel, into the automobile-tire shape of large radius exhibited in figure 4.8.

Many different kinds of STP rollers in water with their tops in air are shown in the photographs in figures 4.8-9 and plates II.4.10-11. The shape of the rollers is associated with minimizing the rigid motions interface potential under constraints. The shapes are corrugated under the action of surface tension, centripetal acceleration, gravity, and geometrical constraints. Many different shapes, even with one Ω can be achieved.

The photographs of isolated rollers in figures 4.8 and plate II.4.11(d) appear to have a nodoid shape of the kind exhibited in figure 2.4 of section II.2. The tops poke into air. Gravity enters into the dynamics of these rollers; the water pushes up on the bottom and the air pulls down at the top. These effects are stabilizing.

Arrays of rollers of STP lubricated on all sides by water can be achieved using the four-roller apparatus of G. I. Taylor sketched in figure 4.9. The apparatus is filled with water up to the plane of symmetry between the upper and lower pairs of rollers, and then STP is added to cover the upper rollers. Rollers of STP, completely surrounded by water, develop out of small sinusoidal disturbances of initially uniform (along rod generators) interfaces; and the water surface between them develops a small wave. This small wave grows in amplitude and results in the many rollers. The interpenetrating rollers that finally develop are steady, stable and lubricated everywhere by water. Photographs of these rollers are exhibited in plate II.4.10. A grown-up version of the small waviness mentioned above can be seen in plate II.4.10(a).

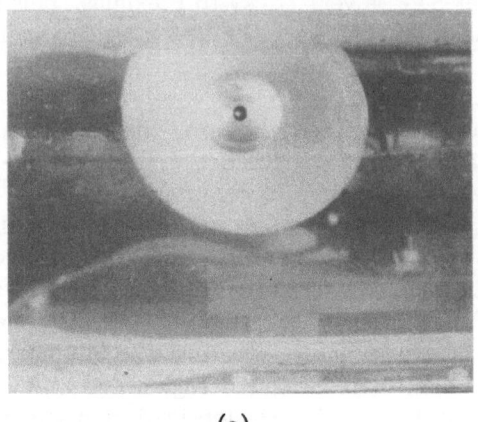

(a)

(b)

Fig. 4.8(a-b). [Figure 14 of Joseph, Nguyen and Beavers, 1984] The same experiment as in plates II.4.6 (b) - (c) after some STP has been removed from the right of the box and then some is added back, and the roller has been moulded with a screwdriver in the shape of an automobile tire. There is a small layer of STP on top of the water. This layer is separated from the roller by a layer of water maintained hydrodynamically. (b) Side view of the roller, looking in from the left side of (a). The roller has detached from the sidewalls under hydrodynamic action.

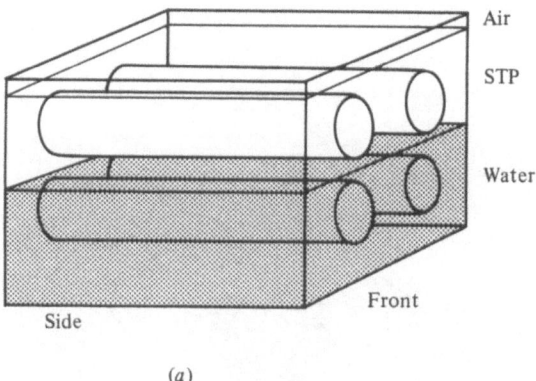

Air

STP

Water

Front

Side

(a)

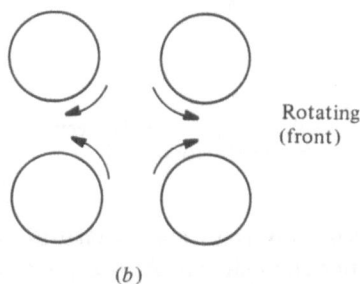

Rotating
(front)

(b)

Fig. 4.9. [Figure 17 of Joseph, Nguyen and Beavers, 1984] Schematic of the experiment with STP and water in the four-roller apparatus shown in plate II.4.10.

II.4(b) Sheet Coatings

When the viscosity of the more viscous liquid is less than about 100 poise in the set of experiments described above, the rollers collapse under gravity and shear. They are too fluid and cannot be maintained. Instead, we get coating in sheets of the type shown in figure 4.2. The material parameters for the liquids for which this type of sheet coating could be maintained in the experiments of JNB are summarized in table 4.1. Photographs showing sheet coating from above and below are exhibited in plates II.4.12-13. An interesting combination of sheet coating and dynamically maintained emulsions is exhibited in plate II.4.14.

A special form of sheet coating with a thin lubricating film of water is shown in plates II.4.15-16. This film appears to have a zero thickness to the naked eye. We use the word monolayer, perhaps inaccurately, to emphasize the apparent thickness of the film.

Recall that the 95000 cP silicone oil wetted the Plexiglas rod and led to the formation of rollers (section II.4 (a)). JNB tried to obtain rollers with

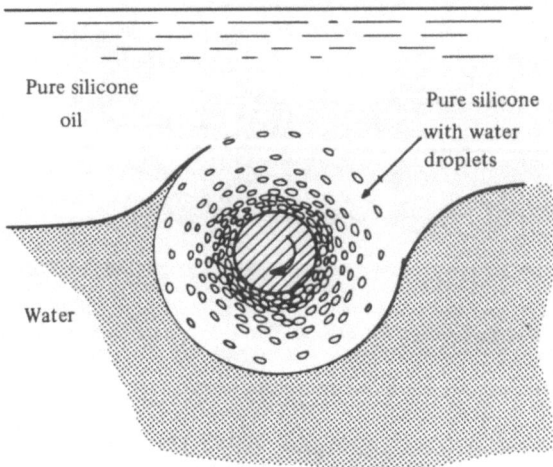

Fig. 4.17. [Figure 16 of Joseph, Nguyen and Beavers, 1984] Fingering of water droplets into high-viscosity (95000 cP) silicone oil.

the same silicone oil in the box (figure 4.1) with an aluminum rod which was wetted by water rather than the oil. At first, over a period of about six hours, an imperfect roller configuration developed, but thereafter, they lost the lubricating water film to a cusp-like water film which is characteristic of fingering instabilities and the formation of emulsions (see figure 4.17). The configuration of components then assumed form as a dilute water-laden emulsion of water doplets in silicone oil. The water droplets were continuously generated from the cusp shown in figure 4.17. The droplets were very effective in reducing the viscosity of the silicone oil. More and more of the water droplets drifted to the rod. After a few days, this collection of drops reached a percolation threshold with rings, like wedding bands, of water around the rod. After three days, all the rings had collected into a sheet of water coating the rod and lubricating the silicone oil. The silicone oil in this configuration appears not to move, though of course there must be some small motion of the oil due to shearing by water. This final configuration is shown in plate II.4.15.

The hydrodynamical history of the silicone oil experiments exhibits roller instabilities, fingering instabilities and the generation of emulsions, and finally a sheet lubrication of high-viscosity silicone oil by a lubricating water layer.

The sheet lubrication illustrated in the idealized sketch of figure 4.2 is achieved in plate II.4.16. In this configuration, there is a water layer on the rod, completely surrounded by silicone oil. This was achieved by withdrawing some water with a syringe after the sheet coating with water, described in the previous paragraph, has completely stabilized. The fully lubricated configuration in plate II.4.16 is robustly stable.

Table 4.1. Experiments in which the low-viscosity fluid coats in sheets. The entries with * indicate sheet coating at high speed. The entries x in the last column give the plate number II.x

Liquid 1	Liquid 2	μ_1 , μ_2	ρ_1, ρ_2	Rod	Rod	Plate
		(cP)	(g/cm^3)	wetted	speed	
				by	(rpm)	
silicone I	glycerol	19, 1761	0.96, 1.25	liquid 1	100	4.13
silicone I	water	19, 1	0.96, 1.0	1	~150	4.12(b)
light machine	glycerol	6.36, 1761	0.831, 1.25	1		
oil						
light machine	corn syrup	6.36, 14	0.831, 1.20	1		
oil	- water					
silicone I	corn syrup	19, 12	0.96, 1.20	1	~150	
	- water					4.12(a)
vegetable oil	glycerol	60, 1761	0.92, 1.25	1		
castor oil	glycerol	700, 1761	0.96, 1.25	1	~50	fig 9,
						JNB
water	dibutyl	1, 18-19	1, 1.045	2		
	phthalate					
polyacrylamide	dibutyl	~700, 18-19	1.02, 1.045	2		
(2% in water)	phthalate					*
castor oil	polyacrylamide	700, 90	0.96, 1.01	1		
	(1% in water)					*
silicone III	water	95000, 1	0.95, 1	2	~65	4.15-
						4.16

II.4(c) Fingering Instabilities and the Formation of Emulsions

Plates II.4.3(c), II.4.14 and figure 4.17 show that relatively large emulsions can be present in shearing flows of two fluids. These bubbles may be called "dynamic" emulsions, since they are dynamically maintained by the shearing and disappear when the apparatus is stopped. There are also smaller bubbles [Drew 1983] which stay in the flow after the apparatus is stopped.

Dynamic emulsions are maintained by motions and disappear when the fluids are laid to rest. These emulsions follow fingering. In many cases, the wetting properties of the rod or the experimental conditions do not allow the formation of lubricating sheets of low-viscosity liquid. In these cases, we get fingering of low-viscosity liquid into high-viscosity liquid. Drops of low-viscosity liquid tear off at the fingertips, leading to the formation of an emulsion of low-viscosity drops in a high-viscosity foam. A type of capillary instability may be associated with drop formation from fingers. Table 4.2 is a summary of the experiments of JNB exhibiting fingering into emulsions.

It is possible to so constrain the dynamics as to form fingers of high-viscosity into low-viscosity liquid; for example, if only a small amount of the high-viscosity liquid is present and there is a lot of the low-viscosity liquid (see the section on "phase inversion" in section II.5). This situation appears to be unusual.

Table 4.2. Experiments in which the low-viscosity fluid coats by fingering in emulsions. Each entry x in the last column gives the plate number II.x. The asterisk * indicates an emulsion at low speed.

Liquid 1	Liquid 2	μ_1, μ_2	ρ_1, ρ_2	Rod	Plate
		(cP)	(g/cm^3)	wetted by	
vegetable oil	water	60, 1	0.92, 1.0	liquid 1	
peanut oil	water	60, 1	0.92, 1.0	1	
castor oil	water	700, 1	0.96, 1.0	1	
light machine oil	water	6.36, 1	0.831, 1.0	1	4.18
castor oil	polyacrylamide	700, 90	0.96, 1.01	1	4.14
	(1% in water)				*
STP	TLA 227	11000, 20000	0.89, 0.895	1	fig.22,
					JNB
STP	polyacrylamide	11000, 90	0.89, 1.01	2	4.21
	(1% in water)				
STP	water	11000, 1	0.89, 1.0	2	
polyacrylamide	dibutyl	700, 19	1.02, 1.045		*
(2% in water)	phthalate				

The first group of emulsification experiments are those in which a light liquid of moderate viscosity, like vegetable oil, light machine oil or castor oil, is floated on top of water, or waterbased polymeric solutions, like poly-acrylamide. In all these cases, the oil wets the Plexiglas rod, and after the whole rod is exposed to oil, the oil clings tenaciously to the rod in a narrow layer, even in a monolayer; when the rotation speed is high enough, then the low-viscosity fluid will migrate next to the rod, as prophesied by the lubrication principle mentioned in section I.3 (f). When the rod rotates slowly, there is a tendency for water to be drawn up onto the rod, but surface tension pulls the water back as shown in plate II.4.18(a). At a higher speed, the water will begin to finger into the oil, depositing droplets as shown in plate II.4.18(b). The fingering instability is sketched in figure 4.19. The continuous formation of droplets leads eventually to emulsification of water droplets in oil foam. The emulsified liquid then coats the rod as shown in plate II.4.18(c). Instead of sheet coating, we get coating by water-laden emulsions. These emulsions have very low viscosities. First, they are water-laden; second, they "tank-tread" like roller bearings (one can think of the emulsion as consisting of drops rolling along the tank-walls, with their treads rotating [Dussan and Davis 1974]) and they seem to be nearly as effective as sheet coats in shielding the high viscosity liquid from shearing.

The photograph of castor oil above polyacrylamide shown in plate II.4.14 is a variant of fingering dynamics leading to drops and emulsions. In this, the polyacrylamide-water droplets are encapsulated at higher speed by a sheet of low-viscosity (polyacrylamide-water) liquid.

A second group of emulsification experiments are generated by conditions which prevent the generation of sheet coating. Some experiments were performed in a box of the type that led to the formation of the rollers shown in plate II.4.6(a)-(c). The only difference was that the box was filled to the top and kept from moving there by a cover-plate. It was expected that the phase configuration of minimum dissipation would lead to the capture of low-viscosity fluid on the rod, with most of it on the rod if the densities were nearly matched. The difference between this sequence of experiments and the ones on the rollers of section II.4 (a) is that the cover-plate forces a kind of hydrodynamic lubrication at the top of the cylinder promoting fingering and the formation of emulsions.

Realizations of the idea of the foregoing paragraph are shown in plates II.4.20(a)-(b). As always, the experiment was started with the static configuration of heavy fluid below. In plate II.4.20 (a), we see the phase configuration of STP (dark) and TLA 227 (light) after a few days with the rod rotating clockwise at about 16 rpm. Both fluids are oil-based polymeric immiscible liquids. The density of TLA 227 is about 0.005 g/cm^3 greater than the density of STP. It can be seen in this figure that after a few days, much of the STP has migrated to the rod, and the streamlines carrying in the STP from remote regions are evident. When the (dark) STP is drawn from remote regions to the rod, it carries with it some (light) TLA 227 by shearing action. After a week, the color differences were very faint and it was not clear if the bulk of the STP was really on the rod. But when the rod was stopped, the STP precipitated out, as illustrated in plate II.4.20 (b).

The effect of emulsions in shielding the high-viscosity fluid is very clearly seen in the experiment with STP (dark) above polyacrylamide in water (clear), shown in plates II.4.21 (a) - (b). The polyacrylamide is heavy and much less viscous (viscosity ratio of 120). The STP has a great affinity for the rod and when the rod is turned on, it pulled along a big annu-

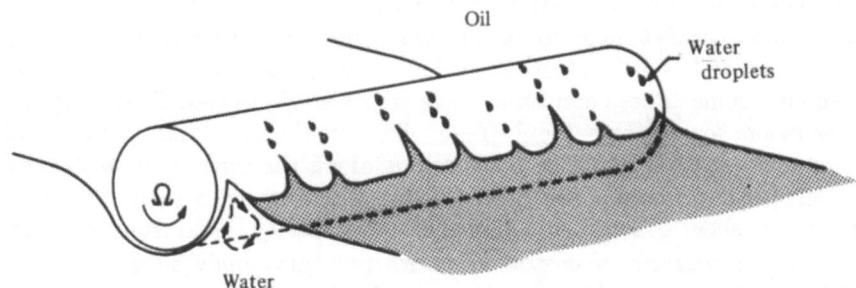

Fig. 4.19. [Figure 20 of Joseph, Nguyen and Beavers, 1984] Fingering instability leading to emulsions of water droplets in oil foam.

lus of the high-viscosity STP, going against the idea that the low-viscosity should migrate to coat the rod. But after a while, the small bubbles of polyacrylamide drifted to the rod and all the shearing was confined to the low-viscosity emulsion near the rod. Then the polyacrylamide in the emulsion near the rod deposited polyacrylamide in almost pure form onto the rod. This appears as a light ring of polyacrylamide on the rotating rod shown in plate II.4.21 (b). The bulk of the STP is completely quiescent, being shielded from the polyacrylamide by the emulsion. We think that this configuration very nearly minimizes dissipation but we did not anticipate that the hydrodynamics would take on such bizarre forms. The rotational speed appeared to have increased a lot (at the same torque) in the week and one half of rod turning, suggesting a big drop in the dissipation of the preferred phase configuration.

A third group of experiments which leads to dynamically maintained emulsions following fingering is described in figures 4.22 - 27. The two fluids are confined to the region between two concentric cylinders (the domain is G of equation (1a.1); see also figure 3.3 of chapter I, and section I.3 (d)), with the outer cylinder stationary and the inner one rotating. We call this a Taylor-Couette apparatus. The liquids are soybean oil (top fluid in figure 4.22, $\mu = 0.6$ p) and silicone oil ($\mu = 0.2$ p). The static configuration shown in figure 4.22 exhibits the effect of surface tension in rounding the interface when the effect of gravity is reduced by a small density difference.

To describe the results, we start with a thought experiment. Suppose that the two fluids have an equal affinity for the Plexiglas cylinders and that the static configuration is as shown in figure 4.22 (a). Now we rotate the inner cylinder. The low-viscosity silicone oil fingers first into the high-viscosity soybean oil, as in figure 4.22 (b). Later, at a higher speed of rotation, the soybean oil fingers into the silicone oil, as in figure 4.22 (c). The end result is an emulsified mixture of silicone oil and soybean oil, and

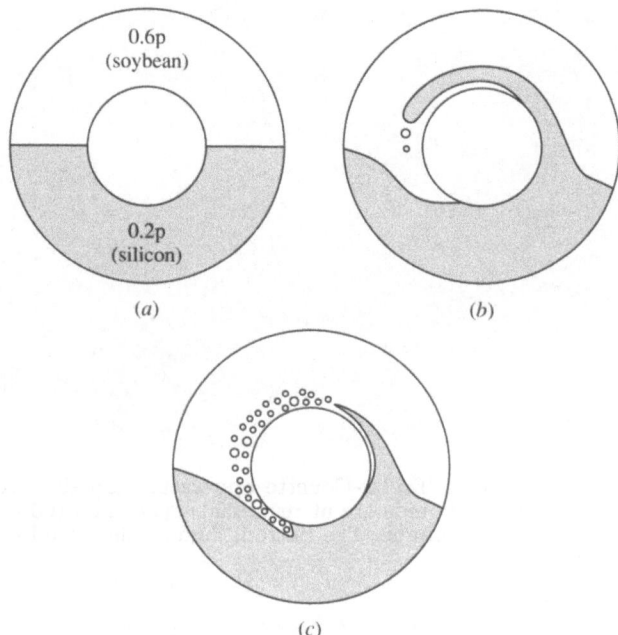

Fig. 4.22. Thought experiment showing fingering when liquids are neutrally wetting. (a) Static configuration with equal volumes; (b) the silicone oil fingers first; (c) the soybean oil fingers.

drops within drops. The number density of silicone oil drops is much larger than the number density of soybean oil drops.

In figures 4.23 - 27, we exhibit photographs of a real experimental situation in which the silicone oil occupies 30% of the total volume. It is clear from the contact angle in figure 4.23, because the preferred fluid wets the solid wall and climbs it, that Plexiglas has a much stronger affinity for soybean than for silicone oil. Nevertheless, for this volume fraction, only the silicone oil fingers. The soybean oil is the disperse phase and does not emulsify. This situation holds for volume fractions less than about 40%.

These experiments establish that typically, the low-viscosity liquid will finger into the high-viscosity liquid. The torque on the inner cylinder exerted by a dynamically maintained emulsion of any volume fraction lies between the torques exerted by the separate high- and low-viscosity constituents.

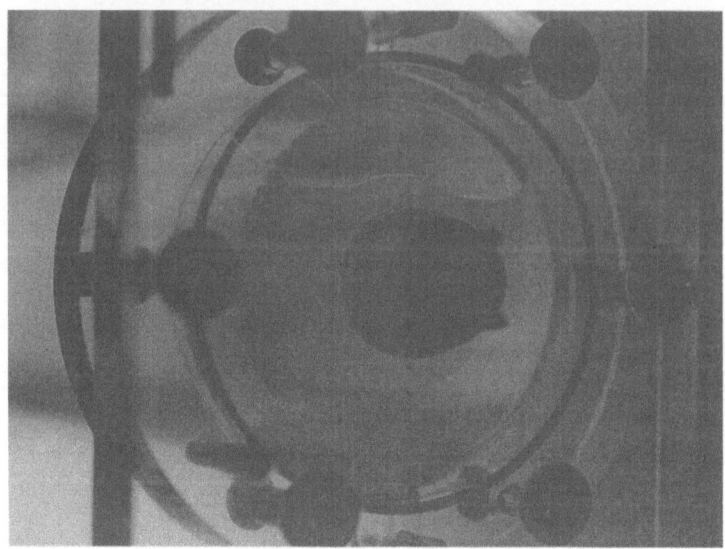

Fig. 4.23. Side view of a Taylor-Couette apparatus, gap size 1.27 cm, when the cylinders and the two liquids are at rest. The soybean oil (60 cP) at the top occupies 70% of the total volume. The bottom fluid is silicone oil (20 cP).

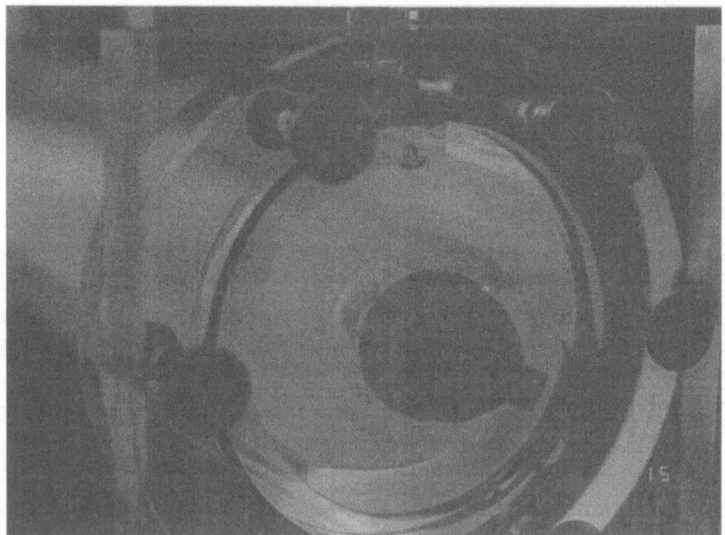

Fig. 4.24. Continuation of the experiment of figure 4.23. The inner cylinder is rotating at a speed just above the stability threshhold for fingering. Silicone oil (20 cP) fingers into soybean oil (60 cP). The formation of silicone oil drops occurs at the end of the fingertips.

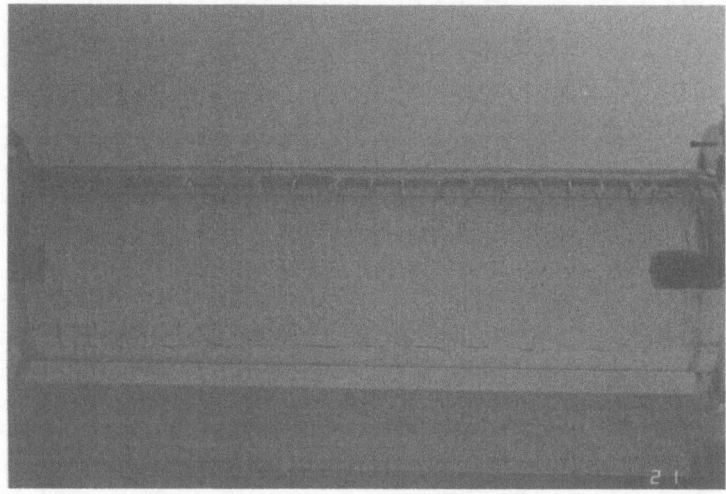

Fig. 4.25. Continuation of the experiment of figure 4.24. Front view of an early stage of fingering. The fingers appear to have a distinct period. They emerge from a sheet of silicone oil which is created by shearing.

Fig. 4.26. Continuation of the experiment of figure 4.25. More fingering.

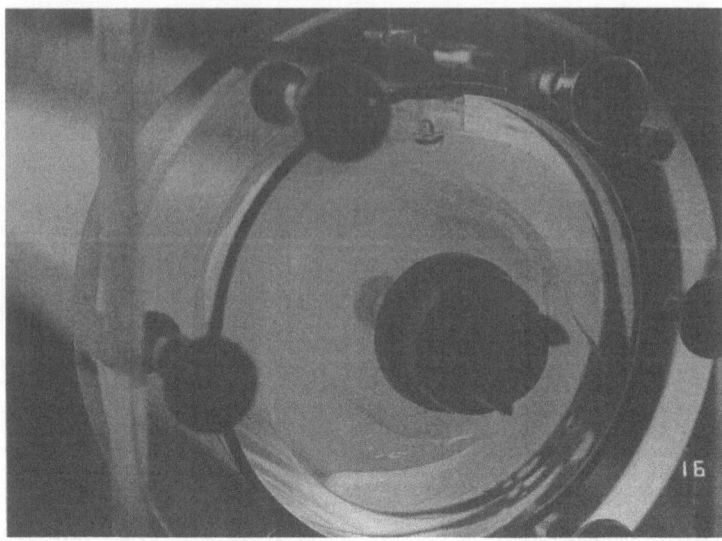

Fig. 4.27. Silicone oil drops migrate to the inner cylinder. These drops are forced to travel in the soybean oil, lying between the main body of silicone oil and the inner cylinder.

II.4(d) Centrifugal Instabilities

The instabilities we have in mind are the ones which are commonly associated with an adverse distribution of angular momentum. For single fluids, an example of such an instability is the formation of Taylor cells. The point of novelty here is the presence of two liquids. Two types of phenomena that occur in our experiments are of interest. The encapsulation instability seems always to position a low-viscosity film or emulsion between the rod and a stagnant body of high-viscosity liquid. Such a configuration is very conducive to the development of an adverse distribution of angular momentum. The tendency for centrifugal forces to throw the heavy, less-viscous liquid outward seems to be resisted by the other more-viscous and still stable portion of the fluid. This type of dynamics may be involved in the flow shown in plate II.4.28. This is reminiscent of the free surface flows exhibited by Moffatt [1977] in his study of viscous films on the outer surface of a rotating cylinder.

A second type of phenomenon develops in the flow of immiscible liquids in a Taylor apparatus set on its side and is described in the next section.

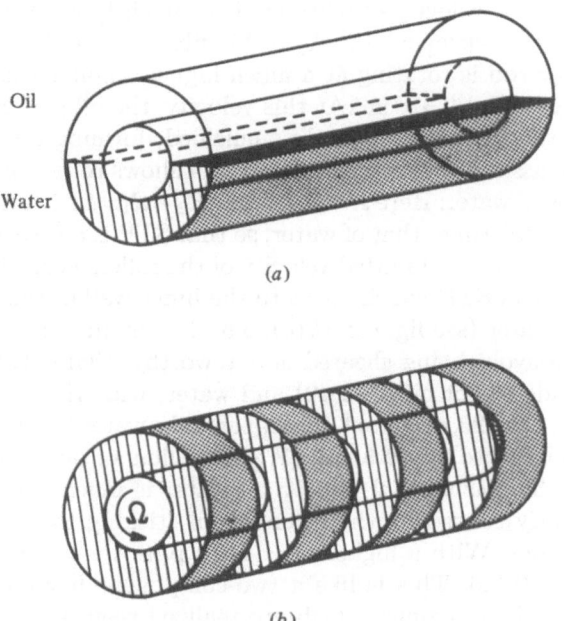

Oil

Water

(a)

(b)

Fig. 5.1. [Figure 25 of Joseph, Nguyen and Beavers, 1984] Sketch of Taylor cells developing from the instability of water. The oil is dynamically passive while the motion in the water is driven by centrifugal instability.

II.5 Taylor-Couette Flow of Two Immiscible Liquids

We begin with a description of plates II.5.2 - II.5.6 which are taken from the paper of Joseph, Nguyen and Beavers [1984]. We shall refer to their paper as JNB. A Taylor apparatus (we have called this a Taylor-Couette apparatus) is set on its side as shown in figure 5.1. The inner diameter of the outer cylinder is 6.35 cm and the length of the space between the inner and outer cylinders is 30.48 cm. The cylinders are made of Plexiglas. The outer is stationary and the inner is free to rotate. Inner cylinders of diameters 5.72cm, 5.08 cm and 3.81 cm were used. Every experiment was carried out with equal volumes of the two liquids, as depicted in figure 5.1(a).

The fluid dynamics of the resulting flow is dominated by a form of Taylor instability which seems to be only weakly influenced by gravity. An idealized sketch of the cells which develop is shown in figure 5.1(b). High- and low-viscosity cells separate each other. Examples of the cells that actually do develop are exhibited in plates II.5.2 - II.5.6.

Plate II.5.2 shows rollers of silicone oil I ($\rho = 0.95$ g/cm^3, $\mu =$19 cP) separating Taylor cells of water. Here, Ω=130 r.p.m., with $R_2 - R_1$=0.635 cm. The azimuthal velocity of silicone oil is much larger than water, presumably because of encapsulation by a thin film of water (see figure 5.8). In plate II.5.3, the rod is rotating at a much higher angular velocity Ω=1810 r.p.m., with $R_2 - R_1$=0.318cm. At this velocity, the oil and water are both unstable and the oil has completely emulsified, forming a single liquid in which we see classical Taylor cells. Plate II.5.4 shows rollers of STP separating Taylor cells of water: Here Ω=86 r.p.m., $R_2 - R_1$=1.27 cm. The viscosity of the STP is 11000 times that of water, so that STP is always stable against Taylor instability. The azimuthal velocity of the rollers is much greater than that of the wave of STP which sticks to the inner wall of the outside cylinder above the water (see figures 5.7(c), 5.8) The manner in which the STP is fractured to avoid being sheared is noteworthy. Plates II.5.5 and II.5.6 show monograde motor oil (SAE40) and water, with $R_2 - R_1 = 0.318$cm. In plate II.5.5, Ω=120 r.p.m.: at this speed only water is unstable, and the dynamics is similar to figure 5.7(c). We see rollers of oil separating Taylor cells of water. The azimuthal velocity of water is much smaller than that of oil, apparently caused by the layers of oil sticking to the inner wall of the outer cylinder. With a higher $\Omega = 440$ r.p.m. we see the phenomenon shown in plate II.5.6. This is like a two-component flow of two different emulsions, and the dynamics which are realized seem to fall under figure 5.7(b).

Encapsulation instabilities enter into the dynamics of the cells shown in plates II.5.2 - II.5.6 in the following way. At low speed, we see a high torque, which very rapidly drops to a lower value. Correlating this with direct observation, we identify the first significant dynamical event as an encapsulation instability in which a sheet of water is pulled around the inner cylinder. After this event, gravity seems less important and the flows tend more to axisymmetry.

The next event is the formation of Taylor cells in the water layer. The dynamics associated with this event may roughly be described as follows. Imagine that a layer of water occupies the region next to the inner cylinder, with $[\![\rho]\!] = 0$ or, equivalently, $g = 0$. As Ω is increased past a critical value, the smooth flow of water gives up stability to Taylor vortices. The oil also moves very slowly in cells, driven not by instability but by shear stresses induced by cellular motion of the dynamically active water. These motions will naturally distort the oil/water interface, as in figure 5.7(a). The large-amplitude limit of the flow in figure 5.7(a) is usually like the flow depicted in figure 5.7(c). In this flow the passive oil cells undergo extremely weak cellular motions driven by shears from active water cells. The oil cells are rollers, and are lubricated at the sides and at the outer cylinder by water. The lubrication of oil rollers by water is enhanced by the fact that water is heavier than oil and will tend to replace the oil layer on the outside of the cylinder. If the oil has a strong adhesion to Plexiglas a small layer of oil

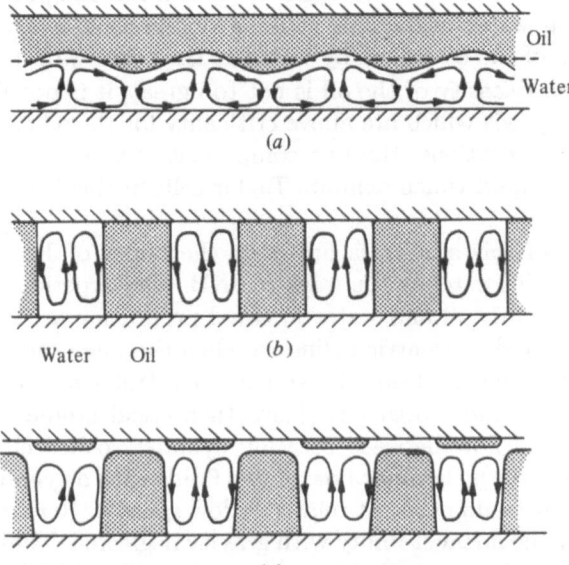

Fig. 5.7. [Figure 31 of Joseph, Nguyen and Beavers, 1984] Sketch of dynamically active water cells and dynamically passive oil cells (shaded) arising from the instability in bicomponent Couette flow between rotating cylinders. The situation in (a) arises near criticality as an instability of layered Couette flow. The situation in (b) could be regarded as the large-amplitude limit of (a) when density differences are negligible. The secondary flow in the oil cells is extremely weak. Instead of (b), we usually see a configuration like (c) with passive oil cells rotating as rigid rollers attached to the inner cylinder and lubricated by water at the outer cylinder.

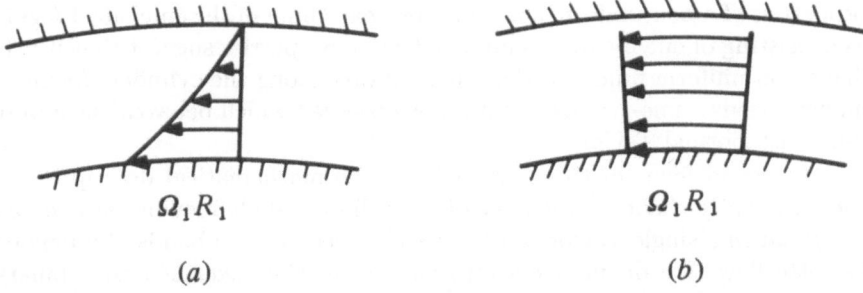

Fig. 5.8. [Figure 32 of Joseph, Nguyen and Beavers, 1984] The azimuthal velocity distribution in (a) water cells, and (b) oil rollers lubricated by water.

will continue to adhere to the outer cylinder at positions above the water
layer where the shears are small (see figure 5.8). In the experiments, it is
easy to see that the azimuthal component of velocity of oil cells near the
outer cylinder is much greater than the azimuthal component of velocity
in neighboring water cells. This striking observation is explained by the
encapsulation of the oil rollers by water at the outer wall.

When the viscosity of the oil is not too great, it is possible to run our
apparatus at speeds which are above criticality for the oil as well as for the
water. In such situations, the two components emulsify strongly, forming
one emulsified liquid which exhibits Taylor cells in classical form (see plate
II.5.3).

The rest of this section is based on the paper of Joseph, Singh and
Chen [1990] which extends the work of JNB. The experimental apparatus
is that of figure 5.1: oil and water in equal proportion are set into motion
between horizontal concentric cylinders when the inner one rotates.

The papers by Y. Renardy and Joseph [1985 a] and by Guillopé,
Joseph, Nguyen, and Rosso [1987] are theoretical studies of the special
form of two-fluid flow between rotating cylinders called circular Couette
flow (this refers to the laminar flow of two fluids with a cylindrical interface
in between that could occur at relatively low speeds). Renardy and Joseph
[1985 a] study the linear stability with gravity neglected. They present a numerical study of the effect of different viscosities, densities and the effect of
surface tension. Asymptotic analyses for short waves in the axial direction,
and also in the azimuthal direction, are presented. The paper by Joseph
and Preziosi [1987], much of which is covered in sections II.1 and II.2, gives
a theoretical explanation of rollers, which is another configuration which
appears in our experiments between cylinders.

Many different flows are realized and described in this section: fingering
flows, coarse and fine emulsions, phase-separated emulsions, lubricated flow,
and banded Couette flows. In one regime, many large bubbles of oil are
formed. In a range of speeds where the water is Taylor unstable and the oil
Taylor stable, there are strange Taylor cells of emulsified fluids whose length
may be three or even four times larger than those normally found in either
oil or water. The length of the cells appears to be associated with effective
properties of a non-uniform emulsion (one can think of the emulsified region
as consisting of one medium, and its effective properties suggest that it acts
like a non-uniform fluid), so the cell sizes vary along the cylinder. At much
higher speeds a fine-grained emulsion is observed which behaves like a pure
fluid with normal Taylor cells.

A second focus of this section is on the mathematical description of
the apparently chaotic trajectory of a small oil bubble moving between an
eddy pair in a single Taylor cell trapped between the oil bands of a banded
Couette flow. We define a discrete autocorrelation sequence on a binary
sequence associated with left and right transitions in the cell to show that
the motion of the bubble is chaotic. A formula for a macroscopic Lyapunov

exponent for chaos on binary sequences is derived and applied to the experiment and to the Lorenz equation to show how binary sequences can be used to discuss chaos in continuous systems. These results and recent results of Feeny and Moon [1989] are used to argue that Lyapunov exponents for switching sequences are not convenient measures for distinguishing between chaos (short range predictability) and white noise (no predictability).

The flows which develop between our rotating cylinders depend strongly on the material properties of the two liquids. A third focus of this section is on dynamically maintained emulsions of two immiscible liquids with nearly matched density. The two fluids are 20 cp silicone oil and soybean oil with a very small density difference and small interfacial tension. The two fluids are vertically stratified by weight when the angular velocity is small. Then one fluid fingers into another. The fingers break into small bubbles driven by capillary instability. The bubbles may give rise to uniform emulsions which are unstable and break up into bands of pure liquid separated by bands of emulsified liquid. It is suggested that the mechanics of band formation is due primarily to the pressure deficit in the wake behind each microbubble.

The nature of fluid-solid interactions takes a more important place in two-fluid dynamics than in single fluid dynamics. There is a competition between the two fluids as to which one will wet a solid boundary. The factors that enter into this competition are not understood. We get some kind of interaction between the physical chemistry of adsorption of fluid at the boundary of a solid with two-fluid dynamics. Wetting is not determined entirely by energy considerations, by contact angles, static or dynamic. The history of the motion also plays an important role in determining the places on the solid which are wet by one liquid or another (see figure 5.11e).

Taylor flows of a single fluid are among the best understood of all fluid phenomena; see, for example, chapter V of Joseph [1976 I], §17 of Drazin and Reid [1982], and chapter 7 of Chandrasekhar [1981]. One reason for this is that the geometry is relatively simple for analysis and perfectly marvelous for experiments. The Taylor apparatus may also evolve as an apparatus of choice in the study of two-fluid dynamics, fluid-solid interaction, and in the study of dynamical properties of emulsions.

II.5(a) Experiments and Parameters

All of the experiments were carried out between two concentric cylinders with axis horizontal, perpendicular to gravity. The outer cylinder and the end plates are plexiglass. The inside diameter of the outer cylinder is 2.495 inches; the outside diameter is $a = 2.986$ inches. The inner cylinder is of aluminum with a diameter of $a = 1.985$ inches. This is a convenient reference for normal Taylor cells which combine two counter-rotating eddies approximately 0.515 inches in length. The length of the cylinder is 11.985 inches. The outer cylinder is fixed and the inner one rotates with angular velocity Ω.

Our Taylor apparatus uses two neoprene lip seals to prevent leakage. The shaft driving the inner cylinder is connected to a torque meter which has a provision for counting rpm. Since the work done by the fluid is equal to the torque × speed, measurements of the torque show how much work the fluid is doing. Also, the more lubricated the flow is, the smaller the torque at any given speed. The torque meter is connected to a mechanical-digital converter which displays the value of the torque and rate of rotation. The digital signal is transferred to a Hewlett-Packard 87 and then sample averaged. Usually there is a fluctuation in the torque before the rotation reaches a new steady state. By monitoring the values displayed on the digital converter, we can determine when the motion is in steady state. The torque was measured for some of the experiments (figure 5.9). For uniformity, each data point was taken one-half hour after establishing a changed condition. Obvious transients were well-delayed after this time. Sometimes there is a slow emulsification which will eventually change the dynamics.

The torque values shown in figures 5.9, 5.13, and 5.17 are not guaranteed because there is an unknown frictional torque due to the neoprene seals which varies with the speed. The decrease in the torque at small values of Ω is almost certainly an effect of seals and not of flow. At higher speed the fraction of the total torque due to the bearings is smaller. Perhaps torques measured at speeds in excess of 80 rpm are reasonably accurate.

The laboratory is temperature-controlled at 25° Celsius. Two different oils were used in the experiments with water and oil: Mobil heavy duty oil with density of 0.97 g/cm³ and viscosity 0.95 poise, and SAE 30 motor oil with density of 0.886 g/cm³ and viscosity of 0.98 poise. The major effect of the density difference occurs in slow or lubricated flow in which the oil floats up. This effect is greater in SAE 30 oil-water systems than in the heavy oil-water system which is more nearly density matched.

A different set of experiments in the same apparatus was carried out with silicone oil and soybean oil (under the brand name of Crisco). The density and viscosity of the silicone oil is $\rho = 0.949$ g/cm³, $\mu = 0.2p$ and of the soybean oil is $\rho = 0.922$ g/cm³, $\mu = 0.46p$.

The interfacial tension between Mobil heavy-duty motor oil and tap water is 30.00 dyne/cm. The interfacial tension between SAE 30 motor oil and tap water is 9.2 dyne/cm. The interfacial tension between $0.2p$ silicone oil and Crisco is 1.4 dyne/cm.

Six dimensionless parameters govern these flows: the viscosity ratio, the density ratio, the volume ratio, a capillary number, a Froude number $U/\sqrt{gL}$ where U is a velocity scale and L is a length scale, and either a Taylor number, usually defined as $4\Omega^2 L^4/\nu^2$ where ν is the kinematic viscosity of one of the fluids, or a Reynolds number UL/ν. It is useful to note that some form of emulsion is usually observed when the Taylor number for the high viscosity constituent is larger than the theoretical critical one for Taylor instability of the corresponding one-fluid problem.

The critical angular velocities Ω in our apparatus are calculated from the linear stability theory as [4.35, 409, 402, 91.7, 217] rpm for [water, SAE 30 motor oil, Mobil heavy-duty motor oil, $0.2Jp$ silicone oil, Crisco] when one fluid rather than two fluids fills the gap. For example, in the range $91.7 < \Omega < 217$ rpm, the one-fluid silicone oil problem is unstable and the Crisco problem is linearly stable. We say that the two-phase flow is *bistable* when one phase is stable and the other is unstable. Bistable flow of water and motor oil occurs when $4.35 < \Omega < 409$ (or 402) rpm. Bistable flow of silicone oil and Crisco occurs when $91.7 < \Omega < 217$ rpm.

II.5(b) Circular Couette Flows

Couette flows of two liquids are here defined as steady axisymmetric flow of two immiscible liquids between infinitely long rotating cylinders of radii $a < b$ and angular velocities Ω_1 and Ω_2 . These flows satisfy the Navier-Stokes equations, no-slip boundary conditions, and classical interface conditions at liquid-liquid interfaces. The Couette flows are a small class of steady and possibly non-axisymmetric flows which could develop between rotating cylinders. In contrast to the one-fluid case, Couette flows of two fluids are not unique even at low speeds. There is a continuum of solutions of at least two different types: layered and banded Couette flows.

II.5(b)(i)   Layered Couette Flows. Layered Couette flows constitute one class of steady solutions (see section on rotating Couette flow in chapter I.3 (f)). For two layers, one is on the inside, the other outside, and the interface between them is at $r = d$. In fact the steady solutions are not unique. There could be any number of layers of any thickness subject to the specification of the volume of each fluid and geometrical constraints. We have shown in section I.3(f) that the layered flow with just two layers, the low viscosity liquid on the inner cylinder, uniquely minimizes the torque when the angular velocity difference is prescribed. We noted that the minimizing torque can be determined by minimization for two layers since we may always consider the problem for adjacent layers. For two layers, the torque M_L per length L is (see equation (3f.17) of chapter I)

$$M_L = \frac{a^2 b^2 (\Omega_2 - \Omega_1)}{(b^2 - a^2)} \frac{\mu_1 \mu_2 (b^2 + ka^2) L}{(b^2 \mu_1 + ka^2 \mu_2)} \tag{5b.1}$$

where

$$k = \frac{V_2}{V_1} \tag{5b.2}$$

is the ratio of volumes. Suppose that we have oil and water, fixing the water volume V_w and oil volume. First we compute M when $\mu_1 = \mu_w$, $V_1 = V_w$; then compute M when $\mu_2 = \mu_w$, $V_2 = V_w$. The case with $\mu_1 = \mu_w$ gives a smaller torque, so the water is on the inner cylinder $r = a$, oil outside.

The stability of layered circular Couette flow was studied by Y. Renardy and Joseph [1985 a] and Guillopé *et al.* [1987]. The linear theory with gravity neglected studied in Renardy and Joseph [1985 a] shows that a thin layer of the less-viscous fluid next to either cylinder is linearly stable and that it is possible to have stability with the less dense fluid lying outside. They also present asymptotic analyses for short waves in the axial direction or in the azimuthal direction. The stable configuration with the less-viscous fluid next to the outer cylinder, layered Couette flows, have never been observed because of effects neglected. However various lubricated flows with less viscous fluid on the wall in the presence of gravity could be regarded as realizations of the torque-minimizing layered Couette flow. For examples of this type we refer the reader to plates II.4.12-13 and II.4.15, and to the description of the flows shown in figures 5.10e-g.

II.5(b)(ii)   Banded Couette Flows. These flows are such that the two fluids are arranged in alternating bands rather than layers. This is an exact solution, the same solution as if there were no bands.

$$\mathbf{u} = \mathbf{e}_\theta V(r) = \mathbf{e}_\theta \left(Ar + \frac{B}{r} \right),$$

$$A = \frac{b^2 \Omega_2 - a^2 \Omega_1}{b^2 - a^2},$$

$$B = (\Omega_1 - \Omega_2) \frac{a^2 b^2}{b^2 - a^2}. \tag{5b.3}$$

The interface between bands are set on the annular areas in the intersection of the gap and planes perpendicular to axis z of the cylinders. The velocity is automatically continuous; the shear stresses $\tau_{z\theta}$ and τ_{zr} are zero and, if the densities are equal, the pressure is continuous across these planes.

Banded Couette flows are not unique, the width of the bands and their number is not determined by stated conditions. The torque on the cylinders is the sum of the torques of each band. The torque M_B on a band of length L, with viscosity μ, is given by

$$M_B = \frac{L\mu}{2} \int_a^b r^3 \left[\frac{d\left(\frac{v}{r}\right)}{dr} \right]^2 dr \tag{5b.4}$$

$$= L\mu a^2 b^2 \frac{(\Omega_2 - \Omega_1)^2}{b^2 - a^2}.$$

Suppose L_1 is the total length of bands of fluid with viscosity μ_1 and L_2 with viscosity μ_2. Then

$$L_1 + L_2 = L,$$

$$\frac{L_2}{L_1} = \frac{V_2}{V_1} = k, \text{and}$$

$$M_B = \frac{(\mu_2 + k\mu_1)}{1 + k} La^2 b^2 \frac{(\Omega_2 - \Omega_1)^2}{b^2 - a^2}. \tag{5b.5}$$

To compare the torques on the cylinders in banded and layered Couette flow we put

$$\frac{M_B}{M_L} = \frac{(b^2\mu_1 + ka^2\mu_2)}{\mu_1\mu_2(b^2 + ka^2)}\frac{(\mu_2 + k\mu_1)}{(1+k)}$$

$$= \frac{\left(m + k\frac{a^2}{b^2}\right)(1+mk)}{m\left(1 + k\frac{a^2}{b^2}\right)(1+k)} \tag{5b.6}$$

where

$$m = \frac{\mu_1}{\mu_2}.$$

The minimum torque in layered Couette flow is when the low viscosity constituent is on the wall ($m < 1$). The function (5b.6) of m decreases from infinity at ($m = 0$) to one at $m = 1$. Hence $M_B > M_L$; layered Couette flow has a smaller torque for the same angular velocity and volume fraction than banded Couette flow for which the oil must necessarily attach itself to both inner and outer layer.

Banded Couette flows are not lubricated. A banded Couette flow without secondary motions has not been seen, but the flows shown in figures 5.11a and 5.18 are banded Couette flows in the water cells which have already become unstable to Taylor vortices. The active water cells in figure 5.18 are the sites for the strange attractor which is described in the section, "Chaotic trajectories of oil bubbles in an unstable water cell".

II.5(c) Rollers

There can be a superficial resemblance between banded Couette flows and rollers. The rollers shown in plates II.5.2 and II.5.4 look exactly like the banded Couette flows shown in figure 5.18 of this section, but the rollers are not attached to the outer cylinder; they rotate nearly as a rigid solid sheared only by water. Truncated rollers are shown in figures 5.10e-g.

II.5(d) Emulsions, Tall Taylor Cells, Cell Nucleation

The generation of emulsion can occur in several ways. The fluids at rest are stratified by gravity. At slow speeds a stable interface of one fluid advances into the other. The advancing interface develops scallops at its leading edge as in figure 5.10a. These scallops become unstable and finger into the host fluid (figures 5.12 and 5.14a). Bubbles form from capillary instability leading directly or eventually to emulsions. The average size of the bubbles in an emulsion decreases as the speed (shear) increases. A coarse emulsion has large oil bubbles (figures 5.10c, d). All these emulsions are maintained by shearing. They collapse to stratification when the motion is stopped.

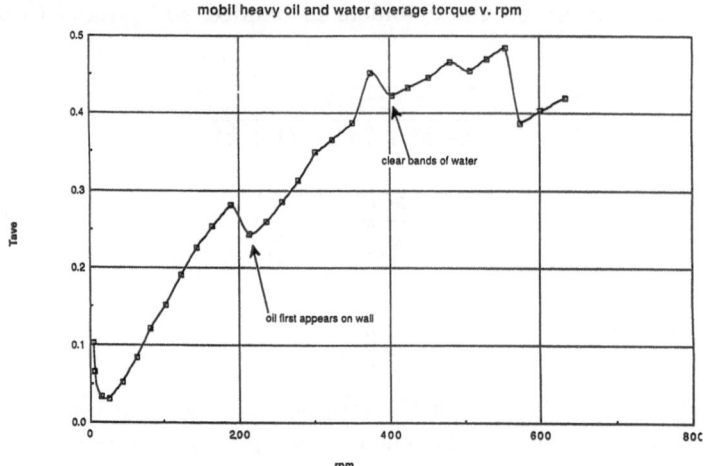

Fig. 5.9. [Figure 1 of Joseph, Singh and Chen, 1990, Plenum Press] Torque versus angular velocity for heavy motor oil and water. The decrease of the torque at the origin is due to the friction of the neoprene seal. We think that the oil foam which appears at $\Omega > 200$ marks the inflow boundary of a Taylor cell for a coarse grained emulsion. The appearance of clear bands is associated with spontaneous clearing of oil from the inner cylinder, increasing lubrication.

Strictly speaking, emulsions are unsteady because the oil-water interfaces are moving. Transient uniform emulsions can occur (figures 5.12 and 5.14b), but they appear to be unstable because they cannot be maintained when the angular velocity of the inner cylinder is fixed (figure 5.14c).

The forms taken by rotating flows of emulsions depend strongly on the fluids used. In the oil-water systems there are very long cells and the cell sizes are variable along the axis of the cylinder (figures 5.10d, 5.11b-c). This nonuniformity may be due to a nonuniformity of the degree of emulsification, the bubble size, along the axis. In any event the long nonuniform cells are robustly stable to changes in the angular velocity. At much higher speeds the degree of emulsification increases and new cell boundaries nucleate, producing more cells. Eventually there is a very fine stable homogeneous emulsion which has square Taylor cells of the usual type (figures 5.11d, 5.14e). These cells are robust; if the angular velocity is reduced the normal cells persist. This shows that fine emulsions have a different response than coarse emulsions even when the gap size and angular velocity are the same.

Typical sequences of flow types are exhibited in the photographs of heavy motor oil in water shown in figure 5.10 and in the photographs of SAE 30 motor oil shown in figure 5.11. Many of the transitions are evident also in the torque graph, figure 5.9, associated with transitions in the heavy oil of figure 5.10.

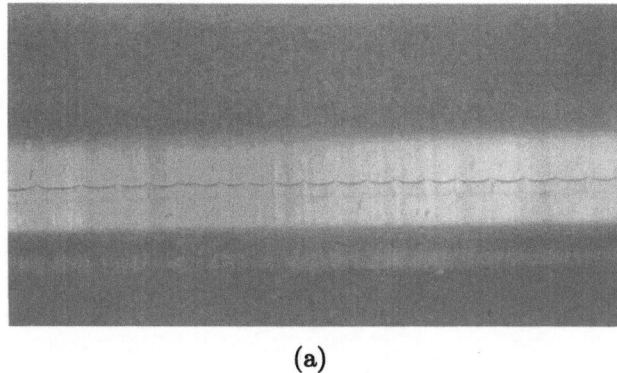

(a)

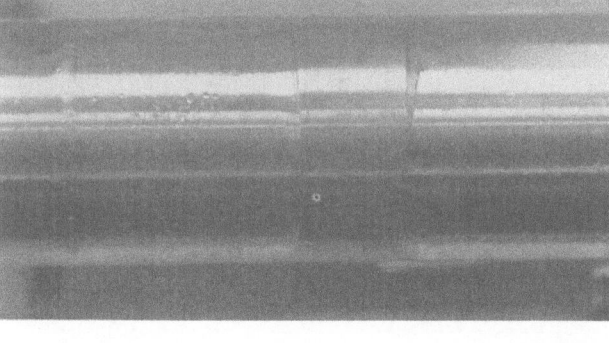

(b)

Fig. 5.10(a-b). [Figure 2 of Joseph, Singh and Chen, 1990, Plenum Press] Heavy motor oil and water. (a) 5 rpm. The inner cylinder is dragging the oil (top) into water (bottom). Scallops are on the leading edge of the oil-water interface which is steady and stable. (b) 31 rpm. The scalloped interface lost stability to a fat finger at 13 rpm. The finger was dragged around the cylinder and lead to the oil band evident in the photograph. There is a small oil roller on the left lubricated by water. Heavy motor oil and water. (Continued)

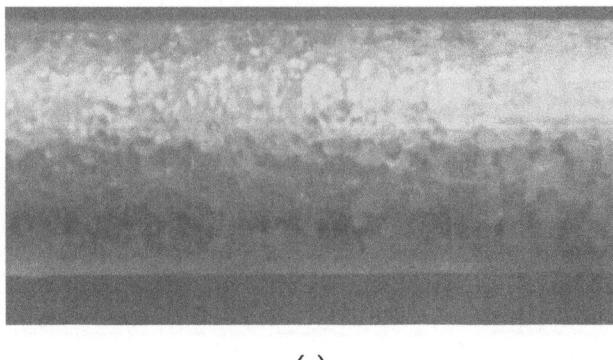

(c)

(d)

Fig. 5.10(c-d). (c) 201 rpm. This marks the beginning of a secondary motion of the oil bubbles. It appears that all of the oil has emulsified. A fine band of foamy oil is deposited on the outer cylinder over the inflow boundary. The cell size may be measured as a fraction of the outside diameter (0.29 inches). Normal Taylor cells are roughly twice the gap, 0.5 inches. These strange cells, which appear to be Taylor cells, are over three times as long as normal Taylor cells. (d) 308 rpm. The oil bubbles are fine markers of the fluid motion elongating themselves in the direction of motion. The foamy oil bands are inflow boundaries. The outflow boundaries are between the inflow boundaries and define a center of orientational symmetry for the fluidized oil bubbles. The length of the cells is not uniform. The cells are three times longer than usual. (Continued)

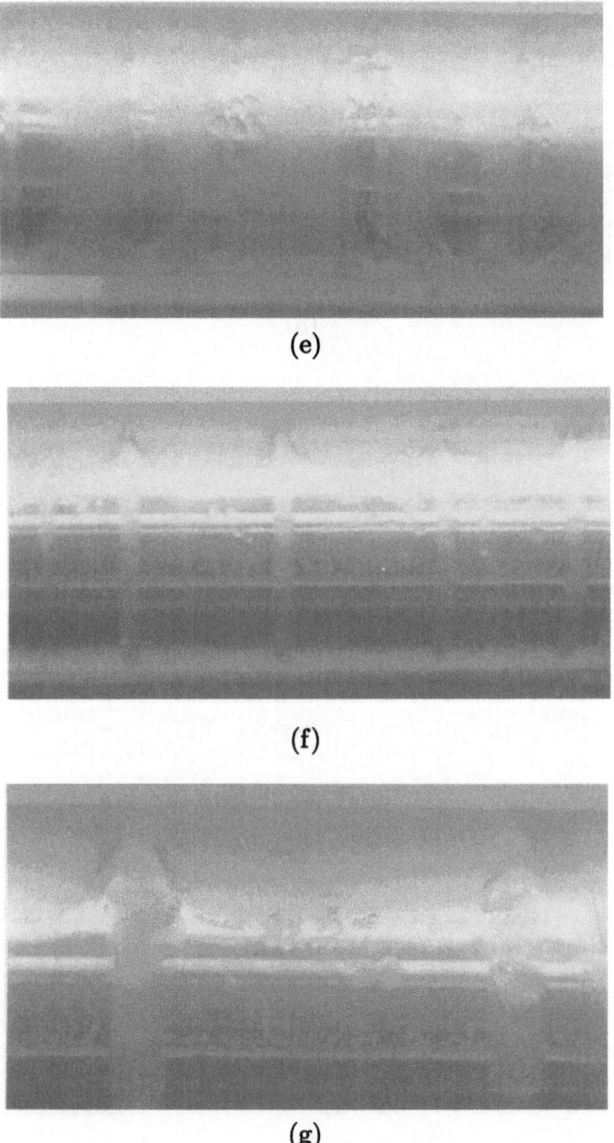

(e)

(f)

(g)

Fig. 5.10(e-g). (e) 500 rpm. The emulsion has almost vanished. The oil on the outside cylinder is lubricated by water on the inside. Truncated rollers are on the inner cylinder. (f) 550 rpm. The oil at the top is shielded from the inner cylinder by a layer of water. It is a lubricated flow, like a layered Couette flow. There are well-lubricated oil rollers on the inner cylinder. (g) 563 rpm. There are bigger rollers and fewer drops. Everything is lubricated by water.

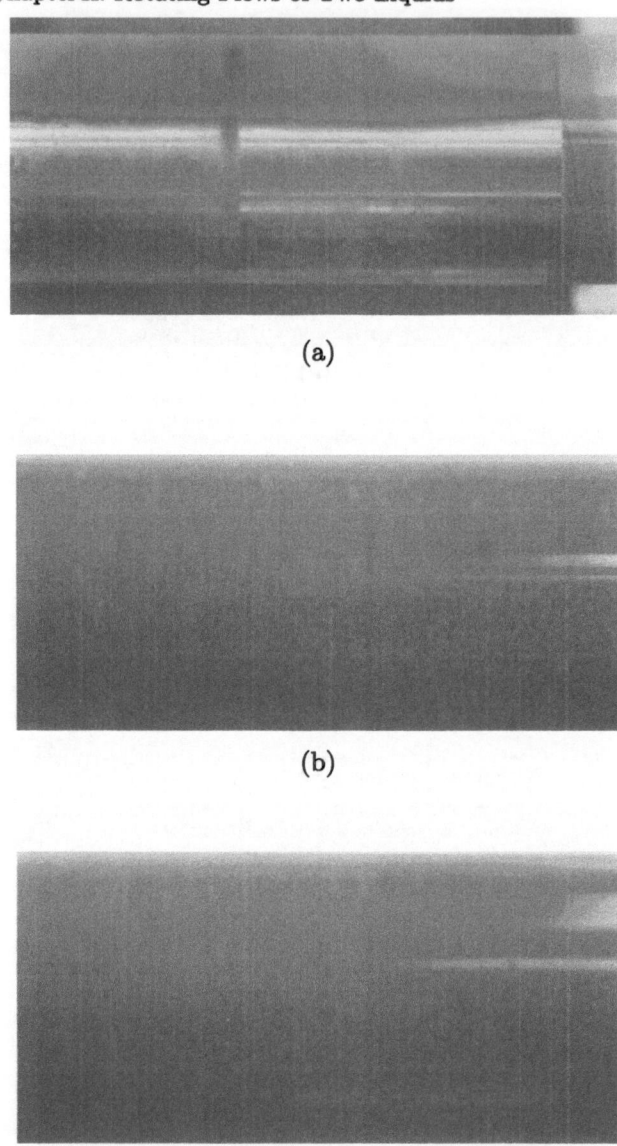

(a)

(b)

(c)

Fig. 5.11(a-c). [Figure 3 of Joseph, Singh and Chen, 1990, Plenum Press] SAE 30 motor oil and water, series 1. This series undergoes development somewhat like the one in plate II.5.10. (a) 99 rpm. A banded Couette flow is on the right and a water-lubricated stratified flow on the left. The oil broke up at about 300 rpm and coarse grained emulsion is formed. (b) 360 rpm. (c) 785 rpm. The secondary motions are beautifully marked by the emulsions. The cell sizes are not uniform but do not change with speed. At higher speeds new cells nucleate and after several adjustments the long cells shorten to the normal length, twice the gap. This adjustment is complete at 2550 rpm. Then the speed is reduced to 360 rpm (d) and the normal cells are retained. Continued.

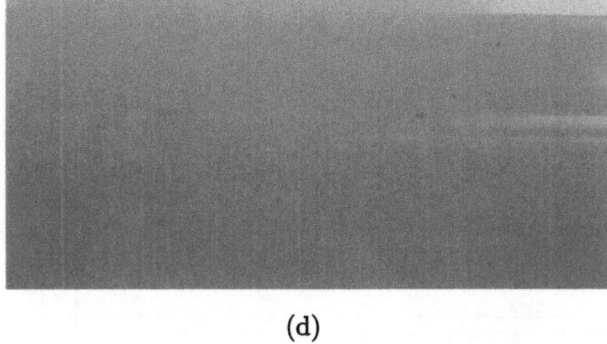

(d)

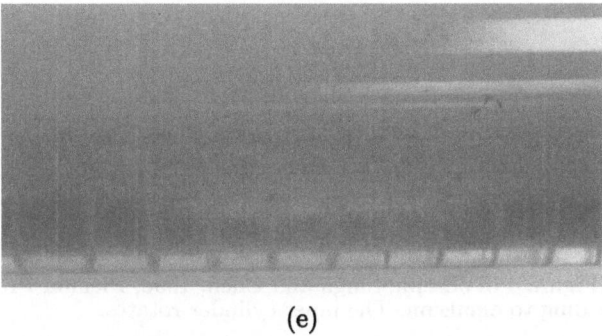

(e)

(f)

Fig. 5.11(d-f). (d) 360 rpm. The difference between (b) and (d) is due possibly to the fact that the emulsion in (d) is finer. (e) 0 rpm. The speed was reduced from 360 rpm to zero. The oil foam at the outside of the inflow boundary is stuck to the wall and it will not come off. This shows clearly that the fluid which wets the wall in different places depends on the history of the motion. (f) 450 rpm. This photograph shows how a banded Couette flow breaks up.

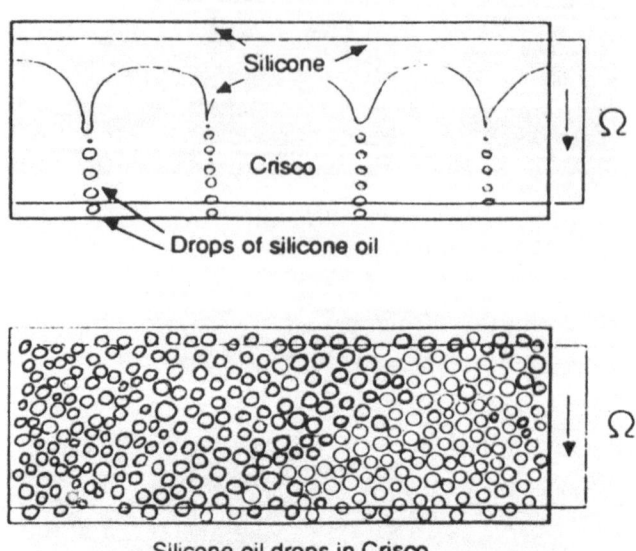

Fig. 5.12. [Figure 4 of Joseph, Singh and Chen, 1990, Plenum Press] Fingering instability leading to emulsion. The inner cylinder rotates.

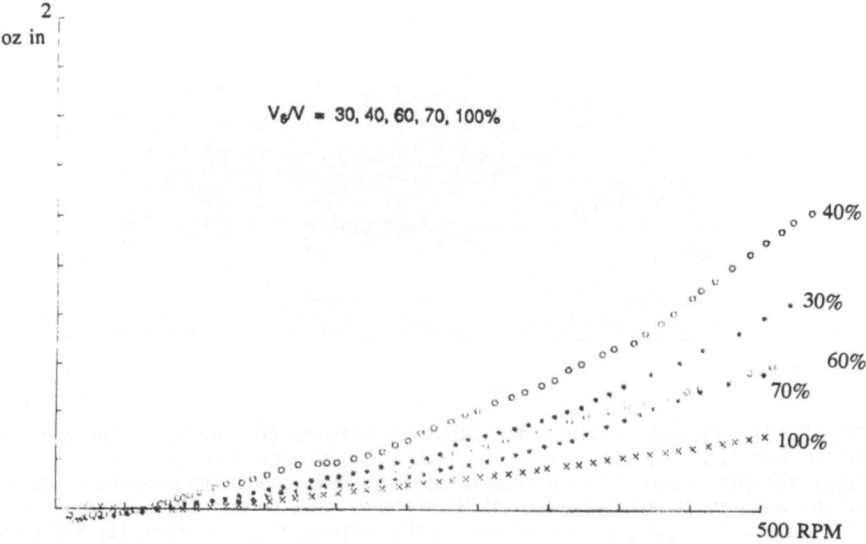

Fig. 5.13. [Figure 5 of Joseph, Singh and Chen, 1990, Plenum Press] Torque versus angular velocity with the volume fraction of silicone oil as a parameter.

II.5(e) Phase Inversion

There is a critical $\phi = \frac{v_1}{v_1 + v_2} = \tilde{\phi}$ for *phase inversion*, which we shall now define. In the experiments $0.6 < \tilde{\phi} < 0.7$. When $\phi < 0.6$, the liquid with the lower viscosity (silicone oil) fingers into the more viscous liquid (Crisco oil) and eventually there are silicone drops in Crisco oil. When $\phi \geq 0.7$, we can sketch the same figure 5.12; but the words Crisco and silicone are interchanged because Crisco, rather than silicone oil, fingers. In either case, the bubbly emulsions which exist before phase separation are uniform without structure. The bubbly emulsions all have smaller torques than the pure low viscosity constant (silicone oil) when Ω is small (see figure 5.12). The torque is an increasing function of volume fraction up to phase inversion; after inversion ($\phi > 0.7$) the torque appears to decrease with increasing ϕ. The *effective viscosity* of bubbly emulsions is smaller than the viscosity of its lowest viscosity constituent in pure form (silicone oil). There is a preference for low viscosity fingers to penetrate into the high viscosity fluid, $\phi < 0.6$. Phase inversion shows that it is also possible to get a more viscous liquid to finger into a less viscous one. (See section VII.15 for a study of phase inversion in vertical pipe flow.)

The torque curves can be used to back out the effective viscosity of an emulsion. The effective viscosity of emulsified Crisco oil is smaller than the effective viscosity of emulsified Crisco oil after phase inversion.

II.5(f) Phase Separation

Apparently the state of uniform emulsification is unstable. At least, we have never been able to maintain a state of uniform emulsification, even after one is created (see figure 5.14). The instability of uniform emulsions in the silicone oil and Crisco oil systems leads to a phase separation. The cause of this phase separation is not understood. It could not be a form of Taylor instability since it occurs at angular velocities well below the critical Taylor number, 91.7, for the silicone oil alone. In fact the phase separation may occur at all finite values of the speed or rotation of the inner cylinder. We think that the state of uniform emulsification is unstable because of wakes which tend to align bubbles in rows. This effect is very clear in beds of spherical particles fluidized by water discussed in the paper by Fortes, Joseph and Lundgren [1987]. In that case there is a scenario called drafting, kissing, and tumbling. Drafting is the mechanism by which one sphere is sucked into the wake of another, as debris is pulled from the side of a road behind a fast-moving truck. The rear sphere accelerates in the wake of the forward sphere and they kiss. The kissing spheres are aligned with the stream.

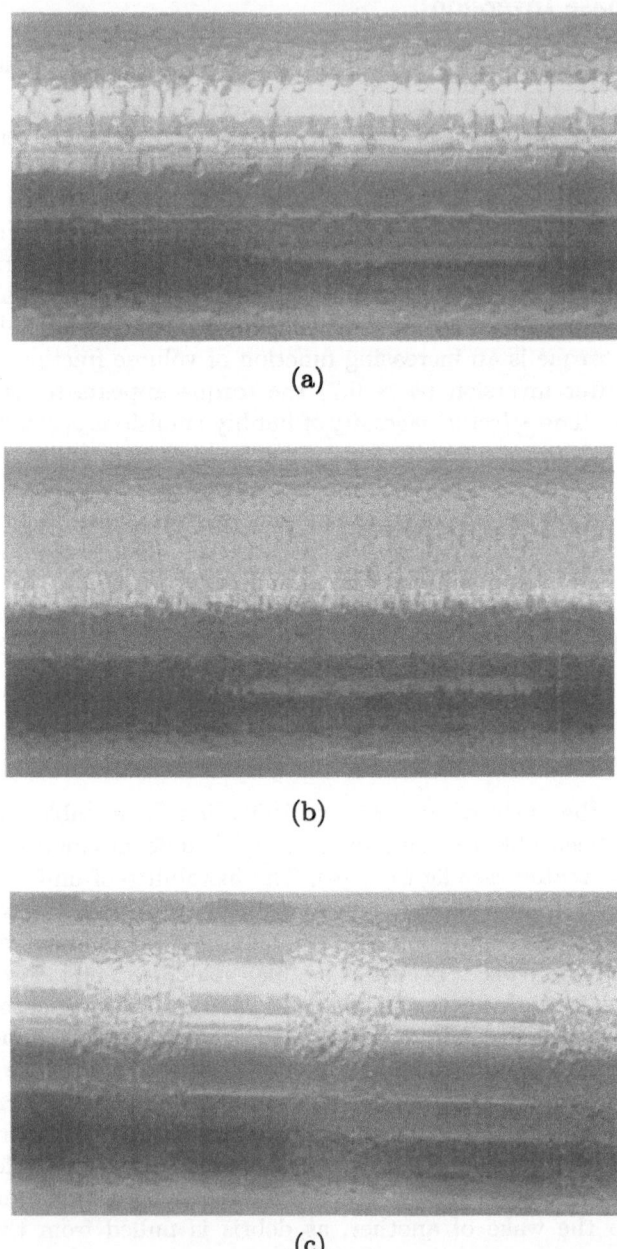

(a)

(b)

(c)

Fig. 5.14(a-c). [Figure 6 of Joseph, Singh and Chen, 1990, Plenum Press] Soybean (Crisco) oil and silicone oil in equal proportions. (a) 16 rpm. Silicone drops form as capillary instability after fingering. (b) "Uniform" emulsion at 24 rpm. (c) Phase separation at 24 rpm. (Continued)

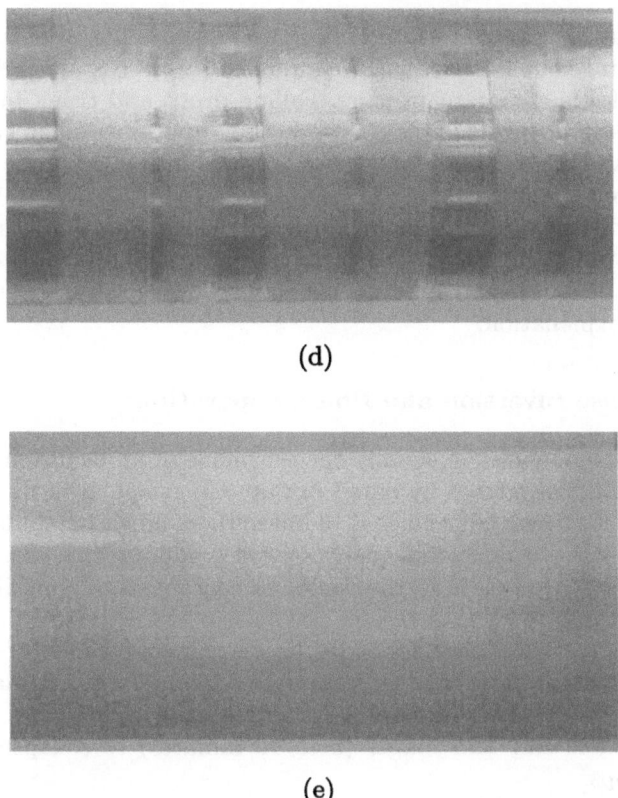

(d)

(e)

Fig. 5.14(d-e). (d) A finer emulsion of silicone oil at 122 rpm. Some of the phase boundaries are very distinct. (e) Stable emulsion with normal Taylor cells at 410 rpm.

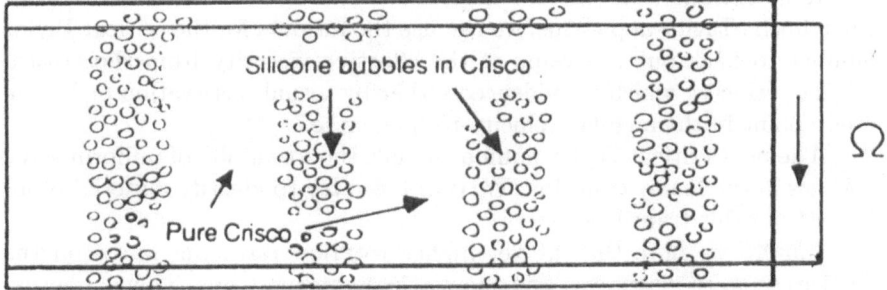

Fig. 5.15. [Figure 8 (upper) of Joseph, Singh, and Chen, 1990, Plenum Press] Phase separation. Bands of emulsified silicone oil are separated by bands of pure Crisco oil, $\phi \leq 0.6$.

The kissing spheres form a long body which is unstable to the same kind of turning couples that cause an aircraft to stall so that spheres tumble into more stable cross-stream pairs. Falling drops and rising bubbles also draft but they do not tumble. Instead, they appear to align, as in drop experiments in which heavy liquids are dropped into a long tube filled with an immiscible lighter liquid.

The drops always align and drafting is obvious. Side bubbles cannot be pulled uniformly in alignment, so there is a tendency to segregate. This explanation is tentative and is not meant to discourage readers from thinking of a better explanation.

II.5(g) Phase Inversion and Phase Separation

When $\phi \leq 0.6$, symmetric equally spaced bands of an emulsion of silicone oil in Crisco are separated by bands of pure Crisco oil, as in figure 5.15.

For $\phi = 0.7$, instead of silicone bubbles, there are Crisco bubbles, again with more or less symmetric equally spaced bands. In this case, a narrow band of pure Crisco oil is in the center of each band of emulsified Crisco oil, as in figure 5.16.

The next event, as Ω increases, is the disappearance of the phase boundaries with uniform mixing, leading to a foamy emulsion that is unlike the grainy emulsion which develops after fingering. The foamy emulsions are more stable and take longer (5 to 10 minutes) to collapse when the rotation stops.

The appearance of foamy emulsions appears to coincide with the appearance of regular Taylor vortices. The secondary motion, which is generated from the instability of the hitherto stable Crisco oil in the bistable phase-separated regime, may be the cause of the mixing leading to foamy emulsions. The critical angular velocity for the onset of Taylor vortices in the composite fluid, which we have called a foamy emulsion, can be determined visually and, when the volume ratio is small, ($\phi < 0.5$) forms a sharp break in the torque curve (see figure 5.17).

If we assume that this emulsion acts basically like a Newtonian fluid when in the Taylor apparatus, we can use the formula for the critical Taylor number to back out the value of the effective viscosity from the critical angular velocity, which is evidenced either by visual observation or by the break point in the angular velocity-torque curve.

The next critical Taylor number signals the formation of uniform wavy vortices in the foamy emulsion. We could also try to identify material properties from this transition.

Finally we note that at yet higher rotation rates the waves on the vortices seem to disappear, and the vortices appear steady. This is unusual because it does not happen in the dynamics of Taylor vortices in single constituent liquids.

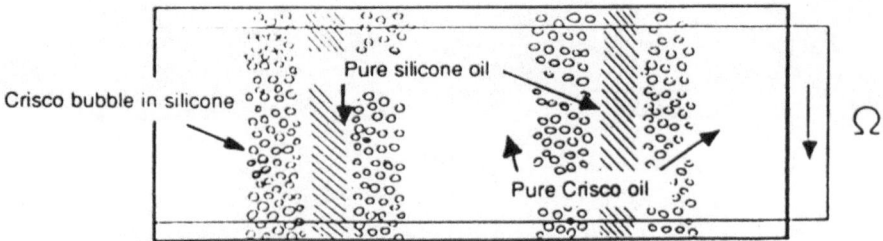

Fig. 5.16. [Figure 8 (lower) of Joseph, Singh and Chen, 1990, Plenum Press]
Phase separation when Crisco oil fingers, $\phi = 0.7$.

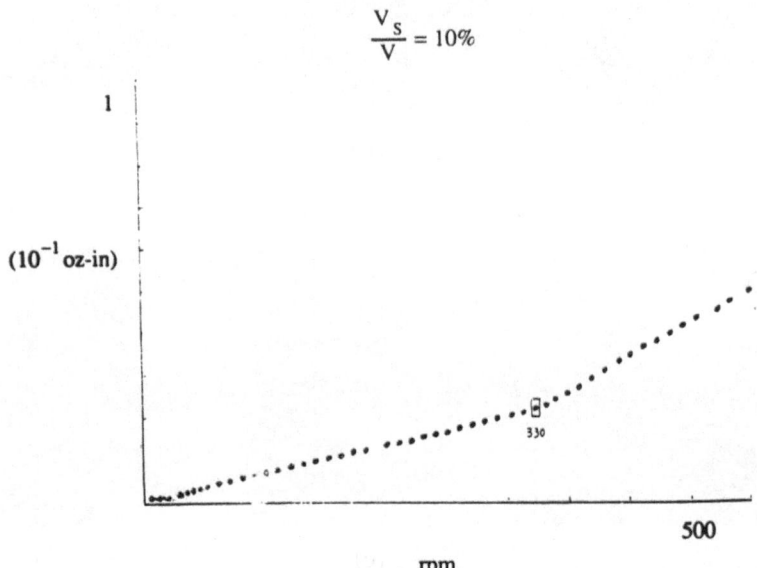

Fig. 5.17. [Figure 9 of Joseph, Singh and Chen, 1990, Plenum Press] Torque
versus angular velocity for an emulsion of 10% silicone oil in Crisco. Taylor cells
appear at 330 rpm.

(a)

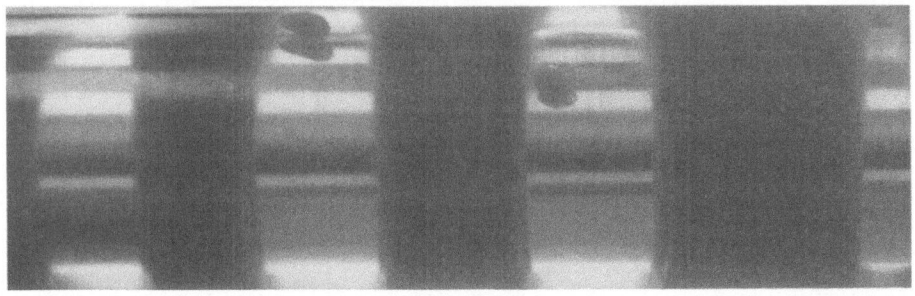

(b)

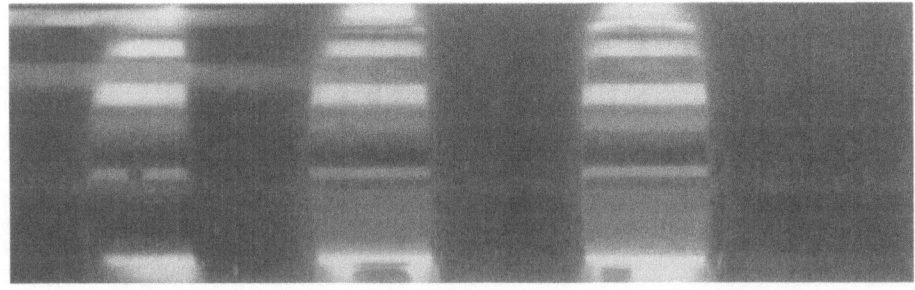

(c)

Fig. 5.18(a-c). [Figure 10 of Joseph, Singh and Chen, 1990, Plenum Press] Banded Couette flows of SAE 30 motor oil and water at 309 rpm. There is a small bubble in the unstable water cell on the left which is visible in (c) but not in (a) or (b). This small bubble undergoes an apparently chaotic motion. The large bubble appears to undergo chaotic switching from right to left.

II.5(h) Chaotic Trajectories of Oil Bubbles in an Unstable Water Cell

In the course of experiments described in section II.5(d), a motion was found which appeared chaotic. At sufficiently high values of the angular velocity prior to emulsification of motor oil, some bubbles of oil are torn away from the oil bands. In some situations it was possible to get one oil bubble into a Taylor cell. This oil bubble is carried round and round by water and is dragged around in the secondary motion due to Taylor instability. A video tape of this was made (see Singh, Mohr and Joseph [1992]) and some still photographs are shown in figure 5.18. The small oil bubble on the left is the one for which the binary sequence is studied. Each time the oil drop goes around it is either in the left eddy or in the right eddy. About 3000 terms were monitored in the sequence LRLLI and assigned the numbers -1 to left and +1 to right. It is difficult to get revealing still photographs of the motion of the small bubble in the leftmost water cell in figure 5.18. However, the large bubble in the center water cell also executes a chaotic motion of a slightly different type as can be seen in the photographs.

Binary sequences. We apply methods of estimation theory (see Singh and Joseph [1989]) to characterize the chaos in the binary number sequence generated by the bubble in our experiments. Consider a sequence $u(n) = \pm 1$ of binary numbers. We assume that the sequence is ergodic so that time averages are the same as ensemble averages. The average is

$$E[u(n)] = \frac{1}{N} \sum_{n=1}^{N} u(n). \qquad (5h.1)$$

In our experiments,

$$E[u(n)] \to 0 \qquad (5h.2)$$

when N is large. Left and right, or ± 1, are equally probable.

Singh and Joseph [1989] showed how to generate a binary sequence for chaotic trajectories of the Lorenz system $[\dot{x}, \dot{y}, \dot{z}] = [\sigma(y-x), rx-y-xz, xy-bz]$ for $(\sigma, b, r) = (10, \frac{8}{3}, 28)$. The binary sequence is generated by projecting the trajectories into the xz plane, as shown in figure 5.19, and monitoring the crossing points of trajectories on the segments AB and CD of the line AD. The crossing times are put into correspondence with the sequence n of integers, left crossings on AB are recorded as $(u(n) = 1)$, and the right crossing of CD, as $u(n) = -1$. The time averages of these sequences vanish for large N, independent of initial condition, so that left and right crossings are equally probable and we may assume that the sequences are ergodic.

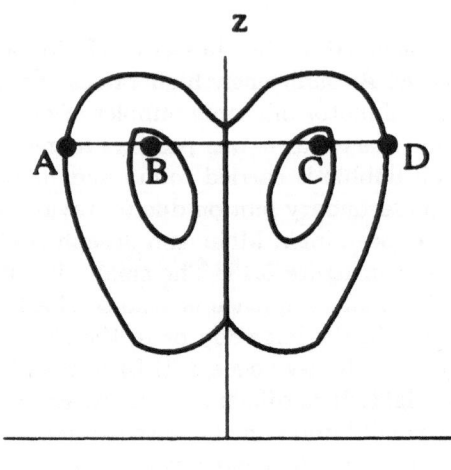

Fig. 5.19. [Figure 11 of Joseph, Singh and Chen, 1990, Plenum Press] The projected trajectories of the Lorenz attractor remain inside the butterfly region and outside the ovals around the fixed points.

Autocorrelations. An estimate of the autocorrelation function on an ergodic binary sequence can be obtained as follows:

$$r(n) = \frac{1}{N} \sum_{k=1}^{N} u(k+n)u(k), \quad n = 1, 2, \ldots \quad N \gg n. \tag{5h.3}$$

The value $r(1)$ represents the correlation between immediate neighbors $(1, 2)$, $(2, 3)$, $(3, 4)$, etc. Value $r(2)$ gives the correlation between separated pairs $(1, 3)$, $(2, 4)$, etc. A chaotic response is one for which $r(1) \neq 0$ and $r(n) \rightarrow 0$ for large n.

For the oil bubble autocorrelation values $r(n)'s$, for large n, are not uniformly close to zero because of the relatively small length of the sequence, $N = 3000$ (figure 5.20). Sequences of different lengths were tried. It was found that $r(n)'s$, for large n's, approached zero uniformly as the length of the sequence was increased.

The Lorenz equations were integrated numerically using the NAG library. Subroutine DO2BBF was used for different tolerance levels in the range 10^{-4} to 10^{-10}. We projected into the xz plane and formed a binary number symbol sequence with 76,000 entries. The autocorrelation function is shown in figure 5.21. The tolerance level in the numerical scheme had absolutely no effect on the nature of autocorrelation sequence, even though sequences generated were quite different for different tolerance levels. For large n, $r(n)$ approached zero uniformly with the increase in length of the sequence, N.

In both cases the decay in the autocorrelations value is very rapid. For large n, autocorrelation values decrease monotonically with the length of the sequence. The decay of autocorrelation for the bubble is essentially complete after $n = 2$, a substantial correlation exists only for $r(1)$. The decay of correlation is slower for the Lorenz system with nonzero $r(n)$ for $n < 6$. We could say that the Lorenz system is less random.

Lyapunov Exponents. Singh and Joseph [1989] derived a macroscopic Lyapunov exponent for binary sequences. Lyapunov exponents for continuous times are locally defined quantities which measure the tendency for chaotic trajectories to diverge exponentially for small time, on the average. One can define the first exponent by

$$\lambda = \frac{1}{t_{N+1} - t_1} \sum_{k=1}^{N} \log_2 \frac{d(t_{k+1})}{d_o(t_k)} \tag{$5h.4$}$$

where $d_0(t_k)$ is the initial distance between two trajectories at time t_k and $d(t_{k+1})$ is the distance between these two trajectories at time $t_{k+1} > t_k$. In the continuous case $d_0(t_k)$ and $d(t_{k+1})$ are infinitesimal and $N \to \infty$.

The concept of distance is not natural to binary sequences. Two trajectories correspond to two strings of binary symbols. We replace the condition that the initial distances between trajectories is small with the condition that we shall only compare strings of symbols which start with the same symbol. We can compare the "distance" between two strings of symbols which both start with $u = 1$ or both with -1, but not with starting values of $+1$ for one string and -1 for the other.

Another condition we need for comparing two strings of symbols is statistical independence. We want uncorrelated sequences so that theorems requiring ergocity, the use of "time" averages, will be appropriate. This requirement is easy to fulfill for our binary sequence symbol string. We compare two strings $u(k), k = 1, 2, \ldots$ with $u(k + M)$ where M is larger than the correlation time for the autocorrelation, for the chaotic bubble, $M > 5$ for the Lorenz attractor.

We can replace (5h.4) with

$$\lambda(t_{N+1} - t_1) = \sum_{k=1}^{N} \log_2 \frac{\bar{d}(k+1)}{\bar{d}_0(k)} \tag{$5h.5$}$$

where $\bar{d}_0(k)$ is the average "distance" between two statistically independent strings at the kth observation. If the two symbols at the $k+1$ st observation have the same sign we say that the "distance" is unchanged, on the average

$$\bar{d}(k+1) = \bar{d}_0(k). \tag{$5h.6$}$$

If the two symbols have different signs after one observation, then

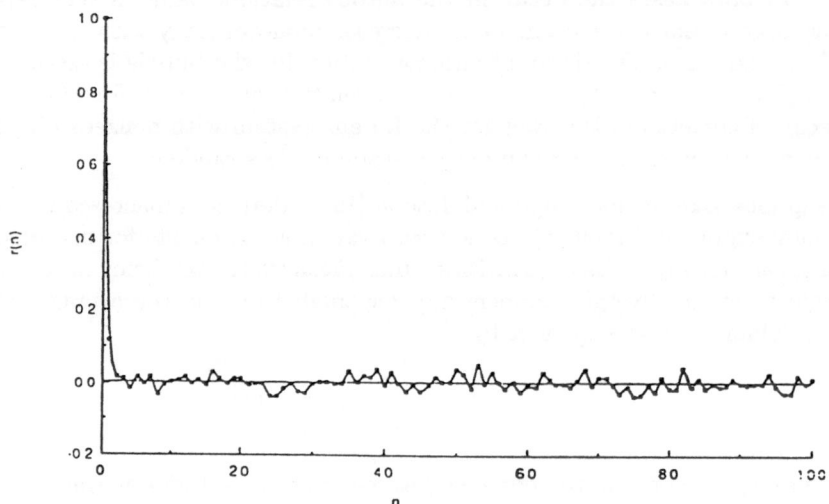

Fig. 5.20. [Figure 12 of Joseph, Singh and Chen, 1990, Plenum Press] The autocorrelation sequence for the oil bubble, $N = 3000$.

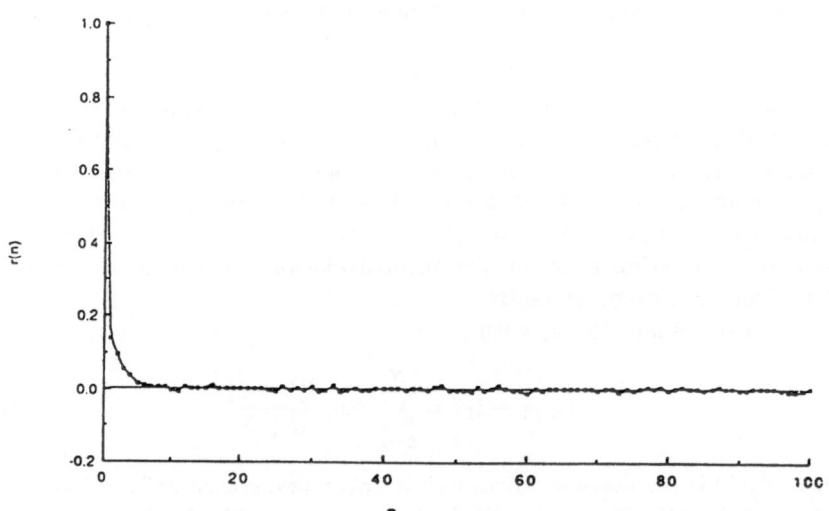

Fig. 5.21. [Figure 13 of Joseph, Singh and Chen, 1990, Plenum Press] The autocorrelation sequence for the Lorenz attractor, $N = 76,000$.

$$\bar{d}(k+1) = c_1 \bar{d}_0(k) \qquad (5h.7)$$

where c_1 is the constant average change of distance. It follows from (5h.6) and (5h.7) that

$$\log_2 \frac{\bar{d}(k+1)}{\bar{d}_0(k)} = \begin{cases} 0 & \text{same sign} \\ \alpha & \text{sign change after iteration} \end{cases} \qquad (5h.8)$$

where $\alpha = \log_2 c_1$.

We now define the set S_1 of ergodic initial distance between strings of symbols

$$S_1 = \{k : u(k)u(k+M) = 1\}. \qquad (5h.9)$$

The complementary set is

$$S_2 = \{k : u(k)u(k+M) = -1\}. \qquad (5h.10)$$

Hence, we may write

$$\log_2 \frac{\bar{d}(k+1)}{\bar{d}_0(k)} = \frac{\alpha}{2}\{1 - u(k+1)u(k+1+M)\}$$

for all symbol sequences which have the same sign at the time k for all $k \in S_1$. Hence,

$$\sum_{k=1}^{N} \log_2 \frac{\bar{d}(k+1)}{\bar{d}_0(k)} = \frac{\alpha}{2} \sum_{k \in S_1} \{1 - u(k+1)u(k+1+M)\}. \qquad (5h.11)$$

The total number of k is N. Let N_1, N_2 be the number of k's in the sets S_1, S_2, and $N_1 + N_2 = N$. We have also that

$$Nr(M) = \sum_{k=1}^{N} u(k)u(k+M) = \sum_{k \in S_1} u(k)u(k+M)$$

$$+ \sum_{k \in S_2} u(k)u(k+M) = N_1 - N_2 = 0, \qquad (5h.12)$$

since $r(M) = 0$ when M is larger than the correlation "time". Hence $N_1 = N_2 = \frac{N}{2}$.

We next define the macroscopic Lyapunov exponent as the average value

$$\lambda_M = \frac{1}{N_1} \sum_{k \in S_1} \log_2 \frac{\bar{d}(k+1)}{\bar{d}_0(k)}$$

$$= \frac{\alpha}{N} \sum_{k \in S_1} \{1 - u(k+1)u(k+1+M)\} \qquad (5h.13)$$

This is related to the average Lyapunov exponents by

$$\frac{\lambda(t_{N+1} - t_1)}{N} = \lambda_M. \tag{5h.14}$$

Singh and Joseph [1989] showed that

$$\lambda_M = \frac{\alpha}{2} \left[1 - r^2(1)\right]. \tag{5h.15}$$

Lyapunov Exponents and White Noise. Singh and Joseph [1989] calculated the macroscopic Lyapunov exponent for the Lorenz system described in the section "Binary sequences". They calculate α as follows. The average distance between starting trajectories on the line $AB(= CD)$ of figure 5.19 is

$$\frac{|AB|}{3} = \bar{d}_0(k).$$

The switching distance is $|AD| - |AB| = \bar{d}(k + 1)$. Hence

$$\frac{\bar{d}(k + 1)}{\bar{d}_0(k)} = 3 \left[\frac{|AD|}{|AB|} - 1\right].$$

They found that $|AD| = 4.31|AB|$. Then from (5h.8) we calculate $\alpha = 3.3$. The relation (5h.14) between the average Lyapunov exponent λ and the macroscopic exponent λ_M may be simplified by putting $t_{N+1} - t_1 = N\Delta T$ where ΔT is the average period. Then

$$\bar{\lambda} = \frac{\lambda_M}{\Delta T} = \frac{\alpha}{2\Delta T} \left(1 - r(1)^2\right)$$

where $\Delta T = 0.7519$ sec. We get

$$\lambda_M = 1.618 \text{bits/period}.$$

The largest Lyapunov exponents computed directly for the Lorenz attractor is

$$\lambda_M = 1.30 \text{bits/period}.$$

Feeny and Moon [1989] have studied a chaotic dry friction oscillator using the method of binary sequences of Singh and Joseph [1989]. They did an experiment with sliding friction in which an imposed change of the normal force caused the slider to stick. They also modeled their experiment with a second order forced ordinary differential equation involving friction coefficient and normal load functions. They did Poincaré sections for the experiments with 2,048 symbols and for the differential equation with 10,000 symbols. The symbols form a string of binary numbers ±1 corresponding to whether the motion is sticking or slipping at each pass through the Poincaré section. They measure distance on the Poincaré plot:

$$\bar{d}_0(k) = \frac{1}{3}, \qquad \bar{d}(k + 1) = 1.$$

Hence, using (5h.15), they get $\alpha = \log_2 3 = 1.585$.

Feeny and Moon studied the tent map and logistic map using the formula (5h.15) with $\alpha = 1.585$. They calculated $r(1)$ for $N = 10^5$ and $N = 2048$. The theoretical value of the largest Lyapunov exponent is $\lambda = 1$ for both the tent map and the logistic map. They compute

$$\lambda_M = \begin{array}{ll} 0.787515 & (10^5 \text{ symbols}) \\ 0.787705 & (2048 \text{ symobls}) \end{array} \Big\} \text{ tent map}$$

$$\lambda_M = \begin{array}{ll} 0.791578 & (10^5 \text{ symbols}) \\ 0.791116 & (2048 \text{ symbols}) \end{array} \Big\} \text{ logistic map}$$

A binary autocorrelation was obtained for their experiments and numerically from the differential equation for a symbol string with $N = 2048$. In both cases the autocorrelation $r(1)$ is very small, less than ± 0.05. They calculate

$$\lambda_M = \begin{cases} 0.79055 & \text{experiment} \\ 0.79219 & \text{numerical integration.} \end{cases}$$

The calculation of the exponent for the Poincaré map from the equations of motion gives

$$\lambda = 0.77.$$

We draw the reader's attention to the fact that for all the calculations done by Feeny and Moon, they get

$$\lambda_M = \frac{\alpha}{2} \left[1 - r(1)^2 \right] = 0.7925 \left[1 - r(1)^2 \right].$$

This shows that $r(1)^2$ is very small in the examples of the tent map, logistic map and experiments.

Short range predictability requires that $r(1), r(2), \ldots, r(M) \neq 0$ for small n, $r(n) \to 0$ for large m. For white noise, we have $r(1) = 0$. The autocorrelation is good for distinguishing short range predictability and white noise. The macroscopic Lyapunov exponent is not useful for making this important distinction. In fact, the macroscopic Lyapunov exponent depends on distance through α, but λ_M/α is universal, does not depend on distance and may be a more intrinsic measure of chaos. Certainly $r(1)$ has a lot less information than the graph of $r(n)$.

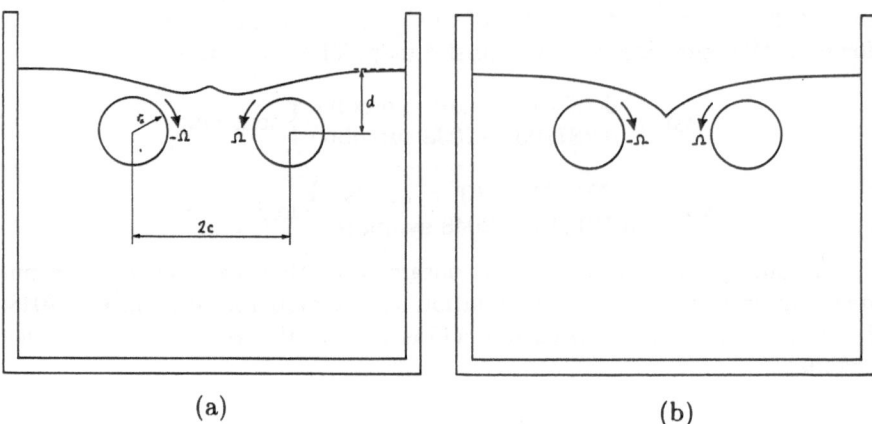

$$(a) \qquad\qquad\qquad\qquad (b)$$

Fig. 6.1(a-b). [Figure 1 of Jeong and Moffatt, 1992] Experimental configuration of Jeong and Moffatt [1992] showing the observed form of the free surface (a) when the rotation rate Ω is very small, and (b) when Ω is larger.

II.6 Two-Dimensional Cusped Interfaces

This section is based on the works of Joseph, Nelson, Renardy and Renardy [1991], Jeong and Moffatt [1992] and Palmquist and Kistler [1992].

Two-dimensional cusped interfaces are line singularities of curvature. It is easy to create a cusp experimentally and there are many ways to do it; for instance, by rotating a cylinder half immersed in liquid. A liquid film is dragged out of the reservoir on one side and is plunged in at the other, where it forms a cusp if the conditions are right. Figure 6.1 shows a pair of counter-rotating cylinders. For slow speeds, the surface is smooth, and the cusp appears to form beyond a critical speed. Both Newtonian and non-Newtonian fluids form cusps, but the transition from a rounded interface to a cusp is gradual in Newtonian liquids and sudden in non-Newtonian liquids.

In section (a), we give an overview of past work relevant to the subject. Section (b) is a description of our experimental results giving data for flows that lead to cusps and those that do not. Section (c) contains a summary of the analysis of Jeong and Moffatt [1992] for Newtonian liquids, and how this compares with our asymptotic analysis near the cusp tip for the case of zero surface tension. In section (d), we present numerical simulations showing the development of a cusp, and the computational work of Palmquist and Kistler [1992] on the flow in the cusp region.

II.6(a) Introduction

Plates II.6.5, II.6.8-9, and figures 6.6-7 were produced in a four-roll mill apparatus (see figure 4.9 of section II.4) of the type introduced by G.I. Taylor to study the deformation of drops and bubbles in a pure straining flow. The cylinders were rotated in the partially filled apparatus shown in the photographs. This was also the way we produced the rollers which are described in the preceding sections of this chapter, and which are an entirely different flow regime. At this point, we ask the reader to inspect the photographs. They show cusped, two-dimensional interfaces. No rounding can be detected, at least not on a visible lengthscale.

Richardson [1968], in his analysis of two-dimensional bubbles in slow flow, considered the situation shown in figure 6.2. This may be viewed as an idealized model of figure 6.3. The origin is a theoretical angular point of a free surface in a two-dimensional Stokes flow. He allowed the angle α to be free and determined its value by requiring that the shear stress condition and the kinematic condition hold at the free surface. He sought the leading contribution to the streamfunction and found

$$\psi = \frac{\sigma}{2\pi\mu} r \log r \sin\theta, \qquad (6a.1)$$

where σ is the surface tension coefficient and μ is the viscosity. This solution also satisfies the normal stress condition. He found that the angle α is π, and proved that *the only possible line singularity of the interface between a viscous liquid and an inviscid two-dimensional bubble is a true cusp.* On the other hand, recent numerical simulations of Palmquist and Kistler [1992] and the analysis of Jeong and Moffatt [1992] indicate that when surface tension is present, there is no line singularity in the interface shape, and hence no genuine cusp, so that pursuing (6a.1) is not going to help us determine the flow. However, this solution does exhibit some intriguing properties. For instance,

(i)   There is a point force 2σ exerted by the free surface on the fluid.
(ii) The velocity near the origin is in the negative x-direction, which is contrary to what it ought to be. Since $\psi = (y/2)\log(x^2 + y^2)$, the velocity in the x-direction is $\psi_y = (1/2)\log(x^2 + y^2) + y^2/(x^2 + y^2)$. If, for instance, we set $y = 0$, then this velocity component is negative for $|x| < 1$.
(iii) The velocity at the origin is infinite.
(iv) The velocity gradient is not square-integrable, leading to an infinite amount of energy dissipation. This implies that an infinite amount of energy is required to maintain the flow and raises questions about the physical realizability of this solution.

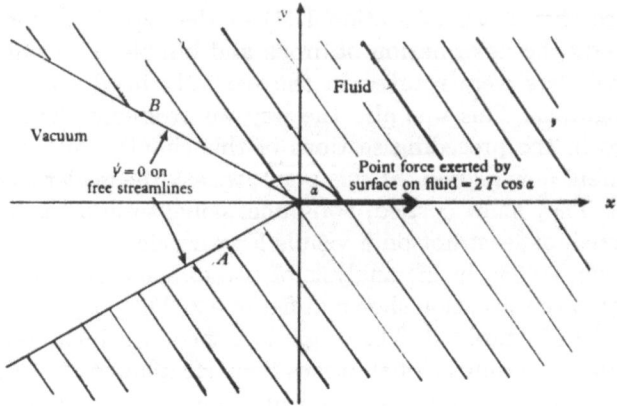

Fig. 6.2. [Figure 7 of Richardson, 1968] Sketch of an angular point at the origin.

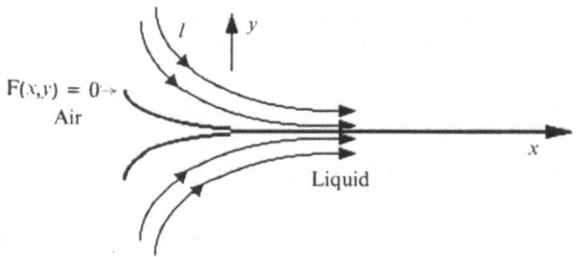

Fig. 6.3. [Figure 1 of Joseph, Nelson, Renardy and Renardy, 1991] Flow past a cusp

In Richardson's solution, the free surface at leading order simply co-incides with the negative x-axis; the actual opening of the cusp remains to be determined. We note in passing that Lamb (p. 607, [1932]) considered the related problem of a point force acting at the center of a circle.

The main features of our experiments cannot be explained in terms of (6a.1) because, as explained in section (c), it turns out that the actual interface is analytic. It is smooth when resolved on a finer lengthscale, ap-proaching a cusp in the limit as surface tension approaches zero, or as the capillary number approaches infinity. When describing the experiments, we therefore speak of a 'cusp' on a visible lengthscale. The experiments suggest that cusped free surfaces are easier to create in non-Newtonian fluids. The

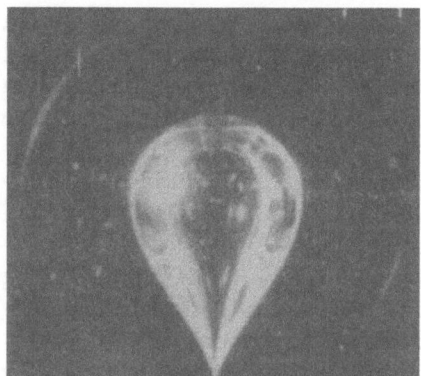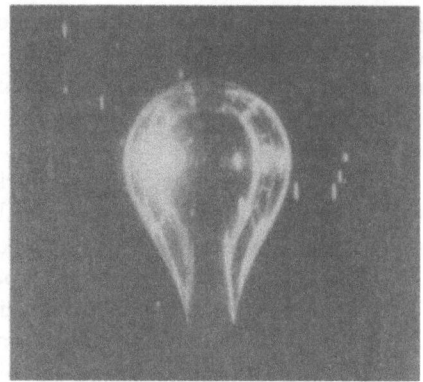

Fig. 6.4. [Figure 2.6-2 in volume 2 of *Dynamics of Polymeric Liquids*, by Bird, Armstrong and Hassager. Copyright ©1977, John Wiley & Sons. Reprinted by permission of John Wiley & Sons] Highly distorted gas bubble of volume approximately 2100 mm^3 rising with velocity 10 mm/s in a polyacrylamide solution. The bubble is seen from two mutually perpendicular directions. Note that the bottom end is not axisymmetric. Photograph by O. Persson, Instituttet for Kemiteknik, Danmarks tekniske Hojskole, Danmark. Bubbles rising in Newtonian liquids do not tail. The shape of the bubble depends strongly on the bubble volume.

cusp does not appear at the slowest speeds; it forms more gradually in Newtonian than in non-Newtonian fluids. This is reminiscent of the transition from rounded to pointed ends in the bubble experiments of Rumscheidt and Mason [1961].

Our asymptotic analysis in section (c) concerns the local behavior near a cusp singularity in the absence of surface tension. In this case, we find an interface shape given at leading order by $y^2 = -cx^3$. This is in agreement with the solution of Jeong and Moffatt [1992] and with experimental data. Our analysis can be extended to linear viscoelastic fluids; in that case the stresses at the cusp are less singular than in the Newtonian case. On the other hand, nonlinear effects in viscoelastic fluids are probably important near the stagnation point that must be present on a smooth interface. The build-up of extensional stresses near such a stagnation point is likely to favor cusping.

The topic of two-dimensional cusps may have some relevance for pointed bubbles (figure 6.4), which are point singularities of the curvature. Such singularities have been described by Taylor [1934], Rumscheidt and Mason [1961], Taylor [1964], Grace [1971], Buckmaster [1972,1973],

Acrivos and Lo [1978], Rallison and Acrivos [1978], Hinch and Acrivos [1979], and Sherwood [1981] for the case of Newtonian fluids. Buckmaster is the only author in this list who discussed Richardson's results. He comments that Richardson has '...shown that such discontinuities, if they exist, must be genuine cusps ..'. He also expresses reservations about the relevance of two-dimensional cusps: 'On the one hand it is doubtful that pointed two-dimensional drops could be stable, and on the other there is a point force associated with corners – which a three-dimensional drop would not generate' These remarks appear to have halted further consideration of two-dimensional cusps, until the recent works we shall discuss in more detail below.

The existence of point singularities for axisymmetric bubbles in Newtonian fluids and the decision whether these might be corner or cusp singularities still seems not to have been decided.

There is a marked difference between the shape of air bubbles rising in Newtonian and non-Newtonian liquids, with a much stronger tendency toward cusping in the non-Newtonian case (see figure 6.4).

The flow at the trailing edge of a rising gas bubble is in a rough way analogous to flow near the cusp in our experiments, though there are obvious and perhaps not so obvious differences between streaming around axisymmetric and plane cusps. The tail shown in figure 6.4 occurs only in non-Newtonian liquids and has been reported in experiments by Philippoff [1937], Warshay, Bogusz, Johnson and Kintner [1959], Mhatre and Kintner [1959], Astarita and Apuzzo [1965], Barnett, Humphrey and Litt [1966], Calderbank [1967], Calderbank, Johnson and Loudon [1970], Leal, Skoog and Acrivos [1971] and Zana and Leal [1978]. Again, it is not certain that the tail can form a true cusp. Astarita and Apuzzo note that 'The bubble is ... unexpected; it is clearly prolate and the lower pole is markedly cuspoidal.' Hassager [1985] reports experiments where the bubble is not axisymmetric but flattened; it appears to have a straight edge at the trailing end.

II.6(b) Experiments

Two equal cylinders whose centerlines lie in the same plane $z = $ constant, perpendicular to gravity, are immersed in liquid approximately up to their centerline. Then the cylinders are put into counter-rotation with the same speed. There are two possibilities: the liquid is drawn up at the outside of the cylinders and forced down in the center (figure 6.1), or vice versa, as in plate II.6.5. Typically more than half of each cylinder is coated with a liquid of relatively uniform thickness. The thickness and covering of the arc of the coating liquid depends on the viscosity of the liquid and the speed of the cylinder.

Cusps form in some liquids and not in others. Once a cusp forms it will persist and sharpen as the angular velocity is increased. Some liquids with sharp cusps are exhibited in plates II.6.5, II.6.8-9, and figures 6.6-7.

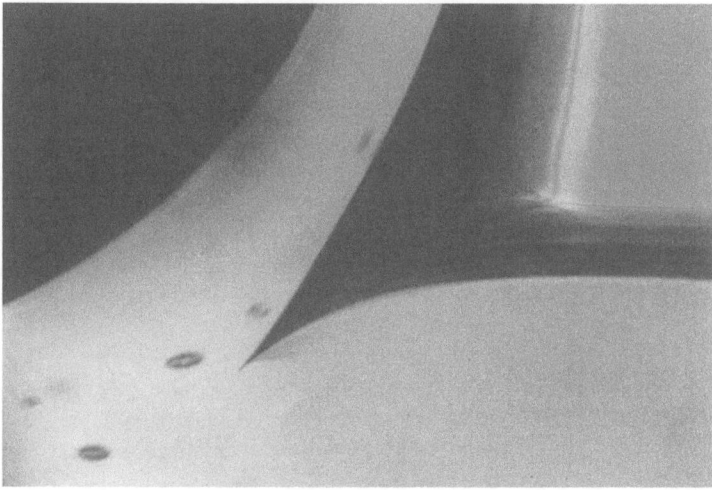

Fig. 6.6. [Figure 4 of Joseph, Nelson, Renardy and Renardy, 1991] Cusp point on the 12500 cs silicone oil as seen through a Nikon microscope.

The cusp tip in plate II.6.9 has been magnified 64 times, and the height of the photograph represents approximately 0.05 inches. Several more photographs of the cusps in the fluids listed in the tables in this section are given in the paper of Joseph, Nelson, Renardy and Renardy [1991]. Liquids which do not cusp have clearly rounded ends as in plate II.6.8(b). Tables 6.1 and 6.2 list fluids which form cusps and table 6.3 lists fluids which do not cusp at the angular speeds we attained. The fluids that do eventually cusp do not do so at the slowest speeds (see plate II.6.9). It therefore appeared at the time these data were collected that cusping is a threshold phenomenon characterized by a threshold parameter which in our experiments is a critical value of the angular velocity of the cylinders. As the rotation speed increases, the shape of the interface changes from rounded to pointed in a continuous manner, until at what we define as the critical speed the interface appears to be a cusp. The critical speed is not distinct for the Newtonian liquids because the changes are gradual. Once a cusp is formed, further increases in the angular velocity will only sharpen the cusp or end in an instability characterized by fingering. When we consider the theoretical evidence we have from sections (c) - (d), the transition to a cusp must be a gradual one for Newtonian liquids and the critical speeds we observed represent the limits of resolution beyond which the curvature of the free surface is so high that it looked like a cusp on a visible lengthscale. Some specific values of this critical speed are given in the figure captions; to convert this rotation speed into a linear speed one has to multiply by the observed radius at the point of cusping which is slightly larger (table II.9) than the radius 1.25 cm of the cylinder.

For non-Newtonian fluids, the critical value of the rotation speed is

(a)

(b)

Fig. 6.7(a-b). [Figure 6 of Joseph, Nelson, Renardy and Renardy, 1991] Cylinder coated with M1 (polyisobutylene in kerosene and polybutene) cusps in air. (a) 19 r.p.m. (b) 75 r.p.m.. At 75 r.p.m., a ribbing instability has developed. The cusped interface has developed scallops.

very distinct. Below this value the interface is round. At the critical speed the interface suddenly looks like a cusp.

The liquids listed in table 6.1 form unambiguous cusps and are non-Newtonian. They have non-zero normal stress differences and they all climb rotating rods. At the same time, these fluids are on the average more viscous than the Newtonian liquids listed in tables 6.2 and 6.3. We ought to say something about honey, because it is a difficult material to work with. It tends to crystallize and the amount of crystallization, which may be undetectable to the eye, determines how the fluid will behave.

We tested the liquids in tables 6.2 and II.8 and none of them will climb a rotating rod. Newtonian fluids which are too mobile will not cusp at the speeds we could attain in our apparatus and may never cusp because of turbulence.

Table 6.1. Representative values of the viscosity and surface tension of cusping non-Newtonian liquids at temperatures in the neighborhood of 23°C.

Fluid	Viscosity (p)	Density (g/cm^3)	Surface Tension (dyn/cm)
1% Aqueous Polyox	61.2	1.000	48.4
2% Aqueous Polyox	600	1.125	44.4
M1	30.0	0.859	29.9
Silicone Oil – 12500cs	122	0.975	21.5
STP	143	0.858	35.0
TLA 510	220	0.868	31.1
Honey	109	1.40	69.3

Table 6.2. Representative values of the viscosity and surface tension of cusping Newtonian liquids at temperatures in the neighborhood of 23°C.

Fluid	Viscosity (p)	Density (g/cm^3)	Surface Tension (dyn/cm)
Castor Oil	8.15	0.960	35.1
Glycerin	8.30	1.265	63.3
Silicone Oil – 500cs	4.86	0.971	21.1
Silicone Oil – 1000cs	9.71	0.971	21.2
Silicone Oil – 5000cs	48.6	0.971	21.3

Table 6.3. Representative values of the viscosity and surface tension of Newtonian liquids that do not cusp at temperatures in the neighborhood of 23°C.

Fluid	Viscosity (p)	Density (g/cm^3)	Surface Tension (dyn/cm)
SAE 30 Motor Oil	2.80	0.886	35.0
Safflower Oil	0.469	0.920	23.0
Silicone Oil – 200cs	1.94	0.970	21.0
Soybean Oil	0.489	0.922	25.7

The critical value of the angular velocity of the cylinders was measured for the transition from rounded to cusped ends in a number of liquid-air systems and in the two-liquid SAE 30 motor oil and water system. A dimensionless capillary number was calculated by the formula

$$Ca = \mu U/\sigma,$$

where μ is the viscosity, σ the surface tension coefficient, and an estimate for the velocity U is ωr, where ω is the angular speed of the cylinder and r is the radius of the fluid at the cusp point. The actual value of r at the cusp point is somewhat larger than the radius of the cylinder. The value of the critical capillary number Ca_c for SAE 30-water was computed using a surface tension constant of 9.22 dyn/cm; the surface tension constant tabulated in table 6.3 is for SAE 30 against air.

Table 6.4. Critical angular velocity and Ca_c at the formation of a cusp.

System	Critical Angular Velocity (rpm)	Critical Radius (cm)	Ca_c
Newtonian Fluids			
Castor Oil	113	1.78	4.89
Glycerin	128	1.49	2.62
Silicone Oil – 500cs	75	1.64	2.96
Silicone Oil – 1000cs	41	1.40	2.75
Silicone Oil – 5000cs	26	1.57	8.21
SAE 30 Motor Oil – Water	51	1.61	2.57
Non-Newtonian Fluids			
Honey	5.58	1.92	1.78
Polyox (1%)	52	1.35	9.24
Polyox (2%)	0.8	1.97	2.10
Silicone Oil – 12500cs	3.0	1.50	2.64
M1	2.4	1.90	0.48
STP	2.9	1.80	2.24

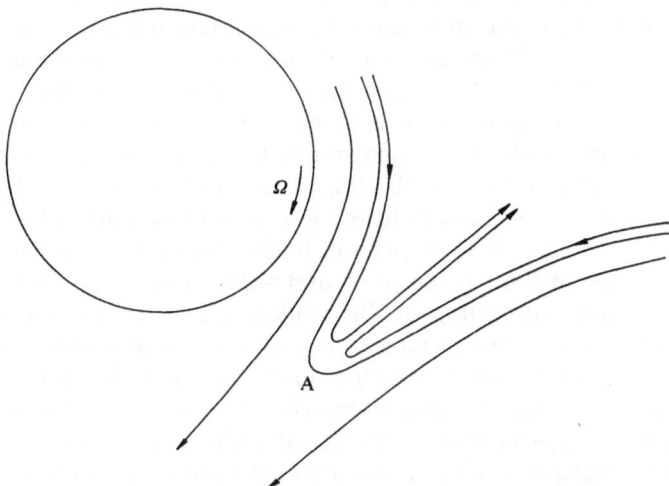

Fig. 6.10. [Figure 16 of Joseph, Nelson, Renardy and Renardy, 1991] Distorted interface in a two-liquid system. The point A is a stagnation point or a cusp. When $\Omega = 0$, each fluid covers one half of the cylinder. We do not know which fluid will finally be on the rod. There is nothing in our equations to tell us; contact angles are not enough.

For all Newtonian fluids, the measured values of Ca_c are of order 1 or larger. When we substitute these values of Ca_c into equation (6c.15), we find that the theoretical radius of curvature is extremely small. These critical situations therefore represent the experimental limit of resolution for the cusp-like paraboloid shape given by (6c.14). Some non-Newtonian fluids lead to lower values of Ca_c. However, this appearance may be deceiving. The 'viscosity' used in computing Ca is the zero shear rate viscosity. The flows studied here are likely to involve significant extensional motion, and extensional viscosities of non-Newtonian fluids increase substantially with the rate of extension.

We have found that it is hard to achieve cusped interfaces in two-liquid systems; the dynamics which would lead to a cusp in air give rise to a fingering flow or migration into rollers in a two-liquid system. Rollers and fingering flows are described in the previous sections of this chapter. Rollers form when one of the two fluids is much more viscous than the other; for example, in STP-water and silicone oil-water systems with oils in excess of 1000p. Therefore, fluids like 12500 cs silicone oil, STP and TLA 510 which give rise to cusps in air lead to rollers in water. On the other hand, 100 cs silicone oil in water gives rise to the fingering of water into silicone oil at higher speeds and not to rollers.

For lower viscosity liquids the interface remains smooth at low speeds, while at higher speeds one liquid will finger into the other. In thinking about this, it is useful to imagine how the streamlines would look at a nearly cusped smooth surface. This is sketched in figure 6.10. It is clear that the nose of the interface must be a stagnation point if the interface is not cusped. For a discussion of the kinematics at a stagnation point, see Goenaga and Higgins [1991]. The internal motion of the fluid inside the nose is yet more complicated. It is being dragged toward the nose near the interface and must either be turned around by the pressure at the stagnation point into the double eddy or else finger into the fluid outside the nose. At higher speed both these possibilities are realized intermittently.

There are evidently some pairs of liquids for which cusp solutions can be realized. One of these is motor oil and water. This cusped surface is not two-dimensional, but scalloped. The scalloping is a fairly common feature and it can be seen in the M1-air interface shown in figure 6.7(b). The crests of the scallops are fingering sites for small water bubbles to enter the oil. At high speeds the air at a cusped interface may also finger into the liquid at these sites. Fingering always appears at scallop sites, really by definition because the scallop is what remains after the bubble has broken away. This kind of fingering leads to emulsions (cf. sections II.4 and II.5).

Questions Posed by the Experiments.

- Is there a real cusp, or just an apparent cusp with a rounded end with rounding that is too small to see?
- Arcs of a cusp, or an apparent cusp, come close together. Is it necessary to take short-range forces, like Van der Waals, to get a correct description of cusping near the apparent tip of the cusp?
- Can the appearance of a cusp, or an apparent cusp, in a Newtonian liquid be correlated with a capillary number alone? Why does the cusping appear to occur at different capillary numbers in the neighborhood of 2.5?
- For some of the very viscous non-Newtonian liquids, we expect surface tension effects to be small: it is a surprise that the capillary number $U\mu/\sigma$, a parameter which should be important for Newtonian liquids, is evidently also relevant for the critical cusping in non-Newtonian liquids where extensional effects should be important. Is the capillary number really an important parameter for cusping in non-Newtonian fluids and how does it enter into cusping dynamics?
- Is the observed difference between cusping in Newtonian and non-Newtonian liquids qualitative or only quantitative? Could the sudden transition to cusping in a non-Newtonian liquid be described as an instability of a stagnation point flow at a free surface?

II.6(c) Theory

II.6(c)(i) Asymptotic Analysis Close to Cusp. We will assume that there is a cusp at the origin of the (x, y)-plane as shown in figure 6.2. The flow is assumed to be a small perturbation of uniform flow with velocity U in the positive x-direction, and the free surface is a small perturbation of the negative x-axis. Let (u, v) denote the perturbations to the velocity and let $y = h(x)$ denote the position of the free surface that is located above the cusp. We analyze Newtonian fluids with zero surface tension. At the cusp region, the highest derivatives govern the singularity of the flow, so that the relevant equation is the Stokes equation. We introduce a streamfunction ψ by

$$u = \psi_y, \quad v = -\psi_x, \qquad (6c.1)$$

and for Stokes flow, ψ satisfies the biharmonic equation $\nabla^4 \psi = 0$.

The positive x-axis is a line of symmetry, we first look for antisymmetric streamfunctions, corresponding to symmetric flow fields. In polar coordinates, Dean and Montagnon [1949] and Michael [1958] have found by separation of variables that the solution

$$\psi = r^\lambda (A \sin(\lambda \phi) + B \sin((\lambda - 2)\phi)) \qquad (6c.2)$$

incorporates the desired symmetry. At $\phi = \pi$, we must satisfy the conditions of zero shear and normal stress,

$$u_y + v_x = 0, \quad 2\mu v_y - p = 0, \qquad (6c.3)$$

which can be shown to translate into

$$\frac{1}{r^2} \psi_{\phi\phi} + \frac{1}{r} \psi_r - \psi_{rr} = 0,$$

$$\frac{3}{r} \psi_{rr\phi} + \frac{1}{r^3} \psi_{\phi\phi\phi} - \frac{3}{r^2} \psi_{r\phi} + \frac{4}{r^3} \psi_\phi = 0. \qquad (6c.4)$$

For the antisymmetric streamfunction, this leads to

$$(\lambda A + (\lambda - 2)B) \sin(\lambda \pi) = 0,$$

$$(A + B) \cos(\lambda \pi) = 0. \qquad (6c.5)$$

We may solve $\sin \lambda \pi = 0$ or $\cos \lambda \pi = 0$ and choose the value of λ which yields a bounded velocity but still describes a cusp. This leads to $\lambda = 3/2$. The linearization about the flat interface $y = 0$ of the kinematic free surface condition leads to

$$Uh' = v = -\psi_x = \psi_r, \qquad (6c.6)$$

and hence $h = -\psi/U$ at $\phi = \pi$:

$$h = \frac{1}{U} r^\lambda (A + B). \qquad (6c.7)$$

Symmetric streamfunctions (i.e. antisymmetric flows) are given by the ansatz

$$\psi = r^\lambda (A \cos(\lambda \phi) + B \cos((\lambda - 2)\phi)). \qquad (6c.8)$$

In this case, (6c.4) leads to

$$(\lambda A + (\lambda - 2)B)\cos(\lambda\pi) = 0,$$

$$(A + B)\sin(\lambda\pi) = 0. \qquad (6c.9)$$

Again, $\lambda = 3/2$ is a solution; in place of (6c.7) we get $h = 0$.

Hence, cusped solutions with $h \sim r^{3/2}$ are possible, giving

$$y = -cx^{3/2}, \qquad (6c.10)$$

but the constant of proportionality is left undetermined at our level of the analysis. We also don't know the range of x for which our perturbation solution is appropriate. When x is larger than the appropriate range, the higher-order corrections may become important. The higher-order corrections would be a composition of symmetric and antisymmetric functions like (6c.2) and (6c.8) in a combination determined globally.

If surface tension is included, we may do a similar asymptotic analysis [Joseph, Nelson, Renardy and Renardy 1991] but it would not be meaningful: the evidence of sections (c)(ii) and (d) below points to the conclusion that a cusp forms only for infinite capillary number. Thus Richardson's solution (6a.1) cannot be put to direct use here.

Joseph [1992] has reasoned that the cusp shape ought to arise from an analytic solution that would probably be generated from the highly elliptic problem governed by the biharmonic equation.[1] This kind of reasoning leads to the cusp shape (6c.10) by the following argument. The shape of the free surface is assumed to be given implicitly by an analytic function $F(x, y) = 0$ before and after cusping. We assume that the cusp is in the least degenerate form in a sense made clear by the following considerations. Let $F(x, y) = 0$ describe a cusp and suppose $F(x, y)$ can be developed in a series of powers of x and y. We are presuming that the cusp is a limiting form of an analytic function $F(x, y) = 0$ for smooth surfaces. Since a cusp is a singular point of the plane curve, the first derivatives vanish at the origin. To have a cusp at the singular point (0,0) the curve can have only one tangent at (0,0). Hence, the discriminant of F, $F_{xx} - F_{xy}$, must vanish at (0,0) and the curvature of the cusp point is singular. For the coordinates of figure 6.3, this is guaranteed if $F_{yy} = 1$ and $F_{xx} = F_{xy} = 0$. Then the form of the cusp is a balance of third-order terms and y^2 where, to lowest order,

$$F(x, y) = F_{yy}(0, 0)y^2 + F_{xxx}(0, 0)x^3 = 0$$

where neglected terms are of the order $x^{7/2}$. Hence to lowest order,

$$y = cx^{3/2}, \qquad c = [-F_{xxx}(0, 0)/F_{yy}(0, 0)]^{1/2}.$$

---

[1] There are theorems to the effect that the solution to an elliptic problem is analytic in the interior of the domain, but here we are addressing the shape of the free boundary, for which there are no theorems concerning analyticity. In addition, the shape of a cusp $F(x, y) = 0$ is not necessarily given by an analytic function.

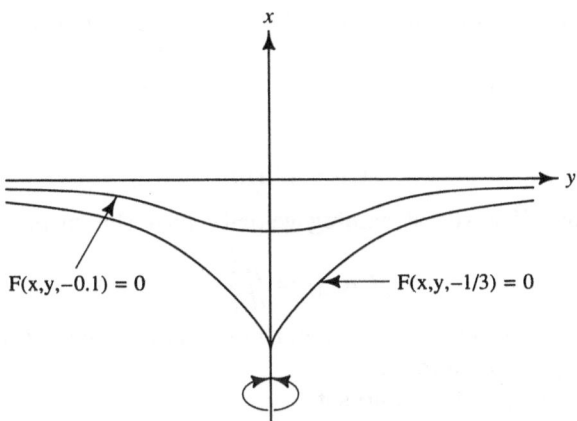

Fig. 6.11. [Joseph, 1992, Understanding cusped interfaces. J Non-Newt. Fluid Mech., to appear, Vol. 1, 152, 153, Elsevier] Cartoon of the solution (50) of Jeong and Moffatt [1992], showing analytic cusping of the form $F(x, y, a) = 0$. There is a vortex dipole placed at $(x, y) = (-d, 0)$.

For obvious reasons, Joseph calls this a generic analytic cusp. This cusp is attained approximately whenever the required derivatives of $F(x, y)$ are small, as is the case already at modestly large values of the capillary number in the solution of the model problem of Jeong and Moffatt [1992], discussed below, in the simulation of the dip coating problem of Palmquist and Kistler [1992], which is discussed in section II.6(d)(ii), and in the model problem discussed in section II.6(d)(i). The generic analytic cusp emerges as an exact solution in all the aforementioned cases as the capillary number tends to infinity (for zero surface tension). Moreover, local solutions of this form also have been obtained for viscoelastic liquids with vanishing surface tension, for the linear viscoelastic fluid discussed in section II.6(c)(iii) and for the second order fluid model discussed by Joseph [1992].

II.6c(ii) Theoretical Analysis of Jeong and Moffatt [1992]. Jeong and Moffatt solve an idealized model of an experiment in which two cylinders are counter-rotated at low Reynolds number about parallel horizontal axes below the free surface of a viscous fluid (see figure 6.1). Figure 6.11 is a sketch of the problem with the vertical axis now being the x-axis as in section II.6(c)(i): the cylinders are represented by a vortex dipole of strength α placed at position $x = -d$ (figure 6.11) and the undisturbed fluid occupies $x < 0$. Surface tension effects are included, but gravity is neglected. The streamfunction satisfies the biharmonic equation, and the solution is obtained by complex variable techniques. Their solution is analytic, with a generic analytic cusp in the limit of zero surface tension.

They find that the free surface is given by

$$F(x,y) = y^2 x - (2a - x)(x + a + 1)^2 = 0 \qquad (6c.11)$$

where a is a complicated function of the capillary number and is such that

$$a + \frac{1}{3} \sim \frac{32}{9} e^{-2\pi Ca} \quad \text{as } Ca \to \infty.$$

Here,

$$Ca = U\mu/\sigma$$

where $U = 16\alpha/d^2$ is the streaming velocity past the cusp. The capillary number

$$Ca_{JM} = \frac{\mu\alpha}{d^2\sigma}$$

used by Jeong and Moffatt is a sixteenth of this. Figure 6.12 displays their results for $Ca_{JM} = 0.1$ and ∞.

If we put $\epsilon = \frac{32}{9} e^{-2\pi Ca}$ and set

$$a = -\frac{1}{3} + \epsilon, \quad X = x + \frac{2}{3}, \qquad (6c.12)$$

for small ϵ, then to lowest order in ϵ, (6c.11) becomes

$$\frac{2}{3}y^2 = X(X^2 - 3\epsilon^2). \qquad (6c.13)$$

Jeong and Moffatt note that this may be put into a universal form

$$\zeta^2 = \frac{3}{2}\eta(\eta + 3)^2 \qquad (6c.14)$$

where $y = \epsilon^{3/2}\zeta$, $X = \epsilon\eta$. This curve exhibits parabolic behavior $\zeta^2 = \frac{27}{2}\eta$ for $\eta \ll 1$, and the cuspidal behavior $\zeta^2 \sim \frac{3}{2}\eta^3$ for $\eta \gg 1$. The points of inflexion at $(\eta, \zeta) = (1, \pm 2\sqrt{6})$ mark the transition between these two regimes. This asymptotic formula agrees with (6c.10) and also provides the constant of proportionality in that result, with $c = \sqrt{3/2}$. In general, the constant c depends on the geometry of the experiment. Figure 6.13 shows a comparison of these formulas with experimental data.

Though the solution given by Jeong and Moffatt is analytic and the free surface is smooth when Ca is finite, the radius of curvature R on the line of symmetry on the free surface is given asymptotically by

$$R \sim \frac{256d}{3} e^{-2\pi Ca}. \qquad (6c.15)$$

This type of exponential behavior can be obtained also from Richardson's solution (6a.1): the range of r over which his solution dominates the uniform-flow solution is given by $(\sigma/2\pi\mu)\log r \sim U$, yielding $\log r \sim 2\pi Ca$.

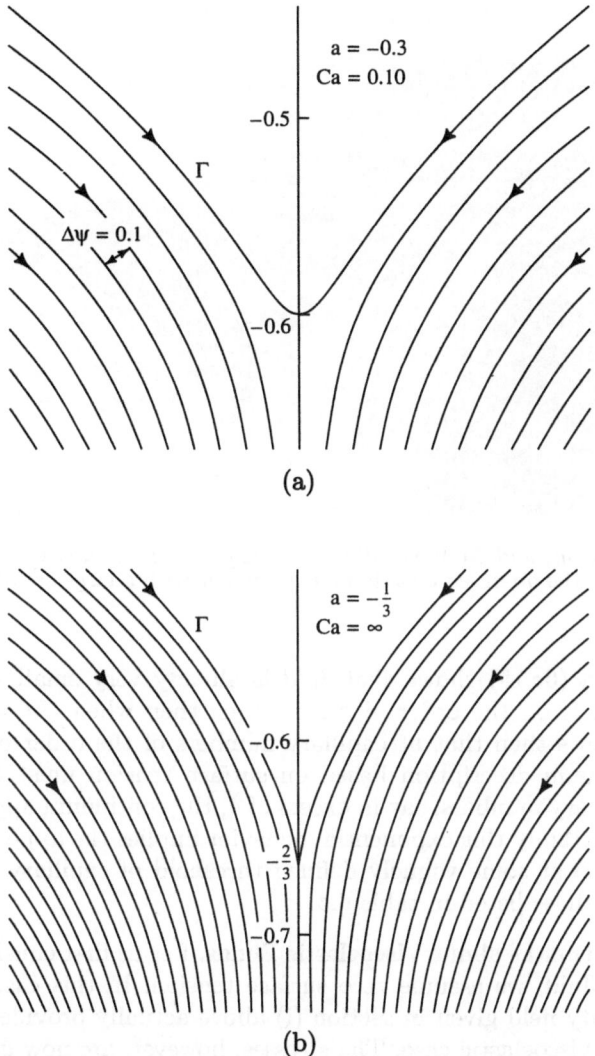

(a)

(b)

Fig. 6.12(a-b). [Figure 4(c)(e) of Jeong and Moffatt, 1992] Streamlines and corresponding free surface shape for various values of the capillary number. All streamlines pass through the singularity at (0, -1), and the streamlines near this point are circles touching the y-axis. Γ denotes the free surface. (a) $Ca = 0.1$. (b) $Ca = \infty$.

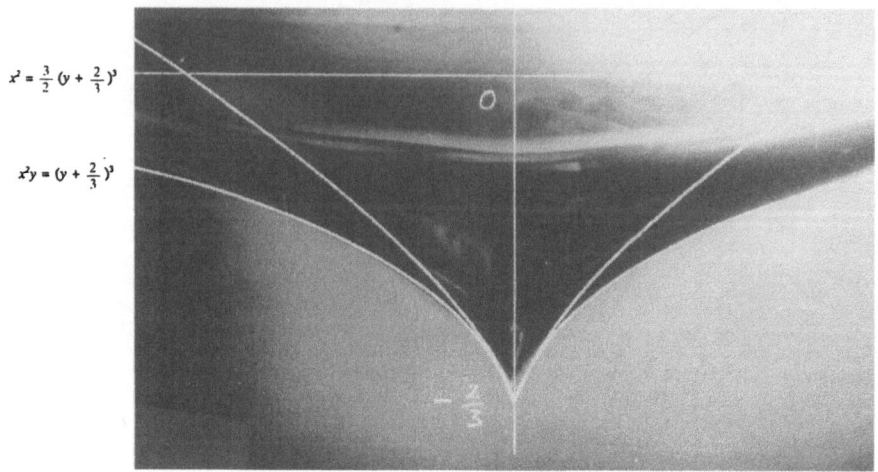

$$x^2 = \frac{3}{2}\left(y + \frac{2}{3}\right)^2$$

$$x^2 y = \left(y + \frac{2}{3}\right)^2$$

Fig. 6.13. [Jeong and Moffatt, 1992] The asymptotic curves (6c.14) - (6c.15) superposed on the observed cusp in an experiment with a large capillary number: $(Ca)_{\mathrm{exp}} = 52.4$.

Equation (6c.15) shows that R/d is already very small when Ca is order 1. If d is on the order of centimeters, then when $Ca = 4$, $R/d \approx 10^{-9}$. It follows then that at capillary numbers of the order of unity, the usual continuum description based on surface tension must break down and for capillary numbers, say in excess of 4, the continuum approximation must fail. Certainly this exponential behavior implies the appearance of an apparent cusp at some vaguely defined threshold of capillary numbers of the order of those listed in table 6.4.

II.6(c)(iii) Remarks about Viscoelastic Fluids. Let us first consider a linear viscoelastic fluid; we assume zero surface tension. In this case, the Newtonian velocity field given in section (i) above actually provides a solution even for the viscoelastic case. The stresses, however, are now given by

$$\mathbf{T}(x, y) = \int_0^\infty G(s)\mathbf{D}(x - Us, y) \, ds, \qquad (6c.16)$$

where G denotes the stress relaxation function and $\mathbf{D}$ the symmetric part of the velocity gradient. We note that the stresses given by (6c.16) would be less singular at the cusp than in the Newtonian case. On the other hand, if we use the second-order fluid approximation, $\mathbf{T} = \mu\mathbf{D} + \alpha_1 U(\partial\mathbf{D}/\partial x)$, where

$$\mu = \int_0^\infty G(s)ds, \quad \alpha_1 = \int_0^\infty sG(s)ds,$$

then the stresses for a given velocity field are more singular than in the Newtonian case. The second-order fluid cannot be a valid approximation close to the cusp, but it may be valid some distance away from it. For further calculations on viscoelastic effects, see Joseph [1992].

It appears doubtful whether linear viscoelasticity is very useful in interpreting experiments in non-Newtonian fluids. The extensional behavior of such fluids is highly nonlinear, leading to extensional stresses much larger than those in Newtonian fluids with the same shear viscosity. We note that a non-cusped interface necessarily has a stagnation point, with the associated potential for a build-up of elongational stresses. Cusping gets rid of the stagnation point and alleviates some of these elongational stresses. This effect should favor cusping in non-Newtonian fluids. Indeed, the experiments show that non-Newtonian fluids form cusps at lower values of Ca than Newtonian fluids. It is to be expected that extensional effects are still important even after cusping and the analysis above on the basis of linear viscoelasticity is probably not valid. We also note that Richardson's analysis does not apply to non-Newtonian fluids. The behavior of non-Newtonian fluids at cusps and in general at corners larger than 180° is not well understood (see Davies [1988] for some partial results).

For non-Newtonian fluids, more dimensionless quantities can be formed, in addition to the capillary number introduced above. The critical capillary numbers displayed in table II.9 are of the same order of magnitude for non-Newtonian as for Newtonian fluids, but there is a definite trend to lower capillary numbers in the non-Newtonian case. To rectify this, one may think of replacing the Newtonian viscosity in the definition of Ca by some non-Newtonian viscosity. We note that the extensional viscosity of non-Newtonian fluids generally increases with elongation rate, so that such a procedure would give higher capillary numbers if we think of the flow as being primarily extensional. We give an alternative dimensional analysis based on the second-order fluid. For this we estimate elongational stresses as $|\alpha_1|U^2/l^2$, where l is a characteristic length, and we estimate stresses resulting from surface tension as σ/l. This leads to the capillary number $Ca = |\alpha_1|U^2/\sigma l$. Next, we estimate l as $U\tau$, where τ is a characteristic timescale, so we have $Ca = |\alpha_1|U/\sigma\tau$. A possible estimate for τ is $|\alpha_1|/\mu$: in this case, we obtain the same capillary number as in the Newtonian case. It may be argued, however, that shorter relaxation times could be important. Using such shorter relaxation times for τ would also raise the value of Ca.

II.6(d) Numerical Results

In part (i), we present numerical results that show the formation of a cusp in a model problem, and in part (ii), we present results of Palmquist and Kistler [1992] which simulate one of our experiments.

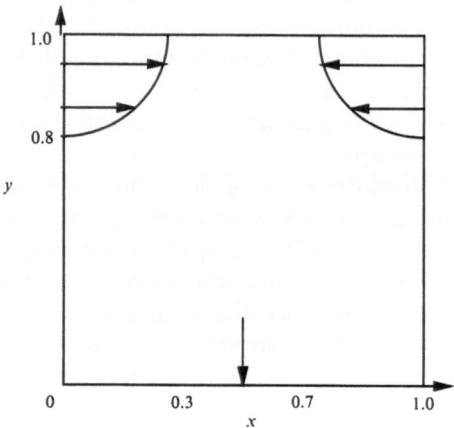

Fig. 6.14. [Joseph, Nelson, Renardy and Renardy, 1991] Flow domain for numerical simulations

II.6(d)(i) Model Problem. Our results are computed with the Fluid Dynamics Package (FIDAP) Version 4.0, which uses the finite element method (cf. Engelman [1982], Engelman and Sani [1986]). The computed free surfaces first appear to have a corner, but as time evolves, this corner sharpens and looks more like a cusp. The solution becomes polluted by wiggles at this point; we shall explain the suspected reasons for this below. We observe that the development of a singularity and the time at which it develops persist under mesh refinement.

The computational domain at time zero is a unit square in the $x - y$ plane, as shown in figure 6.14. The top boundary at $y = 1$, $0 \le x \le 1$, is a free surface. Along the upper parts of the vertical sides ($x = 0$, $0.8 \le y \le 1$ and $x = 1$, $0.8 \le y \le 1$), the boundary condition is zero vertical velocity and the horizontal velocity is half of a parabolic velocity profile. The latter has a maximum value of 1 at $y = 1$. The boundary along $0 \le y \le 0.8$ at $x = 0$ and 1, and along $0 \le x \le 0.3$ at $y = 0$, and $0.7 \le x \le 1.0$ at $y = 0$ is a solid wall. Along $0.3 \le x \le 0.7$ at $y = 0$, the horizontal velocity is zero and the normal traction is 1, so that there is a negative pressure sucking the fluid out. The viscosity and density of the fluid are prescribed to be 1. The value for surface tension is varied, as described below.

A mixed velocity-pressure formulation is used. Interior elements are nine-node isoparametric quadrilaterals with biquadratic velocity and discontinuous linear pressure functions. Three-node quadratic elements are used on the free surface.

An initial velocity field is computed by solving the Stokes problem with the position of the free surface kept fixed. The transient problem is then

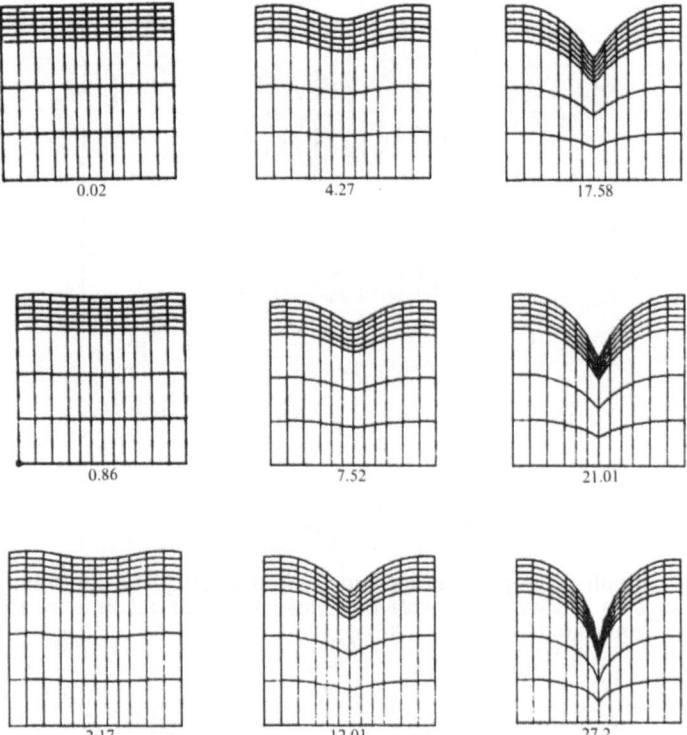

Fig. 6.15. [Joseph, Nelson, Renardy and Renardy, 1991] The top boundary is a free surface. Fluid flows in through the upper parts $0.8 \leq y \leq 1$ of the side boundaries and flows out through the middle part $0.3 \leq x \leq 0.7$ of the bottom boundary. Surface tension is 0.05, which is half the amount in figure 6.18. A cusp forms. Times of free surface evolution are shown below each figure.

computed with the implicit backward Euler integration scheme. At each time-step, the Newton-Raphson scheme is used with a relaxation constant of 0.5. The initial timestep is 0.01, and thereafter, a variable time increment is selected automatically by the code.

We present results for various values of the surface tension. Figure 6.15 displays the formation of a cusp for surface tension 0.05. The mesh size is 21 by 17, and the mesh is displayed in each plot. There are 357 nodes, 80 interior elements and 10 elements on the free surface. The times shown on the plots are 0.02, 0.03, 0.26, 0.86, 2.17, 4.27, 7.52, 12.01, 17.58, 21.01 and 27.2. A cusp starts to form by time 12.01. The velocity vector plot for times 0.01 and 21.01 are shown in figures 6.16 and 6.17, respectively. At later times, a steady state is reached. These computations have been convergence tested on a finer mesh. A computation at zero surface tension resulted in the cusp formation at an earlier time.

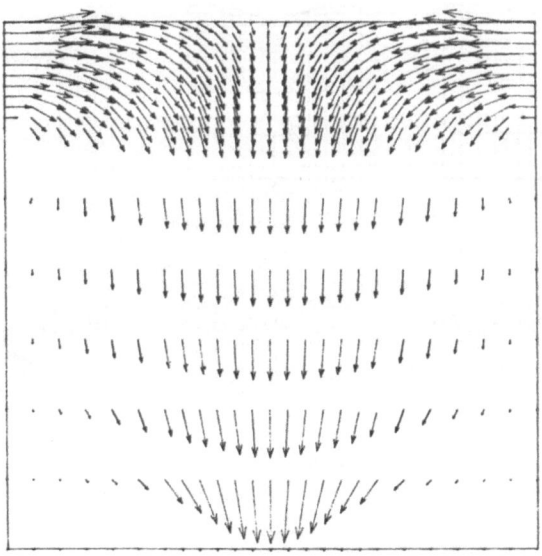

Fig. 6.16. [Joseph, Nelson, Renardy and Renardy, 1991] The top boundary is a free surface. This is a velocity vector plot close to the initial time for the flow in figure 6.15. Time is 0.01.

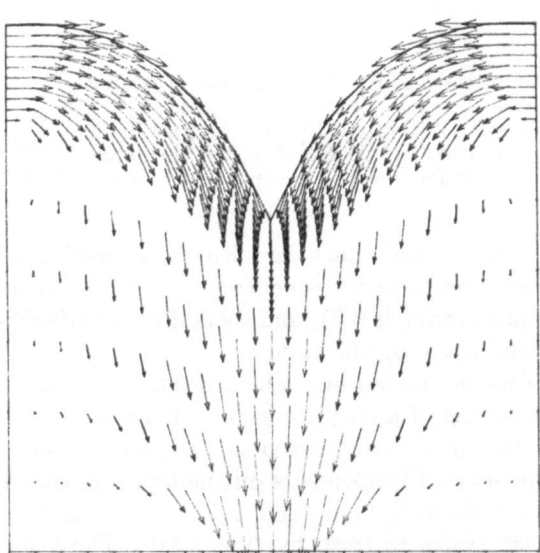

Fig. 6.17. [Joseph, Nelson, Renardy and Renardy, 1991] The top boundary is a free surface. This is a velocity vector plot for the flow in figure 6.15 at the onset of cusp formation. Surface tension is 0.05. Time is 21.01.

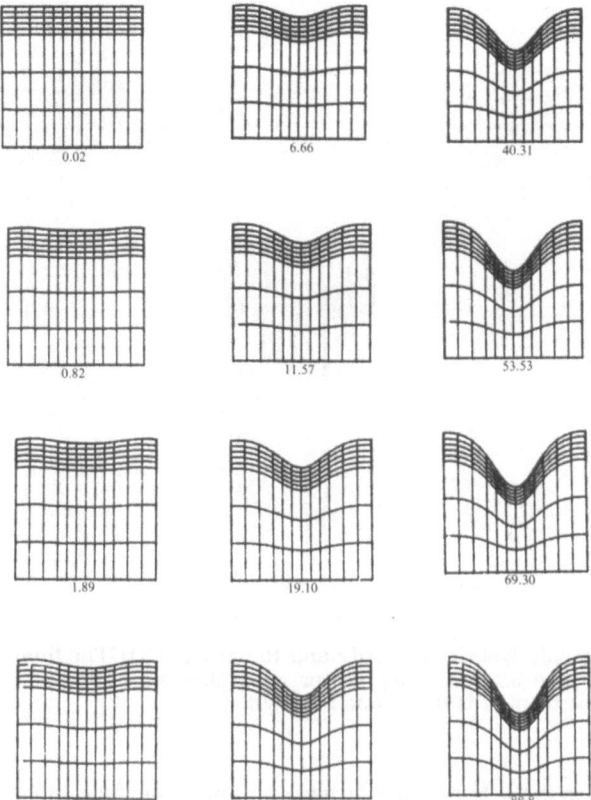

Fig. 6.18. [Joseph, Nelson, Renardy and Renardy, 1991] The top boundary is a free surface. The flow conditions are the same as those of figure 6.15 except for the presence of a sufficiently large surface tension of 0.1 that keeps the free surface smooth. The evolution of the flow is shown. Times of free surface evolution are shown below each figure.

After the formation of the cusp, wiggles appear in the free surface. We believe that this is due to improper handling of the kinematic free surface condition. At a cusp, the kinematic free surface condition must be suspended; fluid particles can move from the free surface into the interior. The code, however, does not know this. The computed results past the development of the cusp can therefore not be trusted.

When surface tension is 0.1, the cusp does not form. Instead, a steady state is attained, with a smooth curve at the free surface. Figure 6.18 shows the evolution of the mesh at times 0.02, 0.03, 0.26, 0.82, 1.89, 3.66, 6.76, 11.57, 19.10, 28.91, 40.31, 53.53, 69.30 and 88.80. By the latter time, a steady state has been reached. The velocity vector plot at time 88.80 is

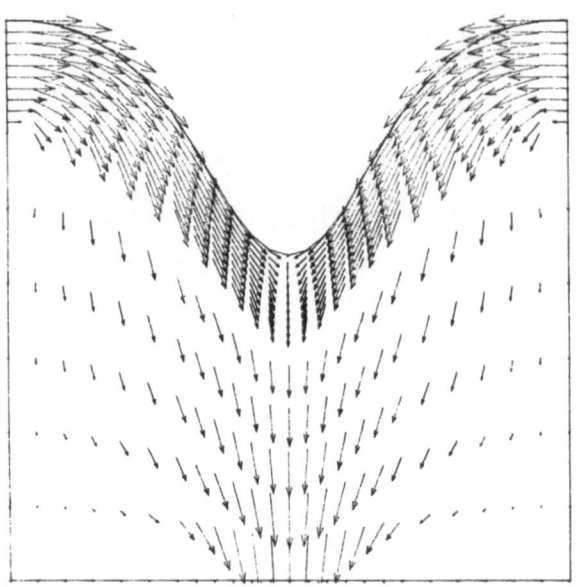

Fig. 6.19. [Joseph, Nelson, Renardy and Renardy, 1991] The final plot of figure 6.18 is shown here as a velocity vector plot, showing that the free surface is smooth. Surface tension is 0.1. Time is 88.80.

shown in figure 6.19. If a much larger amount of surface tension is present (e.g., 1.0), the free surface does not move appreciably from its initial position, and a cusp does not form.

The numerical results of this section show the formation of a cusp in slow viscous flow, for values of the surface tension that are in the order-of-magnitude range observed in the experiments described in the earlier sections.

II.6(d)(ii) Apparent Cusp. In this section, we present parts of the paper of Palmquist and Kistler [1992] on a liquid film plunging into a pool (figure 6.20). They analyze, by means of a finite-element technique, various phenomena that arise in this flow, in particular the formation of a cusp when the wall is moving sufficiently fast (plate II.6.21). Their numerical simulations exhibit corners rather than cusps, but this is the way their code approximates a small region of high curvature. On a macroscopic scale, their results show good agreement with experimental data and the analysis of Jeong and Moffatt [1992]. They call the corner an *apparent cusp*. The apparent cusp approaches a true cusp in the limit of high capillary numbers.

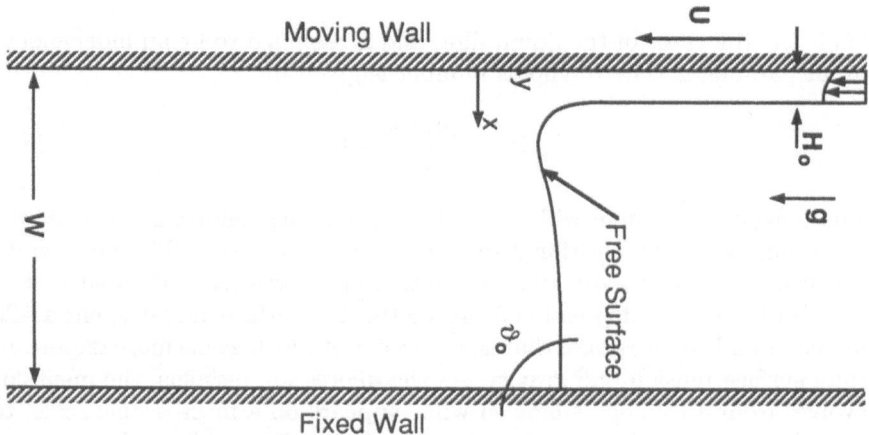

Fig. 6.20. [Palmquist and Kistler, 1992] Flow of a thin liquid film on a moving wall merging with a large body of the same liquid.

They also draw analogies between cusp formation where a large liquid body merges with a thin liquid film, and dynamic wetting where a liquid body advances over previously dry solid, subtending an apparent contact angle. In the latter case, there is a limiting capillary number at which the dynamic contact angle, when measured through the displaced air phase, is zero and the free surface appears to form a cusp with the solid wall.

The key idea in the numerical simulations by Palmquist and Kistler is the assumption that, on a sufficiently small scale, the interface remains smoothly curved and thus incorporates a stagnation point. This assumption excludes *a priori* the formation of a genuine cusp where two interfaces meet at a common tangent. Nevertheless, with considerable mesh refinement, the computations yield meniscus profiles which exhibit a cusp-like singularity on a visible length scale −− as is evident in plate II.6.21. At the onset of cusp formation, the calculated profiles of the plunging film and the interface capping the pool appear to meet at a corner with a measurable angle rather than a genuine cusp. Their computations suggest that the apparent cusp approaches a true cusp in the limit of high capillary numbers, but their finite-element discretization breaks down before this limit is reached.

On a microscopic scale, the region right at the cusp tip is a region of very high curvature, with the maximum curvature increasing exponentially with the capillary number above the critical Ca just as in (6c.15). The parameter space was explored systematically with a continuation method. We refer the reader to their original paper for further details of their steady finite-element code, and for extensive numerical results.

The Stokes number is defined by

$$St = \frac{\rho g H_0^2}{\mu U}. \tag{6d.1}$$

H_0 is the thickness of the liquid film, and is assumed to be an independent input parameter. The Reynolds number is given by

$$Re = \frac{\rho U H_0}{\mu} = 0. \tag{6d.2}$$

The dimensionless pool width $w = W/H_0$ and the contact angle θ_0 at the stationary wall are additional dimensionless parameters. The pool width was chosen to be $w = 10$, and the contact angle was set to $\theta_0 = 90°$.

For high Ca, a depression forms on the free surface, and deepens as Ca increases and St decreases, that is, as viscous effects become more important than surface tension and gravity. As the depression deepens, the meniscus evolves from a rounded shape to what appears on a macroscopic scale, to be a sharply pointed corner. Eventually, a finite-element discretization with a given level of mesh refinement at the bottom of the cusp-like depression fails to resolve the sharply curved meniscus, and the computed free surface profiles exhibit an angular corner discontinuity, sometimes accompanied by spurious wiggles nearby. At sufficiently high Ca, the elements right at the cusp become so distorted that Newton's method fails to converge altogether. A detailed study of this limit indicates that the failure of this refined finite-element discretization coincides with apparent cusp formation in an actual flow.

The numerical results at low St replicate our experimental data for Newtonian fluids: the evolution of the interface from a smooth shape to a cusp is gradual and there is a certain speed, above which the interface appears to be sharply pointed. Cusp-formation thus appears to be a threshold phenomenon. The computed values of critical capillary numbers at the onset of cusp-formation are shown in figure 6.22. They find that at low St, that is, when the viscous drag of the plunging film dominates over gravity effects, the critical capillary number is near $Ca_{\text{cusp}} = 2.5$, a value very similar to that reported by us in section (b) above for several Newtonian liquids (cf. their paper for more documentation). The apparent cusps are deep and sharply pointed, as in figure 6.6.

One of their most interesting results is that *beyond a critical capillary number Ca, the local meniscus curvature increases exponentially with further increases in Ca*. This is consistent with equation (6c.15).

Towards higher Stokes numbers, inertia effects play a key role. In the limit of creeping flow ($Re \ll 1$), Ca_{cusp} increases with St, and the apparent cusp twists in the horizontal direction and becomes less deep and less pointed. Eventually, when gravity dominates over viscous effects at large Stokes numbers, the interface no longer forms a cusp, even when the capillary number becomes large. Instead, it forms a smoothly curved, standing crest similar to that predicted by Hansen [1987] (cf. the paper of Palmquist

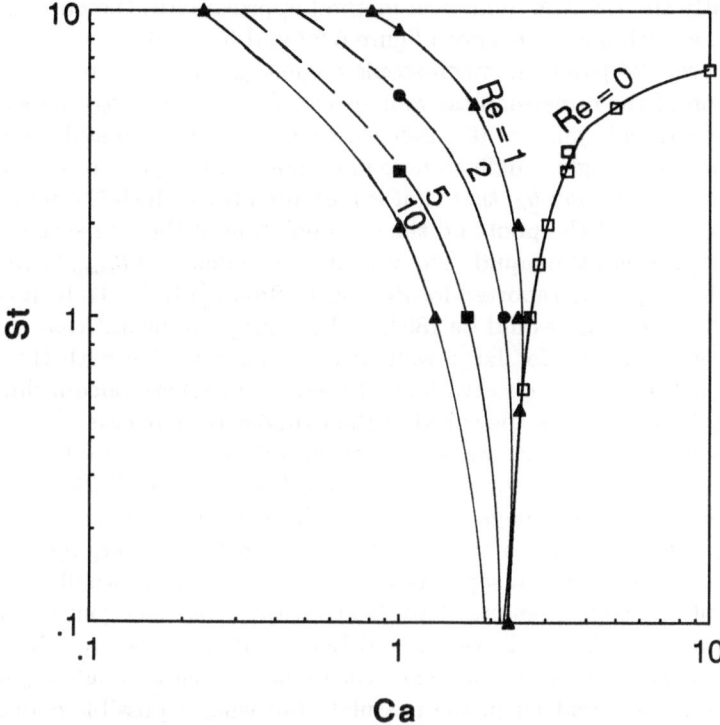

Fig. 6.22. [Palmquist and Kistler, 1992] Critical parameters at the onset of apparent cusp formation. $w = W/H_0 = 10, \theta_0 = 90°$.

and Kistler for further comments on this regime). When $Re \geq O(1)$, on the other hand, cusp formation persists up to very large Stokes numbers. In fact, inertia effects promote very deep and sharply pointed cusps, and Ca_{cusp} can drop well below 2.5. The finite-element results indicate that, near the critical Stokes number above which no cusp forms for $Re \gg 1$, the recirculating eddy that separates the plunging film flow from the stationary wall shrinks and the stagnation line moves away from the depression in the free surface. This finding suggests that the presence of a stagnation line, and the subambient pressure and high deformation rates associated with the flow nearby, are necessary conditions for apparent cusp formation. The proximity of the confining stationary wall and the static contact angle θ_0 on the solid surface can also influence the onset of apparent cusp formation.

Figure 6.23 shows photographs taken by D. D. Joseph *et al.* for a 0.5 Pa.s silicone oil, together with computations by Palmquist and Kistler. The computed profiles replicate the gradual transition from a smooth depression to what macroscopically looks like a cusp, but is more like two flat interfaces meeting at an angle, which may be called an *apparent cusp*. This angle is apparently the way the numerical method approximates the curve (6c.13). There is also similarity between figure 6.22 and figure 6.13.

Figure 6.24 plots the macroscopic angle θ_{cusp} at the apparent cusp as a function of the dimensionless wall speed $1/St$. The analogy between apparent cusps and apparent dynamic wetting lines is most readily evident in this figure. The angle is defined here in a manner analogous to an apparent *dynamic contact angle* θ_d; that is, it is measured through the liquid between the free surface of the pool and the extrapolation of the free surface of the plunging film into the liquid. The velocity dependence of θ_{cusp} is very similar to that typically reported for θ_d (e.g. Hoffman [1975]). To be a genuine cusp, the angle θ_{cusp} would be 180°, and the graph indicates that this may occur asymptotically for large wall speed. This is in line with the theory in section (c) and the observations of section (b) where once a dip starts forming, it sharpens as the speed of the cylinder is increased.

The critical capillary number for air entrainment is found to be consistently higher than that for cusp formation [Bolton and Middleman 1980]. Palmquist and Kistler suggest that 'a genuine cusp is never reached because the displaced air in the narrow wedge between the converging interfaces builds sufficient forces to separate the interfaces, and eventually cause entrainment of visible amounts of air. This postulate is analogous to that put forward much earlier by Deryagin and Levi [1964] for excessive air entrainment at dynamic wetting lines when the dynamic contact angle approaches 180°. At an apparent cusp, the postulate furnishes a possible explanation for a physical mechanism that might alleviate the apparent line singularity (6a.1) which arises when mathematical theories force the two interfaces to meet at a genuine cusp.'

In summary, Palmquist and Kistler suggest that their results −− and the agreement they obtained with our experiments −− provide indirect evidence that what a naked eye, or even an eye aided by a microscope, perceives to be a cusp-shaped corner may be merely a local region of extremely high curvature of the interface. They also hypothesize that, at sufficiently high capillary number, the meniscus may come close to forming a genuine cusp, but that entrainment of visible amounts of air sets on before the limit of two interfaces meeting at a common tangent is reached. Resolving the questions of whether there can be a genuine cusp and at what point excessive air entrainment begins requires more refined analyses of the local dynamics that account for the molecular interaction forces and for the hydrodynamics in the displaced air phase.

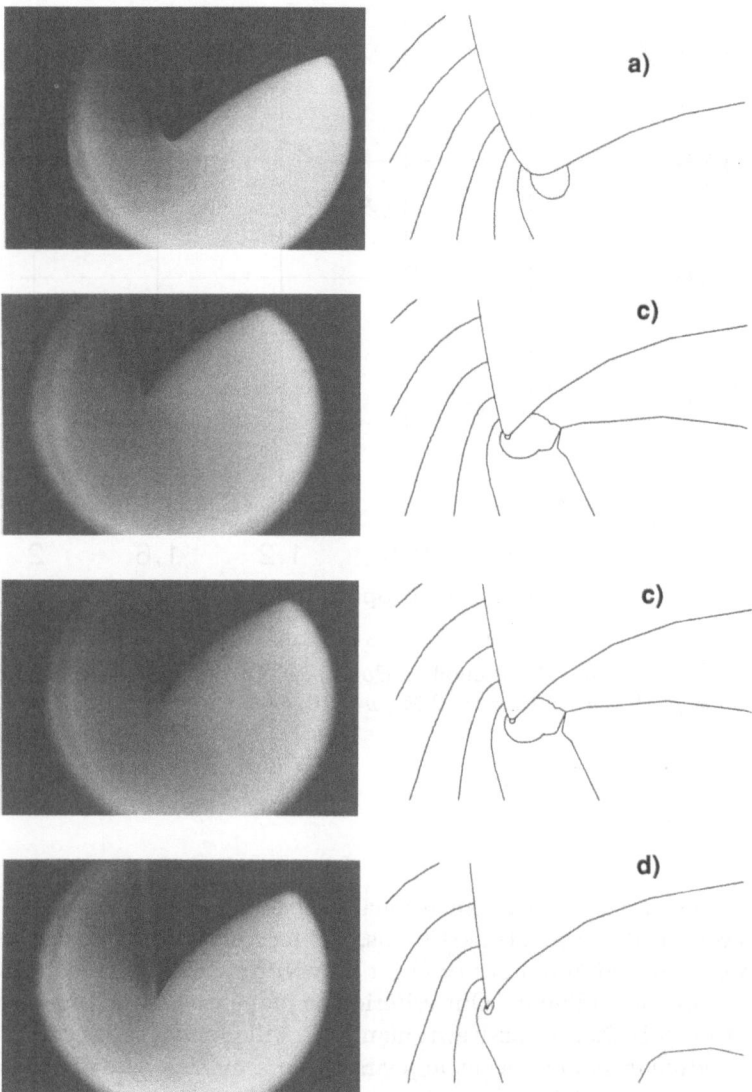

Fig. 6.23. [Palmquist and Kistler, 1992] Comparison between photographs of D. D. Joseph *et al.* and computed predictions of interfacial shapes during cusp formation with a 0.486 mPa.s silicone oil ($\rho = 971$ kg/m^3, $\sigma = 0.0211$ N/m). The computations were performed at a fixed film thickness equivalent to that reported by Joseph *et al.* at the onset of cusping (Bond number $Bo = \rho g H_o^2/\sigma = 6.78$, Galileo number $Ga = \rho^2 g H_0^3/\mu^2 = 2.31$, $w = 10$). (a) $Ca = 2.25$, $St = 2.21$, $Re = 0.65$, (b) $Ca = 2.51$, $St = 2.25$, $Re = 0.77$, (c) $Ca = 2.73$, $St = 2.28$, $Re = 0.88$, (d) $Ca = 2.96$, $St = 2.31$, $Re = 1.00$.

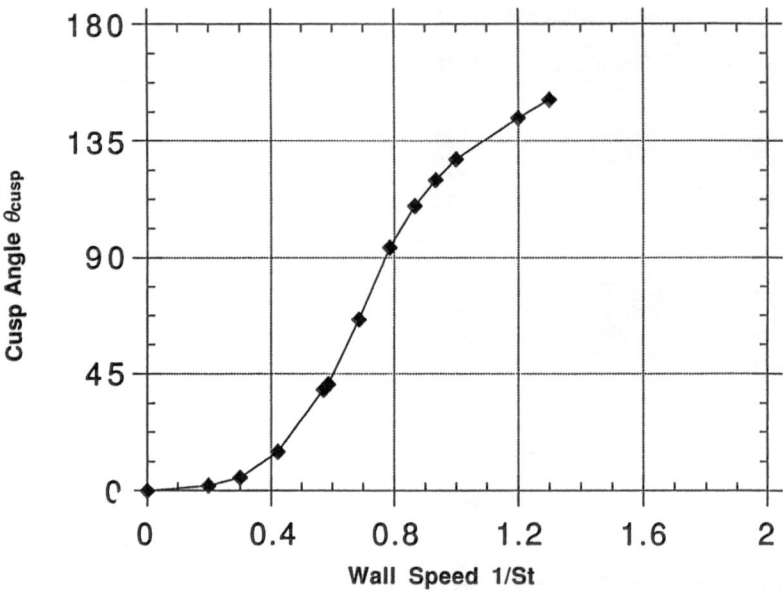

Fig. 6.24. [Palmquist and Kistler, 1992] Apparent cusp angle as a function of velocity of plunging wall. Bond number $Bo = \rho g H_0^2/\sigma = Ca.St = 6.78$, Galileo number $Ga = \rho^2 g H_0^3/\mu^2 = Re.St = 2.31$, $w = 10$, $\theta_0 = 90°$.

II.6(e) Conclusions

1. It is easy to create two-dimensional cusped surfaces at liquid-air interfaces. In the cases studied by us, the formation of cusped surfaces between air and liquid occurred for all non-Newtonian liquids and for some Newtonian liquids. Our criterion to judge whether a fluid is non-Newtonian is that a non-Newtonian fluid will climb a rod.

2. In Newtonian liquids, what appears to be a cusp is actually a small analytic region of high curvature which evolves into a genuine cusp gradually as the capillary number tends to infinity. In experiments, the formation of a cusp appears to be a critical phenomenon because of the limit of resolution of the microscope: other things being equal, the interface is rounded for small values of the streaming speed and 'looks' cusped at a critical speed and beyond. Interpreted in this way, our experimental results agree with the analytical result of Jeong and Moffatt [1992] who solved an idealized problem modeling the experiment with two counter-rotating cylinders. They obtained the formula describing the free surface for their model problem. It is smooth at low

speeds and then develops a paraboloid dip of high curvature, which begins to look like a cusp, for capillary number of order 1 or larger. Their work provides a complete analytical description of the free surface and velocity field for the cusp-like structure.

3. In non-Newtonian fluids the critical speed for cusp formation is very distinct; in Newtonian fluids the change from a rounded to a pointed interface is observed to be gradual.

4. Cusping is unusual at the interface of two liquids. Higher viscosity liquids in water form rollers, which are replaced by fingering flows when the viscosity of the liquid on the cylinder is too low to support a roller.

5. SAE 30 motor oil and water support a cusped interface with periodic scallops along the cusp which form sites for the fingering of water into oil. The motor oil is Newtonian and will not cusp in air.

6. Jeong and Moffatt obtain the formula (6c.14) for the shape of the cusp at infinite capillary number, and our asymptotic formula (6c.10) agrees with this.

7. Linear viscoelasticity would predict stresses at a cusp which are less singular than in the Newtonian case. However, it seems likely that nonlinear extensional behavior of non-Newtonian fluids is quite important. Experimentally, non-Newtonian fluids form cusps at lower speeds than Newtonian fluids. A qualitative explanation may be that cusping avoids the stagnation point inevitably present on a smooth interface and thus relieves extensional stresses.

8. Numerical simulations of Palmquist and Kistler [1992] replicate our experiments. What appears to be a cusp on a macroscopic scale is detected by their numerical scheme as a corner which forms at a critical capillary number. On a microscopic scale, there is a region at the corner of high curvature, and the maximum curvature increases exponentially with the capillary number once the critical value is attained. The corner approaches a true cusp for high capillary numbers. This is consistent with the analytical results of Jeong and Moffatt.

Chapter III
The Two-Layer Bénard Problem

III. 1 Introduction

In this chapter, we will develop a theory of fluid motions arising out of instabilities of heat conduction without motion in two superposed layers of fluid heated from below. For short, we call this a two-layer Bénard problem. Beautiful photographs of the hexagonal cells which develop on a single layer of fluid are exhibited in the frontispiece of the stability book of Chandrasekhar [1961]. However, these cells are generally regarded as arising from surface tension gradients rather than from the instability of conduction, so that the traditional name "Bénard problem" is misinformed.

     We shall restrict our analysis to the case in which the two superposed layers are confined between infinitely extended planes perpendicular to gravity [Renardy and Renardy 1988]. Our analysis is based on the Oberbeck-Boussinesq approximation of the equations of motion and is the two-layer generalization of the one-layer problem which is discussed in books by Chandrasekhar [1961], Joseph [1976], and Drazin and Reid [1981], and in hundreds of research papers. It is instructive to recall some of the results for the one-fluid problem. Rayleigh, in 1916, solved the linear stability problem for one fluid, for "free-free" boundaries (that is, zero normal velocity and zero tangential shear stress). The eigenvalues and eigenfunctions were found in closed form. A countable number of marginal stability curves were found, and are given by the following relationship between the Rayleigh numbers R and the wavenumber α of the disturbance: $R_j = (j^2\pi^2 + \alpha^2)^3/\alpha^2$, $j = 1, 2, ..., R_1 < R_2 < R_3 < ...$ The critical Rayleigh number is $27\pi^4/4$. The Prandtl number does not enter into the criticality condition, but appears in the growth rate. The "rigid-rigid" boundary [Reid and Harris 1958] and "free-rigid" boundary cases [Pellew and Southwell 1940] were solved later: in these cases, the eigenvalues and eigenfunctions cannot be found in closed form but have to be calculated numerically.

Another well-known result for the one-fluid problem, whether the fluid is bounded by walls or by stress-free surfaces, or by a wall below and by a gas above, is the "exchange of stabilities"; that is, the equations are self-adjoint and all the eigenvalues are real. A consequence is that the eigenfunctions at criticality are not oscillatory in time. The exponential growth of small disturbances leads to steady cellular convection (see Figs. 139 - 140 of Van Dyke [1982]). The consideration of additional effects can, however, introduce "overstability", that is, critical eigenfunctions that are oscillatory in time (and critical eigenvalues are a complex conjugate pair). Overstability occurs, for example, if there are temperature-dependent surface tension gradients [Hurle and Jakeman 1971; Smith and Davis 1983], if there is a temperature-dependent solute gradient [Nagata and Thomas 1986; Lee, Lucas and Tyler 1983], in mixtures of superfluids [Schechter, Velarde and Platten 1974], in viscoelastic liquids [Renardy and Renardy 1992], or in the situation where two fluid layers, each having a gradient of a different solute, are superposed [Shirtcliffe 1973]. Another example is free convection between vertical parallel plates [Fujimura 1991b]. When a complex conjugate pair of eigenvalues reaches criticality, a time-periodic solution bifurcates from the rest state. This is called Hopf bifurcation.

The one-fluid convection problem, bounded by a free deformable surface, has been found to yield oscillatory instabilities, as well as the convective ones [Scriven and Sternling 1964; Davis and Homsy 1980]. Benguria and Depassier [1987, 1989] have shown that there are oscillatory instabilities which may occur at values of the Rayleigh number lower than the critical value for the onset of steady convection. Oscillatory instabilities also occur in mixtures of two fluids with a deformable upper surface [Castillo and Velarde 1982; Garcia-Ybarra and Velarde 1987].

In this chapter, we consider the Bénard problem with two fluid layers having different thermal and mechanical properties (see figure 2.1). The novel feature is the presence of the interface, which allows for oscillatory motions. We note some similarities with the types of motions that occur in double diffusion. In double diffusion, a layer of fluid (for example, water) with a dissolved solute (for example, salt) is heated from below [Shirtcliffe 1969; Turner 1979; Knobloch 1980]. The solute gradient (a continuous concentration gradient) has a stabilizing effect while the temperature gradient is destabilizing, and the instability results in convective motions, which can be periodic in time. The interface in the two-layer Bénard problem can play a stabilizing role similar to that of the solute gradient in double diffusion. This stabilization may be due, for example, to surface tension, or stratifications in local densities or thermal conductivities.

Busse [1981] has examined the onset of linear nonoscillatory instability of the arrangement where the unperturbed fluids are static and lie in two layers with a flat horizontal interface. In particular, he focused on the range of parameters pertinent to a model of mantle convection in which the two layers represent the upper and lower mantles. The depth of the upper mantle is thin compared with that of the lower mantle. He shows that the horizontal scale of convection in the lower mantle may determine a horizontal scale of flow in the upper mantle. Hence, it is possible that the horizontal scale of convection in the upper mantle may not necessarily indicate the depth of convection there. He does not distort the interface, and time-periodic motions are excluded in his analysis. The analysis in this chapter includes the perturbations to the interface, since thermal and viscous coupling of the two fluids are coupled with the interface position. Busse's numerical observation that the short-wave response decays exponentially away from the interface is consistent with our asymptotic analysis in chapter III.5.

A slightly different type of oscillatory instability has been studied by Busse [1978] for an idealized model of the Earth's crust and mantle. The crust is modelled as a thin horizontal layer of very viscous fluid floating on a layer of deep, less viscous, heavier fluid. The layers are bounded by free boundaries. In the model, the crust contains homogeneously distributed heat sources and differs from the analysis of this chapter in that the heat sources result in the unperturbed temperature profile depending quadratically on the depth variable instead of linearly. Busse makes assumptions consistent with available data on the Earth in order to obtain analytic expressions in closed form, defining the range of parameters in which oscillatory instabilities would develop. The oscillatory instabilities arise because of the heat sources: the thick regions in the crust are hot and thin regions are cold. Hence, perturbations in the crust thickness lead to alternate hot and cold regions in the crust, which induce convection cells in the mantle. This convection in turn induces a movement in the crust, due to viscous coupling, and after some time, the thick regions become thin and vice versa, so that the mantle convection is reversed. This oscillatory instability is called "crustal thickness instability".

Zeren and Reynolds [1972] studied the linear stability problem for the two-layer Bénard problem for arrangements involving benzene over water, and performed experiments, focusing on the Marangoni effect; i.e., convection is induced by surface tension, which is assumed to depend linearly on the temperature of the free surface. They note that Sternling and Scriven [1959] find purely imaginary eigenvalues in the Marangoni problem where the upper fluid is inviscid (see also Davis and Homsy [1980], Castillo and Velarde [1982], Cerisier, Jamond, Pantaloni and Charmet [1984], Pérez-García, Pantaloni, Occelli and Cerisier [1985]). Zeren and Reynolds chose to compute neutral stability curves corresponding to zero eigenvalues. Y. Renardy and Joseph [1985] showed that even if Marangoni convection is suppressed, instability may occur as a Hopf bifurcation with time periodic convection.

A two-layer Bénard-Couette problem was studied by Gumerman and Homsy [1974]. They begin by analyzing the linear stability of the three-dimensional problem, with the upper boundary moving relative to the lower boundary, so that there is shearing imposed. They do the asymptotic analysis for long waves of the interfacial mode following the method of Yih [1967] (see also Benguria and Depassier [1989]), and check their analysis with a full numerical computation of the original set of equations. In chapter III.4, we present their analysis for long waves for the restricted case of zero shear. They apply their analysis to the water-benzene system of Zeren and Reynolds [1972] and find that criticality is achieved by a one-fluid roll mode: the interfacial mode achieves instability at a higher Rayleigh number. In Part II of Gumerman and Homsy [1974], an analysis is made of the energy equation for the case when surface tension is large and there is a large density difference. Thus, their analysis concerns the case where the interfacial mode is stabilized, and the first instability is due to a one-fluid mode. In addition, the fluid mechanics of the upper fluid is neglected. They present results on global stability and discuss the effects of Marangoni, Rayleigh and Reynolds numbers. In Part III of their work, they present experimental results on systems that satisfy the assumptions made in their Part II. The modes that they observe, including oscillatory ones, are basically those which would occur in the shearing flow of one fluid where the eigenvalues are complex.

The linear stability of the two-layer problem, bounded by a free surface at the top rather than by a solid boundary, was studied numerically by Wahal and Bose [1988]. Their idealized model is relevant to the understanding of coating processes. The Marangoni effect and the effect of surface active agents are taken into account at both interfaces. They present numerical results covering a wide range of the many parameters, yielding convective and oscillatory instabilities. As an aside, they have reproduced our numerical results [Renardy and Joseph 1985b] of chapter III.3(d). They find oscillatory instability when the physical properties of the two liquids are only slightly different from each other, a result that is reminiscent of our findings in this chapter, and in particular, of the results in chapter III.6(a).

We focus on the two-layer Bénard problem without the Marangoni effect. For the linear stability analysis, only two-dimensional disturbances need to be considered because of the rotational symmetry about the vertical axis. In chapter III.3 (b) below, it is shown that the two-layer Bénard problem is not self-adjoint [Renardy 1988a; Renardy and Joseph 1985], so that complex eigenvalues are possible. Indeed, chapter III.3 (d) exhibits an example of a situation where there is an oscillatory onset of instability. The oscillations are essentially due to the competition between the destabilizing temperature gradient and a stable interface between the two fluids. Marginal eigenvalues of this type in the two-dimensional problem are associated with Hopf bifurcations from the motionless state to either a pair of traveling waves or a standing wave [Ruelle 1973]. Both the traveling and standing waves are solutions to the nonlinear problem. If they are both supercritical, then only one of them can be stable; otherwise, they are both unstable. In section III.7, the nonlinear problem is examined in three dimensions [Renardy and Renardy 1988], and in that case, the conclusion about the bifurcating solutions involves more possibilities.

In order to understand the roles of each of the ten parameters characterizing this problem and to aid in developing systematic methods for locating the unstable eigenvalues, it is useful to apply perturbation methods [Renardy and Renardy 1985]. Differences in density, coefficients of cubical expansion, and surface tension are important in the short-wave limit of the interfacial mode (see chapter III.5). In the limit of long waves (see chapter III.4), the volume ratio, and differences in thermal conductivity, density, and coefficients of cubical expansion are important. A related result is that of Yih [1986], which shows that thermal conductivity stratification can destabilize the flow down an inclined heated plane. It is noted that the results of this chapter show that the presence of motion in the fluid is in no way essential for the occurrence of such instabilities.

In general, the eigenvalues and eigenfunctions of the basic unperturbed two-fluid flow cannot be calculated in closed form but must be computed numerically. One exception is the case when the two fluids have similar thermal and mechanical properties. The stability of the interface for fluids with only slightly differing thermal and mechanical properties is presented in chapter III.6. In chapter III.6 (a), we perturb the critical condition of the one-fluid problem [Renardy 1986]. This leads to a regular perturbation problem for a double eigenvalue (interfacial mode and the first critical mode of the one-fluid problem). If slip is allowed at the boundary, then it is possible to solve for the eigenvalues in closed form. The results have helped to elucidate which parameters are important in determining whether oscillatory instabilities occur. The stability of the interface at Rayleigh numbers below criticality of the one-fluid problem is discussed in chapter III.6 (b). A *thin-layer effect* analogous to that of the two-layer shearing flows (see chapter IV) is established, with thermal conductivity assuming the role of viscosity. It is possible to find linearly stable arrangements with the more dense fluid on top!

Experiments on the two-layer Bénard problem have been done by Rasenat [1987] in regimes where the linear stability analysis predicts stability for the interfacial mode, and criticality is attained by a "one-fluid" mode. He gave the following reasons for why it was too difficult to study regimes of *interfacial* instability. For usual fluids, surface tension is high and dominates the interfacial problem: it stabilizes the interfacial mode and the first instability is then due to a "one-fluid" mode, giving rise to convection rolls in one or both of the fluids with a flat or distorted interface. In order to have an oscillatory instability, one may attempt to decrease surface tension, but the fluids would then mix. Thus, the surface tension is left as is, and one seeks to override its effect by another effect. To do this, some other property of the fluids must be numerically higher than the dimensionless surface tension parameter. Of the other fluid properties, viscosity is the only one that can vary over a wide range: from zero to a million. The other fluid properties do not vary greatly. The viscosity should not be small because the dimensionless surface tension parameter is inversely proportional to the viscosity and this parameter must not get too large. Thus, the viscosities of the fluids should be chosen large. On the other hand, the Rayleigh number in each fluid is inversely proportional to the viscosity. One must therefore devise a means of raising the Rayleigh number to criticality. Since the Rayleigh number is also proportional to the temperature difference and the height of the apparatus, there are two choices. One is to raise the temperature difference, but this would run counter to the Boussinesq approximation. The only alternative is to make a sufficiently large apparatus. Thus, the theory in the following sections awaits the parallel development of experimental verification.

Experiments focusing on convective instability due to one-fluid modes have been conducted by Nataf, Moreno and Cardin [1988]. Their two-layer system consists of silicon oil lying above glycerol with both layers being of equal thickness. Convection cells are set up in each layer with the interface stabilized by a large density difference. The two fluids are found to undergo "thermal coupling" (that is, convection cells in each fluid lie above one another and turn in the same directions, with uprising currents above uprisings) rather than "mechanical coupling" (that is, cells of each fluid lie above one another but turn in opposite directions) even though numerical experiments suggest the latter should take place. They conclude that the neglect of the deformation of the interface in the numerical experiments is not responsible for the discrepancy.

In the one-fluid Bénard problem, the most frequently observed convection patterns are rolls and hexagons, although squares are sometimes seen (for an overview on experimental observations, see Golubitsky, Swift and Knobloch [1984]). Rolls and hexagons are both doubly periodic with respect to the hexagonal lattice (this lattice will be described in more detail in section 7). Motivated by this, Buzano and Golubitsky [1983] and Golubitsky, Swift and Knobloch [1984] investigate the bifurcation problem on the hexagonal lattice, of a situation that is more general than the one-fluid Bénard problem, where the instability leads to convection cells. Six modes become unstable via a single eigenvalue passing through zero. The Center Manifold Theorem [Carr 1981; Sijbrand 1981; Chow and Hale 1982; Guckenheimer and Holmes 1983; Golubitsky and Stewart 1985] justifies a description near criticality in terms of a finite number of amplitude functions. The partial differential equations near the bifurcation reduce to three complex amplitude equations for the critical modes. They use group theory to develop a theory of pattern selection in the hexagonal lattice (this does not include many rectangular convection cell patterns). In addition to rolls and hexagons, they find many new solutions, such as triangles. In the steady bifurcation problem (the one-fluid Bénard problem), the results depend on whether there is midplane reflection symmetry (for example, the boundary conditions are the same on the top and bottom plates) or not. In the two-fluid Bénard problem, when criticality is achieved at one isolated zero eigenvalue as the Rayleigh number is increased, the bifurcation theory for the one-fluid problem carries over. That theory would not carry over in the case of a complex conjugate pair of eigenvalues reaching criticality as the Rayleigh number is increased, or when there is instability from some Rayleigh number all the way down to zero Rayleigh number (as in the case of a top-heavy arrangement).

A bifurcation problem closely related to that of the two-layer Bénard problem is that of double diffusion, where a layer of fluid is heated from below and there is a salt gradient. This has been studied by Nagata and Thomas [1986]. For suitable parameter values, Hopf bifurcations of time-periodic solutions occur in roll-like, square and hexagonal convection cell patterns. A version of the Center Manifold Theorem is used to prove the existence of the bifurcating time-periodic solutions, and the stability of these solutions is determined from the Poincaré normal form of the reduced equations on the center manifold. They compute for rolls, squares and hexagons the direction of bifurcation (supercritical or subcritical) and the stability of the bifurcating solutions with respect to small perturbations *having the same cellular structure* (that is, perturbations have the same symmetries as the solutions themselves). For example, their results may show that hexagon pattern solutions are locally asymptotically stable with respect to hexagon pattern disturbances, but their analysis does not handle whether more generally they may be unstable with respect to roll-like disturbances. Thus, in their work, the competition between solutions (or pattern selection) is not addressed. Their analysis would have been more complete if a theory of pattern selection for Hopf bifurcation (such as the work of Roberts, Swift and Wagner [1986]) had been available. The application of the latter theory to double diffusion remains to be done.

The bifurcation problem for the oscillatory onset of instability (that is,

Hopf bifurcation) in the two-fluid Bénard problem is discussed in chapter III.7 [Renardy and Renardy 1988]. The solutions will be assumed to be doubly periodic with respect to a hexagonal lattice. This case is of interest from a physical point of view because such solutions are observed in experiments for the one-fluid problem, and from a mathematical point of view because the symmetry of the hexagonal lattice causes a sixfold degeneracy of the critical eigenvalue. This high level of degeneracy leads to many bifurcating solutions, and to an interesting problem of pattern selection. On the abstract level, Hopf bifurcations with this type of symmetry-induced degeneracy have been investigated by Roberts, Swift and Wagner [1986]. They find eleven qualitatively different types of bifurcating solutions and they identify the parameters which determine the stability of these solutions. The equations they discuss are given in a reduced canonical form, and in order to apply their results to our situation, the reduction process must first be manipulated. This process involves the determination of a center manifold and the theory of normal forms. The partial differential equations are reduced to a system of six ordinary differential equations for the complex-valued amplitude functions (this is twice as many as in the one-fluid problem).

The numerical results in chapter III. 7 (d) concern situations suggested by the linear theory in sections III.3 - 6 to exemplify the mechanisms that maintain oscillations on the interface. In chapter III.6, the linear stability problem is studied for the case when the two fluids have nearly equal properties. The results indicate that the factors which are most essential for the stability of the interface are surface tension, density difference accross the interface, and thermal conductivity stratification. The effects of surface tension and density difference are rather obvious, but the influence of thermal conductivity stratification is more subtle. For disturbances of long to moderate wavelength, thermal conductivity stratification is stabilizing if the less conducting fluid is in a thin layer next to one of the walls and destabilizing if the more conducting fluid is in a thin layer. Indeed, thermal conductivity stratification becomes the dominant influence on stability of the interface as the thickness of the layer tends to zero. These results are very much reminiscent of the role of viscosity stratification in shear flows. Roughly speaking, temperature gradient assumes the role of shear, and thermal conductivity assumes the role of viscosity. For short wave disturbances, on the other hand, thermal conductivity stratification is always destabilizing. This is also the case for viscosity stratification in shear flows. With α denoting the wavenumber of the disturbance, the growth rates of short wave instabilities resulting from thermal conductivity stratification behave like α^{-2}. The effects of density difference and surface tension, on the other hand, are of orders α^{-1} and α, respectively, and will thus dominate for large α if they are present. This motivates the choice of the following three cases for the numerical study of criticality and stability of bifurcating solutions:

1. All fluid properties are equal, but there is surface tension between the two fluids.

2. The thermal expansion coefficients differ in such a way that a stabilizing density difference across the interface results.
3. The two fluids have a thermal conductivity stratification and depth ratio which is stabilizing for disturbances of long to moderate wavelengths. We include enough surface tension to stabilize short wave disturbances, but keep surface tension small enough so that thermal conductivity stratification remains the dominant influence at the critical wavelength.

In chapter III.7 (d), the stability of the bifurcating solutions is discussed for each of these cases. Since the eigenfunctions at criticality are not explicitly known in this problem, a combination of analysis and numerical computation is used.

The two-dimensional counterpart of the analysis of the above problem has been done by Ruelle [1973]. There is a twofold degeneracy of the critical eigenvalue (eigenvalues $\pm iw$ for a critical wavenumber α_c, and also for $-\alpha_c$). The possible bifurcating solutions are standing rolls and travelling rolls. If either is subcritical, then both are unstable. If both are supercritical, then one of them is stable.

An energy stability theory for the Bénard problem for two fluids is given by Joseph [1987]. He derives two energy identities, one from the momentum equation and the other from the heat equation. The derivation is essentially similar to that of Davis and Homsy [1980] for the one-fluid problem with a free surface. Joseph writes that the energy equations cannot be used to prove a condition for global (unconditional) stability because there is an interface term which cannot be estimated. This dilemma is typical of two-layer systems and will be revisited in detail in chapter VI, where the energy analysis for core-annular flow is presented. This is in contrast with the one-fluid Bénard problem of Davis and Homsy [1980] where the energy theory does yield critical Rayleigh numbers for conditional stability in certain limiting situations.

III.2 Formulation of Equations

Two fluids are assumed to fill the space between two parallel boundaries of infinite extent in the (x^*, y^*, z^*) plane (see figure 2.1) [Renardy and Renardy 1988]. Asterisks denote dimensional variables. The upper boundary at $z^* = l^*$ is kept at a constant temperature θ_0^*, and the lower boundary at $z^* = 0$ is kept at a higher constant temperature $\theta_1^* = \theta_0^* + \Delta\theta^*$. The location of the interface is given by $z^* = h^*(x^*, y^*, t^*)$. The average value of h^* is denoted by l_1^*: this is the average height of the lower fluid (fluid 1), and that of the upper fluid (fluid 2) is then $l^* - l_1^* = l_2^*$. The velocity

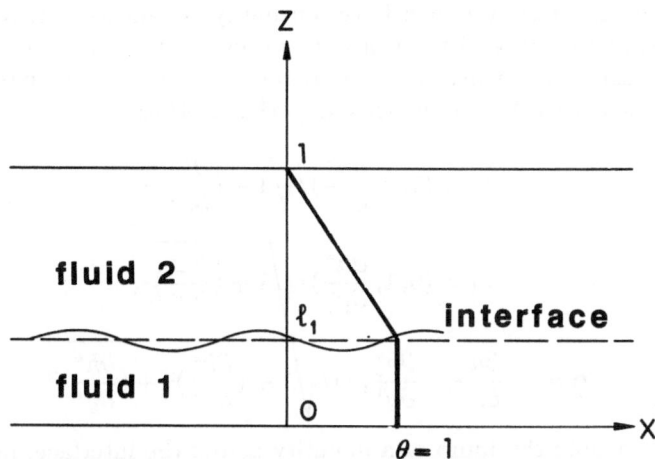

Fig. 2.1. The two-layer Bénard problem in dimensionless variables. The average height of the lower fluid (fluid 1) is l_1, and the average height of the upper fluid is l_2 where $l_1 + l_2 = 1$. This figure shows the basic temperature field in the situation where the thermal conductivity of fluid 1 is much larger than that of fluid 2. The temperature does not vary as much over fluid 1 as it does over fluid 2.

is denoted by $\mathbf{v}^* = (u^*, v^*, w^*)$, the pressure by p^* and the temperature by θ^*.

At temperature θ_0^*, fluid i has coefficient of cubical expansion $\hat{\alpha}_i$, thermal diffusivity κ_i, thermal conductivity k_i, viscosity μ_i, density ρ_i, and kinematic viscosity $\nu_i = \mu_i/\rho_i$. In each fluid, the governing equations are Fourier's law of heat conduction and the Navier-Stokes equations. The Oberbeck-Boussinesq approximation is used; that is, all fluid properties are independent of the temperature, except for the density in the buoyancy term, which is expanded in a Taylor series about the temperature of the top boundary and then truncated. The density here is then approximated as a linear function of temperature, by

$$\rho_i(1 - \hat{\alpha}_i(\theta^* - \theta_0^*)). \tag{2.1}$$

The equations of motion in fluid i read as follows:

$$\dot{\theta}^* + (\mathbf{v}^* \cdot \nabla)\theta^* = \kappa_i \Delta\theta^*,$$

$$\rho_i\big(\dot{\mathbf{v}}^* + (\mathbf{v}^* \cdot \nabla)\mathbf{v}^*\big) = \mu_i\Delta\mathbf{v}^* - \nabla p^* - (\rho_i - \rho_i\hat{\alpha}_i(\theta^* - \theta_0^*))g\mathbf{e}_z, \tag{2.2}$$

$$\mathrm{div}\ \mathbf{v}^* = 0.$$

At the walls, we have the boundary conditions of zero velocity and constant temperature:

$$\mathbf{v}^* = \mathbf{0},\ \theta^* = \theta_1^* \quad \text{at}\ z^* = 0,$$

$$\mathbf{v}^* = \mathbf{0}, \ \theta^* = \theta_0^* \quad \text{at } z^* = l^*. \tag{2.3}$$

At the interface, we must have continuity of velocity, temperature and heat flux, and balance of tractions. To formulate these conditions, we first introduce some notations. Let $\mathbf{t}_1$, $\mathbf{t}_2$ denote two unit vectors parallel to the interface, and let $\mathbf{n}$ be a unit normal to the interface:

$$\mathbf{t}_1 = (1, 0, \frac{\partial h^*}{\partial x^*}) / \sqrt{1 + (\frac{\partial h^*}{\partial x^*})^2},$$

$$\mathbf{t}_2 = (0, 1, \frac{\partial h^*}{\partial y^*}) / \sqrt{1 + (\frac{\partial h^*}{\partial y^*})^2},$$

$$\mathbf{n} = (-\frac{\partial h^*}{\partial x^*}, -\frac{\partial h^*}{\partial y^*}, 1) / \sqrt{1 + (\frac{\partial h^*}{\partial x^*})^2 + (\frac{\partial h^*}{\partial y^*})^2}. \tag{2.4}$$

By $[\![\cdot]\!]$ we denote the jump of a quantity across the interface, i.e. its value in fluid 1 minus its value in fluid 2. Finally, $\mathbf{T}^*$ denotes the stress tensor $\mu[\nabla \mathbf{v}^* + (\nabla \mathbf{v}^*)^T] - p^* \mathbf{1}$.

The conditions to be satisfied across the interface are

$$[\![\mathbf{v}^*]\!] = 0, \ [\![\mathbf{t}_i \cdot \mathbf{T}^* \cdot \mathbf{n}]\!] = 0, \ i = 1, 2,$$

$$[\![\mathbf{n} \cdot \mathbf{T}^* \cdot \mathbf{n}]\!] = \frac{S^* \left\{ \frac{\partial^2 h^*}{\partial x^{*2}}(1 + (\frac{\partial h^*}{\partial y^*})^2) + \frac{\partial^2 h^*}{\partial y^{*2}}(1 + (\frac{\partial h^*}{\partial x^*})^2) - 2\frac{\partial^2 h^*}{\partial x^* \partial y^*}\frac{\partial h^*}{\partial x^*}\frac{\partial h^*}{\partial y^*} \right\}}{(1 + (\nabla h^*)^2)^{3/2}},$$

$$[\![\theta^*]\!] = 0, \ [\![k\mathbf{n} \cdot \nabla \theta^*]\!] = 0,$$

$$\dot{h}^* + u^* \frac{\partial h^*}{\partial x^*} + v^* \frac{\partial h^*}{\partial y^*} = w^*, \tag{2.5}$$

where S^* is the dimensional surface tension coefficient.

There are six dimensionless ratios arising from the fluid properties:

$$m = \frac{\mu_1}{\mu_2},$$

$$r = \frac{\rho_1}{\rho_2},$$

$$\gamma = \frac{\kappa_1}{\kappa_2},$$

$$\zeta = \frac{k_1}{k_2},$$

$$\beta = \frac{\hat{\alpha}_1}{\hat{\alpha}_2},$$

$$l_1 = l_1^*/l^* = 1 - l_2, \quad l_2 = l_2^*/l^*. \tag{2.6}$$

Dimensionless variables (without asterisks) are as follows:

$$(x, y, z) = (x^*, y^*, z^*)/l^*,$$

$$t = \kappa_1 t^*/l^{*2},$$

$$\mathbf{v} = \mathbf{v}^* l^*/\kappa_1,$$

$$\theta = (\theta^* - \theta_0^*)/\Delta\theta^*,$$

$$p = p^* l^{*2}/(\rho_1 \kappa_1^2). \tag{2.7}$$

We define a Rayleigh number

$$R = g\hat{\alpha}_1 \Delta\theta^* l^{*3}/(\kappa_1 \nu_1), \tag{2.8}$$

where g denotes gravitational acceleration, a Prandtl number

$$P = \nu_1/\kappa_1, \tag{2.9}$$

and a dimensionless measure of gravity,

$$G = \frac{g(l^*)^3}{\kappa_1^2}. \tag{2.10}$$

A dimensionless measure of the temperature difference between the plates is

$$\hat{\alpha}_1 \Delta\theta^* = RP/G. \tag{2.11}$$

A dimensionless parameter for surface tension is

$$S = S^* l^*/(\kappa_1 \mu_1). \tag{2.12}$$

Thus, there are ten dimensionless parameters.

Expressed in dimensionless form, the equations in fluid 1 read

$$\dot{\theta} + (\mathbf{v} \cdot \nabla)\theta = \Delta\theta,$$

$$\dot{\mathbf{v}} + (\mathbf{v} \cdot \nabla)\mathbf{v} = P\Delta\mathbf{v} - \nabla p + (RP\theta - G)\mathbf{e}_z, \tag{2.13}$$

$$\operatorname{div} \mathbf{v} = 0.$$

In fluid 2, the equations read:

$$\dot{\theta} + (\mathbf{v} \cdot \nabla)\theta = \frac{1}{\gamma}\Delta\theta,$$

$$\dot{\mathbf{v}} + (\mathbf{v} \cdot \nabla)\mathbf{v} = \frac{r}{m}P\Delta\mathbf{v} - r\nabla p + (\frac{RP}{\beta}\theta - G)\mathbf{e}_z, \tag{2.14}$$

$$\operatorname{div} \mathbf{v} = 0.$$

The boundary conditions at the plates are

$$\mathbf{v} = \mathbf{0}, \ \theta = 1 \text{ at } z = 0,$$

$$\mathbf{v} = \mathbf{0}, \ \theta = 0 \text{ at } z = 1. \tag{2.15}$$

We set

$$h(x, y, t) = h^*(x^*, y^*, t^*)/l^* - l_1. \qquad (2.16)$$

The interface is then at $z = l_1 + h(x, y, t)$, and the conditions to be satisfied at the interface are

$$[\![\mathbf{v}]\!] = 0, \quad [\![\mathbf{t}_i \cdot \mathbf{T} \cdot \mathbf{n}]\!] = 0, \quad i = 1, 2,$$

$$[\![\mathbf{n} \cdot \mathbf{T} \cdot \mathbf{n}]\!] = \frac{PS\left\{ \frac{\partial^2 h}{\partial x^2}\left(1 + \left(\frac{\partial h}{\partial y}\right)^2\right) + \frac{\partial^2 h}{\partial y^2}\left(1 + \left(\frac{\partial h}{\partial x}\right)^2\right) - 2\frac{\partial^2 h}{\partial x \partial y}\frac{\partial h}{\partial x}\frac{\partial h}{\partial y} \right\}}{(1 + (\nabla h)^2)^{3/2}},$$

$$[\![\theta]\!] = 0, \quad [\![k\mathbf{n} \cdot \nabla \theta]\!] = 0,$$

$$\dot{h} + u\frac{\partial h}{\partial x} + v\frac{\partial h}{\partial y} = w. \qquad (2.17)$$

Here we have set the dimensionless stress tensor

$$\mathbf{T} = \begin{cases} P[\nabla \mathbf{v} + (\nabla \mathbf{v})^T] - p\mathbf{1} & \text{in fluid 1,} \\ \frac{P}{m}[\nabla \mathbf{v} + (\nabla \mathbf{v})^T] - p\mathbf{1} & \text{in fluid 2.} \end{cases} \qquad (2.18)$$

A trivial solution to the problem is given by

$$h = 0, \quad \mathbf{v} = \mathbf{0},$$

$$\theta = \begin{cases} 1 - A_1 z & \text{for } 0 \le z \le l_1, \\ A_2(1 - z) & \text{for } l_1 \le z \le 1, \end{cases} \qquad (2.19)$$

$$p = \begin{cases} p_1 - Gz + RPz - RPA_1 z^2/2 & \text{for } 0 \le z \le l_1, \\ p_2 - Gz/r + RPA_2 z/(r\beta) - RPA_2 z^2/(2r\beta) & \text{for } l_1 \le z \le 1, \end{cases}$$

where

$$A_1 = \frac{1}{l_1 + \zeta l_2}, \qquad A_2 = \zeta A_1, \qquad (2.20)$$

and $p_1 - p_2$ must be chosen such that p is continuous at $z = l_1$. Figure 2.1 shows the basic temperature field, and illustrates the role played by the thermal conductivities.

We are concerned with the stability of this trivial solution and with solutions bifurcating from it.

We denote by $\tilde{\theta}$ the difference between θ and the trivial solution (2.19), and by $\tilde{p}$ the difference between p and the trivial solution. This changes the equations in fluid 1 to

$$\dot{\tilde{\theta}} + (\mathbf{v} \cdot \nabla)\tilde{\theta} - A_1 w = \Delta\tilde{\theta},$$

$$\dot{\mathbf{v}} + (\mathbf{v} \cdot \nabla)\mathbf{v} = P\Delta\mathbf{v} - \nabla\tilde{p} + RP\tilde{\theta}\mathbf{e}_z, \qquad (2.21)$$

$$\text{div } \mathbf{v} = 0.$$

In fluid 2, the equations of motion become

$$\dot{\tilde{\theta}} + (\mathbf{v} \cdot \nabla)\tilde{\theta} - A_2 w = \frac{1}{\gamma}\Delta\tilde{\theta},$$

$$\dot{\mathbf{v}} + (\mathbf{v} \cdot \nabla)\mathbf{v} = \frac{r}{m} P \Delta \mathbf{v} - r \nabla \tilde{p} + \frac{RP}{\beta} \tilde{\theta} \mathbf{e}_z, \qquad (2.22)$$

$$\operatorname{div} \mathbf{v} = 0.$$

The boundary conditions on both walls are now

$$\mathbf{v} = \mathbf{0}, \ \tilde{\theta} = 0, \qquad (2.23)$$

and the interface conditions are

$$[\![\mathbf{v}]\!] = \mathbf{0}, \ [\![\mathbf{t}_i \cdot \tilde{\mathbf{T}} \cdot \mathbf{n}]\!] = 0, \ i = 1, 2,$$

$$[\![\mathbf{n} \cdot \tilde{\mathbf{T}} \cdot \mathbf{n}]\!] = -M_1 h + M_2 h^2$$

$$+ \frac{PS\left\{ \frac{\partial^2 h}{\partial x^2} \left(1 + (\frac{\partial h}{\partial y})^2\right) + \frac{\partial^2 h}{\partial y^2} \left(1 + (\frac{\partial h}{\partial x})^2\right) - 2 \frac{\partial^2 h}{\partial x \partial y} \frac{\partial h}{\partial x} \frac{\partial h}{\partial y} \right\}}{(1 + (\nabla h)^2)^{3/2}},$$

$$[\![\tilde{\theta}]\!] = h(A_1 - A_2), \ [\![k \mathbf{n} \cdot \nabla \tilde{\theta}]\!] = 0,$$

$$\dot{h} + u \frac{\partial h}{\partial x} + v \frac{\partial h}{\partial y} = w. \qquad (2.24)$$

Here we have set

$$\tilde{\mathbf{T}} = \begin{cases} P[\nabla \mathbf{v} + (\nabla \mathbf{v})^T] - \tilde{p}\mathbf{1} & \text{in fluid 1,} \\ \frac{P}{m}[\nabla \mathbf{v} + (\nabla \mathbf{v})^T] - \tilde{p}\mathbf{1} & \text{in fluid 2,} \end{cases} \qquad (2.25)$$

$$M_1 = G(1 - \frac{1}{r}) + RPA_2 l_2 (\frac{1}{r\beta} - 1),$$

and

$$M_2 = \frac{RP}{2} \left(\frac{A_2}{r\beta} - A_1 \right). \qquad (2.26)$$

One of the features which make the two-layer problem complicated is the fact that the domain occupied by each fluid is unknown and the interface must be determined as part of the solution. In theoretical studies on free-surface problems, e.g., existence proofs [Beale 1980, 1984; M. Renardy and Joseph 1986], it is customary to introduce a mapping which transforms the domain occupied by each fluid to a fixed domain. Thus the partial differential equations are posed on a known domain; however, the coefficients in the transformed equations become rather complicated. This makes it cumbersome to apply the method in actual calculations. However, the perturbation h of the interface position is small, and we are only concerned with the terms of leading order in an asymptotic expansion which exploits this smallness. This makes a much simpler approach possible. We shall solve the equations on the unperturbed domains $0 \le z \le l_1$ and $l_1 \le z \le 1$, respectively. We think of the solutions on the unperturbed domains as being smoothly extended up to the interface $z = l_1 + h$. We then pose interface conditions at $z = l_1$ which result from a Taylor expansion of the conditions at the interface with respect to h, e.g.

$$[\tilde{\theta}](z = l_1 + h) = [\tilde{\theta}](z = l_1) + h[\tilde{\theta}_z](z = l_1) + \frac{h^2}{2}[\tilde{\theta}_{zz}](z = l_1) + \dots$$

$$= h(A_1 - A_2). \tag{2.27}$$

These interface conditions are then not satisfied exactly, but only to within a certain order of the small parameter. For more information on domain perturbations, see Joseph [1973, 1990 a] and Lebovitz [1982].

For the three-dimensional bifurcation analysis of section 7 below, the equations of motion and interface conditions will need to be restated, with terms ordered according to the degree of nonlinearity. Since only terms up to third degree will be used in the computations, the equations are truncated after the cubic terms. The equations of motion for $0 \le z \le l_1$ are then

$$\dot{\tilde{\theta}} - A_1 w - \Delta\tilde{\theta} = H_1 = -(\mathbf{v} \cdot \nabla)\tilde{\theta},$$

$$\dot{\mathbf{v}} - P\Delta\mathbf{v} + \nabla\tilde{p} - RP\tilde{\theta}\mathbf{e}_z = (H_2, H_3, H_4) = -(\mathbf{v} \cdot \nabla)\mathbf{v}, \tag{2.28}$$

$$\text{div } \mathbf{v} = 0,$$

where the H_i denote the right hand sides. For $l_1 \le z \le 1$ we have

$$\dot{\tilde{\theta}} - A_2 w - \frac{1}{\gamma}\Delta\tilde{\theta} = H_5 = -(\mathbf{v} \cdot \nabla)\tilde{\theta},$$

$$\dot{\mathbf{v}} - \frac{r}{m}P\Delta\mathbf{v} + r\nabla\tilde{p} - \frac{RP}{\beta}\tilde{\theta}\mathbf{e}_z = (H_6, H_7, H_8) = -(\mathbf{v} \cdot \nabla)\mathbf{v}, \tag{2.29}$$

$$\text{div } \mathbf{v} = 0.$$

The boundary conditions at the walls $z = 0$ and $z = 1$ are

$$\mathbf{v} = \mathbf{0}, \ \tilde{\theta} = 0, \tag{2.30}$$

and the interface conditions at $z = l_1$, truncated at third order, are as follows (subscripts 1 or 2 on the velocity components and pressure refer to the values for $z = l_1-$ and $z = l_1+$, respectively).

Continuity of velocity yields

$$[\![\mathbf{v}]\!] = (H_9, H_{10}, H_{11}) = -h[\![\mathbf{v}_z]\!] - \frac{1}{2}h^2[\![\mathbf{v}_{zz}]\!]. \tag{2.31}$$

Balance of shear stresses yields

$$P(\frac{\partial u_1}{\partial z} + \frac{\partial w_1}{\partial x}) - \frac{P}{m}(\frac{\partial u_2}{\partial z} + \frac{\partial w_2}{\partial x}) = H_{12}, \tag{2.32a}$$

$$P(\frac{\partial v_1}{\partial z} + \frac{\partial w_1}{\partial y}) - \frac{P}{m}(\frac{\partial v_2}{\partial z} + \frac{\partial w_2}{\partial y}) = H_{13}, \tag{2.32b}$$

where the nonlinearities are

$$H_{12} = -Ph(\frac{\partial^2 u_1}{\partial z^2} + \frac{\partial^2 w_1}{\partial x\partial z}) + \frac{P}{m}h(\frac{\partial^2 u_2}{\partial z^2} + \frac{\partial^2 w_2}{\partial x\partial z})$$

$$-2P\frac{\partial h}{\partial x}(\frac{\partial w_1}{\partial z} - \frac{\partial u_1}{\partial x})$$

$$+2\frac{P}{m}\frac{\partial h}{\partial x}(\frac{\partial w_2}{\partial z} - \frac{\partial u_2}{\partial x}) + P\frac{\partial h}{\partial y}(\frac{\partial u_1}{\partial y} + \frac{\partial v_1}{\partial x}) - \frac{P}{m}\frac{\partial h}{\partial y}(\frac{\partial u_2}{\partial y} + \frac{\partial v_2}{\partial x})$$

$$+P(\frac{\partial h}{\partial x})^2(\frac{\partial u_1}{\partial z} + \frac{\partial w_1}{\partial x}) - \frac{P}{m}(\frac{\partial h}{\partial x})^2(\frac{\partial u_2}{\partial z} + \frac{\partial w_2}{\partial x})$$

$$+P\frac{\partial h}{\partial x}\frac{\partial h}{\partial y}(\frac{\partial v_1}{\partial z} + \frac{\partial w_1}{\partial y}) - \frac{P}{m}\frac{\partial h}{\partial x}\frac{\partial h}{\partial y}(\frac{\partial v_2}{\partial z} + \frac{\partial w_2}{\partial y})$$

$$-P\frac{h^2}{2}(\frac{\partial^3 u_1}{\partial z^3} + \frac{\partial^3 w_1}{\partial x\partial z^2}) + \frac{P}{m}\frac{h^2}{2}(\frac{\partial^3 u_2}{\partial z^3} + \frac{\partial^3 w_2}{\partial x\partial z^2})$$

$$-2Ph\frac{\partial h}{\partial x}(\frac{\partial^2 w_1}{\partial z^2} - \frac{\partial^2 u_1}{\partial x\partial z}) + 2\frac{P}{m}h\frac{\partial h}{\partial x}(\frac{\partial^2 w_2}{\partial z^2} - \frac{\partial^2 u_2}{\partial x\partial z})$$

$$+Ph\frac{\partial h}{\partial y}(\frac{\partial^2 u_1}{\partial y\partial z} + \frac{\partial^2 v_1}{\partial x\partial z}) - \frac{P}{m}h\frac{\partial h}{\partial y}(\frac{\partial^2 u_2}{\partial y\partial z} + \frac{\partial^2 v_2}{\partial x\partial z}), \qquad (2.33)$$

and

$$H_{13} = -Ph(\frac{\partial^2 v_1}{\partial z^2} + \frac{\partial^2 w_1}{\partial y\partial z}) + \frac{P}{m}h(\frac{\partial^2 v_2}{\partial z^2} + \frac{\partial^2 w_2}{\partial y\partial z})$$

$$-2P\frac{\partial h}{\partial y}(\frac{\partial w_1}{\partial z} - \frac{\partial v_1}{\partial y})$$

$$+2\frac{P}{m}\frac{\partial h}{\partial y}(\frac{\partial w_2}{\partial z} - \frac{\partial v_2}{\partial y}) + P\frac{\partial h}{\partial x}(\frac{\partial u_1}{\partial y} + \frac{\partial v_1}{\partial x}) - \frac{P}{m}\frac{\partial h}{\partial x}(\frac{\partial u_2}{\partial y} + \frac{\partial v_2}{\partial x})$$

$$+P(\frac{\partial h}{\partial y})^2(\frac{\partial v_1}{\partial z} + \frac{\partial w_1}{\partial y}) - \frac{P}{m}(\frac{\partial h}{\partial y})^2(\frac{\partial v_2}{\partial z} + \frac{\partial w_2}{\partial y})$$

$$+P\frac{\partial h}{\partial x}\frac{\partial h}{\partial y}(\frac{\partial u_1}{\partial z} + \frac{\partial w_1}{\partial x}) - \frac{P}{m}\frac{\partial h}{\partial x}\frac{\partial h}{\partial y}(\frac{\partial u_2}{\partial z} + \frac{\partial w_2}{\partial x})$$

$$-P\frac{h^2}{2}(\frac{\partial^3 v_1}{\partial z^3} + \frac{\partial^3 w_1}{\partial y\partial z^2}) + \frac{P}{m}\frac{h^2}{2}(\frac{\partial^3 v_2}{\partial z^3} + \frac{\partial^3 w_2}{\partial y\partial z^2})$$

$$-2Ph\frac{\partial h}{\partial y}(\frac{\partial^2 w_1}{\partial z^2} - \frac{\partial^2 v_1}{\partial y\partial z}) + 2\frac{P}{m}h\frac{\partial h}{\partial y}(\frac{\partial^2 w_2}{\partial z^2} - \frac{\partial^2 v_2}{\partial y\partial z})$$

$$+Ph\frac{\partial h}{\partial x}(\frac{\partial^2 u_1}{\partial y\partial z} + \frac{\partial^2 v_1}{\partial x\partial z}) - \frac{P}{m}h\frac{\partial h}{\partial x}(\frac{\partial^2 u_2}{\partial y\partial z} + \frac{\partial^2 v_2}{\partial x\partial z}). \qquad (2.34)$$

Normal stress balance yields

$$2P\frac{\partial w_1}{\partial z} - 2\frac{P}{m}\frac{\partial w_2}{\partial z} - \tilde{p}_1 + \tilde{p}_2 + M_1 h - PS(\frac{\partial^2 h}{\partial x^2} + \frac{\partial^2 h}{\partial y^2}) = H_{14}, \qquad (2.35)$$

where

$$H_{14} = M_2 h^2 - 2Ph\frac{\partial^2 w_1}{\partial z^2} + 2\frac{P}{m}h\frac{\partial^2 w_2}{\partial z^2} + h\frac{\partial \tilde{p}_1}{\partial z} - h\frac{\partial \tilde{p}_2}{\partial z}$$

$$+2P\frac{\partial h}{\partial x}\left(\frac{\partial w_1}{\partial x}+\frac{\partial u_1}{\partial z}\right)-2\frac{P}{m}\frac{\partial h}{\partial x}\left(\frac{\partial w_2}{\partial x}+\frac{\partial u_2}{\partial z}\right)$$

$$+2P\frac{\partial h}{\partial y}\left(\frac{\partial w_1}{\partial y}+\frac{\partial v_1}{\partial z}\right)-2\frac{P}{m}\frac{\partial h}{\partial y}\left(\frac{\partial w_2}{\partial y}+\frac{\partial v_2}{\partial z}\right)$$

$$-\left(\frac{\partial h}{\partial x}\right)^2\left(2P\frac{\partial u_1}{\partial x}-\tilde{p}_1-2\frac{P}{m}\frac{\partial u_2}{\partial x}+\tilde{p}_2\right)$$

$$-\left(\frac{\partial h}{\partial y}\right)^2\left(2P\frac{\partial v_1}{\partial y}-\tilde{p}_1-2\frac{P}{m}\frac{\partial v_2}{\partial y}+\tilde{p}_2\right)$$

$$-2\frac{\partial h}{\partial x}\frac{\partial h}{\partial y}\left(P\left(\frac{\partial u_1}{\partial y}+\frac{\partial v_1}{\partial x}\right)-\frac{P}{m}\left(\frac{\partial u_2}{\partial y}+\frac{\partial v_2}{\partial x}\right)\right)$$

$$-Ph^2\frac{\partial^3 w_1}{\partial z^3}+\frac{P}{m}h^2\frac{\partial^3 w_2}{\partial z^3}+\frac{h^2}{2}\frac{\partial^2\tilde{p}_1}{\partial z^2}-\frac{h^2}{2}\frac{\partial^2\tilde{p}_2}{\partial^2 z}$$

$$+2Ph\frac{\partial h}{\partial x}\left(\frac{\partial^2 w_1}{\partial x\partial z}+\frac{\partial^2 u_1}{\partial z^2}\right)-2\frac{P}{m}h\frac{\partial h}{\partial x}\left(\frac{\partial^2 w_2}{\partial x\partial z}+\frac{\partial^2 u_2}{\partial z^2}\right)$$

$$+2Ph\frac{\partial h}{\partial y}\left(\frac{\partial^2 w_1}{\partial y\partial z}+\frac{\partial^2 v_1}{\partial z^2}\right)-2\frac{P}{m}h\frac{\partial h}{\partial y}\left(\frac{\partial^2 w_2}{\partial y\partial z}+\frac{\partial^2 v_2}{\partial z^2}\right)$$

$$+\frac{PS}{2}\frac{\partial^2 h}{\partial x^2}\left(\left(\frac{\partial h}{\partial y}\right)^2-\left(\frac{\partial h}{\partial x}\right)^2\right)+\frac{PS}{2}\frac{\partial^2 h}{\partial y^2}\left(\left(\frac{\partial h}{\partial x}\right)^2-\left(\frac{\partial h}{\partial y}\right)^2\right)$$

$$-2PS\frac{\partial^2 h}{\partial x\partial y}\frac{\partial h}{\partial x}\frac{\partial h}{\partial y}-M_1 h\left(\left(\frac{\partial h}{\partial x}\right)^2+\left(\frac{\partial h}{\partial y}\right)^2\right). \tag{2.36}$$

Continuity of temperature yields

$$[\![\tilde{\theta}]\!]-h(A_1-A_2)=H_{15}=-h[\![\tilde{\theta}_z]\!]-\frac{1}{2}h^2[\![\tilde{\theta}_{zz}]\!]. \tag{2.37}$$

Continuity of heat flux yields

$$\frac{\partial\tilde{\theta}_1}{\partial z}-\frac{1}{\zeta}\frac{\partial\tilde{\theta}_2}{\partial z}=H_{16}=-h\left(\frac{\partial^2\tilde{\theta}_1}{\partial z^2}-\frac{1}{\zeta}\frac{\partial^2\tilde{\theta}_2}{\partial z^2}\right)$$

$$+\frac{\partial h}{\partial x}\left(\frac{\partial\tilde{\theta}_1}{\partial x}-\frac{1}{\zeta}\frac{\partial\tilde{\theta}_2}{\partial x}\right)+\frac{\partial h}{\partial y}\left(\frac{\partial\tilde{\theta}_1}{\partial y}-\frac{1}{\zeta}\frac{\partial\tilde{\theta}_2}{\partial y}\right)$$

$$-\frac{1}{2}h^2\left(\frac{\partial^3\tilde{\theta}_1}{\partial z^3}-\frac{1}{\zeta}\frac{\partial^3\tilde{\theta}_2}{\partial z^3}\right)+h\frac{\partial h}{\partial x}\left(\frac{\partial^2\tilde{\theta}_1}{\partial z\partial x}-\frac{1}{\zeta}\frac{\partial^2\tilde{\theta}_2}{\partial z\partial x}\right)+h\frac{\partial h}{\partial y}\left(\frac{\partial^2\tilde{\theta}_1}{\partial z\partial y}-\frac{1}{\zeta}\frac{\partial^2\tilde{\theta}_2}{\partial z\partial y}\right). \tag{2.38}$$

The kinematic free surface condition is

$$\dot{h}-w_1=H_{17}=h\frac{\partial w_1}{\partial z}-\frac{\partial h}{\partial x}u_1-\frac{\partial h}{\partial y}v_1$$

$$+\frac{1}{2}h^2\frac{\partial^2 w_1}{\partial z^2}-h\frac{\partial h}{\partial x}\frac{\partial u_1}{\partial z}-h\frac{\partial h}{\partial y}\frac{\partial v_1}{\partial z}. \tag{2.39}$$

III.3 Linearized Stability Problem

III.3(a) The Governing Equations

The following two-dimensional linearized eigenvalue problem arises from equations (2.28) - (2.39) for the velocity $\mathbf{u} = (u, w)$, perturbation $\Theta(x, z)$ to the temperature, and perturbation p to the pressure. Their dependence on x and t is assumed to be through the factor $\exp(i\alpha x + \sigma t)$.

For $0 \leq z \leq l_1$,

$$\sigma\Theta = wA_1 + \Delta\Theta,$$

$$\sigma u = -\partial p/\partial x + P\Delta u,$$

$$\sigma w = -\partial p/\partial z + RP\Theta + P\Delta w, \qquad (3a.1)$$

$$\frac{\partial u}{\partial x} + \frac{\partial w}{\partial z} = 0.$$

For $l_1 \leq z \leq 1$,

$$\sigma\Theta = wA_2 + \frac{1}{\gamma}\Delta\Theta,$$

$$\sigma u = -r\frac{\partial p}{\partial x} + \frac{r}{m}P\Delta u,$$

$$\sigma w = -r\frac{\partial p}{\partial z} + \frac{RP}{\beta}\Theta + \frac{r}{m}P\Delta w, \qquad (3a.2)$$

$$\frac{\partial u}{\partial x} + \frac{\partial w}{\partial z} = 0.$$

At $z = 0, 1$, the no-slip boundary condition holds and the temperatures are fixed:

$$\Theta = u = w = 0. \qquad (3a.3)$$

The interface conditions linearized about $z=l_1$ are:

continuity of temperature : $[\![\Theta]\!] = h[\![A]\!],$

continuity of heat flux : $[\![k\frac{\partial\Theta}{\partial z}]\!] = 0,$

continuity of velocity : $[\![w]\!] = [\![u]\!] = 0,$

continuity of shear stress : $[\![\mu(\frac{\partial u}{\partial z} + \frac{\partial w}{\partial x})]\!] = 0,$

balance of normal stress:

$$p_2 - p_1 + 2P(\frac{\partial w_1}{\partial z} - \frac{1}{m}\frac{\partial w_2}{\partial z}) + M_1 h - PS\frac{\partial^2 h}{\partial x^2} = 0,$$

and kinematic free surface condition : $\sigma h = w_1$.   (3a.4)

The conservation of volume of the incompressible fluids implies that

$$h(x,t) = h_0 \; exp(i\alpha x + \sigma t)$$

has a zero mean value as a function of x. This is automatic if $\alpha \neq 0$. There is a difference between $\alpha = 0$ and $\alpha \to 0$; the former is disallowed.

In the eigenvalue problem of equations (3a.1) - (3a.4), the perturbation to the interface position is coupled to the eigenvalue through the kinematic interface condition. Thus, if the interface is not allowed to deform, the eigenfunctions would not be expected to be oscillatory in time. An example of such a situation is the Bénard problem with two fluid layers having identical physical properties separated by a rigid plate [Catton and Lienhard V 1984]. The plate location, as well as the nonzero thickness of the plate and fluid layers, is arbitrary. The top and bottom boundaries are rigid and perfectly conducting. They prove that no oscillatory onset is possible. Their proof can be extended to the case where the two fluid layers have different densities, viscosities, thermal diffusivities, thermal expansivities and thermal conductivities. The case of stress-free upper and lower external boundaries is also accomodated. Then, in the limit of vanishingly small plate thickness, their situation is identical to the two-layer problem addressed here, except for the nondeformability of the interface.

III.3(b) The Adjoint Equations

We will show that if we proceed in a natural way, then the adjoint equations for the problem (3a.1) - (3a.4) are not the same as the original equations. This indicates that the eigenvalues need not be real. Let Ω be a strip of width one wavelength $2\pi/\alpha$, covering $0 \leq z \leq 1$. Let Ω_1 be the part of Ω in fluid 1 and Ω_2 be in fluid 2. Let $\bar{\mathbf{u}}^*$, $\bar{\Theta}^*$ and $\bar{p}^*$ be the complex conjugates of the adjoints of $\mathbf{u}$, Θ and p. The asterisks here denote the adjoint and the overbars the complex conjugates.

It is natural to proceed with the integration by parts of

$$\int_{\Omega_1} \bar{\mathbf{u}}^* \cdot (\sigma \mathbf{u} + \nabla p - P \triangle \mathbf{u} - RP\Theta \mathbf{e_z})$$

$$+ \int_{\Omega_2} \bar{\mathbf{u}}^* \cdot (\frac{\sigma}{r}\mathbf{u} + \nabla p - \frac{P}{m} \triangle \mathbf{u} - \frac{RP}{r\beta}\Theta \mathbf{e_z})$$

$$+ \int_{\Omega_1} \bar{\Theta}^*(\sigma\Theta - wA_1 - \triangle\Theta) + \int_{\Omega_2} \frac{\bar{\Theta}^*}{\zeta}(\gamma(\sigma\Theta - wA_2) - \triangle\Theta)$$

$$- \int_{\Omega} \bar{p}^* \nabla \cdot \mathbf{u}. \qquad (3b.1)$$

The last term, arising from the incompressibility condition, actually contributes zero; integrating this by parts yields the integral of $\mathbf{u} \cdot \nabla \bar{p}^*$ over Ω and a boundary integral of $-[\![\bar{p}^* w]\!]$. Equation (3b.1) is equal to

$$\int_{\Omega_1} \mathbf{u} \cdot \left(\sigma \bar{\mathbf{u}}^* - P(\Delta \bar{\mathbf{u}}^* + \nabla(\nabla \cdot \bar{\mathbf{u}}^*)) - A_1 \bar{\Theta}^* \mathbf{e_z} \right)$$

$$- \int_{\Omega} p \nabla \cdot \bar{\mathbf{u}}^*$$

$$+ \int_{\Omega_2} \mathbf{u} \cdot \left(\frac{\sigma}{r} \bar{\mathbf{u}}^* - \frac{P}{m}(\Delta \bar{\mathbf{u}}^* + \nabla(\nabla \cdot \bar{\mathbf{u}}^*)) - \frac{A_2 \gamma}{\zeta} \bar{\Theta}^* \mathbf{e_z} \right)$$

$$+ \int_{\Omega_1} \Theta \left(\sigma \bar{\Theta}^* - PR\bar{w}^* - \Delta \bar{\Theta}^* \right) + \int_{\Omega_2} \Theta \left(\frac{\gamma \sigma}{\zeta} \bar{\Theta}^* - \frac{PR}{r\beta} \bar{w}^* - \frac{1}{\zeta} \Delta \bar{\Theta}^* \right)$$

$$+ \int_{\Omega} \mathbf{u} \cdot \nabla \bar{p}^* - B, \tag{3b.2}$$

where B consists of boundary integrals taken at $z = l_1$ over one wavelength in x. The expression for B is

$$\int -[\![p\bar{w}^*]\!] + P\left(\bar{u}_1^* (\frac{\partial u_1}{\partial z} + \frac{\partial w_1}{\partial x}) - \frac{\bar{u}_2^*}{m}(\frac{\partial u_2}{\partial z} + \frac{\partial w_2}{\partial x}) \right.$$

$$+ \frac{u_2}{m}(\frac{\partial \bar{u}_2^*}{\partial z} + \frac{\partial \bar{w}_2^*}{\partial x}) - u_1(\frac{\partial \bar{u}_1^*}{\partial z} + \frac{\partial \bar{w}_1^*}{\partial x})$$

$$\left. + 2\bar{w}_1^* \frac{\partial w_1}{\partial z} - \frac{2}{m} \bar{w}_2^* \frac{\partial w_2}{\partial z} - 2w_1 \frac{\partial \bar{w}_1^*}{\partial z} + \frac{2}{m} w_2 \frac{\partial \bar{w}_2^*}{\partial z} \right)$$

$$+ \bar{\Theta}_1^* \frac{\partial \Theta_1}{\partial z} - \frac{\bar{\Theta}_2^*}{\zeta} \frac{\partial \Theta_2}{\partial z} + \frac{\Theta_2}{\zeta} \frac{\partial \bar{\Theta}_2^*}{\partial z} - \Theta_1 \frac{\partial \bar{\Theta}_1^*}{\partial z} + [\![\bar{p}^* w]\!] \quad dx$$

$$= \int -(p_1 - 2P\frac{\partial w_1}{\partial z})\bar{w}_1^* + (p_2 - \frac{2P}{m}\frac{\partial w_2}{\partial z})\bar{w}_2^* - 2P(w_1 \frac{\partial \bar{w}_1^*}{\partial z} - \frac{w_2}{m}\frac{\partial \bar{w}_2^*}{\partial z})$$

$$+ P\left([\![\bar{u}^*]\!](\frac{\partial u_1}{\partial z} + \frac{\partial w_1}{\partial x}) + u_1 \left(\frac{1}{m}(\frac{\partial \bar{u}_2^*}{\partial z} + \frac{\partial \bar{w}_2^*}{\partial x}) - (\frac{\partial \bar{u}_1^*}{\partial z} + \frac{\partial \bar{w}_1^*}{\partial x}) \right) \right)$$

$$+ [\![\bar{\Theta}^*]\!]\frac{\partial \Theta_1}{\partial z} - (\Theta_1 \frac{\partial \bar{\Theta}_1^*}{\partial z} - \frac{\Theta_2}{\zeta}\frac{\partial \bar{\Theta}_2^*}{\partial z}) + [\![\bar{p}^* w]\!] \quad dx. \tag{3b.3}$$

The integration in (3b.1) - (3b.2) is facilitated by expressing

$$\int_{\Omega_i} \Delta \mathbf{u} \cdot \bar{\mathbf{u}}^* \tag{3b.4}$$

as

$$\int_{\Omega_i} \bar{\mathbf{u}}^* \cdot \nabla \cdot \left(\nabla \mathbf{u} + (\nabla \mathbf{u})^T \right) \tag{3b.5}$$

where the superscript T denotes the transpose, taking advantage of $\nabla \cdot \mathbf{u} = 0$, to obtain, for example,

$$\int_{\Omega_1} \triangle \mathbf{u} \cdot \bar{\mathbf{u}}^* = \int_{\Omega_1} \mathbf{u} \cdot \triangle \bar{\mathbf{u}}^* + (\mathbf{u} \cdot \nabla)(\nabla \cdot \bar{\mathbf{u}}^*)$$

$$+ \int_{z=l_1,\ x=0}^{x=2\pi/\alpha} \left(\bar{u}^* (\frac{\partial u}{\partial z} + \frac{\partial w}{\partial x}) + 2\bar{w}^* \frac{\partial w}{\partial z} - u(\frac{\partial \bar{u}^*}{\partial z} + \frac{\partial \bar{w}^*}{\partial x}) - 2w \frac{\partial \bar{w}^*}{\partial z} \right) dx.$$

$$(3b.6)$$

Choosing $\mathbf{u}$, Θ, p, and its derivatives to vanish in the neighborhood of the interface, we find

$$\nabla \cdot \bar{\mathbf{u}}^* = 0 \tag{3b.7}$$

and other adjoint equations. Hence, in fluid 1,

$$\sigma \bar{u}^* - P \triangle \bar{u}^* - A_1 \bar{\Theta}^* \mathbf{e_z} = -\nabla \bar{p}^*,$$

$$\sigma \bar{\Theta}^* - PR\bar{w}^* - \triangle \bar{\Theta}^* = 0, \tag{3b.8}$$

and in fluid 2,

$$\frac{\sigma}{r} \bar{\mathbf{u}}^* - \frac{P}{m} \triangle \bar{u}^* - \frac{A_2\gamma}{\zeta} \bar{\Theta}^* \mathbf{e_z} = -\nabla \bar{p}^*,$$

$$\frac{\sigma\gamma}{\zeta} \bar{\Theta}^* - \frac{PR}{r\beta} \bar{w}^* - \frac{1}{\zeta} \triangle \bar{\Theta}^* = 0. \tag{3b.9}$$

We examine B to find the adjoint interface conditions. Together with the continuity of heat flux, velocity and shear stress in equation (3a.4), the latter terms in B in equation (3b.3) yield

$$[\bar{u}^*] = [\bar{\Theta}^*] = [\mu(\frac{\partial \bar{u}^*}{\partial z} + \frac{\partial \bar{w}^*}{\partial x})] = 0. \tag{3b.10}$$

Therefore,

$$B = \int -(p_1 - 2P\frac{\partial w_1}{\partial z})\bar{w}_1^* + (p_2 - \frac{2P}{m}\frac{\partial w_2}{\partial z})\bar{w}_2^*$$

$$+ w_1(\bar{p}_1^* - 2P\frac{\partial \bar{w}_1^*}{\partial z}) - w_2(\bar{p}_2^* - \frac{2P}{m}\frac{\partial \bar{w}_2^*}{\partial z}) - (\Theta_1\frac{\partial \bar{\Theta}_1^*}{\partial z} - \frac{\Theta_2}{\zeta}\frac{\partial \bar{\Theta}_2^*}{\partial z}) \quad dx. \tag{3b.11}$$

The last two terms in the normal stress balance are $M_1 h - PSd^2 h/dx^2$. The h and d^2h/dx^2 are eliminated as follows. By continuity of temperature, $h = [\Theta]/[A]$, and $d^2h/dx^2 = -\alpha^2 h = -\alpha^2[\Theta]/[A] = [\partial^2\Theta/\partial x^2]/[A]$. Thus, the normal stress balance is

$$p_2 - p_1 + 2P(\frac{\partial w_1}{\partial z} - \frac{1}{m}\frac{\partial w_2}{\partial z}) + \frac{1}{[A]}([\Theta]M_1 - PS[\frac{\partial^2\Theta}{\partial x^2}]) = 0. \tag{3b.12}$$

This is used to obtain

$$B = \int w_1 \left((\bar{p}_1^* - 2P\frac{\partial \bar{w}_1^*}{\partial z}) - (\bar{p}_2^* - \frac{2P}{m}\frac{\partial \bar{w}_2^*}{\partial z}) \right)$$

$$-[\bar{w}^*]\left(p_1 - 2P\frac{\partial w_1}{\partial z} - \Theta_1\frac{M_1}{[A]} + \frac{\partial^2\Theta_1}{\partial x^2}\frac{PS}{[A]}\right)$$

$$-\frac{1}{[A]}[\bar{w}^*(\Theta M_1 - \frac{\partial^2\Theta}{\partial x^2}PS)] - \Theta_1\frac{\partial\bar{\Theta}_1^*}{\partial z} + \frac{\Theta_2}{\zeta}\frac{\partial\bar{\Theta}_2^*}{\partial z} \quad dx. \tag{3b.13}$$

We choose

$$[\bar{w}^*] = 0, \tag{3b.14}$$

and use the kinematic free surface condition to write $w_1 = \sigma h = \sigma[\Theta]/[A]$, to find

$$B = \int \frac{[\Theta]}{[A]}\left(\sigma([\bar{p}^*] - 2P\frac{\partial\bar{w}_1^*}{\partial z} + \frac{2P}{m}\frac{\partial\bar{w}_2^*}{\partial z}) - \bar{w}_1^*M_1 + \frac{\partial^2\bar{w}_1^*}{\partial x^2}PS - [A]\frac{\partial\bar{\Theta}_1^*}{\partial z}\right)$$

$$-\Theta_1(\frac{\partial\bar{\Theta}_1^*}{\partial z} - \frac{1}{\zeta}\frac{\partial\bar{\Theta}_2^*}{\partial z}) \quad dx. \tag{3b.15}$$

We choose

$$\frac{\partial\bar{\Theta}_1^*}{\partial z} - \frac{1}{\zeta}\frac{\partial\bar{\Theta}_2^*}{\partial z} = 0 \tag{3b.16}$$

and

$$\sigma([\bar{p}^*] - 2P\frac{\partial\bar{w}_1^*}{\partial z} + \frac{2P}{m}\frac{\partial\bar{w}_2^*}{\partial z}) - \bar{w}_1^*M_1 + \frac{\partial^2\bar{w}_1^*}{\partial x^2}PS - [A]\frac{\partial\bar{\Theta}_1^*}{\partial z} = 0. \tag{3b.17}$$

Equations (3b.7) - (3b.10), (3b.14), (3b.16), (3b.17) are equations adjoint to the linearized system (3a.1) - (3a.4).

III.3(c) Numerical Scheme

A spectral method, namely the Chebyshev-tau method [Orszag 1971; Gottlieb and Orszag 1983; Renardy and Joseph 1985b], is used to discretize the equations in the z-direction. This approximates the eigenvalues for C^∞-eigenfunctions with infinite-order accuracy. To facilitate this method, the variable z is changed to z_i in fluid i, defined by

$$z_1 = \frac{2}{l_1}z - 1, \quad z_2 = \frac{2}{l_2}(z-1) + 1, \tag{3c.1}$$

so that the z_i range over [-1,1] in each fluid.

The number of unknowns is decreased by using the incompressibility condition to eliminate u. In each fluid, we have the heat equation and one momentum equation, resulting in a system that is linear in σ. For $0 \le z \le l_1$,

$$P(\Delta^2 w - \alpha^2 R\Theta) = \sigma \Delta w, \quad wA_1 + \Delta\Theta = \sigma\Theta, \tag{3c.2}$$

where

$$\Delta = \frac{\partial^2}{\partial z^2} - \alpha^2,$$

and for $l_1 \leq z \leq 1$,

$$P(\frac{r}{m} \Delta^2 w - \alpha^2 \frac{R}{\beta} \Theta) = \sigma \Delta w, \quad w A_2 + \frac{1}{\gamma} \Delta \Theta = \sigma \Theta. \qquad (3c.3)$$

The z-dependence of the variables $w(x, z_i)$ and $\Theta(x, z_i)$ are expanded in powers of Chebyshev polynomials $T_m(z_i)$ for $m = 0, ..., N$ giving a total of $4N + 4$ unknown coefficients. Together with the free surface variable h_0 where $h(x, t) = h_0 exp(i\alpha x + \sigma t)$, there are $4N + 5$ unknowns. There are six boundary conditions and seven interface conditions. The term of highest differential order in the momentum equation is $\partial^4 w/\partial z^4$. Since we choose w to be an Nth degree polynomial, the term $\partial^4 w/\partial z^4$ is of degree $N - 4$ and therefore, the momentum equation is truncated at the $N - 4$th degree, yielding $N - 3$ equations in each fluid. Similarly, since the term of highest differential order in the heat equation is $\partial^2 w/\partial z^2$, we truncate this equation at the $N - 2$th degree, yielding $N - 1$ equations in each fluid. The eigenvalues of the resulting $4N + 5$ square matrix were computed in complex double precision on a Vax 11-780.

To check the accuracy and convergence of our computer code, the eigenvalues for the Bénard problem in one fuid were computed, with $P = 1$, $R = 2177.41$ and 47005.6, and $\alpha = 2$. The eigenvalues for this problem are real and are given by Reid and Harris [1958]. The critical eigenvalue in both instances (the critical σ should vanish, with other eigenvalues being negative real) is less than 10^{-5} when $N = 15$. A convergence test with $N = 15$ and 20 showed that several other eigenvalues had converged to at least five figures at $N = 15$.

The computation for two fluids was checked against those of Zeren and Reynolds [1972] by adding an extra term into the shear stress balance at the interface in order to take into account the Marangoni effect. We define a Marangoni number based on fluid 1:

$$Ma = -\left(\frac{dS^*}{d\theta^*}\right) \frac{\Delta\theta^* l^*}{\mu_1 \kappa_1}, \qquad (3c.4)$$

and the shear stress condition at $z = l_1$ is modified to

$$\alpha^2(m - 1)w_1 + m\frac{\partial^2 w_1}{\partial z^2} - \frac{\partial^2 w_2}{\partial z^2} + Ma. \ m\alpha^2(\Theta_1 - A_1 h) = 0. \qquad (3c.5)$$

Their Table 2 was used for the values of the physical variables at 10 degrees C for benzene lying above water. Our eigenvalues were checked against their Table 3 for $l_1 = 0.1$ and 0.6 for heating from below. Note that our definitions of R and Ma are different from theirs. At $l_1 = 0.1$, the conversion of their parameters to ours yields $Ma = 1255.71$, $R = 178.3045$, $\hat{a}_1 \Delta\theta^* = 0.00032537$, $P = 8.1$, $\alpha = 3.5$, and $S = 460320$. We computed $\sigma/P = 0.006186$ using both $N=15$ and 20. This yields 0.00175 for their eigenvalue q. At $l_1=0.6$, their parameters in Table 3 become $Ma =4016.7153$, $R=$

570.3736, $\hat{\alpha}_1 \Delta\theta^* = 0.0010408$, and $\alpha = 2.5$. We computed -0.00436 for their eigenvalue q at $N=15$ and 20. In both cases, stable complex conjugate pairs were found in the spectrum.

III.3(d) Example of a Hopf Bifurcation

To aid the reader in the interpretation of the numerical results [Renardy and Joseph 1985b], we recall some results from the Bénard problem with one fluid. In the simplest case, the layer is bounded at $z = 0$ and $z = 1$ by stress-free conducting boundaries, and [Drazin and Reid 1982]

$$\sigma = -0.5(1+P)(n^2\pi^2+\alpha^2) \pm \left(0.25(P-1)^2(n^2\pi^2+\alpha^2)^2 + \frac{\alpha^2 PR}{(n^2\pi^2+\alpha^2)}\right)^{1/2}$$

$$(3d.1)$$

for $n=1,2,\ldots$. Hence, for $\alpha \to 0$,

$$\sigma \sim -0.5(1+P)n^2\pi^2 \pm 0.5|P-1|n^2\pi^2 \quad < 0, \qquad (3d.2)$$

and as $\alpha \to \infty$,

$$\sigma \sim -\frac{\alpha^2}{2}\left(1 + P \pm |P-1|\right) \quad < 0. \qquad (3d.3)$$

In the critical case where there is one zero eigenvalue with others being negative, the least value of R occurs when

$$\alpha = \pi/\sqrt{2} \quad \text{and} \quad R = (\pi^2+\alpha^2)^3/\alpha^2 = \frac{27\pi^4}{4}. \qquad (3d.4)$$

These formulas are for stress-free surfaces but they give an idea of the variation of σ as a function of R and α in the classical case of one fluid between rigid boundaries.

We now consider the case when there are two fluids with equal properties. This would, at first thought, appear to be a one-fluid problem. However, the presence of the interface introduces an eigensolution with

$$[\Theta] = [A] = 0, \quad h_0 \neq 0, \quad \sigma = 0. \qquad (3d.5)$$

Following Yih [1967], this mode will be called an *interfacial mode*. Thus, the interfacial mode is neutrally stable when the fluids have identical properties, but when the properties differ, it no longer remains neutrally stable and plays an important role in the stability of the arrangement. Besides the interfacial mode, there are other modes which will be called *one-fluid modes*.

The least stable eigenvalues will be tracked as the parameters are varied. The parameters R, P, S, m, r, γ, ζ, β and l_1 are chosen such that there is a critical α, where the least stable eigenvalue satisfies $Re\ \sigma = 0$, with $Re\ \sigma < 0$ for other α. We shall exhibit parameters for which $Im\ \sigma \neq 0$ at criticality. Hence, the linear problem yields oscillations at criticality, and the nonlinear problem can have time-periodic solutions near criticality.

The two fluids are assumed to have equal densities at temperature θ_0^* and the same thermal diffusivities and conductivities: $r=\gamma=\zeta=1$. Let $R=1695.7$, $P=1$, $\hat{\alpha}_1\Delta\theta^* = 0.001$, $S=0$, $Ma=0$, $m=1.1$ and $\beta=0.9$. Thus, if fluid 1 occupies the entire flow, the Rayleigh number is lower than the critical one 1708 [Reid and Harris 1958]. If fluid 2 occupies the entire flow, $R = 2072.52$ and the flow is linearly unstable for a range of α. l_1 is chosen to be 0.4. Figure 3.1 is a graph of the growth rate $Re\ \sigma$ against α. Criticality occurs at approximately $\alpha=3.1$, for which we find $\sigma=0.000072\pm 5.9259i$ when $N=15,20$.

The five numbers next to the curves in figure 3.1 denote branches which display different features. The interfacial mode is associated with branches 1, 3, and 5. Branch 1 can be calculated by tracking the interfacial mode from the case when the properties of the fluids are equal ($\beta= m =1$), letting $\alpha \to 0$, and then by moving β to 0.9 and m to 1.1. The interfacial mode represented by this branch is real-valued.

Branch 2 is associated with the least stable of the one-fluid modes. This branch is approximately -9.87 as $\alpha \to 0$ and would correspond to the largest value of equation (3d.2). The eigenvalue on branch 2 is real-valued.

Branches 1 and 2 coalesce and split into conjugate pairs at $\alpha =1.275$. At $\alpha=6.79$, the conjugate pair again splits into two real-valued eigenvalues on branches 4 and 5. Branch 4 is associated with a one-fluid mode, and remains real, decreasing rapidly as α increases, as indicated in equation (3d.3) for the single fluid problem.

Branch 5 is an interfacial mode. It is real-valued and negative. The stability of branches 4 and 5 for large α is explained by the choice of β and the Boussinesq approximation (equation (2.1)). Consider the densities at the unperturbed interface $z = l_1$, approximated by $\rho_i (1 - \hat{\alpha}_i\Delta\theta^* (\theta - \theta_0))$, with the temperature field $\theta - \theta_0$ given by equation (2.19). Then, with $r=1$ and $\beta =0.9$, we find that $\rho_2(1-0.6\hat{\alpha}_2\Delta\theta^*)$ is the density of fluid 2 at $l_1 = 0.4$ and $\rho_2(1 - 0.54\hat{\alpha}_2\Delta\theta^*)$ is the density of fluid 1. Hence, the heavier fluid is below and gravity may be expected to stabilize short (large α) waves. The interfacial eigenvalue on branch 5 is discussed in section 5(i) below.

Figure 3.2 shows the growth rates versus wavenumber of the disturbance when the Rayleigh number is increased to $R= 2177.41$ while other parameters remain fixed as in figure 3.1. This exhibits an instability arising from the complex conjugate pair of eigenvalues on branch 3. Solutions to the nonlinear three-dimensional problem, close to the onset of this type of oscillatory instability, are investigated in section 7.

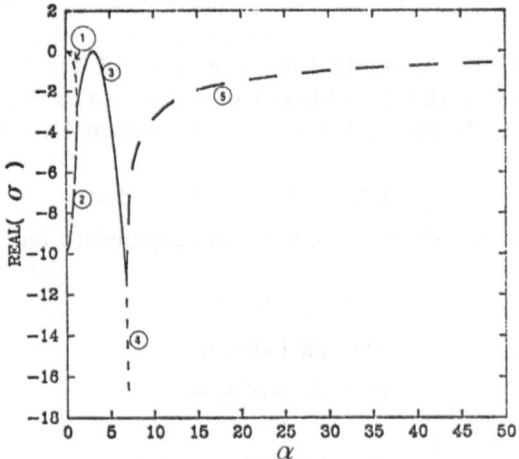

Fig. 3.1. [Renardy and Joseph, 1985b, American Institute of Physics] The growth rate $Re\ \sigma$ is plotted against the wavenumber α for $R = 1695.7$, $P=1$, $\hat{\alpha}_1\Delta\theta^*=0.001$, $S = Ma=0$, $m=1.1$, $\beta= 0.9$, $l_1=0.4$, and $r = \gamma = \zeta=1$. Branches 1, 2, 4 and 5 belong to real-valued eigenvalues. Branch 3 consists of a complex conjugate pair.

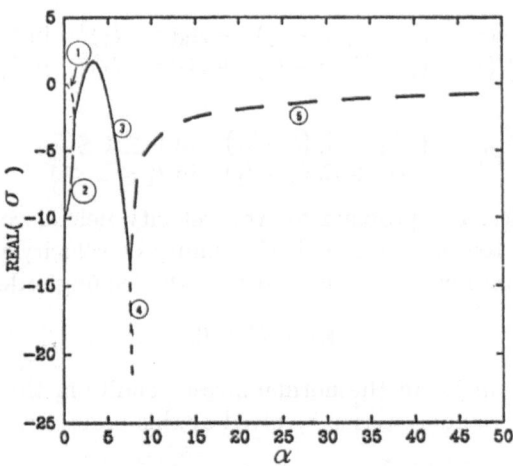

Fig. 3.2. [Renardy and Joseph, 1985b, American Institute of Physics] The growth rate $Re\ \sigma$ is plotted against the wavenumber α for $R = 2177.41$, $P=1$, $\hat{\alpha}_1\Delta\theta^*=0.001$, $S = Ma=0$, $m=1.1$, $\beta= 0.9$, $l_1=0.4$, and $r = \gamma = \zeta=1$. Branches 1, 2, 4 and 5 belong to real-valued eigenvalues. Branch 3 consists of a complex conjugate pair.

III.4 Asymptotic Analysis for Long Waves

Gumerman and Homsy [1974] do the asymptotic analysis for long waves of the interfacial mode in a two-layer Bénard problem with a shearing flow superposed. They use the method for long waves of Yih [1967] to find an expression for the growth rate. This section concerns a subset of their result.

We introduce the streamfunction ψ proportional to $\exp(i\alpha x + \sigma t)$, where

$$\sigma = -i\alpha c, \quad u = \psi_z, \quad w = -i\alpha\psi \tag{4.1}$$

into equations (3a.1) - (3a.4). The following expansions for long waves are assumed:

$$c \sim c_0 + \alpha c_1 + ... \tag{4.2}$$

$$\psi \sim \psi_0 + \alpha\psi_1 + ... \tag{4.3}$$

$$\Theta \sim \Theta_0 + \alpha\Theta_1 + ... \tag{4.4}$$

and

$$h \sim h_0 + \alpha h_1 + ...$$

By eliminating the pressure in the momentum equations, we have

$$D^4\psi_0 = 0, \quad D = \frac{\partial}{\partial z} \tag{4.5}$$

and the heat equation yields and

$$D^2\Theta_0 = 0. \tag{4.6}$$

Thus,

$$\psi_0 = \begin{cases} A_0 + A_1(z - l_1) + A_2(z - l_1)^2 + A_3(z - l_1)^3 & \text{in } 0 \le z \le l_1 \\ B_0 + B_1(z - l_1) + B_2(z - l_1)^2 + B_3(z - l_1)^3 & \text{in } l_1 \le z \le 1 \end{cases} \tag{4.7}$$

and

$$\Theta_0 = \begin{cases} A_4 + A_5(z - l_1) & \text{in } 0 \le z \le l_1 \\ B_4 + B_5(z - l_1) & \text{in } l_1 \le z \le 1 \end{cases} . \tag{4.8}$$

At this order, the problem for the velocity field decouples from the problem for the temperature field. Continuity of velocity yields $A_0 = B_0$ and $A_1 = B_1$. Continuity of shear stress at the leading order is

$$[\![\mu D^2\psi]\!] = 0, \tag{4.9}$$

and yields $B_2 = mA_2$. In the normal stress condition, the pressures scale like $1/\alpha$ and are the only terms involved at this order, or,

$$[\![p]\!] = 0. \tag{4.10}$$

This yields $B_3 = mA_3$. The kinematic condition yields

$$c_0 h_0 = A_0. \tag{4.11}$$

The no-slip condition at the boundaries is $\psi = 0 = D\psi$ at $z = 0, 1$. This yields, expressing the B_i in terms of A_i,

$$A_0 - A_1 l_1 + A_2 l_1^2 - A_3 l_1^3 = 0,$$

$$A_0 + A_1 l_2 + m A_2 l_2^2 + m A_3 l_2^3 = 0,$$

$$A_1 - 2 A_2 l_1 + 3 A_3 l_1^2 = 0,$$

$$A_1 + 2 m A_2 l_2 + 3 m A_3 l_2^2 = 0.$$

The solution is $A_0 = A_1 = A_2 = A_3 = 0$. Equation (4.11) yields that either $c_0 = 0$ or $h_0 = 0$. If $h_0 = 0$, then solving for the temperature field, continuity of temperature yields $A_4 = B_4$. Continuity of heat flux yields $\zeta A_5 = B_5$. Together with the boundary condition at the walls, we have $A_4 = A_5 = B_4 = B_5 = 0$ and there is no eigenfunction. Therefore, $h_0 \neq 0$ and

$$c_0 = 0. \tag{4.12}$$

This result is expected because the problem is symmetric in α, so the growth rate $Re\ \sigma$ $(\sigma = -i\alpha c)$ must be an even power of α.

The temperature field at the leading order is given by

$$A_4 = \frac{l_1 h_0 [\![A]\!]}{l_1 + \zeta l_2}, \tag{4.13}$$

$$A_5 = \frac{A_4}{l_1}, \quad B_4 = A_4 - h_0 [\![A]\!], \quad B_5 = -\frac{B_4}{l_2}. \tag{4.14}$$

At the next order, we have

$$D^4 \psi_1 = \begin{cases} i R \Theta_0 & \text{in fluid 1} \\ \frac{iRm}{r\beta} \Theta_0 & \text{in fluid 2} \end{cases} \tag{4.15}$$

and

$$D^2 \Theta_1 = 0.$$

The problem for the temperature field is again decoupled from that for the velocity field. The solution is

$$\psi_1 = \begin{cases} E_0 + E_1(z - l_1) + E_2(z - l_1)^2 + E_3(z - l_1)^3 \\ \quad + iR(A_4(z - l_1)^4/24 + A_5(z - l_1)^5/120) & \text{in } 0 \leq z \leq l_1 \\ D_0 + D_1(z - l_1) + D_2(z - l_1)^2 + D_3(z - l_1)^3 \\ \quad + \frac{iRm}{r\beta}(B_4(z - l_1)^4/24 + B_5(z - l_1)^5/120) & \text{in } l_1 \leq z \leq 1 \end{cases} \tag{4.16}$$

and

$$\Theta_1 = \begin{cases} E_4 + E_5(z - l_1) & \text{in } 0 \leq z \leq l_1 \\ D_4 + D_5(z - l_1) & \text{in } l_1 \leq z \leq 1 \end{cases}. \tag{4.17}$$

In order to calculate c_1, the problem for the temperature field can be ignored. Continuity of velocity and shear stress at the interface yield $E_0 = D_0$, $E_1 = D_1$ and $m E_2 = D_2$. In the normal stress condition, the jump in the pressure is balanced by the $M_1 h_0$ term to yield $D_3 = mE_3 - imM_1 h_0/(6P)$. The kinematic condition yields

$$c_1 h_0 = E_0. \tag{4.18}$$

The no-slip condition at the walls and the above relations reduce the problem to a system for four unknowns:

$$\begin{pmatrix} 1 & -l_1 & l_1^2 & -l_1^3 \\ 1 & l_2 & ml_2^2 & ml_2^3 \\ 0 & 1 & -2l_1 & 3l_1^2 \\ 0 & 1 & 2ml_2 & 3ml_2^2 \end{pmatrix} \begin{pmatrix} E_0 \\ E_1 \\ E_2 \\ E_3 \end{pmatrix}$$

$$= \begin{pmatrix} -iR\left(\frac{A_4 l_1^4}{24} - \frac{A_5 l_1^5}{120}\right) \\ -\frac{iRm}{r\beta}\left(\frac{B_4 l_2^4}{24} + \frac{B_5 l_2^5}{120}\right) + l_2^3 im\frac{M_1 h_0}{6P} \\ -iR\left(\frac{-A_4 l_1^3}{6} + \frac{A_5 l_1^4}{24}\right) \\ -\frac{iRm}{r\beta}\left(\frac{B_4 l_2^3}{6} + \frac{B_5 l_2^4}{24}\right) + 3l_2^2 im\frac{M_1 h_0}{6P} \end{pmatrix}. \tag{4.19}$$

Here, M_1 is defined by equation (2.25) and represents buoyancy effects.

Thus, $c_1 = a/b$ where

$$a = iR(\frac{A_5}{h_0})l_1^2 l_2 \left[\frac{m\zeta l_2^4}{r\beta}\left(ml_2(\frac{1}{40} + \frac{11 l_1}{120}) + \frac{11 l_1^2}{120}\right) + \frac{l_1^4 ml_2}{15}\right.$$

$$\left. - \frac{11}{120} ml_1^3 l_2(ml_2^2 + l_1(2 - l_1))\right] + \frac{imM_1 l_2^3 l_1^3}{3P}(l_1 + ml_2), \tag{4.20}$$

and

$$b = -m^2 l_2^4 - l_1^4 - 2ml_1 l_2(l_1^2 - l_1 + 2). \tag{4.21}$$

These expressions yield the growth rate, which is the main information that we have sought from the asymptotic analysis for long waves.

Next, consider the situation where l_1 is small. We have that $b \sim -m^2$ and the leading term in a is of order l_1^2, yielding

$$c_1 \sim -\frac{iR l_1^2(1 - \zeta)}{40 r\beta\zeta} - \frac{iM_1 l_1^3}{3P}, \qquad l_1 << 1. \tag{4.22}$$

Since $\sigma \sim -i\alpha^2 c_1$,

$$\sigma \sim -\frac{\alpha^2 R l_1^2(1 - \zeta)}{40 r\beta\zeta} - \frac{\alpha^2 M_1 l_1^3}{3P}. \tag{4.23}$$

Therefore, in the limit as l_1 approaches zero, there is stability if $\zeta < 1$ (and instability if $\zeta > 1$), irrespective of the density stratification. This is the analogue of the thin-layer effect in shearing flows (see chapter IV.6), with the thermal conductivity stratification taking the place of the role played by viscosity stratification in shearing flows. This role is clear if one pictures the basic temperature field, and compares it with the basic velocity field in the shearing case. Figure 2.1 shows the temperature field where the thermal conductivity of the lower fluid is much larger than that of the upper fluid, and is unstable in the long-wave limit if the lower fluid is in a very thin layer. However, in many practical situations involving relatively thin layers, the effect of the buoyancy terms outweigh the importance of the thermal conductivity stratification (because of the smallness of the $\hat{\alpha}_1 \Delta\theta^*$ term in M_1).

III.5 Asymptotic Analysis for Short Waves

III.5(i) The Leading Order Terms. The asymptotic analysis of the interfacial eigenvalue for short waves in a two-layer problem was first investigated by Hooper and Boyd [1983], in the context of unbounded Couette flow (see chapter IV.5). Their formal method is applied below to the two-layer Bénard problem [Renardy and Renardy 1988]. The asymptotic analysis for short waves is a singular perturbation in the sense that one does not simply take the original differential equations and expand them in a series of $1/\alpha$. Rather, the z-variable must first be re-scaled, just like in a boundary-layer analysis. In this process, some of the boundary conditions are discarded, so that the expansion for the eigenfunction cannot be expected to converge to the exact solution on the whole domain. What is sought is an expansion that is valid in a region close to the interface position, for disturbances of rapid variation. The length scale of variation is of the order $1/\alpha$.

The z-variable is replaced by

$$\eta = \alpha(z - l_1), \quad \eta = O(1). \tag{5.1}$$

The equations in fluid 1 are

$$\sigma\Theta - wA_1 = \alpha^2 L^*\Theta, \quad (\sigma - P\alpha^2 L^*)L^*w = -PR\Theta, \tag{5.2}$$

where

$$L^* = \frac{\partial^2}{\partial\eta^2} - 1.$$

In fluid 2,

$$\sigma\Theta - wA_2 = \frac{1}{\gamma}\alpha^2 L^*\Theta, \quad (\sigma - \frac{r}{m}P\alpha^2 L^*)L^*w = -\frac{PR}{\beta}\Theta. \tag{5.3}$$

The interface conditions are

$$w_1 = w_2 = \sigma h, \tag{5.4}$$

$$[\frac{\partial w}{\partial\eta}] = 0, \tag{5.5}$$

$$[\Theta] = [A]h,$$

$$m\frac{\partial^2 w_1}{\partial \eta^2} - \frac{\partial^2 w_2}{\partial \eta^2} + (m-1)w_1 = 0, \tag{5.6}$$

$$\zeta\frac{\partial \Theta_1}{\partial \eta} = \frac{\partial \Theta_2}{\partial \eta},$$

$$\alpha^2\left(\frac{1}{m}\frac{\partial^3 w_2}{\partial \eta^3} - \frac{\partial^3 w_1}{\partial \eta^3}\right) + 3\alpha^2\frac{\partial w_1}{\partial \eta}\left(1 - \frac{1}{m}\right) + S\alpha^3 h$$

$$-hR\alpha\left(\frac{(1/r-1)}{\hat{\alpha}_1\Delta\theta^*} + l_2A_2\left(1 - \frac{1}{r\beta}\right)\right) = (1/r-1)\frac{\sigma}{P}\frac{\partial w_1}{\partial \eta}. \tag{5.7}$$

Since the normal stress condition (5.7) contains both odd and even powers of α, all the variables are formally expanded in powers of $1/\alpha$. To the zeroth and first orders,

$$L^*\Theta = 0 \quad \text{and} \quad L^*w = 0 \tag{5.8}$$

in each fluid. Conditions (5.4) - (5.6) yield

$$w_1 = c_0(1-\eta)e^\eta + O(1/\alpha), \quad w_2 = c_0(1+\eta)e^{-\eta} + O(1/\alpha) \tag{5.9}$$

as $\alpha \to \infty$. Thus, to this and the next order,

$$\frac{\partial w}{\partial \eta} = 0 \tag{5.10}$$

at the interface. Hence, the normal stress condition is

$$\alpha\left(\frac{1}{m}\frac{\partial^3 w_2}{\partial \eta^3} - \frac{\partial^3 w_1}{\partial \eta^3}\right) - \frac{w}{\sigma}R\left(\frac{(1/r-1)}{\hat{\alpha}_1\Delta\theta^*} + l_2A_2\left(1 - \frac{1}{r\beta}\right) - \alpha^2\frac{S}{R}\right) = 0. \tag{5.11}$$

To avoid the trivial solution, we set

$$\sigma = \frac{\sigma_0}{\alpha} + O\left(\frac{1}{\alpha^2}\right) \tag{5.12}$$

for large α. In the normal stress condition, we assume, as in Hooper and Boyd [1983] that

$$\alpha^2\frac{S}{R} = O(1). \tag{5.13}$$

Therefore,

$$\sigma_0 = \frac{R}{2(1/m+1)}\left(\frac{(1/r-1)}{\hat{\alpha}_1\Delta\theta^*} + l_2A_2\left(1 - \frac{1}{r\beta}\right) - \alpha^2\frac{S}{R}\right). \tag{5.14}$$

In the computations for figure 3.1, the asymptotic formula is accurate to 1% for $\alpha > 20$. We computed -1.494 for α=20 using N=15, whereas the asymptotic formula yields -1.48. If we add a surface tension of S=1 to the situation in figure 3.1, the eigenvalue at α=20 is -6.86 whereas the asymptotic formula is -6.72. Equation (5.14) shows that surface tension is always stabilizing for short-wave disturbances and that its effect dominates over those of the other parameters; this action by surface tension is also found in other two-layer flows, such as steady shear flows with two immiscible fluids of different viscosities (see chapter IV.5).

III.5(ii) Thermal Conductivity Stratification. We consider the special case where only the thermal conductivities of the two fluids differ with all other fluid properties equal and surface tension equal to zero [Renardy and Renardy 1988]. We introduce a stream function ψ by setting $w = i\alpha\psi$, $u = -\psi'$. After some algebra, the linear stability problem is reduced to

$$\left\{-P(D^2 - \alpha^2)^3 + (P+1)\sigma(D^2 - \alpha^2)^2 - \sigma^2(D^2 - \alpha^2)\right\}\psi$$

$$= \alpha^2 A_1 RP\psi, \quad 0 \le z \le l_1, \tag{5.15}$$

$$\left\{-P(D^2 - \alpha^2)^3 + (P+1)\sigma(D^2 - \alpha^2)^2 - \sigma^2(D^2 - \alpha^2)\right\}\psi$$

$$= \alpha^2 A_2 RP\psi, \quad l_1 \le z \le 1, \tag{5.16}$$

and the interface conditions become

$$[\![\psi]\!] = [\![\psi']\!] = [\![\psi'']\!] = [\![\psi''']\!] = 0,$$

$$[\![\psi^{iv}]\!] = -i\alpha Rh[\![A]\!], \quad [\![k((D^2 - \alpha^2)^2\psi' - \frac{\sigma}{P}(D^2 - \alpha^2)\psi')]\!] = 0, \quad \sigma h = i\alpha\psi, \tag{5.17}$$

where

$$D = {}' = \partial/\partial z.$$

For large α, we look for eigenfunctions which decay rapidly away from the interface. We ignore the influence of the boundary conditions at the walls, which would only lead to correction terms which are exponentially small in α. We use the rescaled coordinate $\eta = \alpha(z - l_1)$. In the rescaled coordinate, with

$$D = {}' = \partial/\partial\eta,$$

(5.15)-(5.17) read as follows:

$$-P\alpha^6(D^2 - 1)^3\psi + (P+1)\alpha^4\sigma(D^2 - 1)^2\psi - \alpha^2\sigma^2(D^2 - 1)\psi = \alpha^2 A_i RP\psi \tag{5.18}$$

in fluid i, and the interface conditions become

$$[\![\psi]\!] = [\![\psi']\!] = [\![\psi'']\!] = [\![\psi''']\!] = 0,$$

$$[\![\psi^{iv}]\!] = \frac{-iR}{\alpha^3}h[\![A]\!], \quad [\![k(\alpha^4(D^2 - 1)^2\psi' - \frac{\sigma\alpha^2}{P}(D^2 - 1)\psi')]\!] = 0, \quad \sigma h = i\alpha\psi. \tag{5.19}$$

For large α, equations (5.18) and (5.19) are satisfied at leading order if we set $\psi = 0$, $\sigma = 0$ and $h = 1$. From the interface conditions, it is clear that the leading term in ψ is of order α^{-3} and the leading term in σ is of order α^{-2}. To find the leading term in ψ, we must solve the equation

$$(D^2 - 1)^3\psi = 0, \tag{5.20}$$

and we use $h = 1$ in the conditions at the interface. Let $\chi = (D^2 - 1)\psi$, $\xi = (D^2 - 1)^2\psi$. From (5.20) and the condition of decay away from the interface, we find

$$\xi(\eta) = ae^\eta, \ \eta < 0,$$
$$\xi(\eta) = be^{-\eta}, \ \eta > 0. \tag{5.21}$$

The interface conditions yield

$$[\xi] = a - b = -\frac{iR}{\alpha^3}[A], \qquad [k\xi'] = k_1a + k_2b = 0. \tag{5.22}$$

This leads to

$$a = -\frac{iRk_2}{\alpha^3(k_1 + k_2)}[A], \qquad b = \frac{iRk_1}{\alpha^3(k_1 + k_2)}[A]. \tag{5.23}$$

Observing that $(D^2 - 1)\chi = \xi$, we find

$$\chi(\eta) = \frac{a}{2}\eta e^\eta + ce^\eta, \ \eta < 0,$$

$$\chi(\eta) = -\frac{b}{2}\eta e^{-\eta} + de^{-\eta}, \ \eta > 0. \tag{5.24}$$

The interface conditions yield

$$[\chi] = c - d = 0, \qquad [\chi'] = \frac{a}{2} + c + \frac{b}{2} + d = 0, \tag{5.25}$$

which result in

$$c = d = -\frac{a+b}{4} = \frac{iR(k_2 - k_1)}{4\alpha^3(k_1 + k_2)}[A]. \tag{5.26}$$

The equation $(D^2 - 1)\psi = \chi$ now yields

$$\psi(\eta) = \frac{a}{8}\eta^2e^\eta + (\frac{c}{2} - \frac{a}{8})\eta e^\eta + fe^\eta, \qquad \eta < 0,$$

$$\psi(\eta) = \frac{b}{8}\eta^2e^{-\eta} + (\frac{b}{8} - \frac{d}{2})\eta e^{-\eta} + ge^{-\eta}, \qquad \eta > 0. \tag{5.27}$$

From the interface conditions, we find

$$[\psi] = f - g = 0, \ [\psi'] = f + g + \frac{c+d}{2} - \frac{a+b}{8} = 0, \tag{5.28}$$

and hence

$$f = g = \frac{a+b}{16} - \frac{c+d}{4} = \frac{3}{16}(a + b) = \frac{3iR(k_1 - k_2)}{16\alpha^3(k_1 + k_2)}[A]. \tag{5.29}$$

The last of the interface conditions now gives the leading contribution to σ as

$$\sigma \sim i\alpha f = \frac{3R(k_2 - k_1)}{16\alpha^2(k_1 + k_2)}[A] = \frac{3R(k_2 - k_1)^2}{16\alpha^2(k_1 + k_2)(k_2l_1 + k_1l_2)} > 0. \tag{5.30}$$

Hence short waves are unstable if the two fluids differ only in thermal conductivities.

III.6 Liquids with Similar Properties

III.6(a) Close to Criticality for One Fluid

A linear stability analysis is given for the case when the two fluids have only slightly differing mechanical and thermal properties, are allowed to slip at the boundaries (the "free-free" boundary conditions), and the parameters are close to criticality for one fluid. Thus, the differential equations are given by (3a.1) - (3a.2), and the interface conditions by equation (3a.4). The boundary conditions at the external boundaries are

$$\Theta = w = \frac{\partial u}{\partial z} = 0. \tag{6a.1}$$

The first two of these are identical to equation (3a.3), and the latter requires that the tangential shear stress, rather than tangential velocity, be zero. Although this slip condition is physically unrealistic, it has the advantage that the eigenvalues and eigenfunctions at criticality are known in closed form when the two fluids have identical properties. We take advantage of this by considering a perturbation problem [Renardy and Renardy 1985], where the unperturbed problem is the linear stability problem where the fluids have identical properties, and the perturbed problem is the linear stability problem where the fluids have slightly different properties. This allows us to obtain a closed-form expression for the interfacial eigenvalue when the fluids are slightly different, so that the effect of each parameter can be seen. The results help to shed light into the numerical results obtained for the more realistic "rigid-rigid" boundary problem.

In the linear stability problem where the fluids are identical, the one-fluid eigenvalues are real (see, for example, Chandrasekhar [1961], Joseph [1976 II], or Drazin and Reid [1982]) and the presence of the interface introduces a zero eigenvalue. Another zero eigenvalue occurs first at the Rayleigh number and wavenumber α given by (3d.4), which is the first critical value of the one-fluid problem. We fix the Rayleigh number and the wavenumber at this critical value.

The small parameter ϵ in the perturbation analysis is introduced as follows. The quantities $1 - m, 1 - r, 1 - \gamma, 1 - \zeta, 1 - \beta, M_1$ and S are to be regarded as small quantities proportional to ϵ; that is, we set

$$1 - m = \bar{m}\epsilon,$$

$$1 - r = \bar{r}\epsilon,$$

$$1 - \gamma = \bar{\gamma}\epsilon,$$

$$1 - \zeta = \bar{\zeta}\epsilon,$$

$$1 - \beta = \bar{\beta}\epsilon,$$

$$M_1 = \bar{M}_1 \epsilon, \quad \bar{M}_1 = RP(-\frac{\bar{r}}{\hat{\alpha}_1 \Delta \theta^*} + l_2(\bar{r} + \bar{\beta})),$$

where the Rayleigh number

$$R = \frac{27\pi^4}{4},$$

is critical for one fluid, and the surface tension parameter is

$$S = \bar{S}\epsilon. \tag{6a.2}$$

At $\epsilon = 0$, $\sigma = 0$ is an algebraically double eigenvalue, consisting of the interfacial mode and the first critical mode of the one-fluid problem. This eigenvalue is not semi-simple. The Riesz index is two; there is one eigenvector and one generalized eigenvector. Correspondingly, there is one adjoint eigenvector and one generalized adjoint eigenvector (see Iooss and Joseph [1980], chapter IV).

The analysis for small ϵ concerns the regular perturbation of this eigenvalue into two eigenvalues, which can be expanded in powers of $\epsilon^{1/2}$. The purpose of the following calculation is to find the coefficients of $\epsilon^{1/2}$ and ϵ in this expansion.

The leading terms in the perturbation expansion are calculated in closed form. This yields information regarding the following two questions:

(1) Which perturbations lead to eigenvalues with nonvanishing imaginary parts? and

(2) Which parameter perturbations stabilize and which destabilize the flow?

It is found that whether the eigenvalue splits into real values or complex conjugates is independent of the viscosity ratio and the Prandtl number. The oscillatory instability is due to buoyancy instabilities in either one or both of the fluids, countered by a stable interface.

III.6(a)(i) Perturbation of Multiple Eigenvalues

The perturbation expansion for a *double* eigenvalue is not a simple series expansion in ϵ. In addition, eigenfunctions need not have series expansions in powers of ϵ. The perturbation of multiple eigenvalues in the context of matrices is discussed in chapter IV, §1, pp. 285 - 296 of Yakubovich and Starzhinskii [1975]. The procedure involves the generalized eigenspaces belonging to the eigenvalue $\sigma = 0$ for the *unperturbed* problem and its adjoint, and does *not* require finding the eigenspaces of the perturbed $O(\epsilon)$ problem at all.

The pertinent results from Yakubovich and Starzhinskii [1975] are as follows. Suppose σ_0 is an algebraically 2-fold eigenvalue of a matrix L_0. Let $\{a_1, a_2\}$ be a basis for the generalized eigenspace of L_0 with eigenvalue σ_0, and let $\{b_1, b_2\}$ be a basis for the generalized eigenspace of L_0^* (the adjoint of L_0) with eigenvalue $\bar{\sigma}_0$ (the overbar here denotes the complex conjugate). Let L_0 be perturbed into $L(\epsilon) = L_0 + \epsilon L_1 + O(\epsilon^2)$ with L_1 depending smoothly on ϵ. Then the perturbed eigenvalues σ are given by the zeros of the determinant of a matrix $\Psi_{ij}(\epsilon, \sigma), i, j, = 1, 2$, which represents to $O(\epsilon)$ the projection of $L(\epsilon) - \sigma$, first onto the eigenspace of the unperturbed problem and then onto the adjoint eigenspace:

$$\Psi_{ij}(\epsilon, \sigma) = \langle b_i, (L_0 + \epsilon L_1 - \sigma)a_j \rangle + O(\epsilon^2). \qquad (6a.3)$$

Some care must be taken when this result is applied to unbounded operators in infinite dimensional spaces, for example, differential operators. Such an operator has a *"domain"* that is specified not only by smoothness requirements on the function but also by the boundary conditions. If the domain of the operator that is being perturbed depends on ϵ, we cannot apply (6a.3); the domains of $L(\epsilon)$ and L_0 may be different, and their combination would not make sense. We can, however, circumvent this problem by not looking at the differential operator itself, but at its *resolvent* $(L(\epsilon) - \lambda \underline{I})^{-1}$ where λ is not an eigenvalue of $L(\epsilon)$. The domain of this does not depend on ϵ and we will need to redefine Ψ_{ij} accordingly. As will be seen from the following, the resolvent itself does not ever need to be computed.

In order to make these ideas more precise, some notation will be introduced. Let X denote the set of functions (Θ, u, w, h). The following inner product is introduced to generate a Hilbert space:

$$\langle X_1, X_2 \rangle = \int_0^{2\pi/\alpha} \int_{z=0}^{l_1} \bar{\Theta}_1 \Theta_2 + \bar{u}_1 u_2 + \bar{w}_1 w_2 \; dz dx$$

$$+ \int_0^{2\pi/\alpha} \int_{z=l_1}^1 \bar{\Theta}_1 \Theta_2 + \bar{u}_1 u_2 + \bar{w}_1 w_2 \; dz dx$$

$$+ \int_0^{2\pi/\alpha} \bar{h}_1 h_2 \; dx. \qquad (6a.4)$$

In this Hilbert space, we consider the subspace determined by the "Hodge projection" (see space H in Theorem 1.4, [Temam 1979]), that is, by the conditions that the velocity field be divergence-free, that the vertical velocity vanish at the boundaries, and be continuous across the interface.

By $L(\epsilon)$X we denote the right hand sides of equations (3a.1),(3a.2) and the kinematic condition in (3a.4). We regard $L(\epsilon)$ as an operator in the subspace so that the conditions on w at the external boundaries, and the continuity of velocity and the normal stress balance at the interface are an integral part of the definition of $L(\epsilon)$. The domain of definition of $L(\epsilon)$ is determined by the rest of the boundary conditions in (3a.4) and (6a.1), which we write in the form $B(\epsilon)$X=0. The range of the operator $L(\epsilon)$ must satisfy the following conditions in order for the pressure p to be determined as a function of X: The "velocity part" of $L(\epsilon)$X must be divergence free, the vertical velocity must vanish on the walls and be continuous across the interface, and the jump in p across the interface must be given by the normal stress balance. Thus, the problem we wish to solve is: for small ϵ, find σ satisfying

$$L(\epsilon)X = \sigma X, \tag{6a.5}$$

$$B(\epsilon)X = 0,$$

$$L(\epsilon) = L_0 + \epsilon L_1 + O(\epsilon^2),$$

$$B(\epsilon) = B_0 + \epsilon B_1 + O(\epsilon^2).$$

Explicitly,

$$L_0 X = \begin{pmatrix} \Delta\Theta + w \\ -\frac{\partial p}{\partial x} + P\Delta u \\ -\frac{\partial p}{\partial z} + RP\Theta + P\Delta w \\ w \end{pmatrix} \tag{6a.6}$$

in fluids 1 and 2, and

$$L_1 X = \begin{pmatrix} l_2\bar{\zeta}w \\ -\frac{\partial\tilde{p}}{\partial x} \\ -\frac{\partial\tilde{p}}{\partial z} \\ 0 \end{pmatrix} \tag{6a.7}$$

in fluid 1 and

$$\begin{pmatrix} \bar{\gamma}\Delta\Theta - l_1\bar{\zeta}w \\ \bar{r}\frac{\partial p}{\partial x} + (\bar{m}-\bar{r})P\Delta u - \frac{\partial\tilde{p}}{\partial x} \\ \bar{r}\frac{\partial p}{\partial z} + \bar{\beta}RP\Theta + (\bar{m}-\bar{r})P\Delta w - \frac{\partial\tilde{p}}{\partial z} \\ 0 \end{pmatrix} \tag{6a.8}$$

in fluid 2, where $\tilde{p}$ denotes the $O(\epsilon)$-perturbation to the pressure,

$$B_0 X = \begin{pmatrix} \Theta_1 - \Theta_2 \text{ at } z = l_1 \\ \frac{\partial\Theta_1}{\partial z} - \frac{\partial\Theta_2}{\partial z} \text{ at } z = l_1 \\ u_1 - u_2 \text{ at } z = l_1 \\ \frac{\partial u_1}{\partial z} - \frac{\partial u_2}{\partial z} \text{ at } z = l_1 \\ \Theta \text{ at } z = 0, 1 \\ \frac{\partial u}{\partial z} \text{ at } z = 0, 1 \end{pmatrix}, \tag{6a.9}$$

and

$$B_1 X = \begin{pmatrix} -h\bar{\zeta} \text{ at } z = l_1 \\ -\bar{\zeta}\frac{\partial\Theta_1}{\partial z} \text{ at } z = l_1 \\ 0 \text{ at } z = l_1 \\ -\bar{m}(\frac{\partial u_1}{\partial z} + \frac{\partial w_1}{\partial x}) \text{ at } z = l_1 \\ 0 \text{ at } z = 0, 1 \\ 0 \text{ at } z = 0, 1 \end{pmatrix}. \tag{6a.10}$$

With the above definitions, we are now ready to look at the resolvent of $L(\epsilon)$ and then to redefine the matrix in (6a.3). Since ϵ is small, the eigenvalues of $L(\epsilon)$ are close to those of L_0, so that the λ in the resolvent should be chosen well away from zero. We choose $\lambda = 1$. Hence, instead of looking at (6a.5) directly, the equivalent problem

$$(L(\epsilon) - 1)^{-1}X = (\sigma - 1)^{-1}X, \tag{6a.11}$$

where

$$\hat{\sigma} = (\sigma - 1)^{-1},$$

is studied and the problem is perturbed around $\hat{\sigma} = -1$. We note that the definition of $(L(\epsilon)-1)^{-1}$ already incorporates the boundary conditions. The relation (6a.3) is applied to this problem. The determinant of the matrix

$$\Psi_{ij}(\epsilon, \hat{\sigma}) = \langle b_i, ((L(\epsilon) - 1)^{-1} - \hat{\sigma})a_j \rangle + O(\epsilon^2), \ i, j = 1, 2, \tag{6a.12}$$

where the b_i and a_j are as before, is set equal to zero. We will require an expansion of the resolvent in powers of ϵ in order to carry out the calculation of Ψ_{ij}. We note that the inverse of $L_0 - 1$ is defined.

We first have to find the boundary value problem adjoint to (6a.5), for $\epsilon = 0$.

The Adjoint Problem for $\epsilon = 0$. We denote the domains occupied by the two fluids by

$$\Omega_1 = \{0 \leq x \leq 2\pi/\alpha, 0 \leq z \leq l_1\}$$

and

$$\Omega_2 = \{0 \leq x \leq 2\pi/\alpha, l_1 \leq z \leq 1\}.$$

We denote the interface by I, and the lower and upper boundaries by Γ_1 and Γ_2, respectively. Let $X_1 = (\Theta, u, w, h)$ and $X_2 = (\Theta^*, u^*, w^*, h^*)$. Asterisks denote the adjoint. We have

$$\langle X_2, L_0 X_1 \rangle = \int_{\Omega_1} (\bar{\Theta}^* \triangle \Theta + \bar{\Theta}^* w$$

$$-\bar{u}^* \frac{\partial p}{\partial x} + P\bar{u}^* \triangle u - \bar{w}^* \frac{\partial p}{\partial z} + RP\bar{w}^* \Theta + P\bar{w}^* \triangle w)$$

$$+ \int_{\Omega_2} (\bar{\Theta}^* \triangle \Theta + \bar{\Theta}^* w$$

$$-\bar{u}^* \frac{\partial p}{\partial x} + P\bar{u}^* \triangle u - \bar{w}^* \frac{\partial p}{\partial z} + RP\bar{w}^* \Theta + P\bar{w}^* \triangle w)$$

$$+ \int_I \bar{h}^* w_1. \tag{6a.13}$$

This is integrated by parts, using the divergence conditions div $\mathbf{u}$ = div $\mathbf{u}^* = 0$, to obtain

$$\langle X_2, L_0 X_1 \rangle = \int_{\Omega_1} \left(\Theta(\triangle \bar{\Theta}^* + RP\bar{w}^*) \right.$$

$$+ u(P \triangle \bar{u}^* - \frac{\partial \bar{p}^*}{\partial x}) + w(P \triangle \bar{w}^* - \frac{\partial \bar{p}^*}{\partial z} + \bar{\Theta}^*))$$

$$+ \int_{\Omega_2} \left(\Theta(\triangle \bar{\Theta}^* + RP\bar{w}^*) + u(P \triangle \bar{u}^* - \frac{\partial \bar{p}^*}{\partial x}) + w(P \triangle \bar{w}^* - \frac{\partial \bar{p}^*}{\partial z} + \bar{\Theta}^*)) \right.$$

$$+ \int_{\Gamma_1} (- \bar{\Theta}^* \frac{\partial \Theta}{\partial z} + \Theta \frac{\partial \bar{\Theta}^*}{\partial z} - P\bar{u}^* \frac{\partial u}{\partial z} + Pu \frac{\partial \bar{u}^*}{\partial z}$$

$$- P\bar{w}^* \frac{\partial w}{\partial z} + Pw \frac{\partial \bar{w}^*}{\partial z} + \bar{w}^* p - w\bar{p}^*)$$

$$+ \int_{\Gamma_2} (\bar{\Theta}^* \frac{\partial \Theta}{\partial z} - \Theta \frac{\partial \bar{\Theta}^*}{\partial z} + P\bar{u}^* \frac{\partial u}{\partial z} - Pu \frac{\partial \bar{u}^*}{\partial z}$$

$$+ P\bar{w}^* \frac{\partial w}{\partial z} - Pw \frac{\partial \bar{w}^*}{\partial z} - \bar{w}^* p + w\bar{p}^*)$$

$$+ \int_I \bar{h}^* w_1 + [\bar{\Theta}^* \frac{\partial \Theta}{\partial z} - \Theta \frac{\partial \bar{\Theta}^*}{\partial z} + P\bar{u}^* \frac{\partial u}{\partial z} - Pu \frac{\partial \bar{u}^*}{\partial z}$$

$$+ P\bar{w}^* \frac{\partial w}{\partial z} - Pw \frac{\partial \bar{w}^*}{\partial z} - \bar{w}^* p + w\bar{p}^*]. \qquad (6a.14)$$

From this, the adjoint differential operator is found to be

$$L_0^* X_2 = \begin{pmatrix} \triangle \Theta^* + RPw^* \\ P \triangle u^* - \frac{\partial p^*}{\partial x} \\ P \triangle w^* - \frac{\partial p^*}{\partial z} + \Theta^* \\ 0 \end{pmatrix}. \qquad (6a.15)$$

Moreover, since X_1 satisfies the boundary conditions (6a.1) on Γ_1 and Γ_2, the integrals over these boundaries vanish if

$$\Theta^* = w^* = \frac{\partial u^*}{\partial z} = 0. \qquad (6a.16)$$

Into the interface term in (6a.14), we add

$$-P\bar{w}^* (\frac{\partial u}{\partial x} + \frac{\partial w}{\partial z}) + Pw(\frac{\partial \bar{u}^*}{\partial x} + \frac{\partial \bar{w}^*}{\partial z})$$

which is zero. We integrate the x-derivative by parts and use periodicity. This yields

$$\int_I \bar{h}^* w_1 + [\bar{\Theta}^* \frac{\partial \Theta}{\partial z} - \Theta \frac{\partial \bar{\Theta}^*}{\partial z} + P\bar{u}^* (\frac{\partial u}{\partial z} + \frac{\partial w}{\partial x}) - Pu(\frac{\partial \bar{u}^*}{\partial z} + \frac{\partial \bar{w}^*}{\partial x})$$

$$- \bar{w}^* (p - 2P \frac{\partial w}{\partial z}) + w(\bar{p}^* - 2P \frac{\partial \bar{w}^*}{\partial z})]. \qquad (6a.17)$$

From this we find the adjoint interface conditions:

$$[\Theta^*] = 0,$$

$$[\frac{\partial \Theta^*}{\partial z}] = 0,$$

$$[u^*] = 0,$$

$$[w^*] = 0,$$

$$[\frac{\partial u^*}{\partial z} + \frac{\partial w^*}{\partial x}] = 0,$$

$$[p^* - 2P\frac{\partial w^*}{\partial z}] + h^* = 0. \qquad (6a.18)$$

The next step is to determine the generalized eigenvectors of the unperturbed problem ($\epsilon=0$) for both (6a.5) and the adjoint problem (6a.15) - (6a.18).

Generalized Eigenvectors. These are denoted by a_1, a_2 and b_1, b_2, respectively, and satisfy:

$$L_0 a_1 = 0, \quad B_0 a_1 = 0,$$

$$L_0 a_2 = a_1, \quad B_0 a_2 = 0,$$

$$L_0^* b_1 = 0, \quad B_0^* b_1 = 0,$$

$$L_0^* b_2 = b_1, \quad B_0^* b_2 = 0. \qquad (6a.19)$$

If $\epsilon=0$, the variable h does not occur in the right hand sides of (3a.1), (3a.2), or in the interface conditions (3a.4), and we have the eigenfunction [Chandrasekhar 1961; Joseph 1976 II; Drazin and Reid 1982]

$$a_1 = e^{i\alpha x} \begin{pmatrix} 0 \\ 0 \\ 0 \\ 1 \end{pmatrix}. \qquad (6a.20)$$

The equations (3a.1), (3a.2), (3a.4) and (6a.1) are precisely those characterizing the one-fluid Bénard problem. The eigenfunction for this problem now yields a generalized eigenfunction:

$$a_2 = \frac{e^{i\alpha x}}{\sin \pi l_1} \begin{pmatrix} \frac{\sin \pi z}{\pi^2 + \alpha^2} \\ \frac{i\pi}{\alpha} \cos \pi z \\ \sin \pi z \\ 0 \end{pmatrix}. \qquad (6a.21)$$

The adjoint equations agree with the one-fluid Bénard problem if we set $h^* = 0$. Thus the eigenfunction of the one-fluid Bénard problem yields the eigenfunction for the adjoint:

$$b_1 = e^{i\alpha x} \begin{pmatrix} \frac{9}{2} P \pi^2 \sin \pi z \\ \frac{i\pi}{\alpha} \cos \pi z \\ \sin \pi z \\ 0 \end{pmatrix}. \qquad (6a.22)$$

The generalized eigenvector b_2 of the adjoint is found from the last equation in (6a.19). Setting

$$b_2 = \begin{pmatrix} \Theta^* \\ u^* \\ w^* \\ h^* \end{pmatrix},$$

this leads to the equations:

$$\Delta\Theta^* + RPw^* = \frac{9}{2}P\pi^2 e^{i\alpha x} \sin\pi z,$$

$$P\Delta u^* - \frac{\partial p^*}{\partial x} = \frac{i\pi}{\alpha}e^{i\alpha x}\cos\pi z, \qquad (6a.23)$$

$$P\Delta w^* + \Theta^* - \frac{\partial p^*}{\partial z} = e^{i\alpha x}\sin\pi z,$$

$$\frac{\partial u^*}{\partial x} + \frac{\partial w^*}{\partial z} = 0.$$

We set $w^* = w_0^* e^{i\alpha x}$ etc., and obtain by combining the equations:

$$\left(\frac{d^2}{dz^2} - \alpha^2\right)^3 w_0^* + \frac{27}{8}\pi^6 w_0^* = \frac{9}{4}\pi^4\left(1 + \frac{1}{P}\right)\sin\pi z. \qquad (6a.24)$$

The general solution of this equation is

$$w_0^* = c_1 \sinh Q_1 z + c_2 \cosh Q_1 z$$

$$+ c_3 \sinh Q_2 z + c_4 \cosh Q_2 z$$

$$+ c_5 \sin\pi z + c_6 \cos\pi z - \frac{1}{6\pi}\left(1 + \frac{1}{P}\right)z\cos\pi z \qquad (6a.25)$$

in fluid 1, and

$$w_0^* = d_1 \sinh Q_1(z-1) + d_2 \cosh Q_1(z-1)$$

$$+ d_3 \sinh Q_2(z-1) + d_4 \cosh Q_2(z-1)$$

$$+ d_5 \sin\pi(z-1) + d_6 \cos\pi(z-1) - \frac{1}{6\pi}\left(1 + \frac{1}{P}\right)z\cos\pi z \qquad (6a.26)$$

in fluid 2, where

$$Q_1 = \frac{\pi}{2}\sqrt[4]{52}\, e^{i\phi/2}, \quad Q_2 = \frac{\pi}{2}\sqrt[4]{52}\, e^{-i\phi/2}, \qquad (6a.27)$$

and ϕ is determined by $\cos\phi = 5/\sqrt{52}$, $\sin\phi = 3\sqrt{3/52}$. The coefficients c_1 - c_6 and d_1 - d_6 must be determined such that the boundary conditions are satisfied. By using (6a.23), we can show that the conditions (6a.16) at the walls reduce to

$$w_0^* = \frac{d^2}{dz^2}w_0^* = \frac{d^4}{dz^4}w_0^* = 0. \qquad (6a.28)$$

At z=0, this yields

$$c_2 + c_4 + c_6 = 0,$$
$$Q_1^2 c_2 + Q_2^2 c_4 - \pi^2 c_6 = 0,$$
$$Q_1^4 c_2 + Q_2^4 c_4 + \pi^4 c_6 = 0. \tag{6a.29}$$

From this, we obtain

$$c_2 = c_4 = c_6 = 0. \tag{6a.30}$$

At $z=1$, we find

$$d_2 + d_4 + d_6 + \frac{1}{6\pi}(1 + \frac{1}{P}) = 0,$$

$$Q_1^2 d_2 + Q_2^2 d_4 - \pi^2 d_6 - \frac{\pi}{6}(1 + \frac{1}{P}) = 0,$$

$$Q_1^4 d_2 + Q_2^4 d_4 + \pi^4 d_6 + \frac{\pi^3}{6}(1 + \frac{1}{P}) = 0. \tag{6a.31}$$

This yields

$$d_2 = d_4 = 0, \quad d_6 = -\frac{1}{6\pi}(1 + \frac{1}{P}). \tag{6a.32}$$

The first five of the conditions (6a.18) lead, after eliminating u^* and Θ^* from (6a.23), to the conditions

$$[w_0^*] = [\frac{dw_0^*}{dz}] = [\frac{d^2 w_0^*}{dz^2}] = [\frac{d^4 w_0^*}{dz^4}]$$

$$= [\frac{d^5 w_0^*}{dz^5} - \pi^2 \frac{d^3 w_0^*}{dz^3}] = 0. \tag{6a.33}$$

We can set the coefficient d_5 to zero for the following reason. The coefficient d_5 multiplies $w = \sin \pi(z - 1)$ in the generalized eigenvector. The eigenvector b_1 has $w = \sin \pi z$. Since any multiple of b_1, added to b_2, i.e. $b_2 + Cb_1$, is also a generalized eigenvector, we choose C to be d_5. This essentially gets rid of d_5 in Ω_2, replaces c_5 in Ω_1 by $c_5 + d_5$, and we rename $c_5 + d_5$ as c_5. We thus obtain the following system of equations.

$$c_1 \sinh Q_1 l_1 + c_3 \sinh Q_2 l_1 + c_5 \sin \pi l_1$$

$$= -d_1 \sinh Q_1 l_2 - d_3 \sinh Q_2 l_2 + d_6 \cos \pi l_2,$$

$$c_1 Q_1 \cosh Q_1 l_1 + c_3 Q_2 \cosh Q_2 l_1 + \pi c_5 \cos \pi l_1 = Q_1 d_1 \cosh Q_1 l_2$$

$$+ Q_2 d_3 \cosh Q_2 l_2 + \pi d_6 \sin \pi l_2,$$

$$c_1 Q_1^2 \sinh Q_1 l_1 + c_3 Q_2^2 \sinh Q_2 l_1 - \pi^2 c_5 \sin \pi l_1 = -Q_1^2 d_1 \sinh Q_1 l_2$$

$$- Q_2^2 d_3 \sinh Q_2 l_2 - \pi^2 d_6 \cos \pi l_2,$$

$$c_1 Q_1^4 \sinh Q_1 l_1 + c_3 Q_2^4 \sinh Q_2 l_1 + \pi^4 c_5 \sin \pi l_1 = -Q_1^4 d_1 \sinh Q_1 l_2$$

$$- Q_2^4 d_3 \sinh Q_2 l_2 + \pi^4 d_6 \cos \pi l_2, \tag{6a.34}$$

$$c_1 (Q_1^5 - \pi^2 Q_1^3) \cosh Q_1 l_1 + c_3 (Q_2^5 - \pi^2 Q_2^3) \cosh Q_2 l_1 + 2c_5 \pi^5 \cos \pi l_1$$

$$= d_1 (Q_1^5 - \pi^2 Q_1^3) \cosh Q_1 l_2 + d_3 (Q_2^5 - \pi^2 Q_2^3) \cosh Q_2 l_2 + 2d_6 \pi^5 \sin \pi l_2.$$

We eliminate c_1 from the third and fourth equations by using the first equation.

$$(Q_2^2 - Q_1^2) \sinh Q_2 l_1 c_3 - (\pi^2 - Q_1^2) \sin \pi l_1 c_5 + (Q_2^2 - Q_1^2) \sinh Q_2 l_2$$
$$= -(\pi^2 + Q_1^2) \cos \pi l_2 d_6,$$
$$(Q_2^3(Q_2^2 - \pi^2) - Q_1^2 Q_2(Q_1^2 - \pi^2)) \cosh Q_2 l_1 c_3 + (2\pi^5 - \pi Q_1^2(Q_1^2 - \pi^2)) \cos \pi l_1 c_5$$
$$+ (-Q_2^3(Q_2^2 - \pi^2) + Q_2 Q_1^2(Q_1^2 - \pi^2)) \cosh Q_2 l_2 d_3$$
$$= (2\pi^5 - Q_1^2 \pi(Q_1^2 - \pi^2)) \sin \pi l_2 d_6,$$
$$(Q_2^4 - Q_1^4) \sinh Q_2 l_1 c_3 + (\pi^4 - Q_1^4) \sin \pi l_1 c_5 + (Q_2^4 - Q_1^4) \sinh Q_2 l_2 d_3$$
$$= (\pi^4 - Q_1^4) \cos \pi l_2 d_6. \tag{6a.35}$$

The first and third equations of (6a.35) yield

$$c_5 = d_6 \cot \pi l_2, \tag{6a.36}$$

and

$$c_3 \sinh Q_2 l_1 + d_3 \sinh Q_2 l_2 = 0. \tag{6a.37}$$

The second equation of (6a.35) yields

$$c_3 \cosh Q_2 l_1 - c_3 \cosh Q_2 l_2 = \frac{\pi d_6(1 + \sqrt{3}i)}{2Q_2 \sin \pi l_2}.$$

Hence,

$$c_3 = \frac{d_6 \pi(1 + \sqrt{3}i) \sinh Q_2 l_2}{2Q_2 \sin \pi l_2 \sinh Q_2}, \tag{6a.38}$$

and

$$d_3 = -\frac{d_6 \pi(1 + \sqrt{3}i) \sinh Q_2 l_1}{2Q_2 \sinh Q_2 \sin \pi l_2}. \tag{6a.39}$$

From the first and second equations of (6a.34), we find that

$$c_1 = \bar{c}_3, \quad \text{and} \quad d_1 = \bar{d}_3. \tag{6a.40}$$

Having calculated a_1, a_2, b_1, b_2, we return to the evaluation of (6a.12).

Calculation of Some Inner Products. In order to apply formula (6a.12), we must determine the expressions

$$\langle b_i, (L(\epsilon) - 1)^{-1} a_j \rangle \tag{6a.41}$$

to first order in ϵ. To facilitate this calculation, we introduce x_j^0 and x_j^1 defined by

$$(L(\epsilon) - 1)^{-1} a_j = x_j^0 + \epsilon x_j^1 + O(\epsilon^2). \tag{6a.42}$$

Equating the coefficients of equal powers of ϵ, we find the equations governing x_j^0 and x_j^1

$$(L_0 - 1)x_j^0 = a_j,$$

$$B_0 x_j^0 = 0, \tag{6a.43}$$

and

$$L_1 x_j^0 + (L_0 - 1)x_j^1 = 0,$$

$$B_1 x_j^0 + B_0 x_j^1 = 0. \tag{6a.44}$$

From (6a.43), we immediately find $x_1^0 = -a_1$, $x_2^0 = -a_2 - a_1$. We will not need the solutions x_j^1 to the perturbation problem (6a.44) but only certain inner products involving them, namely $\langle b_i, x_j^1 \rangle$. This is seen from (6a.12) and (6a.42):

$$\Psi_{ij}(\epsilon, \hat\sigma) = \langle b_i, x_j^0 \rangle + \epsilon \langle b_i, x_j^1 \rangle - \hat\sigma \langle b_i, a_j \rangle + O(\epsilon^2). \tag{6a.45}$$

We calculate $\langle b_i, x_j^1 \rangle$ from (6a.44):

$$\langle b_i, x_j^1 \rangle = \langle b_i, L_1 x_j^0 \rangle + \langle b_i, L_0 x_j^1 \rangle, \tag{6a.46}$$

and an integration by parts:

$$\langle b_i, L_0 x_j^1 \rangle = \langle L_0^* b_i, x_j^1 \rangle + boundary\ integrals, \tag{6a.47}$$

where the boundary integrals are evaluated using the second part of (6a.44). (The boundary integrals would vanish if $B_0 x_j^1$ were zero.) Details of these calculations are given next.

The boundary integrals that arise in (6a.47) and the entries of the matrix Ψ_{ij} defined in (6a.45) are calculated. In addition to the boundary terms arising in (6a.47), we also integrate the term arising from $\tilde p$ in $\langle b_i, L_1 x_j^0 \rangle$ by parts. This yields another integral over the interface, which we combine with those from (6a.47) into an expression Γ_{ij}. The form of Γ_{ij} can be read off from the calculation of the adjoint in equations (6a.13) -(6a.18). The terms remaining in $\langle b_i, L_1 x_j^0 \rangle$ will be denoted by

$$\langle\langle b_i, L_1 x_j^0 \rangle\rangle. \tag{6a.48}$$

We thus have

$$\langle b_i, L_0 x_j^1 \rangle + \langle b_i, L_1 x_j^0 \rangle = \langle L_0^* b_i, x_j^1 \rangle + \langle\langle b_i, L_1 x_j^0 \rangle\rangle + \Gamma_{ij}, \tag{6a.49}$$

where

$$\Gamma_{ij} = \int_I \bar\Theta^* \zeta \frac{\partial \Theta}{\partial z} - \zeta h \frac{\partial \bar\Theta^*}{\partial z} + P \bar u^* \bar m \left(\frac{\partial u}{\partial z} + \frac{\partial w}{\partial x} \right)$$

$$- \bar w^* \left(-2P\bar m \frac{\partial w}{\partial z} + h(\bar M_1 + \frac{\pi^2}{2} P\bar S) \right) dx, \tag{6a.50}$$

$$b_i = \begin{pmatrix} \Theta^* \\ u^* \\ w^* \\ h^* \end{pmatrix} \quad \text{and} \quad x_j^0 = \begin{pmatrix} \Theta \\ u \\ w \\ h \end{pmatrix}. \tag{6a.51}$$

Here, the interval of integration I extends over one wavelength in x, at $z = l_1$. Hence, equations (6a.46) becomes:

$$\langle b_1, x_1^1 \rangle = \Gamma_{11} + \langle\langle b_1, L_1 x_1^0 \rangle\rangle,$$

$$\langle b_1, x_2^1 \rangle = \Gamma_{12} + \langle\langle b_1, L_1 x_2^0 \rangle\rangle,$$

$$\langle b_2, x_1^1 \rangle = \Gamma_{21} + \langle\langle b_2, L_1 x_1^0 \rangle\rangle + \langle b_1, x_1^1 \rangle, \tag{6a.52}$$

$$\langle b_2, x_2^1 \rangle = \Gamma_{22} + \langle\langle b_2, L_1 x_2^0 \rangle\rangle + \langle b_1, x_2^1 \rangle,$$

where

$$\Gamma_{11} = 2\sqrt{2}(9P\frac{\pi^3}{2}\bar{\zeta}\cos\pi l_1 + (\bar{M}_1 + \frac{\pi^2}{2}P\bar{S})\sin\pi l_1),$$

$$\Gamma_{12} = 2\sqrt{2}(P\pi\cos\pi l_1((-3 + 9\frac{\pi^2}{2})\bar{\zeta} - \bar{m}) + (\bar{M}_1 + \frac{\pi^2}{2}P\bar{S})\sin\pi l_1),$$

and

$$\Gamma_{21} = 2\frac{\pi}{\alpha}(\bar{\zeta}\frac{\partial\bar{\Theta}^*}{\partial z} + \bar{w}_1^*(\bar{M}_1 + \frac{\pi^2}{2}P\bar{S}))_{z=l_1}$$

$$= 2\sqrt{2}(\frac{\pi}{4}(1 - 11P)\bar{\zeta}\cos\pi l_1$$

$$+2P\frac{\bar{\zeta}}{\pi^2}(c_1(Q_1^2 - \frac{\pi^2}{2})^2 Q_1\cosh Q_1 l_1 + c_3(Q_2^2 - \frac{\pi^2}{2})^2 Q_2\cosh Q_2 l_1)$$

$$+(1 + P)\bar{\zeta}\frac{3}{4}\pi^2(\frac{cos^2\pi l_1}{sin\ \pi l_1} + l_1\sin\pi l_1)$$

$$+(\bar{M}_1 + \frac{\pi^2}{2}P\bar{S})(c_1\sinh Q_1 l_1 + c_3\sinh Q_2 l_1 + (1 + \frac{1}{P})\frac{l_2}{6\pi}\cos\pi l_1)).$$

(Here Θ^* etc. denote components of b_2 without the factor $e^{i\alpha x}$, which has already been integrated out.)

We will see later that Γ_{22} is not required. Since $L_1 x_1^0$ contains only the $\tilde{p}$-term in both fluids, the first equation in (6a.52) reduces to

$$\langle b_1, x_1^1 \rangle = \Gamma_{11}. \tag{6a.53}$$

Since $x_1^0 = -a_1$,

$$\Psi_{11}(\epsilon, \hat{\sigma}) = -(1 + \hat{\sigma})\langle b_1, a_1 \rangle + \epsilon\Gamma_{11} + O(\epsilon^2). \tag{6a.54}$$

We find that $\langle b_1, a_1 \rangle = 0$ so that

$$\Psi_{11}(\epsilon, \hat{\sigma}) = \epsilon\Gamma_{11} + O(\epsilon^2). \tag{6a.55}$$

Since $x_2^0 = -a_1 - a_2$,

$$\Psi_{12}(\epsilon, \hat{\sigma}) = -(1 + \hat{\sigma})\langle b_1, a_2 \rangle + \epsilon(\Gamma_{12} + \langle\langle b_1, L_1 x_2^0 \rangle\rangle) + O(\epsilon^2). \tag{6a.56}$$

We set

$$\hat{\sigma} = -1 + \tau_1\epsilon^{1/2} + \tau_2\epsilon + O(\epsilon^{3/2}) \tag{6a.57}$$

so that

$$\Psi_{12}(\epsilon, \hat{\sigma}) = -\sqrt{\epsilon}\tau_1\langle b_1, a_2\rangle + \epsilon(-\tau_2\langle b_1, a_2\rangle + \Gamma_{12} + \langle\langle b_1, L_1 x_2^0\rangle\rangle) + O(\epsilon^{3/2}).$$
$$(6a.58)$$

We have

$$\langle b_1, a_2\rangle = \frac{1}{\sin\pi l_1}\frac{2\pi}{\alpha}\int_0^1 \frac{9P\pi^2\sin^2\pi z}{2(\pi^2+\alpha^2)} + \frac{\pi^2\cos^2\pi z}{\alpha^2} + \sin^2\pi z dz$$

$$= \frac{3\sqrt{2}(1+P)}{\sin\pi l_1}, \qquad (6a.59)$$

and

$$\langle\langle b_1, L_1 x_2^0\rangle\rangle = \frac{2\pi}{\alpha\sin\pi l_1}\left(\bar{\zeta}l_2\int_0^{l_1}\frac{9}{2}P\pi^2(-\sin^2\pi z)dz\right.$$

$$+\int_{l_1}^1 \sin^2\pi z(\frac{9}{2}P\pi^2\bar{\zeta}l_1 + \frac{9}{2}P\pi^2\bar{\gamma} + \frac{3}{2}P\pi^2\bar{m} - \frac{9}{2}P\pi^2(\bar{r}+\bar{\beta}))$$

$$\left.+3P\bar{m}\pi^2\cos^2\pi z dz\right)$$

$$= \frac{2\sqrt{2}}{\sin\pi l_1}(\frac{9}{4}P\pi^2 l_2(\bar{\gamma}+\bar{m}-\bar{r}-\bar{\beta}) + \frac{9}{8}P\pi\sin 2\pi l_1(\bar{\zeta}+\bar{\gamma}-\bar{r}-\bar{\beta}-\frac{\bar{m}}{3})). \quad (6a.60)$$

Next,

$$\Psi_{21}(\epsilon, \hat{\sigma}) = -(1+\hat{\sigma})\langle b_2, a_1\rangle + \epsilon(\Gamma_{21} + \langle b_1, x_1^1\rangle) + O(\epsilon^2). \qquad (6a.61)$$

We note that

$$\langle b_2, a_1\rangle = \langle b_2, L_0 a_2\rangle = \langle L_0^* b_2, a_2\rangle = \langle b_1, a_2\rangle. \qquad (6a.62)$$

Hence,

$$\Psi_{21}(\epsilon, \hat{\sigma}) = -\sqrt{\epsilon}\tau_1\langle b_1, a_2\rangle + \epsilon(\Gamma_{21} + \langle b_1, x_1^1\rangle) + O(\epsilon^{3/2}). \qquad (6a.63)$$

We find
$$\Psi_{22}(\epsilon, \hat{\sigma}) = \langle b_2, x_2^0\rangle - (-1 + \sqrt{\epsilon}\tau_1)\langle b_2, a_2\rangle + O(\epsilon)$$
$$= -\langle b_2, a_1\rangle - \sqrt{\epsilon}\tau_1\langle b_2, a_2\rangle + O(\epsilon). \qquad (6a.64)$$

Collecting the $O(\epsilon)$-terms from the equation $det\Psi_{ij} = 0$, we find

$$\tau_1^2 = -\frac{\Gamma_{11}}{\langle b_2, a_1\rangle}. \qquad (6a.65)$$

Collecting the $O(\epsilon^{3/2})$-terms, we find

$$\tau_2 = \frac{\Gamma_{12} + \Gamma_{21} + \langle\langle b_1, L_1 x_2^0\rangle\rangle}{2\langle b_1, a_2\rangle} + \frac{\tau_1^2}{2}(-1 + \frac{\langle b_2, a_2\rangle}{\langle b_1, a_2\rangle}). \qquad (6a.66)$$

We now need to calculate $\langle b_2, a_2\rangle$, where a_2 is given by (6a.21) and b_2 by (6a.23) - (6a.40):

$$\langle b_2, a_2 \rangle = \frac{2\pi}{\alpha \sin \pi l_1} \{ \int_0^{l_1} + \int_{l_1}^1 \} \frac{\bar{\Theta}^* \sin \pi z}{(\pi^2 + \alpha^2)} + \frac{i\pi \bar{u}^* \cos \pi z}{\alpha} + \bar{w}^* \sin \pi z \, dz.$$

$$(6a.67)$$

We express the integrand in terms of w^* by using equations (6a.23):

$$i\bar{u}^* = \frac{\bar{w}_z^*}{\alpha} \quad \text{and} \quad \bar{\Theta}^* = 3 \sin \pi z + \frac{2P}{\pi^2} \Delta^2 \bar{w}^*. \qquad (6a.68)$$

Hence,

$$\langle b_2, a_2 \rangle = \frac{2\sqrt{2}}{\sin \pi l_1} \{ \frac{1}{\pi^2} + (\int_0^{l_1} + \int_{l_1}^1) \frac{4P}{3\pi^4} \Delta^2 \bar{w}^* \sin \pi z + 3\bar{w}^* \sin \pi z dz \}.$$

$$(6a.69)$$

Integration by parts simplifies the first integral as follows:

$$\int \Delta^2 \bar{w}^* \sin \pi z dz = \sin \pi l_1 [\frac{\partial^3 \bar{w}^*}{\partial z^3}] + \frac{9\pi^4}{4} \int \bar{w}^* \sin \pi z dz,$$

so that we are left with having to evaluate $[\frac{\partial^3 \bar{w}^*}{\partial z^3}]$ and $\int \bar{w}^* \sin \pi z dz$ and substituting them into

$$\langle b_2, a_2 \rangle = \frac{2\sqrt{2}}{\sin \pi l_1} \{ \frac{1}{\pi^2} + \frac{4P \sin \pi l_1}{3\pi^4} [\frac{\partial^3 \bar{w}^*}{\partial z^3}] + 3(1 + P) \int \bar{w}^* \sin \pi z dz \}.$$

$$(6a.70)$$

The former is facilitated by multiplying equation (6a.24) by $\sin \pi z$ and integrating by parts. This leads to

$$[\frac{\partial^3 \bar{w}^*}{\partial z^3}] = -\frac{3\pi^2(1 + \frac{1}{P})}{4 \sin \pi l_1}. \qquad (6a.71)$$

From (6a.25) - (6a.26), we have

$$\bar{w}^* = c_1 \sinh Q_1 z + c_3 \sinh Q_2 z + c_5 \sin \pi z - \frac{(1 + \frac{1}{P})}{6\pi} z \cos \pi z \qquad (6a.72)$$

in fluid 1 and

$$\bar{w}^* = d_1 \sinh Q_1(z - 1) + d_3 \sinh Q_2(z - 1) + \frac{1}{6\pi}(1 + \frac{1}{P})(1 - z) \cos \pi z$$

$$(6a.73)$$

in fluid 2, with the coefficients given by (6a.30), (6a.32), (6a.36), (6a.38) - (6a.40). After some algebra, we find

$$\int_0^1 \bar{w}^* \sin \pi z dz = -\frac{(1 + \frac{1}{P})}{12\pi}(l_1 \cot \pi l_2 + \frac{1}{2\pi}). \qquad (6a.74)$$

Using (6a.70) - (6a.74),

$$\langle b_2, a_2 \rangle = -\frac{2\sqrt{2}}{\pi \sin \pi l_1} \{ \frac{P}{\pi} + \frac{(1 + P)^2}{4P}(l_1 \cot \pi l_2 + \frac{1}{2\pi}) \}. \qquad (6a.75)$$

In the following section, we discuss the solution of the eigenvalue perturbation problem, which now reads $\det \Psi_{ij} = 0$, where Ψ is given by (6a.45).

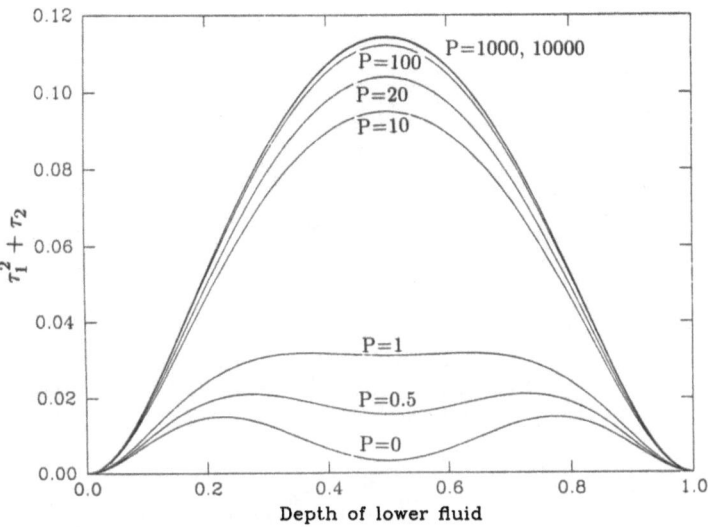

Fig. 6.1. [Renardy and Renardy, 1985, American Institute of Physics] Graph of $\tau_1^2 + \tau_2$ versus l_1 for various Prandtl numbers; $\bar{S}=1$, $\bar{\zeta}= \bar{r} = \bar{\beta} =\bar{m}= \bar{\gamma}=0$, $\hat{\alpha}_1 \Delta\theta^*=0.001$.

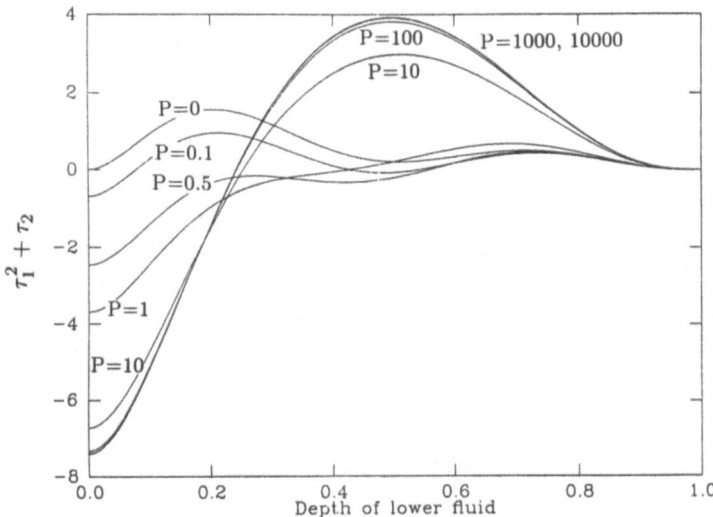

Fig. 6.2. [Renardy and Renardy, 1985, American Institute of Physics] Graph of $\tau_1^2 + \tau_2$ versus l_1 for various Prandtl numbers; $\bar{\beta}=1$, $\bar{\zeta}= \bar{r} = \bar{S} =\bar{m}= \bar{\gamma}=0$, $\hat{\alpha}_1 \Delta\theta^*=0.001$.

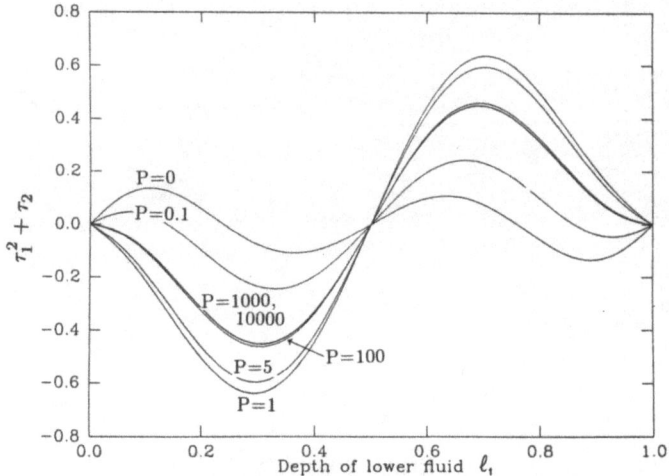

Fig. 6.3. [Renardy and Renardy, 1985, American Institute of Physics] Graph of $\tau_1^2 + \tau_2$ versus l_1 for various Prandtl numbers; $\bar{\zeta}=1$, $\bar{\beta}= \bar{r} = \bar{S} =\bar{m}= \bar{\gamma}=0$, $\hat{\alpha}_1 \Delta\theta^*=0.001$.

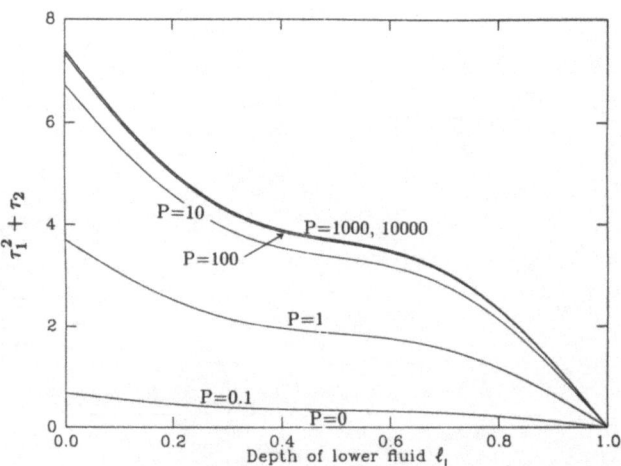

Fig. 6.4. [Renardy and Renardy, 1985, American Institute of Physics] Graph of $\tau_1^2 + \tau_2$ versus l_1 for various Prandtl numbers; $\bar{m}=1$, $\bar{\zeta}= \bar{r} = \bar{S} =\bar{\beta}= \bar{\gamma}=0$, $\hat{\alpha}_1 \Delta\theta^*=0.001$.

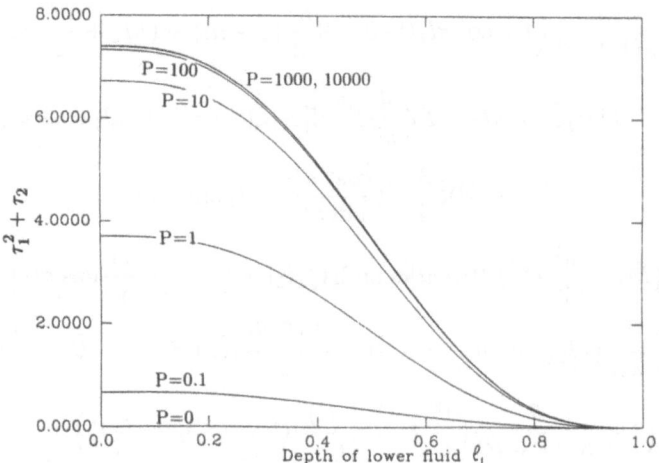

Fig. 6.5. [Renardy and Renardy, 1985, American Institute of Physics] Graph of $\tau_1^2 + \tau_2$ versus l_1 for various Prandtl numbers; $\bar{\gamma}=1$, $\bar{\zeta} = \bar{r} = \bar{S} = \bar{\beta} = \bar{m} = 0$, $\hat{\alpha}_1 \Delta\theta^* = 0.001$.

III.6(a)(ii) Results and Discussion

Since we set $\hat{\sigma} = -1 + \tau_1 \epsilon^{1/2} + \tau_2 \epsilon + O(\epsilon^{3/2})$ (see equation (6a.57)), and $\sigma = 1 + 1/\hat{\sigma}$,

$$\sigma = -\tau_1 \sqrt{\epsilon} - (\tau_1^2 + \tau_2)\epsilon + O(\epsilon^{3/2}). \qquad (6a.76)$$

For the first-order term, we find (see equations (6a.52), (6a.59), (6a.62), (6a.65))

$$\tau_1^2 = -\frac{P\pi^2 \sin^2 \pi l_1}{3(P+1)} \left(9\pi\bar{\zeta} \cot \pi l_1 + \frac{27}{2}\pi^2[-\frac{\bar{r}}{\hat{\alpha}_1 \Delta T^*} + l_2(\bar{\beta}+\bar{r})] + \bar{S}\right). \qquad (6a.77)$$

The sign of this quantity determines whether the eigenvalue splits into real values or complex conjugates. We note that, in the example of a Hopf bifurcation given in section 3(d), $\bar{\zeta}$, $\bar{r}$ and $\bar{S}$ were 0, and $\bar{\beta}$ was positive. (The boundary conditions there are "rigid- rigid" rather than "free-free"; nevertheless, some qualitative similarity is expected). Complex eigenvalues are found, in agreement with the prediction of (6a.77). It is also worth noting that (6a.77) does not involve the viscosity ratio, and that the sign does not depend on the Prandtl number.

If τ_1^2 is positive, we always have instability. However, if τ_1^2 is negative, we have to find $\tau_1^2 + \tau_2$ in order to determine whether the perturbations are stabilizing or destabilizing. Unfortunately, the formula for τ_2 is not nearly as simple as that for τ_1 (see equations (6a.59), (6a.60), (6a.63), (6a.75)). We find

$$
\tau_1^2 + \tau_2 = \frac{\sin \pi l_1}{3(1+P)} \Big(P\pi \cos \pi l_1 ([-3 + 9\frac{\pi^2}{2}]\bar{\zeta} - \bar{m}) + (\bar{M}_1 + \frac{\pi^2}{2} P\bar{S}) \sin \pi l_1
$$

$$
+\frac{\pi}{4}(1 - 11P)\bar{\zeta} \cos \pi l_1 + P\bar{\zeta}\frac{4}{\pi^2} Real[c_1(Q_1^2 - \pi^2/2)^2 Q_1 \cosh Q_1 l_1]
$$

$$
+(1+P)\bar{\zeta}\frac{3}{4}\pi^2(\frac{\cos^2 \pi l_1}{\sin \pi l_1} + l_1 \sin \pi l_1)
$$

$$
+(\bar{M}_1 + \frac{\pi^2}{2}P\bar{S})(2Real[c_1 \sinh Q_1 l_1] + (1 + \frac{1}{P})\frac{l_2}{6\pi}\cos \pi l_1)
$$

$$
+\frac{9\pi P}{4\sin \pi l_1}\Big(\pi l_2(\bar{\gamma} + \bar{m} - \bar{r} - \bar{\beta}) + \frac{\sin 2\pi l_1}{2}(\bar{\zeta} + \bar{\gamma} - \bar{r} - \bar{\beta} - \frac{\bar{m}}{3}))\Big)
$$

$$
+\frac{\tau_1^2}{2}\Big(1 - \frac{2}{3\pi(1+P)}(\frac{P}{\pi} + \frac{(1+P)^2}{4P}(l_1 \cot \pi l_2 + \frac{1}{2\pi}))\Big)\Big). \tag{6a.78}
$$

where $\bar{M}_1$ is defined in (6a.2), c_1 and Q_1 are defined in equations (6a.27), (6a.38), and (6a.40).

We examine some limiting cases below. If the Oberbeck-Boussinesq approximation is to be justified, then $\hat{\alpha}_1 \Delta\theta^*$ must be small, e.g. $\hat{\alpha} = 5 \times 10^{-4}K^{-1}$ and $\Delta\theta^* \leq 10K$ [Drazin and Reid 1982; Joseph 1976 II]. For the purpose of computing graphs, we have taken

$$
\hat{\alpha}_1 \Delta\theta^* = 0.001. \tag{6a.79}
$$

As $P \to \infty$, both τ_1^2 and $\tau_1^2 + \tau_2$ approach finite limits. Concerning cases v and vi below, we note that the expansion for $\hat{\sigma}$ in powers of $\epsilon^{1/2}$ is derived under the assumption that $\tau_1 \neq 0$ and breaks down if $\tau_1 = 0$. However, the sum of the two eigenvalues does not behave singularly [Yakubovich and Starzhinskii 1975] but like $-2(\tau_1^2 + \tau_2)\epsilon$, even when $\tau_1 = 0$. In cases v and vi below, one of the eigenvalues remains zero under the perturbation of viscosity or thermal diffusivity because the static solution, with the linear temperature profile, is a solution to the problem wherever the interface is. Therefore, the other eigenvalue is equal to $-2\tau_2\epsilon + O(\epsilon^2)$.

Case i. $\bar{S} \neq 0, \bar{\zeta} = \bar{r} = \bar{\beta} = \bar{m} = 0.$. If $\bar{S} < 0$, then $\tau_1^2 > 0$ so the instability is exponential in time and leads to mixing. If $\bar{S} > 0$, then $\tau_1^2 < 0$ so the surface tension opposes the tendency for convection. However, whether this leads to a growth or not depends on the sign of (6a.78) which depends nonlinearly on l_1 and the Prandtl number. Figure 6.1 shows this dependence to be $\tau_1^2 + \tau_2 > 0$ so that σ is stable and oscillatory.

Case ii. $\bar{r} \neq 0, \bar{S} = \bar{\zeta} = \bar{\beta} = \bar{m} = 0.$. Since $\hat{\alpha}_1 \Delta\theta^*$ is small, τ_1^2 and $\bar{r}$ have the same signs: if Fluid 1 is the less dense fluid, a convective instability takes place; if Fluid 1 is the more dense fluid, stable wavy modes occur. The graph of $\tau_1^2 + \tau_2$ versus l_1 for $\bar{r} < 0$ is similar to figure 6.1.

Case iii. $\bar{\beta} \neq 0, \bar{S} = \bar{r} = \bar{\zeta} = \bar{m} = 0$. . The signs of τ_1^2 and $\bar{\beta}$ are opposite. A heuristic explanation is as follows. In the Boussinesq approximation (see equation (2.1)), the density at any depth l_1 in the buoyancy term of the momentum equations is approximated by

$$\rho_i(1 - \hat{\alpha}_i \Delta \theta^* (\theta - \theta_0)), \ i = 1, 2$$

where

$$(\theta - \theta_0) = 1 - z + \Theta. \tag{6a.80}$$

Hence, at a depth z which is close to the interface $z = l_1$, the densities are approximately $\rho_i(1 - l_2\hat{\alpha}_i\Delta\theta^*)$. Therefore, if $\bar{\beta} < 0$, the lower fluid is less dense than the upper fluid, locally at the interface, so that $\tau_1^2 > 0$. On the other hand, if $\bar{\beta} > 0$, then the lower fluid is the more dense and $\tau_1^2 < 0$. Figure 6.2 presents $\tau_1^2 + \tau_2$ versus l_1 for $\bar{\beta} = 1$, showing that for a wide range of l_1 close to 0 and of Prandtl numbers greater than 0, there are unstable oscillatory modes.

Case iv. $\bar{\zeta} \neq 0, \bar{S} = \bar{r} = \bar{\beta} = \bar{m} = 0$. The dependence of τ_1^2 to l_1 is through $-\bar{\zeta} \sin 2\pi l_1$. Hence, if the thicker layer has the lesser coefficient of conductivity, (i.e. if $\bar{\zeta} > 0$ and $0.5 < l_1 < 1$; or if $\bar{\zeta} < 0$ and $0 < l_1 < 0.5$) then convective instability results. If the thicker layer has the greater coefficient of conductivity, then oscillatory modes occur, but whether these would be stable or not depends on the sign of $\tau_1^2 + \tau_2$. Figure 6.3 shows a wide range of unstable oscillatory modes for $0 < l_1 < 0.5$ when $\bar{\zeta} = 1$. This graph is antisymmetric about $l_1 = 0.5$, as can be deduced non-trivially from the equations. We note that when the fluid with the greater conductivity occupies most of the flowfield, the stability of the eigenvalue σ with nonzero imaginary part depends on the Prandtl number: if the Prandtl number is very small, the oscillatory modes are stable, and if the Prandtl number is well away from 0, the oscillatory modes are unstable.

Case v. $\bar{m} \neq 0, \bar{\zeta} = \bar{S} = \bar{r} = \bar{\beta} = 0$. We find that $\tau_1^2 = 0$, and figure 6.4 shows $\tau_1^2 + \tau_2$ versus l_1 for various Prandtl numbers. If Fluid 1 is the less viscous fluid, then there is stability, and if Fluid 1 is the more viscous fluid then there is convective instability, regardless of the depth ratio, in agreement with heuristic expectation.

Case vi. $\bar{\gamma} \neq 0, \bar{m} = \bar{\zeta} = \bar{S} = \bar{r} = \bar{\beta} = 0$. The ratio γ of thermal diffusivities plays a similar role here to the ratio m of viscosities, which measures the rates of momentum diffusion. As in Case v above, $\tau_1^2 = 0$ and the graph of $\tau_1^2 + \tau_2$ versus l_1 is shown in figure 6.5. If Fluid 1 has the lesser coefficient of thermal diffusivity, then σ is stable. If Fluid 1 has the greater coefficient of thermal diffusivity, then convective instability ensues, regardless of the depth ratio, as expected.

III.6(b) Low Rayleigh Numbers

The above study has shown that when the Rayleigh number is close to the critical value of the one-fluid problem, the two-fluid problem can be overstable. When the Rayleigh number is less than this, the unperturbed one-fluid problem is below criticality, so that the analysis for two similar fluids involves the perturbation of a simple zero eigenvalue. Since the equations are real, the perturbed eigenvalue is also real, and overstability does not occur [Renardy 1986]. We have a steady onset of instability.

The short-wave asymptotic formulas for σ are identical to those obtained for rigid-rigid boundaries (see section 5), and agree with the short-wave asymptotics of our perturbation formula. Differences in density, coefficients of cubical expansion and surface tension are important in the short-wave limit. In the long-wave limit, the volume ratio, and differences in thermal conductivity, density and coefficients of cubical expansion are important.

By making appropriate choices for the parameters, it is possible to find a linearly stable arrangement with the more dense fluid on top. In fact, when the depth of either fluid is relatively small, it is possible to find linearly stable arrangements with the more dense fluid on top [Renardy 1986]. This is a "thin-layer effect", reminiscent of lubricating effects in shear flows in two layers of different viscosity (see chapter IV), where linearly stable arrangements with the heavier fluid on top are possible if the conditions are right.

III.6(b)(i) Perturbation Analysis

The Rayleigh number is arbitrary but less than the critical value of the one-fluid problem, given by equation (3d.4). The wavenumber α is fixed but arbitrary. At $\epsilon = 0$, there is a simple eigenvalue $\sigma = 0$, arising from the presence of the interface. This eigenvalue is perturbed into a power series in ϵ. One can either write the eigenvalue and eigenvector as a series in ϵ and use the Fredholm alternative, or expand the resolvent of the differential operator as done in the previous section 6(a), and also in this section. The two methods are equivalent and lead to the same results.

The procedure for finding the coefficient of ϵ in the series expansion is like the one used in section (a). Suppose σ_0 is an algebraically simple eigenvalue of a matrix L_0. Let A be an eigenvector of L_0 with eigenvalue σ_0, and let C be an eigenvector of L_0^* (the adjoint of L_0) with eigenvalue $\bar{\sigma}_0$ (the overbar here denotes the complex conjugate). Let L_0 be perturbed into $L(\epsilon) = L_0 + \epsilon L_1 + O(\epsilon^2)$ with L_1 depending smoothly on ϵ. Then the perturbed eigenvalue σ is analytic in ϵ and is given by setting the following expression $\Psi(\epsilon, \sigma)$, which represents to $O(\epsilon)$ the projection of $L(\epsilon) - \sigma$, first onto the eigenspace of the unperturbed problem and then onto the adjoint eigenspace, to zero:

$$\Psi(\epsilon, \sigma) = \langle C, (L_0 + \epsilon L_1 - \sigma)A \rangle + O(\epsilon^2). \tag{6b.1}$$

These results for matrices will be applied to an infinite-dimensional operator below. In order to do this, as in the previous section, we will need to redefine Ψ.

The problem to be solved is given by equation (6a.5) - (6a.10), and like (6a.12), Ψ is redefined to be:

$$\Psi(\epsilon, \hat{\sigma}) = \langle C, ((L(\epsilon) - 1)^{-1} - \hat{\sigma})A \rangle + O(\epsilon^2), \tag{6b.2}$$

$$\hat{\sigma} = (\sigma - 1)^{-1}$$

where the C and A are as before.

The boundary value problem adjoint to (6a.5) is found from equations (6a.15), (6a.16), and (6a.18). The eigenvectors A and C are determined below. They satisfy:

$$L_0 A = 0, \ B_0 A = 0,$$
$$L_0^* C = 0, \ B_0^* C = 0. \tag{6b.3}$$

Equation (6a.20) yields the eigenfunction

$$A = e^{i\alpha x} \begin{pmatrix} 0 \\ 0 \\ 0 \\ 1 \end{pmatrix}. \tag{6b.4}$$

The adjoint equations in (6b.3), together with (6a.15), (6a.16) and (6a.18), give rise to

$$C = \begin{pmatrix} \Theta^* \\ u^* \\ w^* \\ h^* \end{pmatrix} \tag{6b.5}$$

where

$$\Delta \Theta^* + RPw^* = 0,$$

$$P \Delta u^* - \frac{\partial p^*}{\partial x} = 0,$$

$$P \Delta w^* + \Theta^* - \frac{\partial p^*}{\partial z} = 0,$$

$$\frac{\partial u^*}{\partial x} + \frac{\partial w^*}{\partial z} = 0. \tag{6b.6}$$

Boundary and interface conditions are given by (6a.16) and (6a.18):

$$\Theta^* = w^* = \frac{\partial u^*}{\partial z} = 0 \quad \text{at} \quad z = 0, 1, \tag{6b.7}$$

$$\text{at} \ z = l_1, \quad [\Theta^*] = 0, \tag{6b.8}$$

$$[\frac{\partial \Theta^*}{\partial z}] = 0,$$

$$[u^*] = 0,$$
$$[w^*] = 0,$$
$$[\frac{\partial u^*}{\partial z} + \frac{\partial w^*}{\partial x}] = 0,$$
$$[p^* - 2P\frac{\partial w^*}{\partial z}] + h^* = 0.$$

We set $w^* = w_0^* e^{i\alpha x}$ etc., and manipulate the equations to find that

$$(\frac{d^2}{dz^2} - \alpha^2)^3 w_0^* + R\alpha^2 w_0^* = 0. \tag{6b.9}$$

The general solution of this equation is

$$w_0^* = c_1 \sinh Q_1 z + c_2 \cosh Q_1 z + c_3 \sinh Q_2 z + c_4 \cosh Q_2 z$$

$$+ c_5 \sinh Q_3 z + c_6 \cosh Q_3 z \tag{6b.10}$$

in fluid 1, and

$$w_0^* = d_1 \sinh Q_1(z - 1) + d_2 \cosh Q_1(z - 1)$$

$$+ d_3 \sinh Q_2(z - 1) + d_4 \cosh Q_2(z - 1)$$

$$+ d_5 \sinh Q_3(z - 1) + d_6 \cosh Q_3(z - 1) \tag{6b.11}$$

in fluid 2, where Q_1 is the complex conjugate of Q_3,

$$Q_1^2 = \alpha^2 + (R\alpha^2)^{1/3} e^{i\pi/3},$$

$$Q_2^2 = \alpha^2 - (R\alpha^2)^{1/3},$$

$$Q_3^2 = \alpha^2 + (R\alpha^2)^{1/3} e^{-i\pi/3}. \tag{6b.12}$$

The case when $Q_2 = 0$ is considered later (see equations (6b.21) - (6b.24)). When $Q_2^2 = -\pi^2$, $\sinh Q_2 = 0$, and the problem reduces to that considered in section (a), i.e. the perturbation of a double zero eigenvalue. This occurs first at $R = \frac{27\pi^4}{4}$ and $\alpha = \pi/\sqrt{2}$, and the Q_1 and Q_3 above reduce to Q_1 and Q_2 of equation (6a.27). In fact, $\sinh Q_2 = 0$ for $R = (j^2\pi^2 + \alpha^2)^3/\alpha^2$, j =non-zero integer, so that when the basic one-fluid problem is at neutral stability, our c_3 and d_3 are ∞.

The coefficients c_1 - c_6 and d_1 - d_6 must be determined such that the boundary conditions are satisfied. The conditions at the walls are given by equation (6a.28). At z=0, this yields

$$c_2 + c_4 + c_6 = 0,$$

$$Q_1^2 c_2 + Q_2^2 c_4 + Q_3^2 c_6 = 0, \tag{6b.13}$$

$$Q_1^4 c_2 + Q_2^4 c_4 + Q_3^4 c_6 = 0.$$

From this, we obtain

$$c_2 = c_4 = c_6 = 0. \tag{6b.14}$$

At z=1, we find

$$d_2 + d_4 + d_6 = 0,$$
$$Q_1^2 d_2 + Q_2^2 d_4 + Q_3^2 d_6 = 0, \qquad (6b.15)$$
$$Q_1^4 d_2 + Q_2^4 d_4 + Q_3^4 d_6 = 0.$$

This yields

$$d_2 = d_4 = d_6 = 0. \qquad (6b.16)$$

The first five of conditions (6a.18) lead, after eliminating u^* and Θ^*, to the conditions

$$[w_0^*] = [\frac{dw_0^*}{dz}] = [\frac{d^2 w_0^*}{dz^2}] = [\frac{d^4 w_0^*}{dz^4}]$$
$$= [\frac{d^5 w_0^*}{dz^5} - 2\alpha^2 \frac{d^3 w_0^*}{dz^3}] = 0. \qquad (6b.17)$$

We thus obtain the following system of equations.

$$c_1 \sinh Q_1 l_1 + c_3 \sinh Q_2 l_1 + c_5 \sinh Q_3 l_1$$
$$= -d_1 \sinh Q_1 l_2 - d_3 \sinh Q_2 l_2 - d_5 \sinh Q_3 l_2,$$
$$c_1 Q_1 \cosh Q_1 l_1 + c_3 Q_2 \cosh Q_2 l_1 + c_5 Q_3 \cosh Q_3 l_1 = Q_1 d_1 \cosh Q_1 l_2$$
$$+ Q_2 d_3 \cosh Q_2 l_2 + Q_3 d_5 \cosh Q_3 l_2,$$
$$c_1 Q_1^2 \sinh Q_1 l_1 + c_3 Q_2^2 \sinh Q_2 l_1 + c_5 Q_3^2 \sinh Q_3 l_1 = -Q_1^2 d_1 \sinh Q_1 l_2$$
$$- Q_2^2 d_3 \sinh Q_2 l_2 - Q_3^2 d_5 \sinh Q_3 l_2,$$
$$c_1 Q_1^4 \sinh Q_1 l_1 + c_3 Q_2^4 \sinh Q_2 l_1 + c_5 Q_3^4 \sinh Q_3 l_1 = -Q_1^4 d_1 \sinh Q_1 l_2$$
$$- Q_2^4 d_3 \sinh Q_2 l_2 - Q_3^4 d_5 \sinh Q_3 l_2, \qquad (6b.18)$$
$$c_1 (Q_1^5 - 2\alpha^2 Q_1^3) \cosh Q_1 l_1 + c_3 (Q_2^5 - 2\alpha^2 Q_2^3) \cosh Q_2 l_1$$
$$+ c_5 (Q_3^5 - 2\alpha^2 Q_3^3) \cosh Q_3 l_1 = d_1 (Q_1^5 - 2\alpha^2 Q_1^3) \cosh Q_1 l_2$$
$$+ d_3 (Q_2^5 - 2\alpha^2 Q_2^3) \cosh Q_2 l_2 + d_5 (Q_3^5 - 2\alpha^2 Q_3^3) \cosh Q_3 l_2.$$

From these, we find the following relations that will be useful later:

$$c_1 = -d_1 \frac{\sinh Q_1 l_2}{\sinh Q_1 l_1},$$
$$c_3 = -d_3 \frac{\sinh Q_2 l_2}{\sinh Q_2 l_1}, \qquad (6b.19)$$
$$c_5 = -d_5 \frac{\sinh Q_3 l_2}{\sinh Q_3 l_1}.$$

We express the coefficients in terms of d_5:

$$c_1 = d_5 \frac{Q_3 \sinh Q_3 \sinh Q_1 l_2}{Q_1 \sinh Q_1 \sinh Q_3 l_1} \frac{(1 + i\sqrt{3})}{2}$$

$$c_3 = d_5 \frac{Q_3 \sinh Q_3 \sinh Q_2 l_2}{Q_2 \sinh Q_2 \sinh Q_3 l_1} \frac{(1 - i\sqrt{3})}{2}$$

$$c_5 = -d_5 \frac{\sinh Q_3 l_2}{\sinh Q_3 l_1} \qquad (6b.20)$$

$$d_1 = -d_5 \frac{Q_3 \sinh Q_3 \sinh Q_1 l_1}{Q_1 \sinh Q_1 \sinh Q_3 l_1} \frac{(1 + i\sqrt{3})}{2}$$

$$d_3 = -d_5 \frac{Q_3 \sinh Q_3 \sinh Q_2 l_1}{Q_2 \sinh Q_2 \sinh Q_3 l_1} \frac{(1 - i\sqrt{3})}{2}.$$

When $\alpha^2 = (R\alpha^2)^{1/3}$, $Q_2 = 0$. The general solution of (6b.9) is

$$w_0^* = c_1 \sinh Q_1 z + c_3 z + c_5 \sinh Q_3 z \qquad \text{in fluid 1} \qquad (6b.21)$$

and

$$w_0^* = d_1 \sinh Q_1 (z-1) + d_3 (z-1) + d_5 \sinh Q_3 (z-1) \qquad \text{in fluid 2.} \quad (6b.22)$$

As expected, from taking the limit as $Q_2 \to 0$ in (6b.19),

$$c_3 = -\frac{l_2}{l_1} d_3. \qquad (6b.24)$$

Hence, (6b.20) holds for c_1, c_5 and d_1, and

$$c_3 = d_5 \frac{Q_3 l_2 \sinh Q_3}{\sinh Q_3 l_1} \frac{(1 - i\sqrt{3})}{2}. \qquad (6b.24)$$

Having found A and C, we return to the calculation of (6b.2), where we must determine the expressions

$$\langle C, (L(\epsilon) - 1)^{-1} A \rangle \qquad (6b.25)$$

to first order in ϵ. To facilitate this calculation, we introduce x^0 and x^1 defined by

$$(L(\epsilon) - 1)^{-1} A = x^0 + \epsilon x^1 + O(\epsilon^2). \qquad (6b.26)$$

Equating the coefficients of equal powers of ϵ, we find the equations governing x^0 and x^1

$$(L_0 - 1)x^0 = A,$$
$$B_0 x^0 = 0, \qquad (6b.27)$$

and

$$L_1 x^0 + (L_0 - 1)x^1 = 0,$$
$$B_1 x^0 + B_0 x^1 = 0. \qquad (6b.28)$$

From (6b.27), we find $x^0 = -A$. We will not need the solution x^1 to the perturbation problem (6b.28) but only the inner product $\langle C, x^1 \rangle$ involving x^1. This is seen from (6b.2) and (6b.26):

$$\Psi(\epsilon, \hat{\sigma}) = \langle C, x^0 \rangle + \epsilon \langle C, x^1 \rangle - \hat{\sigma} \langle C, A \rangle + O(\epsilon^2). \tag{6b.29}$$

We calculate $\langle C, x^1 \rangle$ from (6b.28):

$$\langle C, x^1 \rangle = \langle C, L_1 x^0 \rangle + \langle C, L_0 x^1 \rangle, \tag{6b.30}$$

and an integration by parts:

$$\langle C, L_0 x^1 \rangle = \langle L_0^* C, x^1 \rangle + \Gamma, \tag{6b.31}$$

where the boundary integrals Γ are evaluated using the second part of (6b.28). (The boundary integrals would vanish if $B_0 x^1$ were zero.)

We note that $x^0 = -A$ and since $L_0 A = 0$, $\frac{\partial p}{\partial x} = \frac{\partial p}{\partial z} = 0$ so that

$$L_1 A = \begin{pmatrix} 0 \\ -\frac{\partial \tilde{p}}{\partial x} \\ -\frac{\partial \tilde{p}}{\partial z} \\ 0 \end{pmatrix} \tag{6b.31}$$

in both fluids so that

$$\langle C, L_1 x^0 \rangle = -\langle C, L_1 A \rangle$$

$$= \int_{\Omega_1} \bar{u}^* \cdot \nabla \tilde{p} + \int_{\Omega_2} \bar{u}^* \cdot \nabla \tilde{p}$$

$$= \bar{w}^* [\![\tilde{p}]\!], \tag{6b.32}$$

and

$$[\![\tilde{p}]\!] = (\bar{M}_1 + P \bar{S} \alpha^2) e^{i\alpha x}. \tag{6b.33}$$

The form of Γ can be read off from the calculation of the adjoint in equation (6a.17):

$$\Gamma = \int_I \bar{\zeta} e^{i\alpha x} \frac{\partial \bar{\Theta}^*}{\partial z}, \tag{6b.34}$$

where the interval of integration I extends over one wavelength in x, at $z = l_1$, and $\Theta_z^* = \frac{P}{\alpha^2} \Delta^2 w_z^*$. From setting $\Psi = 0$, we have

$$\sigma = -\epsilon \frac{(\Gamma + \bar{w}^* [\![\tilde{p}]\!])}{\langle C, A \rangle} + O(\epsilon^2). \tag{6b.35}$$

We next evaluate $\langle C, A \rangle$:

$$\langle C, A \rangle = \int_I \bar{h}^* e^{i\alpha x} \tag{6b.36}$$

where

$$h^* = -[\![p^*]\!] = -\frac{P}{\alpha^2} [\![w_{zzz}^*]\!]. \tag{6b.37}$$

III.6(b)(ii) Results and Discussion

When $\sinh Q_2 \neq 0$, we have, using (6b.34) and equations leading up to it

$$\Gamma + \bar{w}^*[\tilde{p}] = \frac{2\pi}{\alpha}\left(\bar{\zeta}\frac{P}{\alpha^2}\left[\bar{c}_1 Q_3(Q_3^2 - \alpha^2)^2\cosh Q_3 l_1 + \bar{c}_3 \bar{Q}_2(Q_2^2 - \alpha^2)^2\cosh\bar{Q}_2 l_1\right.\right.$$

$$\left.+\bar{c}_5 Q_1(Q_1^2 - \alpha^2)^2\cosh Q_1 l_1\right]$$

$$+(\bar{M}_1 + \alpha^2 P\bar{S})\left[\bar{c}_1\sinh Q_3 l_1 + \bar{c}_3\sinh\bar{Q}_2 l_1 + \bar{c}_5\sinh Q_1 l_1\right]\right) \qquad (6b.38)$$

where Q_i are defined in equation (6b.12), c_i are defined in equation (6b.19), and

$$[w_{0zzz}^*] = Q_1^3(c_1\cosh Q_1 l_1 - d_1\cosh Q_1 l_2) + Q_2^3(c_3\cosh Q_2 l_1 - d_3\cosh Q_2 l_2)$$

$$+Q_3^3(c_5\cosh Q_3 l_1 - d_5\cosh Q_3 l_2), \qquad (6b.39)$$

which, on using equations (6b.19), becomes

$$= d_5 Q_3\frac{3}{2}(R\alpha^2)^{1/3}(-1 + i\sqrt{3})\frac{\sinh Q_3}{\sinh Q_3 l_1}. \qquad (6b.40)$$

Hence,

$$\langle C, A\rangle = \frac{2\pi}{\alpha}\frac{P}{\alpha^2}\bar{d}_5 Q_1\frac{3}{2}(R\alpha^2)^{1/3}(1 + i\sqrt{3})\frac{\sinh Q_1}{\sinh Q_1 l_1}, \qquad (6b.41)$$

and

$$\sigma \sim -\frac{\epsilon}{6}\left(\bar{\zeta}(R\alpha^2)^{1/3}\left[-(1 - i\sqrt{3})\frac{\sinh Q_3 l_2\cosh Q_3 l_1}{\sinh Q_3} + 2\frac{\sinh\bar{Q}_2 l_2\cosh\bar{Q}_2 l_1}{\sinh\bar{Q}_2}\right.\right.$$

$$\left.-(1 + i\sqrt{3})\frac{\sinh Q_1 l_2\cosh Q_1 l_1}{\sinh Q_1}\right]$$

$$+\frac{\alpha^2}{(R\alpha^2)^{1/3}}(\bar{M}_1/P + \alpha^2\bar{S})\left[-(1 + i\sqrt{3})\frac{\sinh Q_3 l_2\sinh Q_3 l_1}{Q_3\sinh Q_3}\right.$$

$$+2\frac{\sinh\bar{Q}_2 l_2\sinh\bar{Q}_2 l_1}{\bar{Q}_2\sinh\bar{Q}_2}$$

$$\left.\left.-(1 - i\sqrt{3})\frac{\sinh Q_1 l_2\sinh Q_1 l_1}{Q_1\sinh Q_1}\right]\right) + O(\epsilon^2) \quad\text{as}\quad \epsilon \to 0. \qquad (6b.42)$$

Since Q_1 and Q_3 are complex conjugates and Q_2 is either purely real or purely imaginary, we find that σ is real, independent of the Prandtl number of the basic one-fluid flow, the perturbation of the viscosity and the perturbation of thermal diffusivity.

For fixed α, $\alpha_1 \Delta\theta^*$ and R, σ has the following symmetry properties. The coefficient of $\epsilon\bar{\zeta}$ in the expression for σ is antisymmetric about $l_1 = 0.5$, as can be seen after some algebra by interchanging l_1 and l_2 in (6b.42). Figure 6.6 shows the coefficient of $\epsilon\bar{\zeta}$ in the expression for σ as a function of l_1 with $\bar{r} = \bar{\beta} = \bar{S} = 0$. It shows that if the thicker layer is the more conductive fluid, σ is stable. This is consistent with what would be expected intuitively when most of the fluid is conductive. In this case, the basic temperature profile would be almost constant in the bulk of the flow whilst most of the temperature variation would be in the less conductive fluid. This arrangement would be expected to be stable in analogy with results known for two-layer shearing flows. There is a parallel between the role of conductivity here and the role of viscosity in shearing flows. For example, in plane Couette flow with two layers (see chapter IV), the viscosities enter into the expression for the basic velocity profile in the same way that the conductivities enter into the basic temperature profile: when there is a large viscosity difference, linearly stable arrangements tend to be those where there is a thin layer of the less viscous fluid.

The coefficient of $\epsilon\bar{r}$ in the expression for σ is not symmetric about $l_1 = 0.5$, despite appearances. Since $\hat{\alpha}_1 \Delta\theta^*$ must be chosen small compared with 1, we have set $\hat{\alpha}_1 \Delta\theta^* = 0.001$ in the numerical computations. This means that in $\bar{M}_1/P$, l_2 is small compared with the constant $1/\alpha_1\Delta\theta^*$. Hence figure 6.7, which plots the coefficient of $\epsilon\bar{r}$ in σ, appears to be almost symmetric about $l_1 = 0.5$. Figure 6.7 shows that when the lower fluid is the less dense, σ is positive, indicating instability.

The coefficient of $\epsilon\bar{\beta}$ in the expression for σ is the product of l_2 and a term symmetric about $l_1 = 0.5$. In the Boussinesq approximation, the densities of fluids 1 and 2 are $\rho_1(1 - \hat{\alpha}_1\Delta\theta^*(1 - z + \Theta))$ and $\rho_1(1 - \hat{\alpha}_1\Delta\theta^*(1 - z + \Theta)/(1 - \bar{\beta}\epsilon))$ respectively. At the interface, these are approximately $\rho_1(1 - \hat{\alpha}_1\Delta\theta^* l_2)$ and $\rho_1(1 - \hat{\alpha}_1\Delta\theta^* l_2/(1 - \bar{\beta}\epsilon))$ respectively. Hence, when $\bar{\beta} > 0$, the upper fluid is the less dense locally at the interface and σ would be expected to be negative, as in figure 6.8.

The coefficient of $\epsilon\bar{M}_1/P$ or $\epsilon\bar{S}$ (see figure 6.9 where $\bar{\zeta} = \bar{r} = \bar{\beta} = 0$) is symmetric about $l_1 = 0.5$ as can be seen after an interchange of l_1 with l_2 in (6b.42). Figure 6.9 shows that when surface tension is positive, the interface is linearly stable and when surface tension is negative, there is instability.

If $Q_2 = 0$, then the two terms in (6b.38) involving $\bar{c}_3$ are replaced by $\alpha^4\bar{c}_3$ and $l_1\bar{c}_3$ respectively, and the term involving Q_2 in (6b.39) vanishes. Equation (6b.41) is unchanged. In (6b.42), the two terms containing $\bar{Q}_2$ in the square brackets are to be replaced by $2l_2$ and $2l_2l_1$ respectively. The symmetries discussed in the preceding paragraphs still hold.

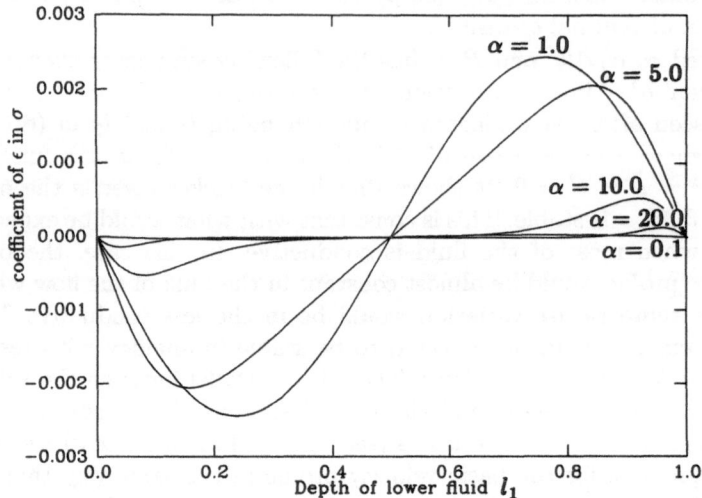

Fig. 6.6. [Renardy, 1986, American Institute of Physics] Graph of the coefficient of the ϵ in (6b.42) for σ versus l_1. Here $\hat{\alpha}_1 \Delta\theta^* = 0.001$, $R=1$, $\bar{\zeta}=1$, $\bar{r} = \bar{S} = \bar{\beta}=0$, $\alpha=0.1$, 1.0, 5.0, 10.0, 20.0.

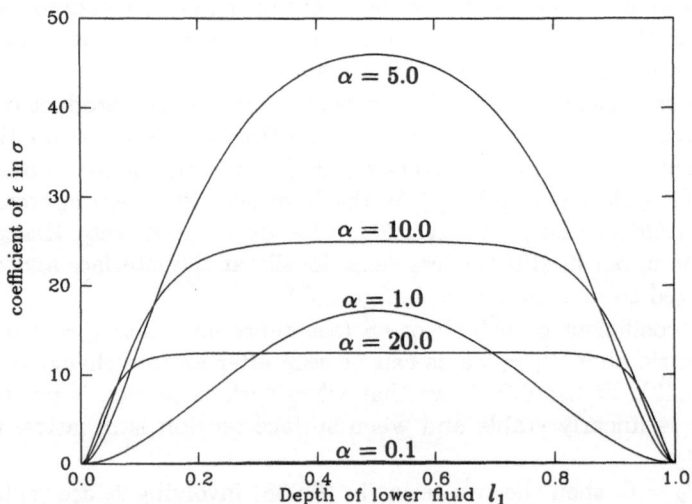

Fig. 6.7. [Renardy, 1986, American Institute of Physics] Graph of the coefficient of ϵ in the expression (6b.42) for σ versus l_1. Here $\hat{\alpha}_1 \Delta\theta^* = 0.001$, $R=1$, $\bar{r}=1$, $\bar{\zeta} = \bar{S} = \bar{\beta}=0$, $\alpha=0.1$, 1.0, 5.0, 10.0, 20.0.

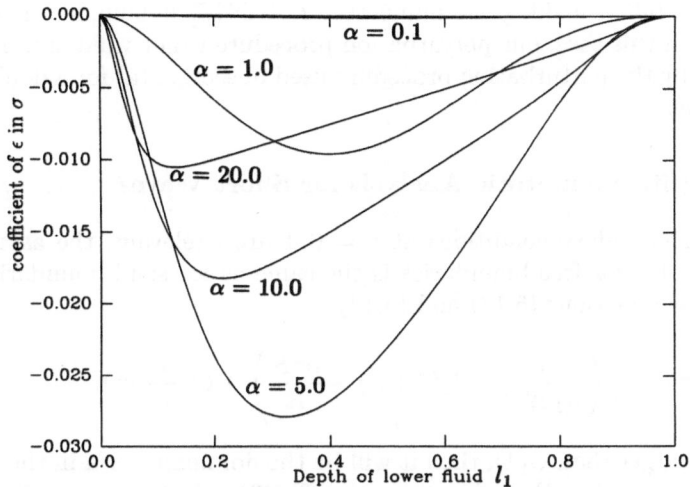

Fig. 6.8. [Renardy, 1986, American Institute of Physics] Graph of the coefficient of ϵ in the expression (6b.42) for σ versus l_1. Here $\hat{\alpha}_1 \Delta\theta^* =0.001$, $R=1$, $\bar{\beta}=1$, $\bar{\zeta} = \bar{S} =\bar{r}=0$, $\alpha=0.1, 1.0, 5.0, 10.0, 20.0$.

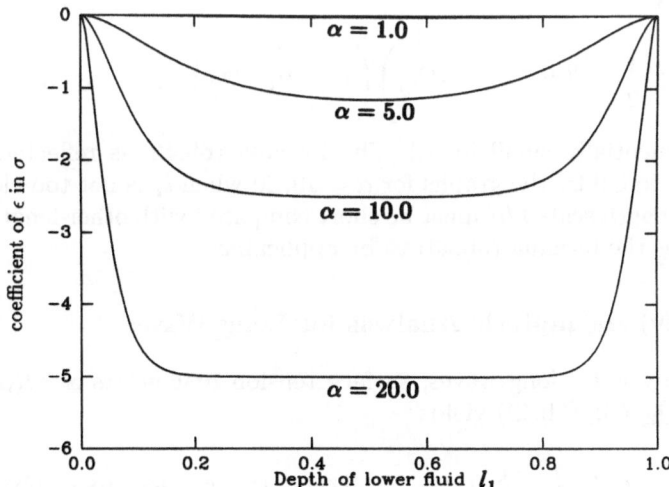

Fig. 6.9. [Renardy, 1986, American Institute of Physics] Graph of the coefficient of ϵ in the expression (6b.42) for σ versus l_1. Here $\hat{\alpha}_1 \Delta\theta^* =0.001$, $R=1$, $\bar{S}=1$, $\bar{\zeta} = \bar{r} =\bar{\beta}=0$, $\alpha=0.1, 1.0, 5.0, 10.0, 20.0$.

If $Q_2^2 = -j^2\pi^2$, $j =$ non-zero integer, i.e., $\sinh Q_2 = 0$, then as pointed out after equation (6b.12), c_3 and d_3 are infinite. In $\langle C, A\rangle$, d_3 is multiplied by $\sinh Q_2$ so $\langle C, A\rangle$ is bounded when the basic one-fluid problem is at neutral stability, making the numerator $\Gamma + \tilde{w}^* [\![\tilde{p}]\!]$ unbounded due to the $\bar{c}_3$-term. In this case, our perturbation procedure is not valid, and must be replaced by the perturbation procedure used in section (a) for a double zero eigenvalue.

III.6(b)(iii) Asymptotic Analysis for Short Waves

Since the boundary conditions at $z = 0, 1$ are irrelevant, the asymptotic behavior of σ for free boundaries is the same as for solid boundaries, and is given by equations (5.12) and (5.14):

$$\sigma \sim \frac{R\epsilon}{4\alpha}\left(\frac{\bar{r}}{\hat{a}_1 \Delta T^*} - l_2(\bar{r} + \bar{\beta}) - \frac{\alpha^2 \bar{S}}{R}\right) + O(\frac{1}{\alpha^2}, \epsilon^2). \qquad (6b.43)$$

If $\bar{S}\alpha^2$ is larger than $O(1)$, then it will be the dominant term in the asymptotic expansion but the other terms in (6b.43) will not necessarily be the correct next-order terms.

The expansion of (6b.42) for large α agrees with this formula. We note that the coefficient of $-\frac{\epsilon}{6}\bar{\zeta}(R\alpha^2)^{1/3}$ in equation (6b.42) behaves as

$$\left(-(1 - i\sqrt{3})\frac{1}{2}(e^{-2l_1 Q_3} - e^{-2l_2 Q_3}) + e^{-2l_1 \bar{Q}_2} - e^{-2l_2 \bar{Q}_2}\right.$$

$$\left.-(1 + i\sqrt{3})\frac{1}{2}(e^{-2l_1 Q_1} - e^{-2l_2 Q_1})\right)\left(1 + O(e^{-2\alpha})\right) \quad \text{as} \quad \alpha \to \infty \quad (6b.44)$$

so is exponentially small in $|\alpha|$. The formula (6b.43) is reflected in the figures 6.6 to 6.9 by the graphs for $\alpha = 10, 20$ when l_1 is not too close to 0 or 1. The length scale $1/\alpha$ must be short compared with other length scales in order for the formula (6b.43) to be applicable.

III.6(b)(iv) Asymptotic Analysis for Long Waves

In the analysis for long waves, surface tension first enters at $O(\alpha^4)$ and, provided $Q_2 \neq 0$, (6b.42) yields

$$\sigma \sim \epsilon R\alpha^2 l_2\left(-\frac{\bar{\zeta}}{360}(15l_1^4 + 30l_1^2(l_2^2 - 1) + 7 + l_2^2(-10 + 3l_2^2))\right.$$

$$\left. + \frac{l_1}{6}\left(-\frac{\bar{r}}{\hat{a}_1 \Delta T^*} + l_2(\bar{r} + \bar{\beta}))(l_1^2 + l_2^2 - 1)\right)\right)$$

$$+O(\alpha^4) \qquad \text{as} \qquad \alpha \to 0. \tag{6b.45}$$

When $Q_2 = 0$, the above dominant terms are $O(\alpha^6)$ and

$$\sigma \sim \epsilon\alpha^4 \frac{l_2 l_1}{6}(l_1^2 + l_2^2 - 1)\bar{S} + O(\alpha^6) \qquad \text{as} \quad \alpha \to 0. \tag{6b.46}$$

III.6(b)(v) Thin-layer Effects

In the linear stability analysis of parallel shear flows of two fluids with different viscosities and densities, the effect of gravity on the density stratification can be countered by the effects of viscosity stratification and surface tension (see chapter IV). A linearly stable arrangement is possible with light fluid below if it is much less viscous and is in a thin layer. In the present problem, the basic flow has no shear, but we again find the stability of thin layers counter to intuition. Here, the role of viscosity stratification is taken by the thermal conductivity stratification.

The value of σ for l_1 small is essentially $l_1 \frac{\partial \sigma}{\partial l_1}(l_1 = 0)$ where, from equation (6b.42),

$$\frac{\partial \sigma}{\partial l_1}(l_1 = 0)$$

$$= -\frac{\epsilon}{3}\bar{\zeta}(R\alpha^2)^{1/3}\left(Real[(1 + i\sqrt{3})Q_1 \coth Q_1] - \bar{Q}_2 \coth \bar{Q}_2 \right) + O(\epsilon^2). \tag{6b.47}$$

Hence, in the presence of a thin layer, the difference in thermal conductivity dominates the stability criterion for long and order one waves. Moreover, we can determine the sign of the quantity in (6b.47). The coefficient of $-\epsilon\bar{\zeta}$ in (6b.47) is positive for Rayleigh numbers up to the first critical value $27\pi^4/4$ for the one-fluid problem, and typically looks like figure 6.10. At the first critical value of the Rayleigh number, the coefficient has a pole and is infinite at the critical value of $\alpha = \pi/\sqrt{2}$. As the Rayleigh number is increased above the first critical value, there will be two values of α for which the one-fluid problem is neutrally stable, and the coefficient will have two poles. Our perturbation scheme is valid away from the neutral curve. Hence, the arrangement with a thin layer of a less dense fluid below is stable to long and order one wavelength disturbances, if the lower fluid has a small thermal conductivity.

For short wave disturbances, however, the effect of the thermal conductivities is exponentially small, and the density difference and surface tension dominate, so that in order to counteract an adverse density stratification, the density difference should not be large whilst the surface tension should be large. An example of the thin-layer effect is displayed in figure 6.11. The conduction solution is linearly stable at low Rayleigh numbers, e.g. $R = 0.1$, even though fluid 1 is the less dense.

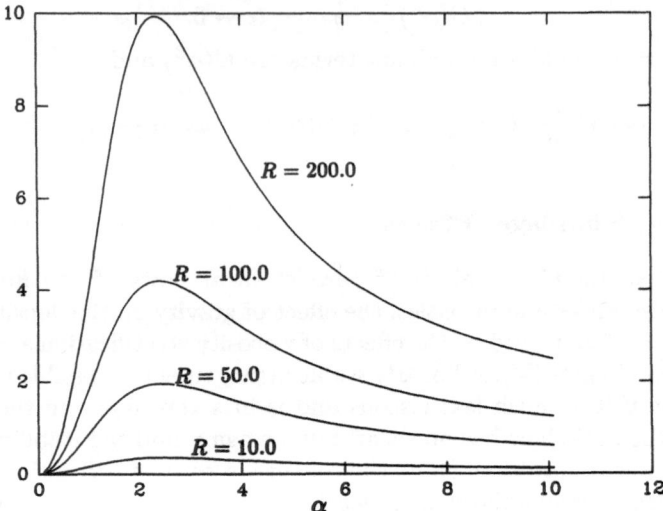

Fig. 6.10. [Renardy, 1986, American Institute of Physics] Graph of the coefficient of $-\epsilon\bar{\zeta}$ in $(\partial\sigma/\partial l_1)$ $(l_1 = 0)$ versus α for R=10.0, 50.0, 100.0, 200.0. The amplitudes decay to zero for large α. As R approaches $27\pi^4/4$, the peak amplitude approaches infinity and the location of the peak approaches $\alpha = \pi/\sqrt{2}$.

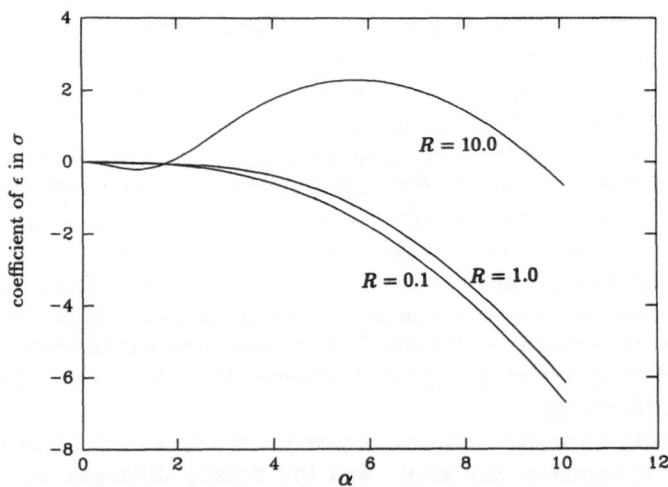

Fig. 6.11. [Renardy, 1986, American Institute of Physics] Graph of the coefficient of ϵ in the expression (6b.42) for σ versus α. Here $\bar{S}$=10.0, $\bar{\zeta}$ =100.0, $\bar{r}$ = 0.1, $\bar{\beta}$=1.0, l_1=0.05, R=0.1, 1.0, 10.0. Fluid 1 is the less dense fluid. At R=0.1 and 1.0, the arrangement is linearly stable for all α. As R increases, the arrangement will become less stable. For example, at R=10.0, the arrangement is unstable.

III.6(c) Summary

We summarize the results we have obtained for the stability of the conduction solution in similar liquids:

1. If the Rayleigh number is below criticality, then the interfacial eigenvalue σ is real. In addition, we have a-d below.

    a. Surface tension is stabilizing.

    b. If the fluids differ only in thermal conductivity, then σ is positive if the thicker layer is less conducting and stable if the thicker layer is more conducting.

    c. A density difference across the interface which can be achieved by stratification in coefficients of cubical expansion is stabilizing if the lower fluid is heavier and destabilizing if the upper fluid is heavier. A dimensionless measure of the density difference across the interface is given by the quantity M_1, which involves the ratio r of the densities (at the temperature of the upper plate) and the ratio β of the coefficients of thermal expansion. In applications where the Boussinesq approximation holds, G/RP is usually large, and therefore the first term in M_1 is dominant unless r is very close to 1.

    d. The stratification in viscosity and thermal diffusivity does not affect the growth rate σ at leading order.

2. If the Rayleigh number is critical for one fluid, then both the interfacial eigenvalue and one of the one-fluid eigenvalues are zero. When the problem is perturbed so that the two fluids differ slightly, this double eigenvalue zero may split up into real eigenvalues or into a complex conjugate pair. The following discussion applies to the case where the properties of fluid 1 are held fixed and fluid 2 is perturbed.

    a. The introduction of surface tension leads to stable, complex conjugate eigenvalues.

    b. When the fluids differ only in thermal conductivity and the thicker layer is less conducting, the eigenvalues become real and positive. When the thicker layer is more conducting, the eigenvalues become complex. Stability depends on the Prandtl number and depth ratio.

    c. A density difference across the interface results in real eigenvalues and instability if the upper fluid is heavier. If the lower fluid is heavier, the eigenvalues become complex. If $r > 1$ and G is large, the conduction solution is stable. If the density difference at the interface is due to a difference in the coefficients of thermal expansion, the sign of σ depends on the Prandtl number and depth ratio.

    d. When the fluids differ only in viscosity or in thermal diffusivity, the eigenvalues are real. If the upper fluid (which is the one that is perturbed) is made less viscous or less conducting, then instability results. This is expected because the introduction of a less viscous or less diffusive fluid increases the "effective" Rayleigh number.

From this discussion of similar liquids we conclude that the generation of complex eigenvalues leading to Hopf bifurcation may be achieved by manipulating surface tension or the differences of density or thermal conductivity across the layer.

III.7 Nonlinear Bifurcation Analysis

Bifurcation from the trivial state (2.19) will be investigated for the situation where criticality is achieved at an isolated value of the Rayleigh number by a complex conjugate pair of eigenvalues [Renardy and Renardy 1988]. The oscillatory nature of the instability is due to the opposing influences of a destabilizing temperature gradient and a stable interface.

We use the notation of section III.2. In the x-and y-directions, we assume that the solution is doubly periodic, i.e. we have $\tilde{\theta}(\mathbf{x} + n_1\mathbf{x}_1 + n_2\mathbf{x}_2, t) = \tilde{\theta}(\mathbf{x}, t)$ for every pair of integers (n_1, n_2). Here $\mathbf{x} = (x, y, z)$ and the vectors $\mathbf{x}_1$ and $\mathbf{x}_2$ span a hexagonal lattice of period W:

$$\mathbf{x}_1 = W \cdot (\frac{\sqrt{3}}{2}, \frac{1}{2}, 0), \quad \mathbf{x}_2 = W \cdot (0, 1, 0). \tag{7.1}$$

Figure 7.1 illustrates the spanning vectors with respect to a hexagonal lattice.

The lattice obtained from this double periodicity is invariant under the symmetries of the hexagon; that is, rotation by multiples of 60 degrees, reflection across $\mathbf{a_i}$ defined in equations (7.2) - (7.3) below, reflection across axes perpendicular to the $\mathbf{a_i}$.

The same periodicity condition holds for $\mathbf{v}$, $\tilde{p}$ and h. We can then expand $\tilde{\theta}$ in a Fourier series with modes proportional to $e^{i\mathbf{a}\cdot\mathbf{x}}$. Here, $e^{i\mathbf{a}\cdot\mathbf{x}}$ must satisfy the same periodicity condition as $\tilde{\theta}(x, y, z, t)$. Thus, $e^{i\mathbf{a}\cdot\mathbf{x}_1}$ and $e^{i\mathbf{a}\cdot\mathbf{x}_2}$ must be 1, and therefore $\mathbf{a}\cdot\mathbf{x}_1$ and $\mathbf{a}\cdot\mathbf{x}_2$ must be multiples of 2π. We set up the vectors $\mathbf{a}_1$ and $\mathbf{a}_2$ to span such vectors (the reciprocal lattice) as follows: $\mathbf{a}_i \cdot \mathbf{x}_j = 2\pi\delta_{ij}$. This leads to

$$\mathbf{a}_1 = \frac{4\pi}{W\sqrt{3}}(1,0,0), \quad \mathbf{a}_2 = \frac{4\pi}{W\sqrt{3}}(-\frac{1}{2}, \frac{\sqrt{3}}{2}, 0), \tag{7.2}$$

$\mathbf{a} = k\mathbf{a}_1 + l\mathbf{a}_2$, where k, l are integers. We need to define:

$$\mathbf{a}_3 = -\mathbf{a}_1 - \mathbf{a}_2. \tag{7.3}$$

The Fourier expansion of $\tilde{\theta}$ reads

$$\tilde{\theta}(x, y, z, t) = \sum_{k,l=-\infty}^{\infty} \tilde{\theta}_{kl}(z, t)e^{ik\mathbf{a}_1 \cdot \mathbf{x} + il\mathbf{a}_2 \cdot \mathbf{x}}. \tag{7.4}$$

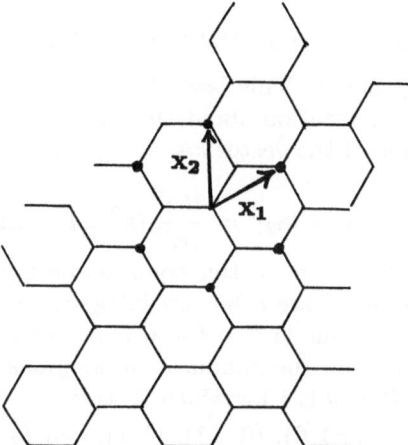

Fig. 7.1. The spanning vectors $\mathbf{x_1} = W(\sqrt{3}/2, 1/2, 0)$ and $\mathbf{x_2} = W(0, 1, 0)$ in the $x - y$ plane. The hexagons drawn in the background are an example of a pattern that satisfies the required double periodicity.

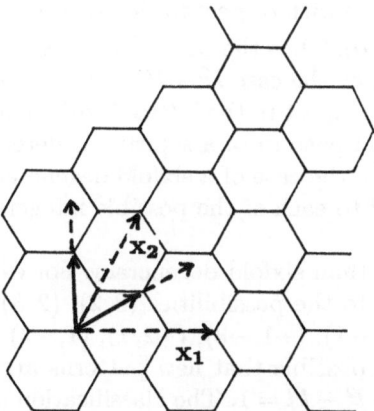

Fig. 7.2. The bold arrows are the spanning vectors for $k^2 + l^2 - kl = 1$ and are identical to those of figure 7.1 for the lattice of period W. The broken arrows emanating over them are the spanning vectors for the case $k^2 + l^2 - kl = 3$. When these are rotated by 30 degrees anticlockwise, we see that the background hexagons again possess the periodicity required by these broken spanning vectors.

We next consider the linearization of problem (2.28)-(2.39). In this linearized problem, we can separate variables, i.e. we look for solutions in the form

$$\tilde{\theta}(x, y, z, t) = e^{\sigma t} e^{ik\mathbf{a_1 \cdot x} + il\mathbf{a_2 \cdot x}} \hat{\theta}(z), \qquad (7.5)$$

and similarly for $\mathbf{v}$, $\tilde{p}$ and h. This leads to an eigenvalue problem for σ. Due to symmetry under rotation about the z-axis, the eigenvalues do not depend on the direction of the vector $k\mathbf{a_1} + l\mathbf{a_2}$, but only on its magnitude α:

$$\alpha^2 = |k\mathbf{a_1} + l\mathbf{a_2}|^2 = \frac{16\pi^2}{3W^2}(k^2 + l^2 - kl). \qquad (7.6)$$

For a specific wavenumber α, this equation determines the period W of the lattice. Since k and l are arbitrary integers, the factor $k^2 + l^2 - kl$ can be $0, 1, 3, 4, 7, ...$ The mode $k = l = 0$ is not of interest in the linear problem, but will enter into the nonlinear interactions later. The smallest nonzero value of $k^2 + l^2 - kl$ is 1 for which W is $W_1 = \frac{4\pi}{\sqrt{3}\alpha}$. This occurs for six possible pairs (k, l): $(\pm 1, 0)$, $(0, \pm 1)$, $(1, 1)$, and $(-1, -1)$. Thus, there is a sixfold degeneracy of the corresponding eigenvalue. We note that the solutions with period W_1 are also permitted as solutions on lattices with larger periods. At a value of $k^2 + l^2 - kl$ larger than one, two possibilities exist:

(i)   There is again a sixfold degeneracy of the eigenvalue. For example, $k^2 + l^2 - kl = 3$ occurs for $(k, l) = (1, 2)$, $(2, 1)$, $(-1, -2)$, $(-2, -1)$, $(1, -1)$, and $(-1, 1)$. Here, the value W from equation (7.6) is $W_1\sqrt{3} = \frac{4\pi}{\alpha}$. However, the solutions with lattice period W_1 are also solutions with lattice period $W_1\sqrt{3}$ with respect to the spanning vectors $W_1\sqrt{3}(1, 0, 0)$ and $W_1\sqrt{3}(\frac{1}{2}, \frac{\sqrt{3}}{2}, 0)$. Thus the case $k^2 + l^2 - kl = 3$ will yield the same physical solutions as the case $k^2 + l^2 - kl = 1$; only the orientation of the patterns with respect to the lattice is different (see figure 7.2). The size of the physical pattern of a solution is determined by the critical wavenumber α_c. In the case of a sixfold degeneracy, the solutions with $k^2 + l^2 - kl$ equal to each of the possible integers are identical after a rotation.

(ii) There is a higher than sixfold degeneracy. For example, the case $k^2 + l^2 - kl = 7$ leads to the possibilities $(3, 2)$, $(2, 3)$, $(-3, -2)$, $(-2, -3)$, $(3, 1)$, $(1, 3)$, $(-3, -1)$, $(-1, -3)$, $(-2, 1)$, $(1, -2)$, $(2, -1)$, and $(-1, 2)$. In this case, it is possible that new patterns arise, in addition to the solutions for $k^2 + l^2 - kl = 1$. The classification of such patterns is an open problem.

We now set $k^2 + l^2 - kl = 1$ and consider the eigenvalue problem for the case $l = 0$, $k = 1$. Here, $\mathbf{a_1} = (\alpha, 0, 0)$. This is the linearized problem of section III.3, where disturbances are proportional to $\exp(\sigma t + i\alpha x)$. The linearized equations for v, arising from equations (2.28) - (2.39), are decoupled from the rest of the problem:

$$\sigma v - P(v'' - \alpha^2 v) = 0 \quad \text{in fluid 1,}$$

$$\sigma v - \frac{rP}{m}(v'' - \alpha^2 v) = 0 \quad \text{in fluid } 2, \tag{7.7}$$

$$v = 0 \quad \text{at} \quad z = 0, 1, \tag{7.8}$$

and at $z = l_1$,

$$\llbracket v \rrbracket = 0, \quad v_1' - \frac{1}{m}v_2' = 0. \tag{7.9}$$

This problem leads to negative eigenvalues because

$$\sigma\left(\int_0^{l_1} r|v|^2 \, dz + \int_{l_1}^1 |v|^2 \, dz\right)$$

$$= Pr\left(\int_0^{l_1} \bar{v}(v'' - \alpha^2 v) \, dz + \int_{l_1}^1 \frac{1}{m}\bar{v}(v'' - \alpha^2 v) \, dz\right)$$

$$= -rP\alpha^2\left(\int_0^{l_1} |v|^2 \, dz + \frac{1}{m}\int_{l_1}^1 |v|^2 \, dz\right) + rP\left(\int_0^{l_1} \bar{v}v'' \, dz + \frac{1}{m}\int_{l_1}^1 \bar{v}v'' \, dz\right). \tag{7.10}$$

The latter bracket is integrated by parts, and the boundary terms at $z = l_1$ vanish. Therefore,

$$\sigma\left(\int_0^{l_1} r|v|^2 \, dz + \int_{l_1}^1 |v|^2 \, dz\right)$$

$$= -rP\left(\int_0^{l_1} \alpha^2|v|^2 + |v'|^2 \, dz + \frac{1}{m}\int_{l_1}^1 \alpha^2|v|^2 + |v'|^2 \, dz\right). \tag{7.11}$$

Hence, $\sigma < 0$.

III.7(a) Problem in Finite Dimension

In this section, the relevant results of Roberts, Swift and Wagner [1986] are summarized. They consider a reduced system of the form

$$\frac{d\mathbf{u}}{dt} + \mathbf{F}(\mathbf{u}, \lambda) = \mathbf{0}, \tag{7a.1}$$

where $\mathbf{u} \in \mathbb{R}^{12}$ and λ is a real parameter. The reduction of the Bénard problem near criticality to such a system will be discussed in detail in the next sections. The parameter λ corresponds to the difference between the Rayleigh number and its critical value, $R - R_c$, and, roughly speaking, the vector $\mathbf{u}$ consists of amplitude factors that multiply the critical eigenfunctions. These critical eigenfunctions are waves propagating in the directions of $\mathbf{a}_1, \mathbf{a}_2, \mathbf{a}_3 = -\mathbf{a}_1 - \mathbf{a}_2$ and $-\mathbf{a}_1, -\mathbf{a}_2, -\mathbf{a}_3$, where these vectors are defined by equations (7.2) - (7.3). We denote the (complex) amplitude of the

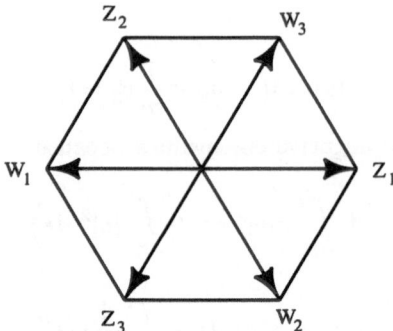

Fig. 7.3. The displayed hexagon is constructed from vertices that are the wave vectors $\mathbf{a_1}$, $\mathbf{a_2}$, $\mathbf{a_3}$, $-\mathbf{a_1}$, $-\mathbf{a_2}$ and $-\mathbf{a_3}$. The corresponding wave amplitudes are marked as z_1, z_2, z_3, w_1, w_2 and w_3 respectively.

wave propagating in the direction of $\mathbf{a}_i$ by z_i and the amplitude of the wave propagating in the direction of $-\mathbf{a}_i$ by w_i. See figure 7.3.

We set $\mathbf{u} = (z_1, z_2, z_3, w_1, w_2, w_3)$. The system (7a.1) can be written in the form

$$\frac{dz_i}{dt} + F_i(z_1, z_2, z_3, w_1, w_2, w_3, \lambda) = 0, \; i = 1, 2, 3,$$

$$\frac{dw_i}{dt} + F_{i+3}(z_1, z_2, z_3, w_1, w_2, w_3, \lambda) = 0, \; i = 1, 2, 3. \tag{7a.2}$$

The original equations governing the problem are invariant under any rotation in the $x - y$ plane, translations in the $x - y$ plane, and reflections across the axes: $x \to -x$, $y \to -y$. The symmetries of the "problem" are those of the equations plus the imposed double periodicity. The problem is invariant under any translations in the $x - y$ plane, and the symmetries of the hexagon. A symmetry of the "solution" is the symmetry operation that, applied to the solution, reproduces the solution; this involves a more restricted class of symmetry operations than those of the problem.

It follows from some of the properties of the hexagonal symmetry (namely, invariance after rotation by $2\pi/3$, by $4\pi/3$, by π, by $5\pi/3$ and by $\pi/3$ radians, respectively) that

$$F_2(z_1, z_2, z_3, w_1, w_2, w_3, \lambda) = F_1(z_2, z_3, z_1, w_2, w_3, w_1, \lambda),$$

$$F_3(z_1, z_2, z_3, w_1, w_2, w_3, \lambda) = F_1(z_3, z_1, z_2, w_3, w_1, w_2, \lambda),$$

$$F_4(z_1, z_2, z_3, w_1, w_2, w_3, \lambda) = F_1(w_1, w_2, w_3, z_1, z_2, z_3, \lambda), \tag{7a.3}$$

$$F_5(z_1, z_2, z_3, w_1, w_2, w_3, \lambda) = F_1(w_2, w_3, w_1, z_2, z_3, z_1, \lambda),$$

$$F_6(z_1, z_2, z_3, w_1, w_2, w_3, \lambda) = F_1(w_3, w_1, w_2, z_3, z_1, z_2, \lambda).$$

Color Insert

Color Insert

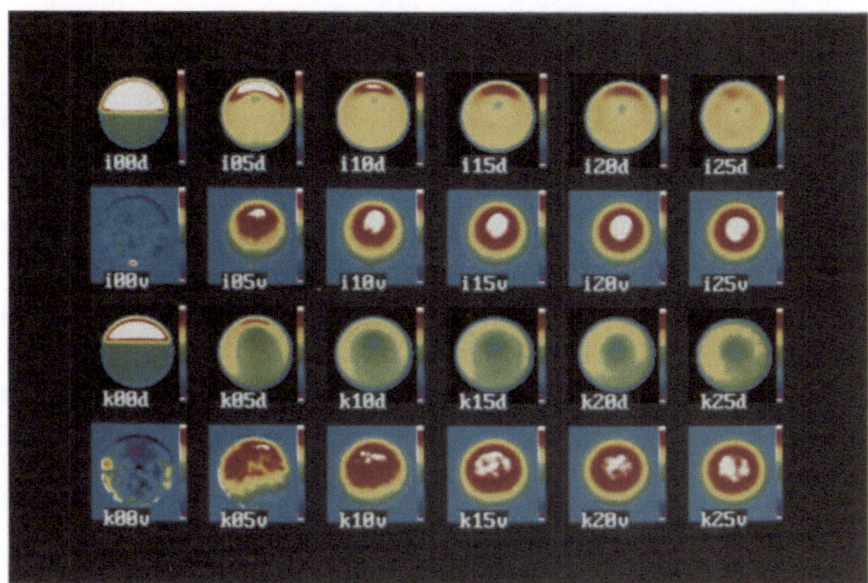

Plate I.1.2. [Altobelli, Givler and Fukushima, 1991, John Wiley & Sons] Inhomogeneous transverse distribution of axial fluid velocity and concentration for negatively buoyant suspensions (0.76mm diameter plastic beads in SAE80W oil with density ratio of 1.18) in a horizontal tube (2.5cm diameter) obtained by nuclear magnetic resonance imaging (NMRI). The images in rows with black backgrounds, marked ...d show the concentration distribution of the liquid according to the color bar. The concentration of solids is the complement of the liquid concentration: that is, white is the region with no solids. The rows with cyan backgrounds, ...v, show the normalized axial velocity distribution with cyan being zero velocity and white representing the maximum velocity measured. The top two rows, marked i..., are for solids concentration by volume of 30% while the bottom rows, k..., are for 38%. The leftmost images, ..00.., are with no flow while the rightmost images, ..25.., are with the fastest flows: that is, with average liquid velocities of 25 and 17 cm/s for the 30% and 38% solids suspensions, respectively. The flow creates an island of high solids concentration which moves closer to the center of the tube as the flow velocity increases. The velocity profile is not parabolic but quite blunt at these concentrations. The NMRI method used is described in Majors, Givler and Fukushima [1989].

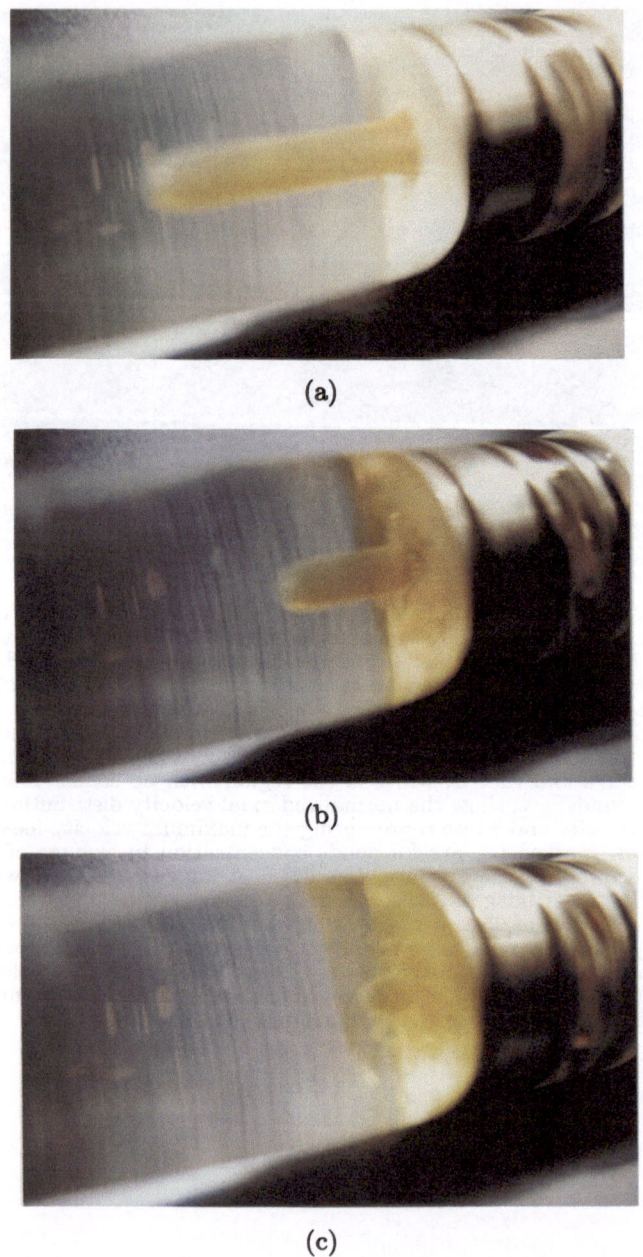

(a)

(b)

(c)

Plate I.1.7(a-c). The sequence (a) - (f) depicts spin-up of SAE motor oil in water.

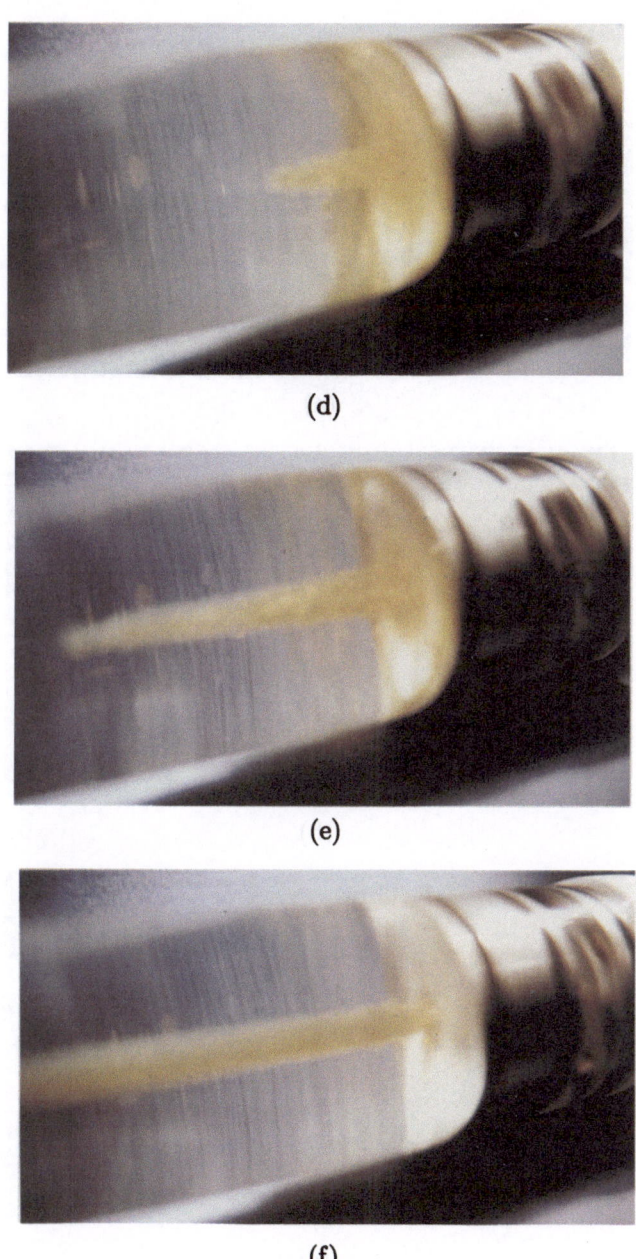

(d)

(e)

(f)

Plate I.1.7(d-f). Continued.

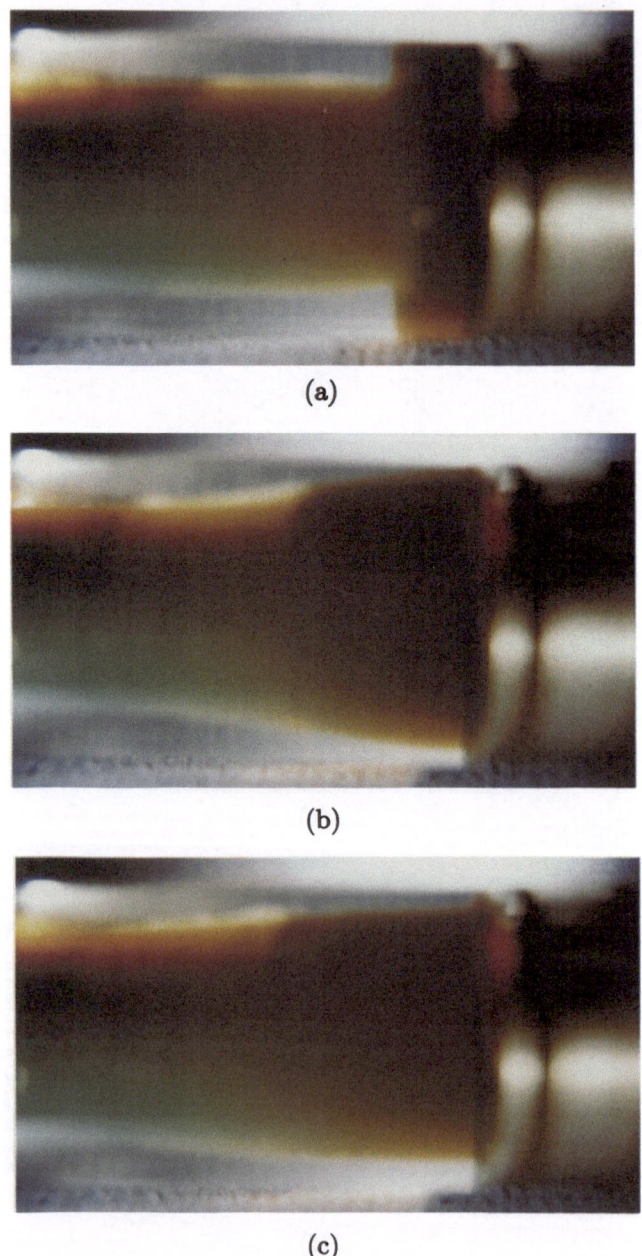

(a)

(b)

(c)

Plate I.1.8(a-c). The sequence of photographs I.1.8 (a) - (c) depicts spin-up of SAE motor oil in 80% glycerin plus 20% water solution.

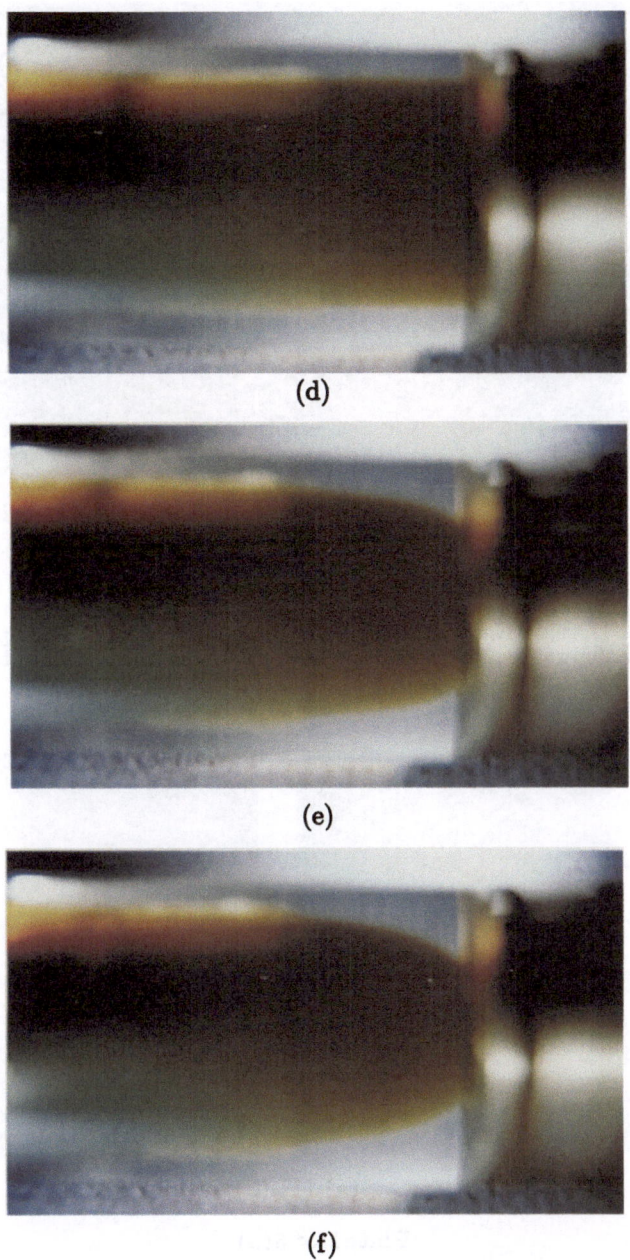

(d)

(e)

(f)

Plate I.1.8(d-f). The sequence of photographs I.1.8 (d) - (f) depicts spin-down.

Plate I.3.1. [Joseph, Nguyen and Beavers, 1984] This is a photograph of one of the static configurations of dibutyl phthalate bubbles in glycerol/water solution.

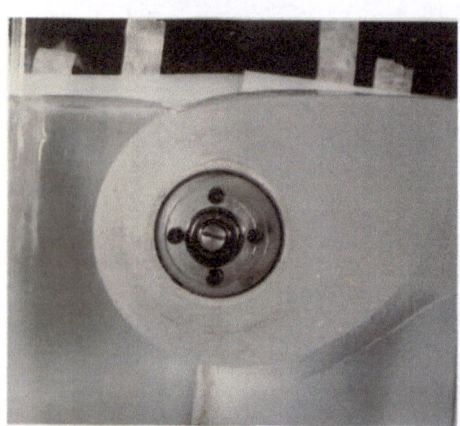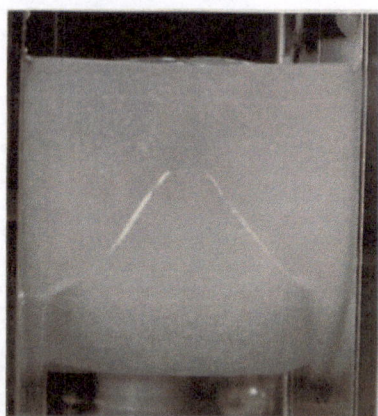

(a) (1)-(2)

Plate II.4.3(a).

(a) (3)-(4)

(b) (1)-(2)

Plate II.4.3(a-b). [Figure 15 of Joseph, Nguyen and Beavers, 1984] (a) The formation of a roller of silicone oil III (ρ =0.95 g/cm^3, μ = 950 P) in water. In the beginning the rod rotates counterclockwise at 10 rpm. The torque is 1.4 lb in. Fingering instabilities lead to an emulsion of water droplets in silicone oil. (1) Front view; (2) side view. Three days later as more water droplets are formed, the effective viscosity of silicone oil decreases and the rod rotates at a higher speed (16 rpm). At this speed, the roller is formed but does not rotate as a solid body since it is still attached to the sidewall. (3) Front view; (4) side view. (b) The speed of the rod is increased. The torque goes up to above 5 lb in. The two photographs (1) and (2), which are taken 5 seconds apart, show the dynamics through which the roller detaches itself from the sidewall. After detaching from the wall, the torque goes down to 0.6 lb in.

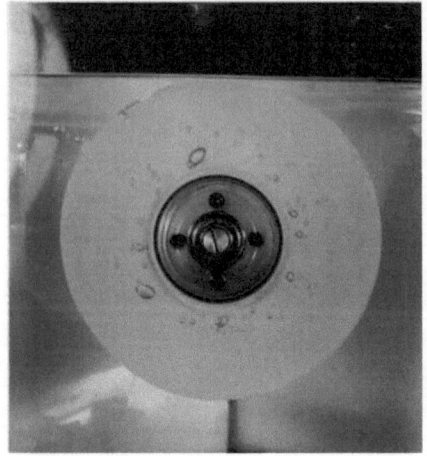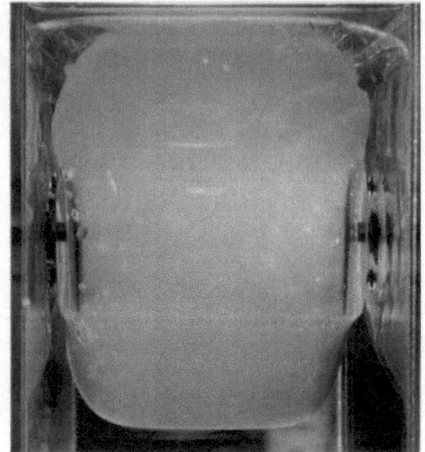

(c) (1)-(2)

(c) (3)-(4)

Plate II.4.3(c). After detaching from the wall, the roller becomes unstable. To restore stability, the speed of the rod is reduced to 12 rpm. The roller is stable but the shape is irregular. (1) Front view; (2) side view. After a few hours the roller moulded itself into the cylindrical shape shown in (3) and (4). The speed of the rod and the speed on the surface of the roller are almost the same showing that the roller is very nearly in a solid body rotation with small shearing by water at the roller rim. The torque is zero on our meter. The cylindrical shape shown in (4) is due to the large effect of gravity at the top of the roller.

(a)

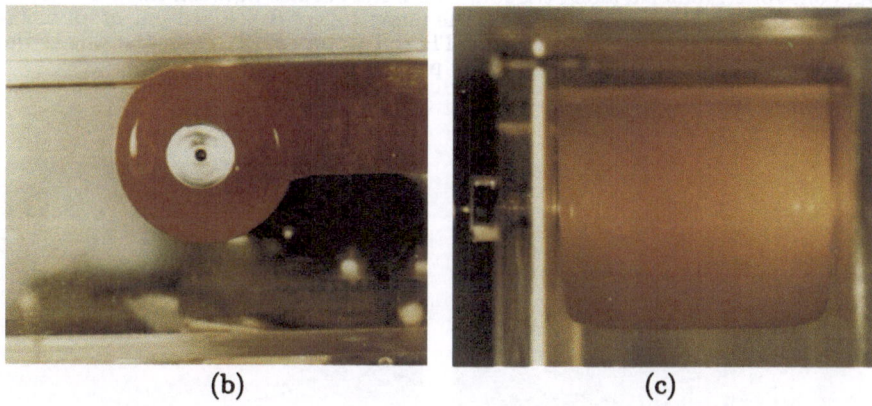

(b)   (c)

Plate II.4.6(a-c). [Figure 12 of Joseph, Nguyen and Beavers, 1984] (a) The static configuration of STP (density ρ=0.89 g/cm^3, viscosity μ= 11000 cp, dark) on water (μ=1 cp). (b) Front view of the box. The rod rotates clockwise at about 40 rpm. The STP on the right is nearly stationary and is shielded from shearing by a thin layer of water. (c) Side view of the box looking in from the left of (b). The STP roller is also shielded from shearing against the Plexiglas walls by a layer of water.

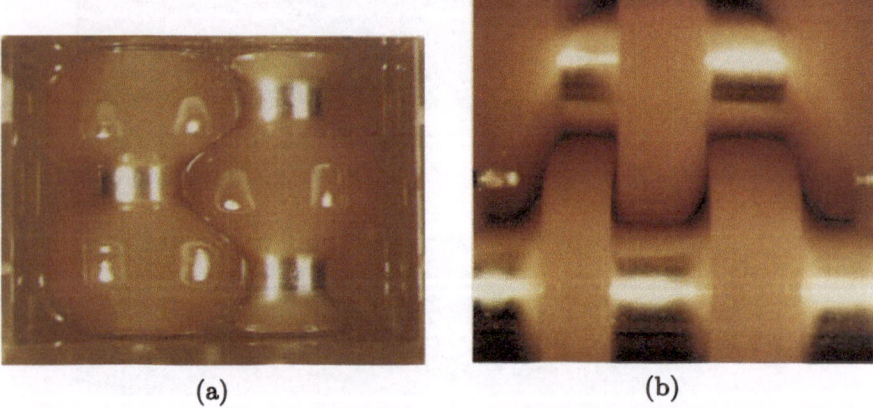

(a)　　　　　　　　　　　(b)

Plate II.4.10(a-b). [Figure 18 of Joseph, Nguyen and Beavers, 1984] (a) Inter-penetrating rollers of STP and water as seen from the top of the box sketched in figure II.4.9. There is water on every side of the STP rollers. Initially, each rod had one roller, uniform along the rod. The water surface between them developed a small wave, which grew, and then the many rollers developed out of an insta-bility. (b) Interpenetrating rollers of STP and water as seen from the side of the box sketched in figure II.4.9. The clear parts are water.

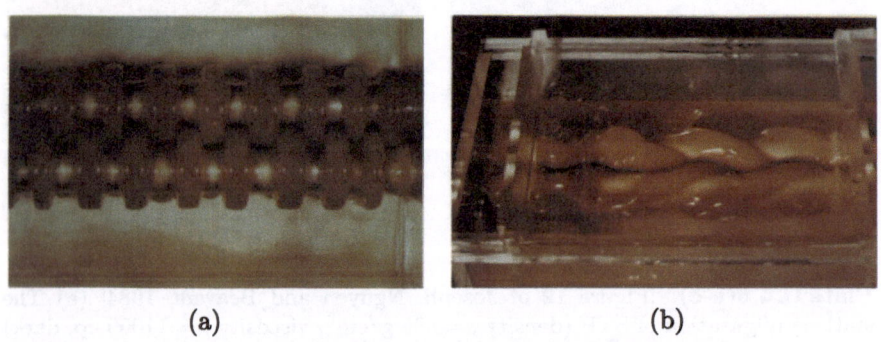

(a)　　　　　　　　　　　(b)

Plate II.4.11(a-d). [Figures 3,2,4 and 5 of Joseph, Nguyen and Beavers, 1986] Different types of rollers: (a) rollers on two cylinders; (b) "pretzels" on two cylin-ders; (c) "diamonds" on two cylinders; (d) fat rollers on one cylinder.

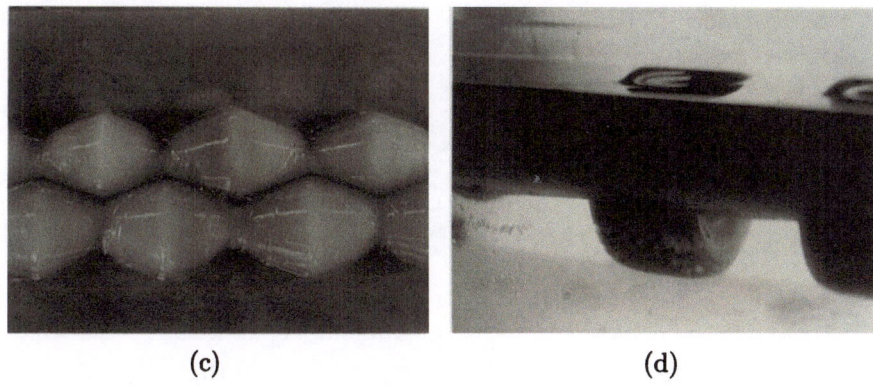

(c)   (d)

Plate II.4.11(c-d). Continued.

(a)   (b)

Plate II.4.12(a-b). [Figure 7 of Joseph, Nguyen and Beavers, 1984] Encapsulation of silicone oil I (ρ=0.96 g/cm^3, $\mu = 19$cP) by (a) corn syrup/water solution ($\rho = 1.2$ g/cm^3, $\mu = 12$ cP, blue); and (b) water (blue). The speed of the rod is about 150 rpm.

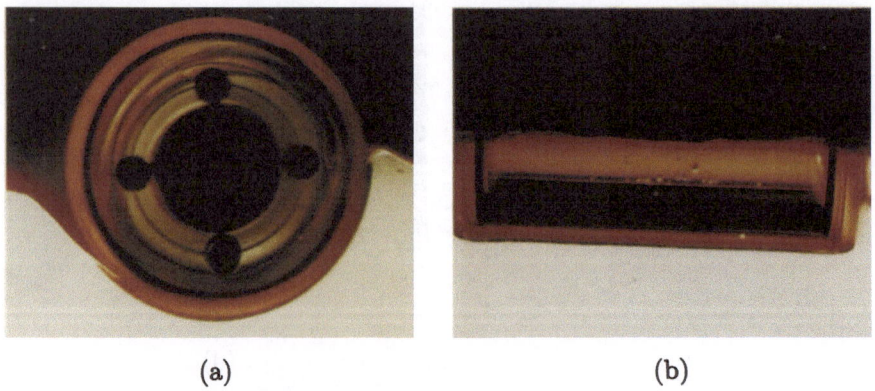

(a)   (b)

Plate II.4.13(a-b). [Figure 8 of Joseph, Nguyen and Beavers, 1984] Encapsulation of glycerol ($\rho = 1.25$ g/cm^3, $\mu = 1760$ cP, clear) by silicone oil I ($\rho = 0.96$ g/cm^3, $\mu = 19$ cP, red); (a) front view; (b) side view.

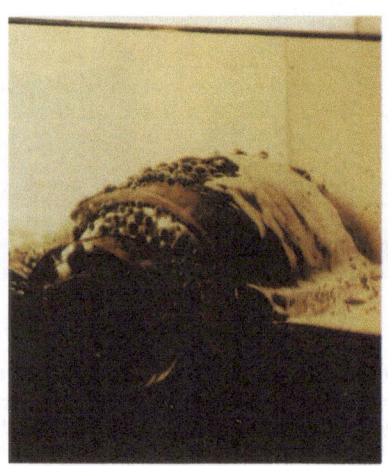

Plate II.4.14. [Figure 21 of Joseph, Nguyen and Beavers, 1984] A combination of emulsion and sheet coating of 1% polyacrylamide/water ($\mu = 90$ cP, $\rho = 1.01$ g/cm^3, blue) in castor oil ($\mu = 700$ cP, $\rho = 0.96$ g/cm^3, yellow). The rod rotates clockwise at 1 rpm. The rod is covered by polyacrylamide/water droplets and these droplets in turn are encapsulated by a sheet of polyacrylamide/water shielding them from shearing against the high-viscosity castor oil. The large droplets and the stability of the sheets of polyacrylamide are a non-Newtonian effect.

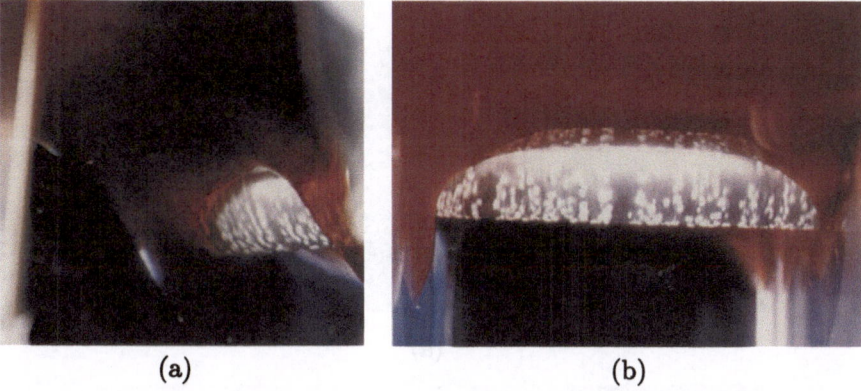

(a)   (b)

Plate II.4.15(a-b). [Figure 10 of Joseph, Nguyen and Beavers, 1984] Lubrication of silicone oil III (ρ =0.95 g/cm^3, μ = 95000 cP, red) by water (clear). The speed of the rod is about 65 rpm. The rod is lubricated entirely by water; (a) back view, looking at an angle from below; (b) side view.

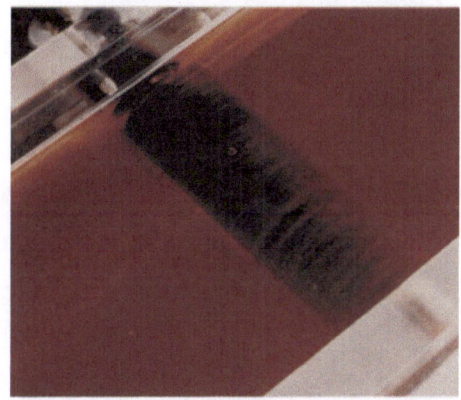

Plate II.4.16. [Figure 11 of Joseph, Nguyen and Beavers, 1984] The same experiment as in figure II.4.15. After the configuration in figure II.4.15 is achieved, some water is withdrawn from the box so as to lower the level of silicone oil below the rod. The rod is surrounded by a very thin film of water and the water is surrounded by oil. The rod is completely lubricated by the water and the oil is sheared only by water. In the type of experiment shown here, it is possible to obtain a water-lubricated aluminum rod in which the thickness of the water layer is confined to a monolayer on the rod. In such configurations, the rod rotates at high speeds, but the silicone oil is dead still. The rod "slips" on the silicone oil completely, as in an inviscid fluid.

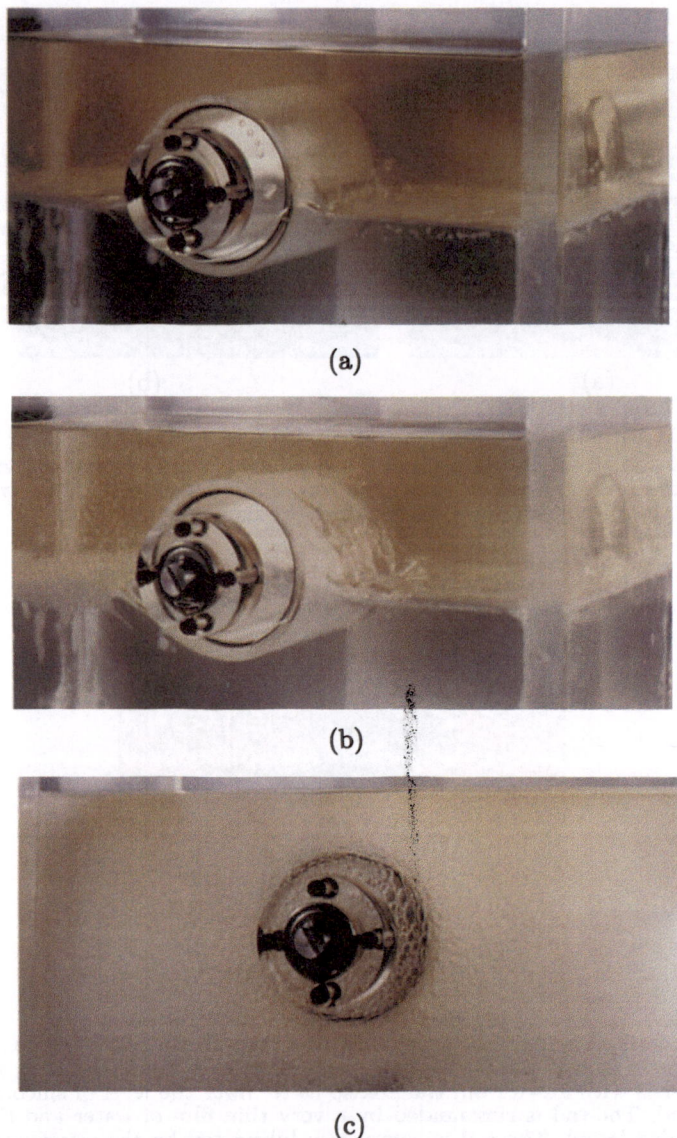

(a)

(b)

(c)

Plate II.4.18(a-c). [Figure 19 of Joseph, Nguyen and Beavers, 1984] (a) Light machine oil ($\rho = 0.831$ g/cm^3, $\mu = 6.36$ cP, yellow) on water. The rod rotates counter clockwise at 115 rpm. The contact angle in this experiment seems to be fixed with the contact line slipping on the rod in such a way as to stay fixed in space. The configuration of the contact line and the tenacity of the contact angle even under pressure from the intense water circulation under the free surface are noteworthy.   (b) The rod rotates at 195 rpm. This plate shows the fingering instability in mature form and how water droplets are torn off the fingers.   (c) Emulsion of water in light machine oil at 300 rpm. The water-laden emulsion shields the main body of oil (on the top left and right) from shearing.

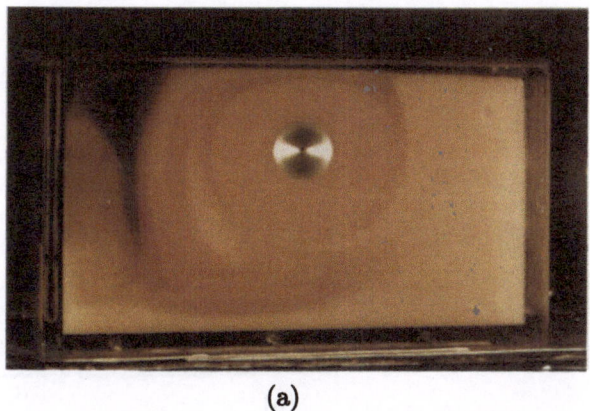

(a)

(b)

Plate II.4.20(a-b). [Figure 22 of Joseph, Nguyen and Beavers, 1984] (a) Emulsion of STP ($\mu = 11000$ cP, $\rho = 0.89$ g/cm^3, dark brown) in TLA 227 ($\mu = 20000$ cP, $\rho = 0.895$ g/cm^3, light brown). The rod rotates clockwise at about 16 rpm. Streamlines show STP being drawn into the rod.   (b) Emulsion of STP in TLA 227. The motor is stopped. A ring of STP (dark brown) precipitates out of TLA (light brown) around the rod.

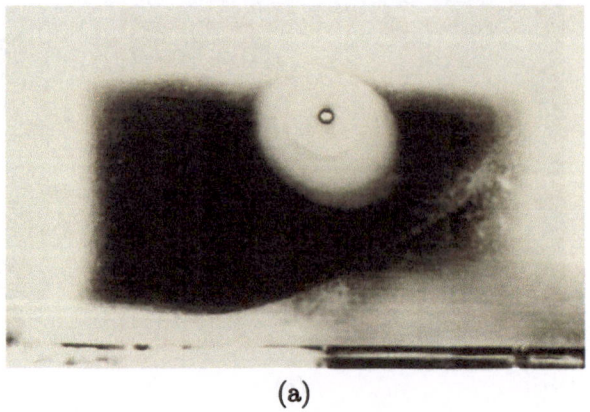

(a)

(b)

Plate II.4.21(a-b). [Figure 23 of Joseph, Nguyen and Beavers, 1984] Emulsion of 1% polyacrylamide/water ($\mu = 90$ cP, $\rho = 1.01$ g/cm^3) in STP ($\mu = 11000$ cP, $\rho = 0.890$ g/cm^3, dark). The polyacrylamide/water solution is on the bottom and on the rod, which rotates clockwise at about 45 rpm. (b) Close-up view showing the emulsion and the ring of polyacrylamide/water solution around the rod.

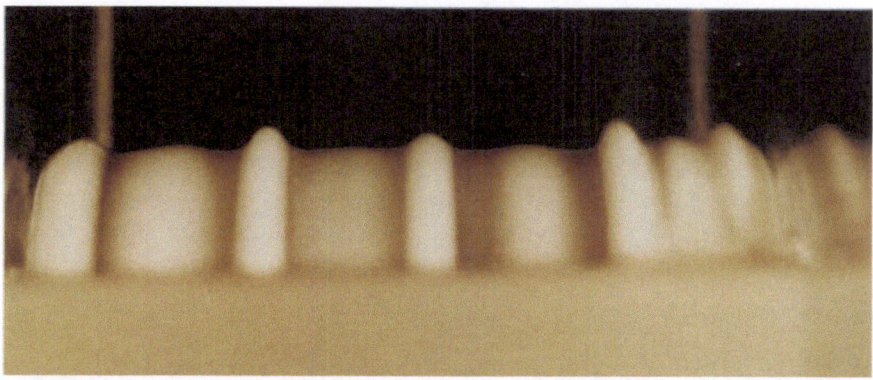

Plate II.4.28. [Figure 24 of Joseph, Nguyen and Beavers, 1984] Centrifugal instability of multigrade motor oil (10W40) in water. The layer of oil on the rod is separated from the main body of oil by a sheet of water. At the speed of about 300 r.p.m., the oil-water interface becomes wavy.

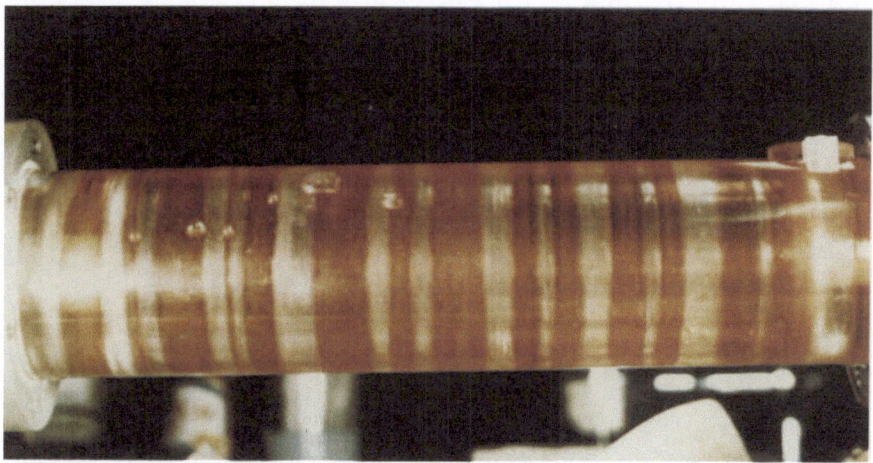

Plate II.5.2. [Figure 26 of Joseph, Nguyen and Beavers, 1984] Rollers of silicone oil I (red) separating Taylor cells of water (clear).

Plate II.5.3. [Figure 27 of Joseph, Nguyen and Beavers, 1984] As in plate II.5.2, but with much higher angular velocity.

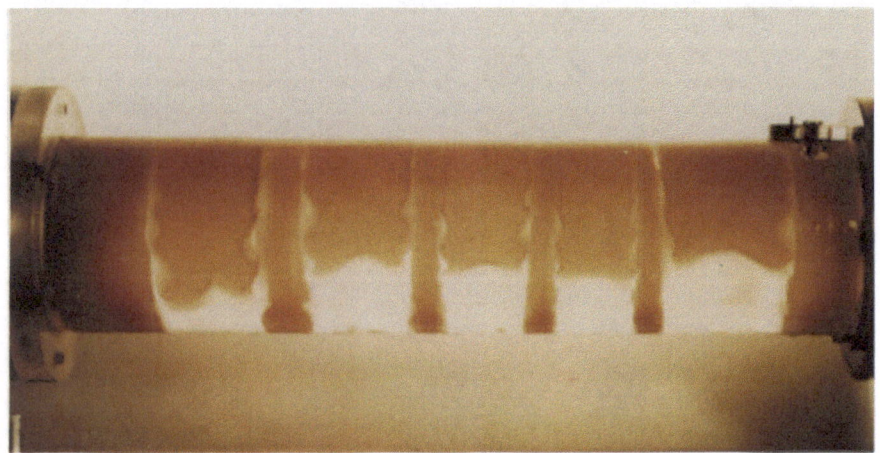

Plate II.5.4. [Figure 28 of Joseph, Nguyen and Beavers, 1984] Rollers of STP (orange) separating Taylor cells of water (clear).

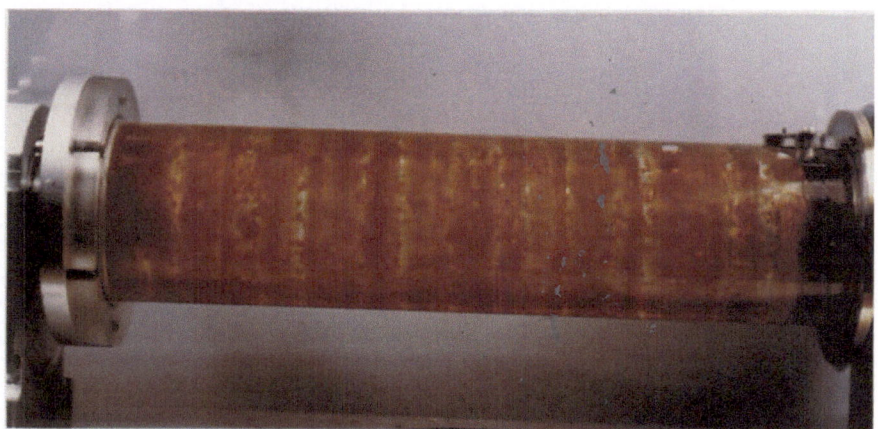

Plate II.5.5. [Figure 29 of Joseph, Nguyen and Beavers, 1984] Monograde motor oil (brown) and water (clear).

Plate II.5.6. [Figure 30 of Joseph, Nguyen and Beavers, 1984] Monograde motor oil (SAE40) and water at $\Omega=440$ r.p.m. The light cells are emulsions of oil with small droplets of water. The dark cells are of water with many large drops of oil.

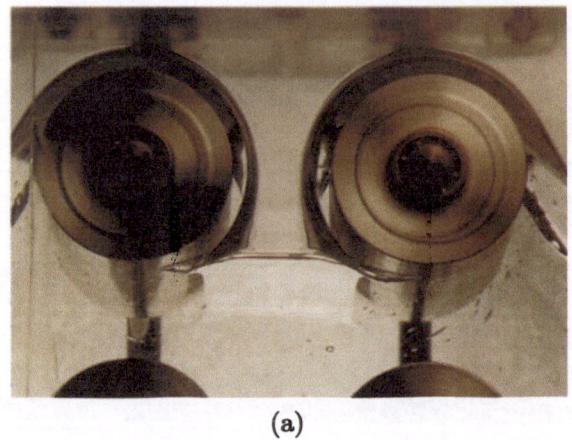

(a)

(b)

Plate II.6.5(a-b). [Joseph, Nelson, Renardy and Renardy, 1991] Cylinder coated with 12500 cs silicone oil cusps in air. (a) Liquid is dragged down between the cylinders. (b) Liquid is dragged up between two cylinders.

(a)

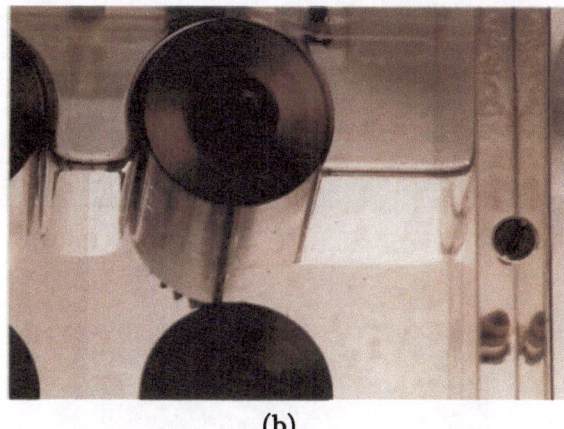

(b)

Plate II.6.8(a-b). [Joseph, Nelson, Renardy and Renardy, 1991] Low-molecular-weight silicone oils do not cusp at 200 r.p.m. (a) 500 cs silicone oil is marginal, but a Nikon microscope shows it does cusp. (b) 200 cs silicone oil does not cusp at 200 r.p.m.

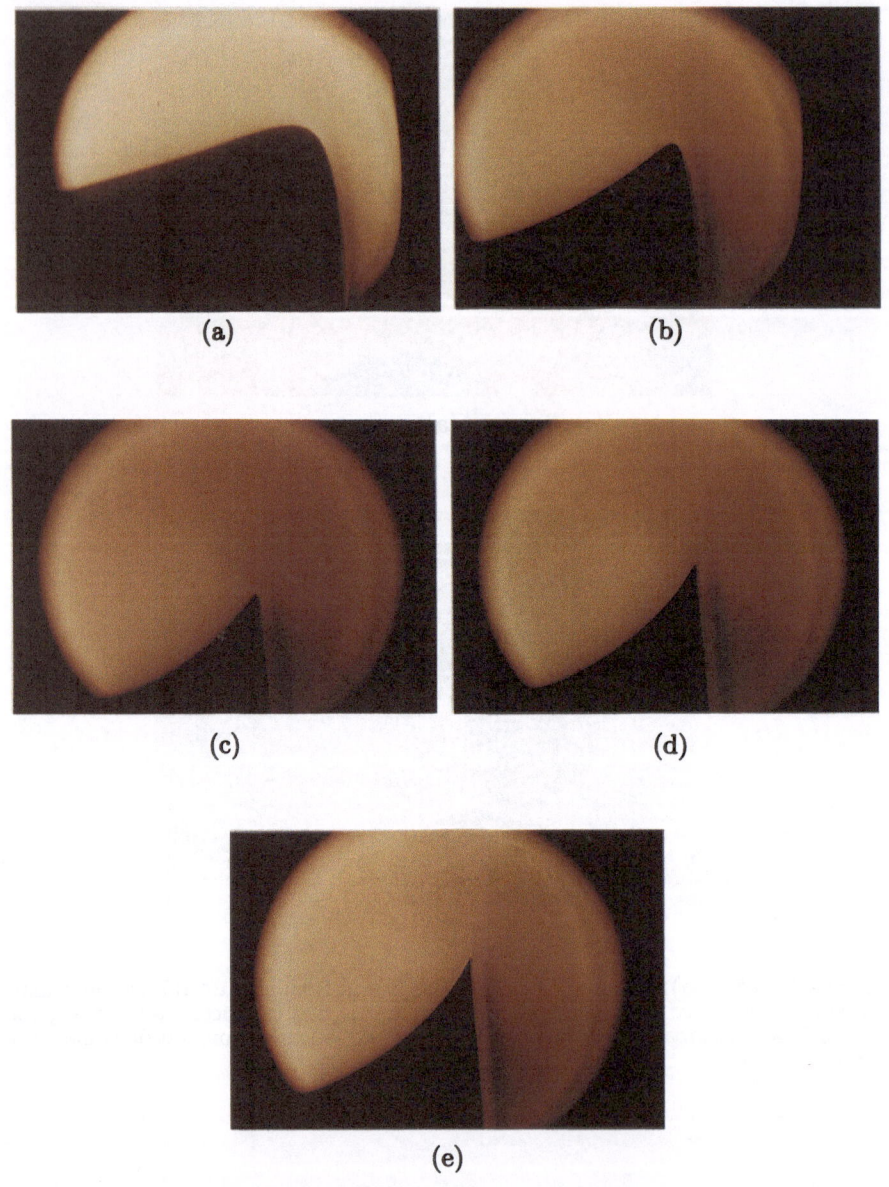

Plate II.6.9(a-e). [Joseph, Nelson, Renardy and Renardy, 1991] The critical speed for 500 cs silicone oil is approximately 75 r.p.m. (a) 44 r.p.m. (b) 59 r.p.m. (c) 65 r.p.m. (d) 70 r.p.m. (e) 75 r.p.m.

Plate II.6.21. [Palmquist and Kistler, 1992] Color contour plot of a flow field with an apparent cusp.

Standing Rolls

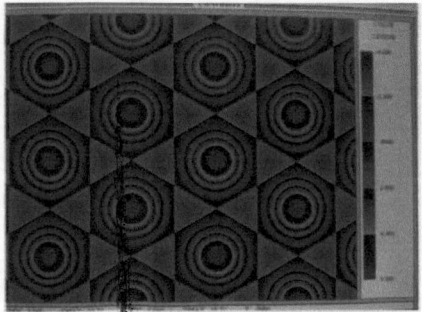

Standing Hexagons

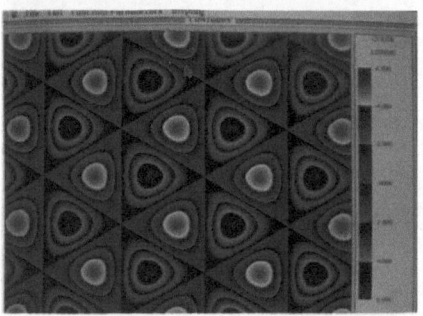

Standing Regular Triangles

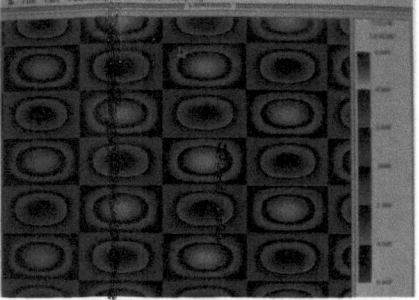

Standing Patchwork Quilt

Plate III.7.4. [Photographs for the plates III.7.4 were taken by Prof. David Wagner, Department of Mathematics, University of Houston, Texas.] On this page, we display the four standing wave solutions. The upper patterns are: standing rolls, standing hexagons. The lower patterns are: standing regular triangles, standing patchwork quilt. These are contour plots for an amplitude function, e.g., the vertical component of the velocity. Continued.

Travelling Rolls

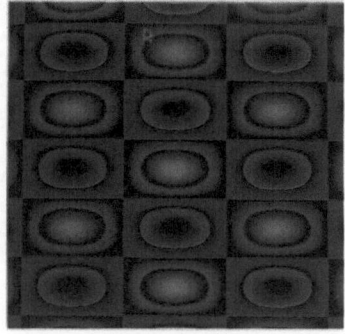

Travelling Patchwork Quilt (1)

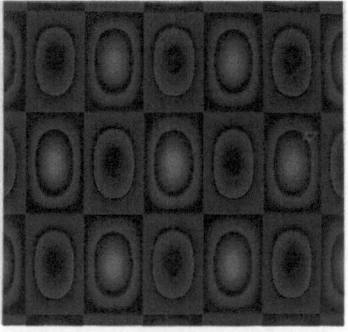

Travelling Patchwork Quilt (2)

Plate III.7.4. [These photographs were taken by Prof. David Wagner.] We display the three travelling wave solutions. The upper pattern shows travelling rolls. The lower patterns are the travelling patchwork quilt (1) and the travelling patchwork quilt (2). The arrows indicate the direction of travel. Continued.

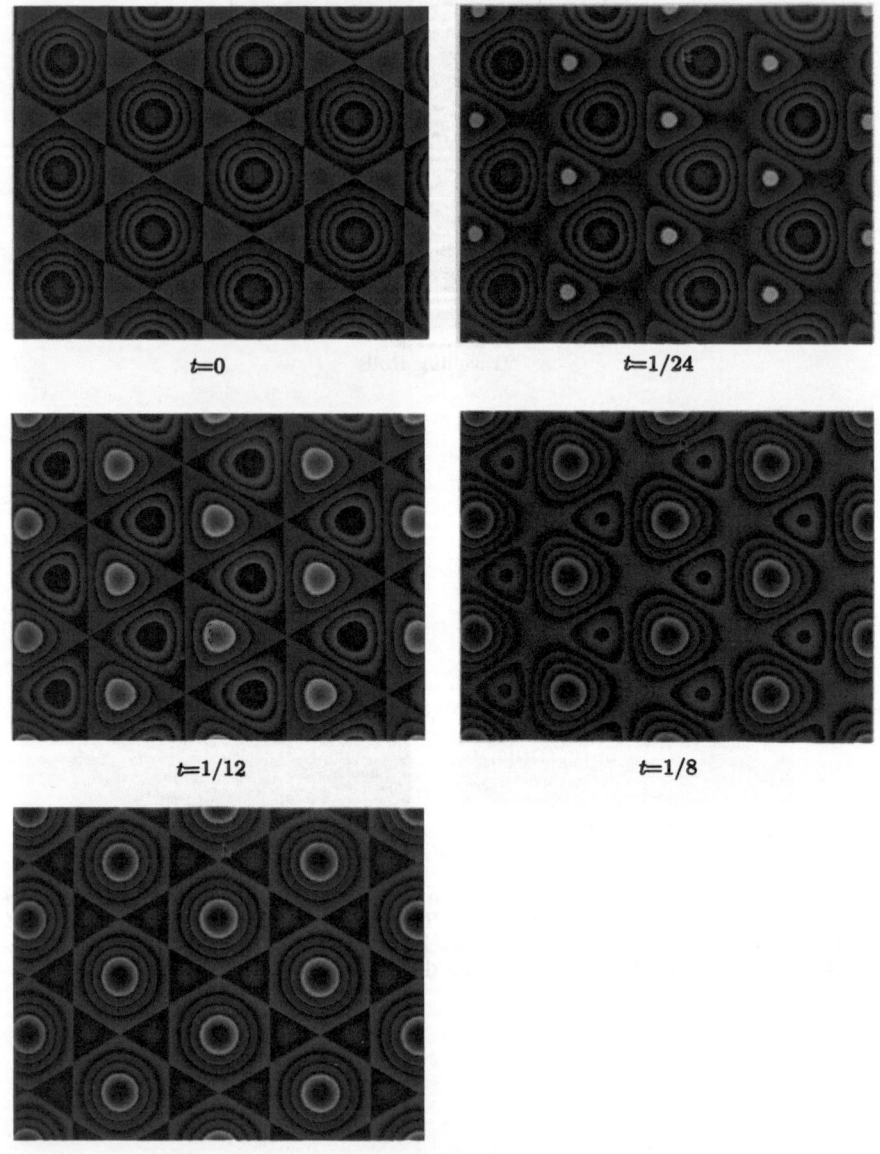

$t=0$

$t=1/24$

$t=1/12$

$t=1/8$

$t=1/6$

Oscillating Triangles

Plate III.7.4. [These photographs were taken by Prof. David Wagner.] The next four solutions are neither standing nor travelling waves. This page describes the oscillating triangles. Upper patterns: $t = 0$, $t = 1/24$. Middle patterns: $t = 1/12$, $t = 1/8$. Lower pattern: $t = 1/6$. Continued.

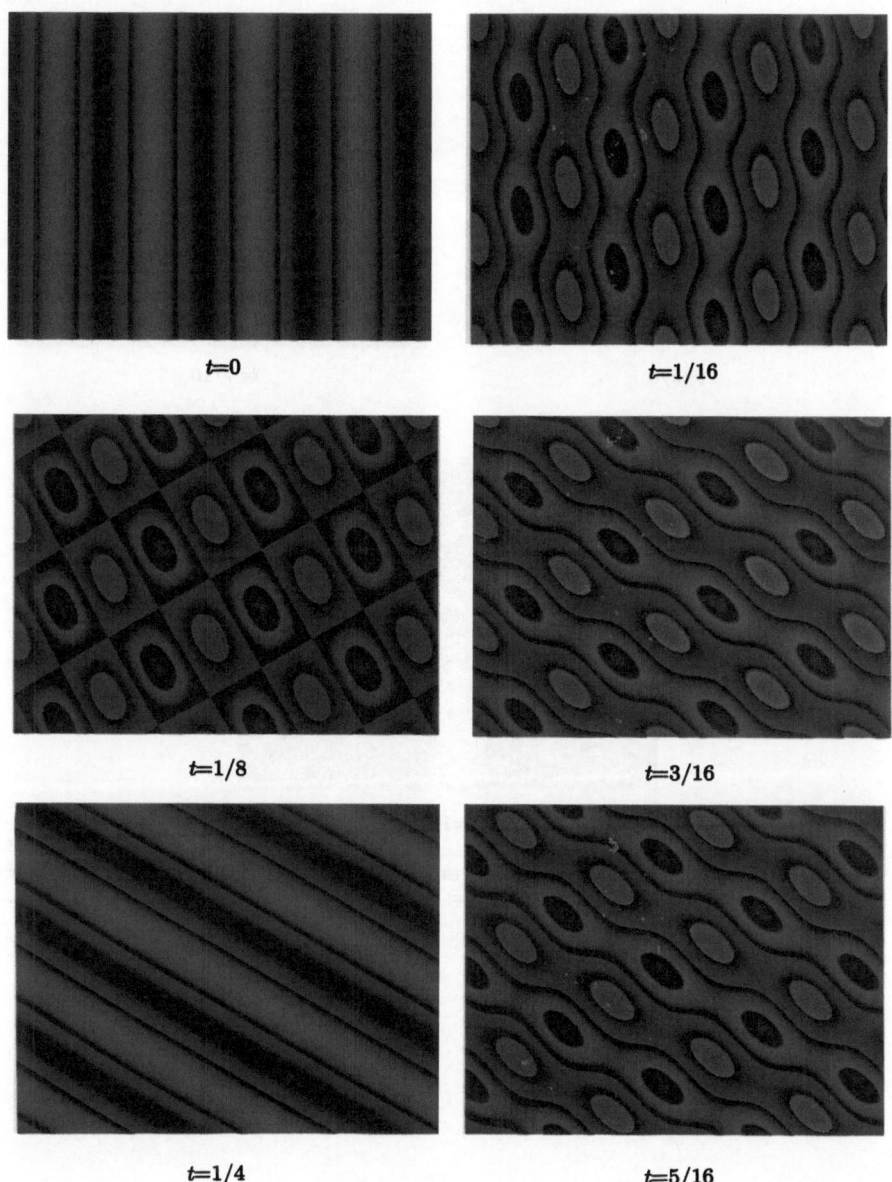

$t=0$

$t=1/16$

$t=1/8$

$t=3/16$

$t=1/4$

$t=5/16$

Plate III.7.4. [These photographs were taken by Prof. David Wagner.] Wavy rolls (1). Upper patterns: $t = 0, t = 1/16$. Middle patterns: $t = 1/8, t = 3/16$. Lower pattern: $t = 1/4, t = 5/16$. Continued.

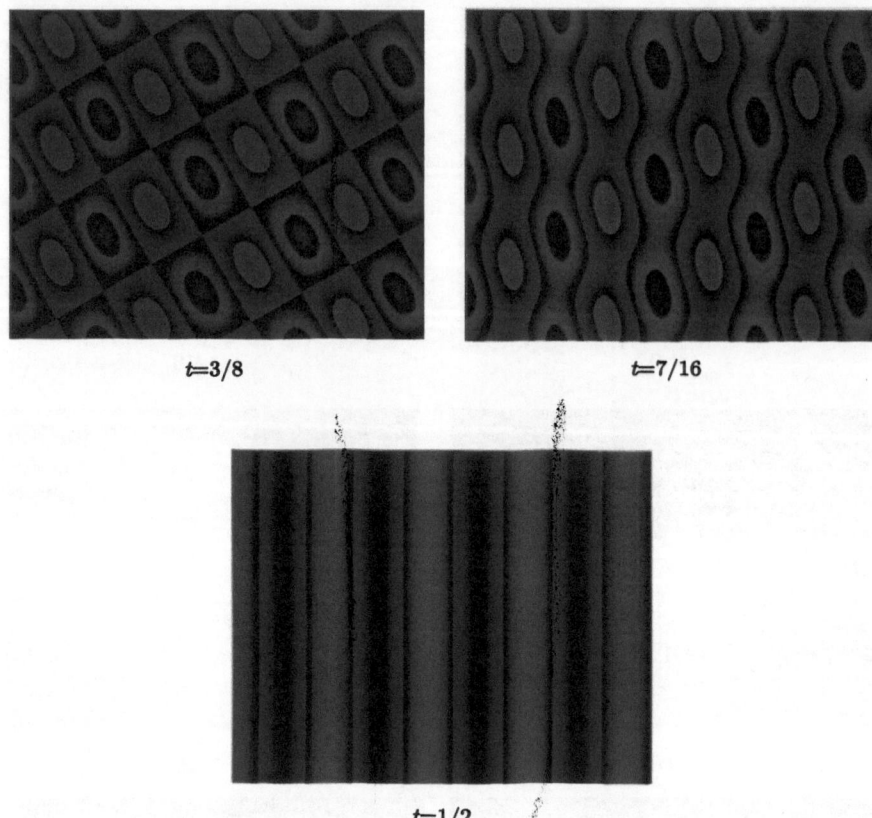

$t=3/8$

$t=7/16$

$t=1/2$

Plate III.7.4. [These photographs were taken by Prof. David Wagner.] Wavy rolls (1) continued. Upper pattern: $t = 3/8, t = 7/16$. Lower pattern: $t = 1/2$. Continued.

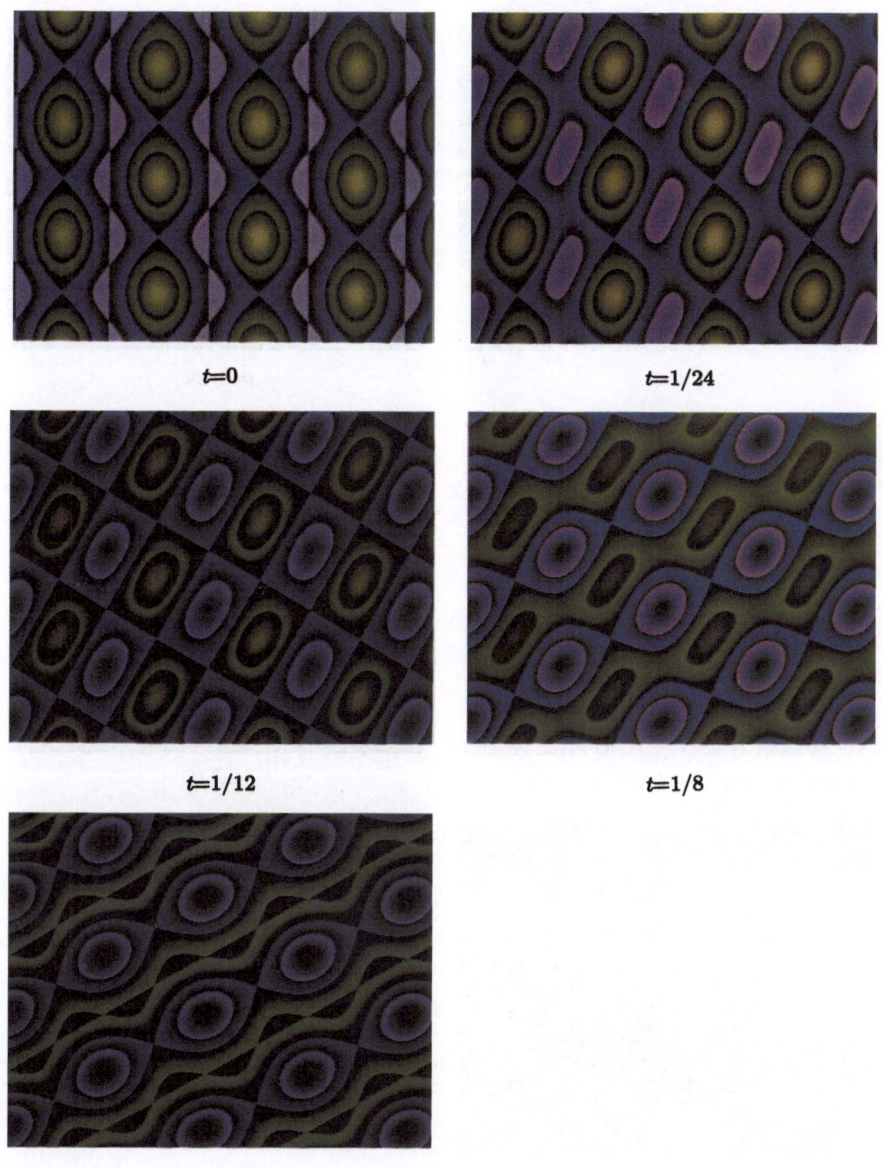

t=0

t=1/24

t=1/12

t=1/8

t=1/6

Twisted Patchwork Quilt

Plate III.7.4. [These photographs were taken by Prof. David Wagner.] Twisted patchwork quilt. Upper patterns: $t = 0, t = 1/24$. Middle patterns: $t = 1/12, t = 1/8$. Lower pattern: $t = 1/6$. Continued.

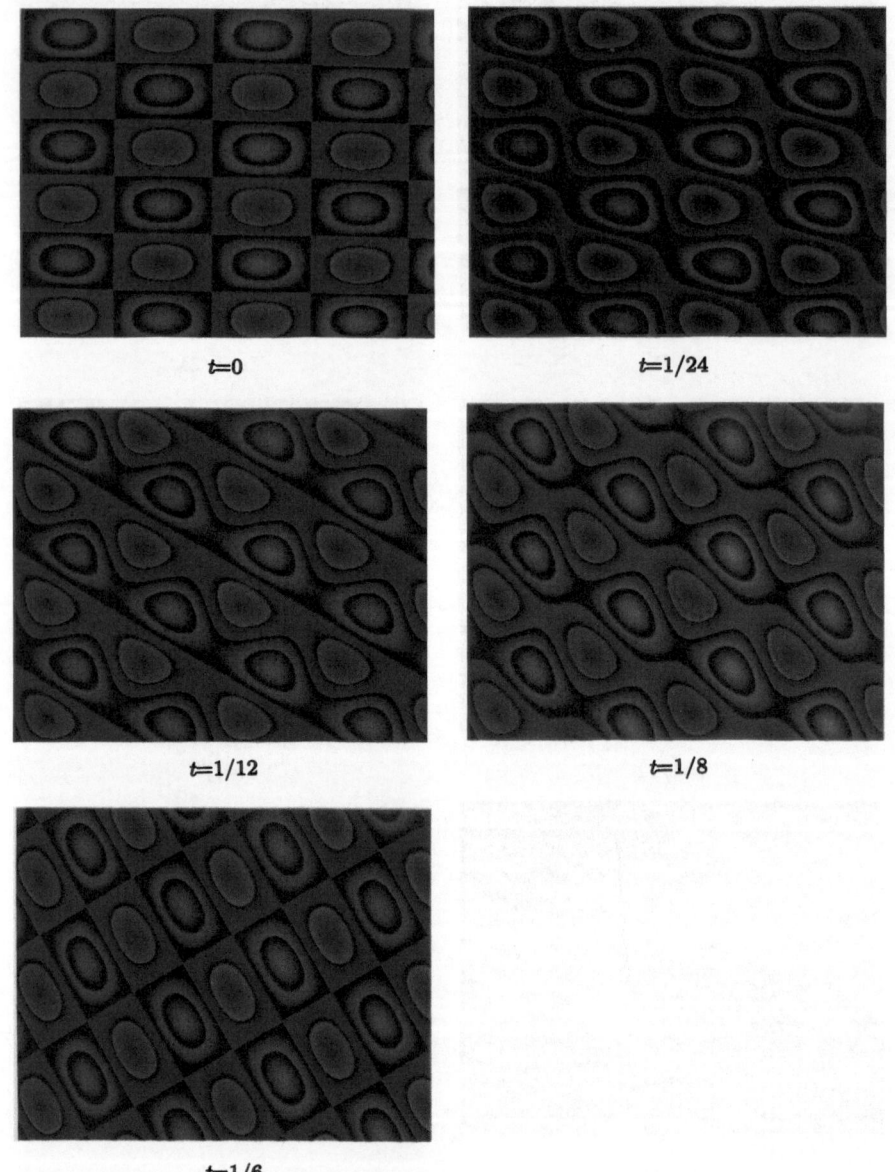

$t=0$ $t=1/24$

$t=1/12$ $t=1/8$

$t=1/6$

Wavy Rolls (2)

Plate III.7.4 . [These photographs were taken by Prof. David Wagner.] Wavy rolls (2). Upper patterns: $t = 0, t = 1/24$. Middle patterns: $t = 1/12, t = 1/8$. Lower pattern: $t = 1/6$.

The symmetries of the hexagonal lattice and the theory of normal forms can be used to reduce F_1 to the form

$$F_1(z_1, z_2, z_3, w_1, w_2, w_3, \lambda) = \mu(\lambda)z_1 + \alpha_1(\lambda)|z_1|^2 z_1 + \alpha_2(\lambda)|w_1|^2 z_1$$

$$+\alpha_3(\lambda)(|z_2|^2 + |z_3|^2)z_1 + \alpha_4(\lambda)(|w_2|^2 + |w_3|^2)z_1 + \alpha_5(\lambda)(z_2 w_2 + z_3 w_3)\bar{w}_1 + ...,$$
$$(7a.4)$$

where the dots denote terms of higher than third degree. It is assumed that criticality is at $\lambda = 0$ and that instability occurs for $\lambda > 0$, i.e. $\mu(0)$ is purely imaginary and Re $\mu'(0) < 0$. Equation (7a.4) is the ultimate form of the problem that is achieved in section III.7 (c), and this is the equation in "normal form".

There are twelve degrees of freedom for the eigenfunctions. Roberts *et al.* find all time-periodic solutions (a periodic solution is also called a "limit cycle") that are given by two degrees of freedom rather than twelve. In other words, they consider those solutions which have enough symmetries that the twelve-dimensional eigenspace reduces to two dimensions if the symmetries are imposed. This means that just one of the z_i and w_i is free while the others are related to each other. For example, when they impose invariance under rotations by multiples of 60 degrees and reflections across the axes of the hexagon, they obtain a two-dimensional solution set where the z_i are equal and get the standing hexagon (see solution number 2 in the list below). Then, in the two-dimensional case, they are back to the standard *Hopf Bifurcation Theorem* [Iooss and Joseph 1980]:

Let
$$\dot{x} = f(x, \lambda), \quad x \in \mathbb{R}^2, \quad \lambda \in \mathbb{R},$$

and let f be smooth (i.e., has a number of derivatives). Assume that

$$f(0, \lambda) = 0.$$

(This implies that $x = 0$ is a solution.) At $\lambda = 0$, suppose the Jacobian $D_x f(0,0)$, defined by

$$D_x f = \left(\frac{\partial f_i}{\partial x_j}\right) = \begin{pmatrix} \frac{\partial f_1}{\partial x_1} & \frac{\partial f_1}{\partial x_2} \\ \frac{\partial f_2}{\partial x_1} & \frac{\partial f_2}{\partial x_2} \end{pmatrix},$$

has eigenvalues $\pm iw$ and that

$$\frac{d}{d\lambda} Re \text{ (eigenvalues)} > 0.$$

(This ensures that eigenvalues cross the imaginary axis with non-vanishing speed. Also, if this quantity is negative, one needs to just change the definition of λ.) Then there is a family of periodic solutions $x(t, \epsilon)$, $\lambda(\epsilon)$ with period $T(\epsilon)$, which are smooth funcions of ϵ and satisfy

$$\lim_{\epsilon \to 0} \lambda(\epsilon) = 0, \quad \lim_{\epsilon \to 0} x(t, \epsilon) = 0, \quad \lim_{\epsilon \to 0} T(\epsilon) = 2\pi/w.$$

(In generic situations, i.e., when certain quantities are not zero, we can take $\lambda(\epsilon)$ to be one of the two values $\pm\epsilon^2$, depending on whether the bifurcated solution is subcritical or supercritical.) The bifurcating periodic solution is stable if it is supercritical, and unstable if it is subcritical.

Roberts *et al.* find eleven qualitatively different classes of bifurcated solutions. The period ($T(\epsilon)$ in the above notation) of a bifurcating solution is $2\pi/(1+\tau)$, where τ is an extra parameter, the "perturbed period", which is like the ϵ^2 or λ above: $\lim_{\tau\to 0}$ of the period is then 2π, with their eigenvalues taken as $\pm i$ rather than the $\pm iw$ in the preceding paragraph.

Each of the eleven classes of solutions can be characterized by a certain relationship which gives z_2, z_3, w_1, w_2 and w_3 in terms of z_1. For example, if $z_1 = w_1$ and $z_2 = z_3 = w_2 = w_3 = 0$, then there are waves of equal amplitudes travelling in opposite directions, so that the pattern is a standing wave. The eleven classes of solutions are as follows:

1. Standing rolls: $z_1 = w_1$, $z_2 = z_3 = w_2 = w_3 = 0$.
2. Standing hexagons: $z_1 = z_2 = z_3 = w_1 = w_2 = w_3$.
3. Standing regular triangles: $z_1 = z_2 = z_3 = -w_1 = -w_2 = -w_3$.
4. Standing patchwork quilt: $z_1 = z_2 = w_1 = w_2$, $z_3 = w_3 = 0$.
5. Travelling rolls: $z_2 = z_3 = w_1 = w_2 = w_3 = 0$.
6. Travelling patchwork quilt (1): $z_1 = z_3$, $z_2 = w_1 = w_2 = w_3 = 0$.
7. Travelling patchwork quilt (2): $z_1 = w_3$, $z_2 = z_3 = w_1 = w_2 = 0$.
8. Oscillating triangles; $z_1 = z_2 = z_3$, $w_1 = w_2 = w_3 = 0$.
9. Wavy rolls (1): $z_1 = z_3 = w_1 = -w_3$, $z_2 = w_2 = 0$.
10. Twisted patchwork quilt: $z_2 = e^{2\pi i/3} z_1$, $z_3 = e^{4\pi i/3} z_1$, $w_i = z_i$.
11. Wavy rolls (2): $z_2 = e^{2\pi i/3} z_1$, $z_3 = e^{4\pi i/3} z_1$, $w_i = -z_i$.

Of course, other solutions can be obtained from these by a symmetry of the hexagonal lattice; e.g., we can rotate the standing rolls by 120 degrees to obtain another solution. This corresponds to the choice $z_2 = w_2$ and $z_1 = z_3 = w_1 = w_3 = 0$. Some patterns get turned as time progresses (for example, the wavy rolls in plate III.7.4 are turned by 60 degrees at time $1/6$, and then are back to the original position after a period) so there are spatio-temporal symmetries as well that are considered by Roberts et al. to get the whole picture.

Roberts *et al.* illustrate in their figure 3 the patterns corresponding to the solutions above (see plate III.7.4). The variable u, of which these diagrams give contour plots, is a suitably chosen component of the solution to the Bénard problem, we may for example choose the interface position. The plots represent the leading-order linear terms in the solution Φ given by equation (7b.22), and involve the eigenfunctions.

In plate III.7.4, solutions 1-4 are standing waves, i.e. the patterns simply oscillate with time. These plots can be thought of as pictures taken at time zero. Solutions 5-7 are travelling waves, i.e. the pattern moves in the direction indicated by the arrow without changing its shape. The time dependence of the remaining solutions is indicated by a series of pictures.

The above list of bifurcated solutions is complete in the set of solutions restricted by symmetry to a two-dimensional subspace of $\mathbb{R}^{12}$. To our knowledge, there is no proof that this list is complete in $\mathbb{R}^{12}$. A complete discussion of the stability of the eleven solutions is given by Roberts et al. They use Floquet theory which is a linear theory of stability for solutions that depend periodically on time. The stability is with respect to arbitrary perturbations, restricted only by the double periodicity. Their results are summarized in their Table 3.2, which we reproduce below.

Each branch is subcritical (unstable) if $\Delta_0 < 0$ and supercritical (stable or unstable depending on the other Δ_ks) if $\Delta_0 > 0$. A branch is stable if all Δ_k, $k = 0, 1, ..., r$ are positive (a stable solution is also called an "attractor") and unstable if any of them is negative. It is interesting to note that a supercritical branch can be unstable in this situation where the critical eigenvalue is not simple. (An example of a critical situation with a simple eigenvalue is the two-layer plane Couette flow (see chapter IV), where there is a critical eigenvalue iw for wavenumber α_c and $-iw$ for $-\alpha_c$.) Nagata and Thomas [1986] studied the stability of some of the above solutions with respect to perturbations with the same symmetries as the solutions themselves; this also yields, as in the case of a simple critical eigenvalue, that a supercritical branch is stable.

Table 3.2 of Roberts et. al is as follows:

Solution	$\Delta_0, \Delta_1, ..., \Delta_r$
1	$\mathrm{Re}(\alpha_1 + \alpha_2)$
	$\mathrm{Re}(\alpha_1 - \alpha_2)$
	$\mathrm{Re}(-\alpha_1 - \alpha_2 + \alpha_3 + \alpha_4)$
	$\lvert -\alpha_1 - \alpha_2 + \alpha_3 + \alpha_4 \rvert^2 - \lvert\alpha_5\rvert^2$
2,3	$\mathrm{Re}(\alpha_1 + \alpha_2 + 2\alpha_3 + 2\alpha_4 + 2\alpha_5)$
	$\mathrm{Re}(\alpha_1 - \alpha_2 + 2\alpha_3 - 2\alpha_4 - 2\alpha_5)$
	$\mathrm{Re}(\alpha_1 - \alpha_2 - \alpha_3 + \alpha_4 - 2\alpha_5)$
	$\mathrm{Re}(\alpha_1 + \alpha_2 - \alpha_3 - \alpha_4 - 4\alpha_5)$
	$\lvert \alpha_1 + \alpha_2 - \alpha_3 - \alpha_4 - 4\alpha_5 \rvert^2 - \lvert \alpha_1 + \alpha_2 - \alpha_3 - \alpha_4 + 2\alpha_5 \rvert^2$
	$\pm\psi_1$
4	$\mathrm{Re}(\alpha_1 + \alpha_2 + \alpha_3 + \alpha_4 + \alpha_5)$
	$\mathrm{Re}(\alpha_1 - \alpha_2 - \alpha_3 + \alpha_4 - \alpha_5)$
	$\mathrm{Re}(\alpha_1 - \alpha_2 + \alpha_3 - \alpha_4 - \alpha_5)$
	$\mathrm{Re}(-\alpha_1 - \alpha_2 + \alpha_3 + \alpha_4 - \alpha_5)$
	$\lvert -\alpha_1 - \alpha_2 + \alpha_3 + \alpha_4 - \alpha_5 \rvert^2 - 4\lvert\alpha_5\rvert^2$
	$\mathrm{Re}(\alpha_1 + \alpha_2 - \alpha_3 - \alpha_4 - 3\alpha_5)$
	$\lvert \alpha_1 + \alpha_2 - \alpha_3 - \alpha_4 - 3\alpha_5 \rvert^2 - \lvert \alpha_1 + \alpha_2 - \alpha_3 - \alpha_4 + \alpha_5 \rvert^2$
5	$\mathrm{Re}(\alpha_1)$
	$\mathrm{Re}(-\alpha_1 + \alpha_2)$
	$\mathrm{Re}(-\alpha_1 + \alpha_3)$

$$\mathrm{Re}(-\alpha_1 + \alpha_4)$$

6
$$\mathrm{Re}(\alpha_1 + \alpha_3)$$
$$\mathrm{Re}(\alpha_1 - \alpha_3)$$
$$\mathrm{Re}(-\alpha_1 + \alpha_3)$$
$$\mathrm{Re}(-\alpha_1 + \alpha_2 - \alpha_3 + \alpha_4 + \alpha_5)$$
$$\mathrm{Re}(-\alpha_1 + \alpha_2 - \alpha_3 + \alpha_4 - \alpha_5)$$
$$\mathrm{Re}(-\alpha_1 - \alpha_3 + 2\alpha_4)$$

7
$$\mathrm{Re}(\alpha_1 + \alpha_4)$$
$$\mathrm{Re}(\alpha_1 - \alpha_4)$$
$$\mathrm{Re}(-\alpha_1 + \alpha_3)$$
$$\mathrm{Re}(-\alpha_1 + \alpha_2 + \alpha_3 - \alpha_4 + \alpha_5)$$
$$\mathrm{Re}(-\alpha_1 + \alpha_2 + \alpha_3 - \alpha_4 - \alpha_5)$$

8
$$\mathrm{Re}(\alpha_1 + 2\alpha_3)$$
$$\mathrm{Re}(\alpha_1 - \alpha_3)$$
$$\mathrm{Re}(-\alpha_1 + \alpha_2 - 2\alpha_3 + 2\alpha_4 + 2\alpha_5)$$
$$\mathrm{Re}(-\alpha_1 + \alpha_2 - 2\alpha_3 + 2\alpha_4 - \alpha_5)$$

9
$$\mathrm{Re}(\alpha_1 + \alpha_2 + \alpha_3 + \alpha_4 - \alpha_5)$$
$$\mathrm{Re}(\alpha_1 - \alpha_2 + \alpha_3 - \alpha_4 + \alpha_5)$$
$$\mathrm{Re}(\alpha_1 - \alpha_2 - \alpha_3 + \alpha_4 + \alpha_5)$$
$$\mathrm{Re}(-\alpha_1 - \alpha_2 + \alpha_3 + \alpha_4 + \alpha_5)$$
$$\mathrm{Re}(\alpha_1 + \alpha_2 - \alpha_3 - \alpha_4 + 3\alpha_5)$$
$$|\alpha_1 + \alpha_2 - \alpha_3 - \alpha_4 + 3\alpha_5|^2 - |\alpha_1 + \alpha_2 - \alpha_3 - \alpha_4 - \alpha_5|^2$$

10,11
$$\mathrm{Re}(\alpha_1 + \alpha_2 + 2\alpha_3 + 2\alpha_4 - \alpha_5)$$
$$\mathrm{Re}(\alpha_1 - \alpha_2 - \alpha_3 + \alpha_4 + \alpha_5)$$
$$\mathrm{Re}(\alpha_1 - \alpha_2 + 2\alpha_3 - 2\alpha_4 + \alpha_5)$$
$$\mathrm{Re}(T + \sqrt{T^2 - D})$$
$$\mathrm{Re}(T - \sqrt{T^2 - D}), \text{ where}$$
$$T = \mathrm{Re}(\alpha_1 + \alpha_2 - \alpha_3 - \alpha_4 - \alpha_5) + 3\bar\alpha_5,$$
$$D = 6\bar\alpha_5(\alpha_1 + \alpha_2 - \alpha_3 - \alpha_4 - \alpha_5)$$
$$\pm\psi_2$$

Here ψ_1 and ψ_2 are terms that can only be determined if (7a.4) is expanded to higher order, and we shall not do so here. It is conceivable, however, that there are situations where the higher order terms would have to be computed. It turns out that for the computations presented in this section, this need does not arise. The sign in front of ψ_1 is opposite for the standing hexagons and standing regular triangles and likewise the sign in front of ψ_2 is opposite for the twisted patchwork quilt and wavy rolls (2).

Numerical values for the α_i are presented in section (d), together with the implications for stability. Without the numerical results, some observations can already be made. For the travelling patchwork quilt (1) (number 6), Δ_1 and Δ_2 have opposite signs, and hence this solution is never stable. Also there are some mutually exclusive cases. For example, if the travelling rolls are stable, then the standing rolls, oscillating triangles and travelling patchwork quilt (2) must all be unstable. The oscillating triangles and the twisted patchwork quilt or wavy rolls (2) cannot simultaneously be stable. Also, the wavy rolls (1) and travelling patchwork quilt (2) cannot both be stable.

III.7(b) Reduction to Finite Dimension

In order to obtain a system of the form (7a.1), we must reduce the partial differential equations governing the Bénard problem to a finite set of ordinary differential equations. This is achieved by the center manifold approach. Roughly speaking, the Center Manifold Theorem says that in the neighborhood of criticality the dynamics is governed by the (finitely many) critical modes. There is no version of the Center Manifold Theorem in the literature which can be applied in a straightforward manner to our problem. A full justification of the center manifold approach would be quite tedious, due to the variable domain occupied by each fluid and the nonlinear nature of the interface conditions. We shall not attempt a proof here, although it can be done. The coercive PDE estimates required for this were derived in M. Renardy and Joseph [1986] for a problem involving parallel shear flow of two fluids, and they can easily be extended to the Bénard problem. In the following, we shall take the existence of a center manifold for granted and do formal calculations.

To describe the center manifold approach, we must first introduce some notations. The unknowns in (2.28)-(2.39) are $\mathbf{v}$, $\tilde{p}$, $\tilde{\theta}$ and h. We denote this set of unknowns by Φ. Equations (2.28)-(2.39) can be written in the schematic form

$$L\Phi = N_2(\Phi, \Phi) + N_3(\Phi, \Phi, \Phi). \qquad (7b.1)$$

Here the operator L incorporates the linear terms, while N_2 and N_3 stand for the quadratic and cubic terms. We can write the real linear operator L in the form $A + \frac{d}{dt}B$. We also introduce the notation $L(\sigma) = A + \sigma B$.

The definitions of N_2 and N_3 are extended in a symmetric fashion to the case when their arguments are different:

$$N_2(\Phi, \Psi) := \frac{1}{4}\Big(N_2(\Phi + \Psi, \Phi + \Psi) - N_2(\Phi - \Psi, \Phi - \Psi)\Big), \qquad (7b.2)$$

$$N_3(\Phi, \Psi, \Xi) := \frac{1}{24}\Big(N_3(\Phi + \Psi + \Xi, \Phi + \Psi + \Xi, \Phi + \Psi + \Xi)$$

$$+ N_3(\Phi - \Psi - \Xi, \Phi - \Psi - \Xi, \Phi - \Psi - \Xi)$$

$$+ N_3(-\Phi + \Psi - \Xi, -\Phi + \Psi - \Xi, -\Phi + \Psi - \Xi)$$

$$+ N_3(-\Phi - \Psi + \Xi, -\Phi - \Psi + \Xi, -\Phi - \Psi + \Xi)\Big). \qquad (7b.3)$$

For example, if $N_2(\Phi, \Phi) = \Phi(d\Phi/dz)$, then $N_2(f, g) = 0.5(fdg/dz + gdf/dz)$. This represents the average of the two 'permutations' of the quadratic expression. Similarly, $N_3(f, g, h)$ is defined as $1/6$ of the six possible permutations of the cubic expression.

We look for solutions Φ with the periodicity of the hexagonal lattice, where the period W (see equation (7.6)) is given by $4\pi/\alpha\sqrt{3}$ and α is the critical wavelength found in the linear analysis, e.g., of section 3 above. We regard all fluid properties as fixed and we view the Rayleigh number as a bifurcation parameter. We set $\lambda = R - R_c$, where R_c is the critical value of the Rayleigh number.

The reduction to the center manifold will involve the solution of problems of the form $L(\sigma)\Phi = f$. We note that in (2.28)-(2.39) no nonlinear terms appear in the incompressibility condition and the boundary conditions on the walls, and we shall therefore always assume that the corresponding components of f are zero. The other components of f are labeled f_1 through f_{17} in accordance with the labeling of the components of H in (2.28)-(2.39): here, f denotes a general form for the nonlinearities and H denotes the particular ones in (2.28) - (2.39).

A digression is made to consider the $(0,0)$ Fourier components (i.e., no x, y - dependence or, $k = l = 0$ in equation (7.5)) of equations (2.28) - (2.39): the purpose is to make remarks on $L(\sigma)$. With d/dt replaced by σ, the problem for w is, from the incompressibility condition,

$$w' = 0 \quad \text{in fluids 1 and 2,}$$

$$w = 0 \quad \text{at} \quad z = 0 \text{ and } 1,$$

$$[\![w]\!] = H_{11} \quad \text{at} \quad z = l_1. \tag{7b.4}$$

The first two conditions yield $w = 0$ and this is consistent with the third only if $H_{11}=0$. The problem for u is

$$\sigma u - Pu'' = H_2 \quad \text{in fluid 1,}$$

$$\sigma u - \frac{rP}{m}u'' = H_6 \quad \text{in fluid 2,}$$

$$u = 0 \quad \text{at} \quad z = 0 \text{ and } 1, \tag{7b.5}$$

and at $z = l_1$,

$$[\![u]\!] = H_9 \quad Pu'_1 - \frac{P}{m}u'_2 = H_{12}. \tag{7b.6}$$

Likewise, the problem for v is

$$\sigma v - Pv'' = H_3 \quad \text{in fluid 1,}$$

$$\sigma v - \frac{rP}{m}v'' = H_7 \quad \text{in fluid 2,}$$

$$v = 0 \quad \text{at} \quad z = 0 \text{ and } 1, \tag{7b.7}$$

and at $z = l_1$,

$$[\![v]\!] = H_{10} \quad Pv'_1 - \frac{P}{m}v'_2 = H_{13}. \tag{7b.8}$$

The problems for u and v yield negative eigenvalues: the proof follows from equations (7.7) - (7.11) with α set to 0.

The kinematic free surface condition is $\sigma h = H_{17}$. If $\sigma \neq 0$, then we find h from this, and then $\tilde{\theta}$ from

$$\sigma\tilde{\theta} - \tilde{\theta}'' = H_1, \quad \text{in fluid 1,}$$

$$\sigma\tilde{\theta} - \frac{1}{\gamma}\tilde{\theta}'' = H_5, \quad \text{in fluid 2,}$$

$$\tilde{\theta} = 0 \quad \text{at } z = 0 \text{ and } 1, \tag{7b.9}$$

and at $z = l_1$,

$$[\![\tilde{\theta}]\!] - h[\![A]\!] = H_{15} \quad \tilde{\theta}_1' - \frac{1}{\zeta}\tilde{\theta}_2' = H_{16}, \tag{7b.10}$$

and finally $\tilde{p}$ (up to an undetermined constant) from

$$\tilde{p}' - RP\tilde{\theta} = H_4 \quad \text{in fluid 1,}$$

$$r\tilde{p}' - \frac{RP}{\beta}\tilde{\theta} = H_8 \quad \text{in fluid 2,} \tag{7b.11}$$

and at $z = l_1$,

$$-[\![\tilde{p}]\!] - M_1 h = H_{14}. \tag{7b.12}$$

If $\sigma = 0$, then the kinematic free surface condition requires that $H_{17} = 0$ and h can be arbitrary: for any given h, the rest of the problem is solved as before.

The above equations show that there are nullspaces of $L(\sigma)$ associated with the $(0,0)$ Fourier component. For any σ, the operator $L(\sigma)$ has a trivial one-dimensional nullspace ($u = v = w = \tilde{\theta} = h = 0$, $\tilde{p} = const.$) due to the fact that we can add an arbitrary constant to the pressure; to make the pressure unique we can normalize it; for example, we can set $\tilde{p}$ for fluid 1 equal to 0 at $z = l_1$. In the following, we impose such a restriction on the solution Φ. The corresponding solvability condition is found by noting that for the $(0,0)$ Fourier component, $w = 0$ in both fluids. Thus the solvability condition is that the $(0,0)$ component of f_{11} must be zero.

In addition, $L(\sigma)$ has an eigenvalue zero with an eigenfunction that has no x, y-dependence. This eigenfunction simply corresponds to a constant vertical shift of the interface and is given by $h = const.$, $u = v = w = 0$, and $\tilde{\theta}$ and $\tilde{p}$ adjusted accordingly. We assume that the volume ratio of the two fluids is given, and we therefore have to rule out this eigenfunction; i.e. we require that the average of h ($\int \int_A h \, dx dy$, where A is the lattice cell as in (7b.40)) must be zero. The corresponding solvability condition is that the average of (or the $(0,0)$ Fourier component of) f_{17} has to be zero. These solvability conditions are always satisfied below because the nonlinear terms in this particular problem are such that they automatically satisfy them. For example, look at H_{17}. Its average, or its $(0,0)$ Fourier component, is

$$\int\int_A H_{17}dxdy$$

$$= \int\int_A h\frac{\partial w_1}{\partial z} - \frac{\partial h}{\partial x}u_1 - \frac{\partial h}{\partial y}v_1 + \frac{1}{2}h^2\frac{\partial^2 w_1}{\partial z^2} - h\frac{\partial h}{\partial x}\frac{\partial u_1}{\partial z} - h\frac{\partial h}{\partial y}\frac{\partial v_1}{\partial z}dxdy.$$

We integrate the second and fifth terms with respect to x by parts, and the third and last terms with respect to y by parts. This yields the boundary terms hu_1, hv_1, $\frac{1}{2}h^2\frac{\partial u_1}{\partial z}$ and $\frac{1}{2}h^2\frac{\partial v_1}{\partial z}$ on the boundaries of the lattice cell A, and these vanish due to periodicity. Thus, the average of H_{17} is

$$= \int\int_A h\frac{\partial w_1}{\partial z} + h\frac{\partial u_1}{\partial x} + h\frac{\partial v_1}{\partial y} + \frac{1}{2}h^2\frac{\partial^2 w_1}{\partial z^2} + \frac{1}{2}h^2\frac{\partial^2 u_1}{\partial z\partial x} + \frac{1}{2}h^2\frac{\partial^2 v_1}{\partial z\partial y}dxdy$$

$$= \int\int_A h\ \text{div}\ \mathbf{v} + \frac{1}{2}h^2\frac{\partial}{\partial z}\ \text{div}\ \mathbf{v}\quad dxdy,$$

and this is zero since div $\mathbf{v}$ is zero.

At criticality ($\lambda = 0$), we have a pair of complex conjugate eigenvalues $\pm i\omega$ of the linearized problem, and each of them has six eigenfunctions. For λ near 0, we denote by $-\mu(\lambda)$ the eigenvalue which arises from perturbing $-i\omega$. We denote by $\zeta_k(\lambda)$, $k = 1, 2, ..., 6$ the eigenfunctions belonging to $-\mu(\lambda)$, i.e.

$$L(-\mu(\lambda))\zeta_k(\lambda) = 0, \tag{7b.13}$$

and those belonging to $-\bar{\mu}(\lambda)$ are their complex conjugates.

The eigenfunctions ζ_k are related to each other by rotations. Let P_ϕ denote the transformation of rotation through angle ϕ:

$$P_\phi := \begin{pmatrix} \cos\phi & -\sin\phi \\ \sin\phi & \cos\phi \end{pmatrix}. \tag{7b.14}$$

With $\Phi = (u, v, w, \tilde{p}, \tilde{\theta}, h)$, where (u, v, w) is the velocity at (x, y, z), we define a transformation Q_ϕ which consists of two rotations, one on the velocity and one on the point in space:

$$Q_\phi\Phi(x, y, z) := (P_\phi(u, v), w, \tilde{p}, \tilde{\theta}, h)(P_{-\phi}(x, y), z). \tag{7b.15}$$

The physical meaning of the transformation $Q_\phi\Phi$ is that we take a given pattern Φ and rotate it by an angle ϕ.

The eigenfunction computed in section 3 has the form $\zeta_1 = \tilde{\zeta}(z)\exp(i\mathbf{a}_1 \cdot \mathbf{x})$, where $\mathbf{a}_1 = (\alpha, 0, 0)$, $\mathbf{x} = (x, y, z)$, and α is the critical wavenumber. The other eigenfunctions are obtained as follows:

$$\zeta_2 = Q_{2\pi/3}\zeta_1, \ \zeta_3 = Q_{4\pi/3}\zeta_1, \ \zeta_4 = Q_\pi\zeta_1, \ \zeta_5 = Q_{5\pi/3}\zeta_1, \ \zeta_6 = Q_{\pi/3}\zeta_1.$$
$$\tag{7b.16}$$

These correspond to the values of (k, l) in (7.5) of $(0, 1)$, $(-1, -1)$, $(-1, 0)$, $(0, -1)$ and $(1, 1)$. These transformations can be visualized as follows. The vectors $\pm\mathbf{a}_1$, $\pm\mathbf{a}_2$ and $\pm\mathbf{a}_3$ defined by equations (7.2) - (7.3) emanate from the center of a hexagon and terminate at its six vertices (see figure 7.3). The vertex given by the end of the vector $\mathbf{a_1}$ is identified with the eigenfunction ζ_1. The values of $\Phi(u, v, w, \tilde{p}, \tilde{\theta}, h)$ at that vertex belongs to ζ_1. Now rotate ζ_1 by π radians to get to the opposite vertex. The x - and y - components of the velocity for ζ_4 are then facing the other way from that of ζ_1, so that the set of variables in Φ for ζ_4 is that of ζ_1 except for $(-u, -v)$. In addition, the coordinates of the vertex given by the end of $\mathbf{a_2}$ is $(-x, -y)$ of that given by $\mathbf{a_1}$. The other transformations in (7b.16) can be visualized in similar ways.

The dependence of ζ_k on x and y is as follows. ζ_1 evaluated at a particular coordinate (x^*, y^*) in the plane depends on x^* and y^* through $\exp(i\alpha x^*)$ and has no y^*-dependence; in other words, the exponential dependence is only on the first component of the coordinate. Let us look at one of the variables, say w. Since $\zeta_2 = Q_{2\pi/3}\zeta_1$, the w of ζ_2 at the point (x^*, y^*) is the w of ζ_1 at the point $P_{-2\pi/3}(x^*, y^*)$, where $P_{-2\pi/3}(x^*, y^*)$

$$
= \begin{pmatrix} \cos(-2\pi/3) & -\sin(-2\pi/3) \\ \sin(-2\pi/3) & \cos(-2\pi/3) \end{pmatrix} \begin{pmatrix} x^* \\ y^* \end{pmatrix} = \begin{pmatrix} -\frac{1}{2}x^* + \frac{\sqrt{3}}{2}y^* \\ -\frac{\sqrt{3}}{2}x^* - \frac{1}{2}y^* \end{pmatrix}.
$$

Hence ζ_2 at (x^*, y^*) is ζ_1 at $(-\frac{1}{2}x^* + \frac{\sqrt{3}}{2}y^*, -\frac{\sqrt{3}}{2}x^* - \frac{1}{2}y^*)$ and the exponential dependence is again on the first coordinate. Thus, dropping the asterisks for generality, $\zeta_2 \sim \exp(i\alpha(-\frac{1}{2}x + \frac{\sqrt{3}}{2}y))$. Similarly, the dependence of ζ_3, ζ_4, ζ_5 and ζ_6 on x and y is through $\exp(i\alpha(-\frac{1}{2}x - \frac{\sqrt{3}}{2}y))$, $\exp(-i\alpha x)$, $\exp(i\alpha(\frac{1}{2}x - \frac{\sqrt{3}}{2}y))$ and $\exp(i\alpha(\frac{1}{2}x + \frac{\sqrt{3}}{2}y))$, respectively.

Corresponding to the sixfold eigenvalue we have six solvability conditions for the equation $L(-\mu(\lambda))\Phi = f$: $(b_k, f) = 0$ for $k = 1, ..., 6$. The adjoint eigenfunctions b_k are calculated from

$$
(b_k, L(-\mu(\lambda))\Phi) = 0, \quad k = 1, ..., 6, \tag{7b.17}
$$

for every Φ, and have the same dependence on x and y as the respective eigenfunctions ζ_k. Here $(\cdot, \cdot)$ denotes an appropriate inner product, e.g. the L^2 inner product (see after equation (7b.38) for further discussion on the inner product). In the actual computations, we are dealing with a discretized problem where $L(-\mu(\lambda))$ becomes a finite-dimensional matrix. We can then choose $(\cdot, \cdot)$ to be the Euclidean inner product in the finite-dimensional space (see equations (7b.39) - (7b.40)) and the b_k simply become the null-vectors of the adjoint matrix. The adjoint eigenvector computed in section 3 has the form

$$
b_1 = \tilde{b}(z)\exp(i\mathbf{a_1} \cdot \mathbf{x}). \tag{7b.18}
$$

In analogy with the ζ_k, the other b_k's are obtained as follows:

$$
b_2 = R_{2\pi/3}b_1, \; b_3 = R_{4\pi/3}b_1, \; b_4 = R_\pi b_1, \; b_5 = R_{5\pi/3}b_1, \; b_6 = R_{\pi/3}b_1, \tag{7b.19}
$$

where

$$R_\phi f(x,y,z) = (f_1, P_\phi(f_2, f_3), f_4, f_5, P_\phi(f_6, f_7), f_8, P_\phi(f_9, f_{10}),$$

$$f_{11}, P_\phi(f_{12}, f_{13}), f_{14}, f_{15}, f_{16}, f_{17})(P_{-\phi}(x,y), z). \tag{7b.20}$$

Here, the transformation R_ϕ is analogous to Q_ϕ so that the rotation P_ϕ is applied to the components f_2, f_3, f_6, f_7, f_9, f_{10}, f_{12} and f_{13} that arise from the (u, v) components of the velocity.

The normalization of $\tilde{\zeta}$ and $\tilde{b}$ is given by

$$(b_i, B\zeta_j) = \delta_{ij}, \ i,j = 1, ..., 6. \tag{7b.21}$$

For convenience, we denote w_1, w_2 and w_3 in (7a.2) by z_4, z_5 and z_6.

Any real-valued function Φ can be decomposed (in a manner analogous to the finite-dimensional case of the decomposition of C^n by the n linearly independent eigenfunctions of an n by n matrix with distinct eigenvalues) in the form

$$\Phi = \sum_{i=1}^{6} z_i \zeta_i + \sum_{i=1}^{6} \bar{z}_i \bar{\zeta}_i + \Psi, \tag{7b.22}$$

where z_i are complex numbers and Ψ represents a linear combination of eigenvectors (and possibly generalized eigenvectors) belonging to stable eigenvalues. This decomposition of Φ is like decomposing a function as a Fourier series on an interval. To use the Center Manifold Theorem, we need to have the fact that Ψ consists of stable modes. The expression on the right hand side of (7b.22) converges to Φ in the mean square, like a Fourier series expansion of a function. The ζ_i and the functions in Ψ form a complete set.

Suppose Ψ represents an eigenvector belonging to an eigenvalue s, not equal to $-\mu(\lambda)$. Then $L(s)\Psi = 0$ and hence $(b_i, (A+sB)\Psi) = 0$. But $A^*b_i = \bar{\mu}(\lambda)B^*b_i$ and thus $(b_i, (A + sB)\Psi) = (b_i, A\Psi) + s(b_i, B\Psi) = (A^*b_i, \Psi) + s(b_i, B\Psi) = (\mu(\lambda) + s)(b_i, B\Psi)$. Therefore,

$$(b_i, B\Psi) = 0, \ i = 1, ..., 6. \tag{7b.23}$$

Since $(b_i, (A - \mu(\lambda)B)\Psi) = 0$, we also have

$$(b_i, A\Psi) = 0, \ i = 1, ..., 6. \tag{7b.24}$$

Since $(A - \bar{\mu}(\lambda)B)\bar{\zeta}_i = 0$, we have $(b_j, (A - \bar{\mu}(\lambda)B)\bar{\zeta}_i) = 0$ and we conclude as before that $(b_j, B\bar{\zeta}_i) = 0$ for $i,j = 1, ..., 6$. By taking the inner product of (7b.22) with b_i, we therefore obtain

$$z_i = (b_i, B\Phi), \ i = 1, ..., 6. \tag{7b.25}$$

A projection operator Π is defined such that it picks out the components of Φ that consist of the critical modes and annihilates the stable modes:

$$\Pi\Phi = 2 \, \text{Re} \sum_{i=1}^{6} (b_i, B\Phi)\zeta_i, \tag{7b.26}$$

so that

$$\Psi = (I - \Pi)\Phi \tag{7b.27}$$

in the above decomposition.

The decomposition of a real-valued function f motivated by the normalization conditions (7b.21) and the decomposition (7b.22) is

$$f = \sum_{i=1}^{6}(b_i, f)B\zeta_i + \sum_{i=1}^{6}(\bar{b}_i, f)B\bar{\zeta}_i + g. \tag{7b.28}$$

Here

$$(b_i, g) = 0, \quad i = 1, ..., 6. \tag{7b.29}$$

With the projection operator $\tilde{\Pi}$ defined as

$$\tilde{\Pi}f = 2 \text{ Re } \sum_{i=1}^{6}(b_i, f)B\zeta_i, \tag{7b.30}$$

we have

$$g = (I - \tilde{\Pi})f. \tag{7b.31}$$

The inner product of (7b.1) with b_i is formed, where $L\Phi = (A + B\frac{d}{dt})\Phi$. Using $(b_i, A\Phi) = \mu(\lambda)(b_i, B\Phi)$, together with (7b.25), we obtain

$$\frac{dz_i}{dt} + \mu(\lambda)z_i = (b_i, N_2(\Phi, \Phi)) + (b_i, N_3(\Phi, \Phi, \Phi)), \quad i = 1, ..., 6. \tag{7b.32}$$

Next, the projection $I - \tilde{\Pi}$ is applied to (7b.22). Since $A\zeta_i = \mu(\lambda)B\zeta_i$,

$$L\Phi = \sum_{i=1}^{6}(\frac{d}{dt} + \mu)z_iB\zeta_i + \sum_{i=1}^{6}(\frac{d}{dt} + \bar{\mu})\bar{z}_iB\bar{\zeta}_i + L\Psi. \tag{7b.33}$$

The application of $I - \tilde{\Pi}$ to the terms under summations yields zero. Hence $(I - \tilde{\Pi})L\Phi = (I - \tilde{\Pi})L\Psi$. Using (7b.23) and (7b.24), $\tilde{\Pi}L\Psi = 0$ and thus

$$(A + B\frac{d}{dt})\Psi = N_2(\Phi, \Phi) + N_3(\Phi, \Phi, \Phi)$$

$$-2 \text{ Re } \left(\sum_{i=1}^{6}(b_i, N_2(\Phi, \Phi) + N_3(\Phi, \Phi, \Phi))B\zeta_i\right). \tag{7b.34}$$

The Center Manifold Theorem states that, in a neighborhood of $\Phi = 0$, there is a manifold Γ (called the center manifold) of the form $\Psi = \tau(z_1, z_2, z_3, z_4, z_5, z_6)$ with the following properties:

1. All solutions with initial data on the center manifold remain on the center manifold as long as they remain small.
2. All small periodic solutions lie on the center manifold.

3.   The stability of a small periodic solution is determined by its stability within the center manifold, in other words, all Floquet multipliers corresponding to directions outside the center manifold are stable.

The center manifold is usually not uniquely determined but, even if there are many center manifolds, their asymptotic expansions agree to all orders. The above properties hold for every center manifold.

In order to make use of the results of section 7(a) at the leading nonlinearities, only quadratic terms in the asymptotic approximation to the center manifold are required. Thus, the first property above is used to write the following expression

$$\Psi = 2 \operatorname{Re} \left(\sum_{i,j=1}^{6} z_i z_j \psi_{ij} + z_i \bar{z}_j \chi_{ij} \right) + ..., \qquad (7b.35)$$

where the dots indicate terms of higher than quadratic order.

The second property above says that the solutions we are interested in (the small periodic solutions, taking into account the nonlinear terms) are found on the center manifold. These are solutions close to the linear eigenfunctions. There need not be a center manifold if amplitudes become large.

Without loss of generality, the symmetry conditions

$$\psi_{ij} = \psi_{ji}, \ \chi_{ij} = \bar{\chi}_{ji}, \qquad (7b.36)$$

are assumed. Equation (7b.35) is inserted into (7b.34) and (7b.32) is used to express the time derivatives dz_i/dt. By comparing quadratic terms, we obtain

$$(A - 2\mu(\lambda)B)\psi_{ij} = N_2(\zeta_i, \zeta_j) - \sum_{k=1}^{6}(b_k, N_2(\zeta_i, \zeta_j))B\zeta_k$$

$$- \sum_{k=1}^{6}(\bar{b}_k, N_2(\zeta_i, \zeta_j))B\bar{\zeta}_k, \qquad (7b.37)$$

$$(A - (\mu(\lambda) + \bar{\mu}(\lambda))B)\chi_{ij} = N_2(\zeta_i, \bar{\zeta}_j) - \sum_{k=1}^{6}(b_k, N_2(\zeta_i, \bar{\zeta}_j))B\zeta_k$$

$$- \sum_{k=1}^{6}(\bar{b}_k, N_2(\zeta_i, \bar{\zeta}_j))B\bar{\zeta}_k. \qquad (7b.38)$$

In our actual computations, the inner product of two functions g_1 and g_2, defined in fluid j and expressed in terms of Chebyshev polynomials as follows,

$$g_1 = \sum_{i=0}^{N} g_{1i} T_i(z_j) \exp(i\alpha_1 x + i\beta_1 y),$$

$$g_2 = \sum_{i=0}^{N} g_{2i} T_i(z_j) \exp(i\alpha_2 x + i\beta_2 y), \qquad (7b.39)$$

$$z_1 = \frac{2}{l_1} z - 1, \ z_2 = \frac{2}{l_2}(z-1) + 1,$$

is the following Euclidean inner product:

$$(g_1, g_2) = \frac{\int \int_A \exp[i(-\alpha_1 + \alpha_2)x + i(-\beta_1 + \beta_2)y] \ dx \ dy}{\int \int_A dx \ dy} \times \sum_{i=0}^{N} \bar{g}_{1i} g_{2i}.$$

$$(7b.40)$$

Here A represents a cell of the hexagonal lattice, i.e. the parallelogram spanned by the two basis vectors x_1 and x_2 of (7.1). Thus, unless $-\alpha_1 + \alpha_2 = -\beta_1 + \beta_2 = 0$, the inner product vanishes (because of periodicity).

We note that many of the inner products in (7b.37)- (7b.38) vanish. For example, in the equation for ψ_{11}, $N_2(\zeta_1, \zeta_1)$ is proportional to $\exp(2i\alpha x)$ and none of the b_k have this x, y-dependence. Moreover, we only need to evaluate ψ_{11} at $\lambda = 0$, and hence we compute it from $(A - 2\mu(0)B)\psi_{11} = N_2(\zeta_1, \zeta_1)$. Thus, the components $\tilde{\theta}$, u, v, w, $\tilde{p}$ and h of ψ_{11} satisfy the following equations (with the f_i denoting the quadratic parts of H_i):

$$-A_1 w - \Delta\tilde{\theta} + 2\sigma\tilde{\theta} = f_1 \text{ of } N_2(\zeta_1, \zeta_1),$$

$$-P\Delta u + \frac{\partial\tilde{p}}{\partial x} + 2\sigma u = f_2 \text{ of } N_2(\zeta_1, \zeta_1),$$

$$-P\Delta v + \frac{\partial\tilde{p}}{\partial y} + 2\sigma v = f_3 \text{ of } N_2(\zeta_1, \zeta_1),$$

$$-P\Delta w + \frac{\partial\tilde{p}}{\partial z} + 2\sigma w - RP\tilde{\theta} = f_4 \text{ of } N_2(\zeta_1, \zeta_1),$$

$$-A_2 w - \frac{1}{\gamma}\Delta\tilde{\theta} + 2\sigma\tilde{\theta} = f_5 \text{ of } N_2(\zeta_1, \zeta_1),$$

$$-\frac{r}{m}P\Delta u + r\frac{\partial\tilde{p}}{\partial x} + 2\sigma u = f_6 \text{ of } N_2(\zeta_1, \zeta_1),$$

$$-\frac{r}{m}P\Delta v + r\frac{\partial\tilde{p}}{\partial y} + 2\sigma v = f_7 \text{ of } N_2(\zeta_1, \zeta_1),$$

$$-\frac{r}{m}P\Delta w + r\frac{\partial\tilde{p}}{\partial z} + 2\sigma w - \frac{RP}{\beta}\tilde{\theta} = f_8 \text{ of } N_2(\zeta_1, \zeta_1),$$

$$[\![u]\!] = f_9 \text{ of } N_2(\zeta_1, \zeta_1),$$

$$[\![v]\!] = f_{10} \text{ of } N_2(\zeta_1, \zeta_1),$$

$$[\![w]\!] = f_{11} \text{ of } N_2(\zeta_1, \zeta_1),$$

$$P\left(\frac{\partial u_1}{\partial z} + \frac{\partial w_1}{\partial x}\right) - \frac{P}{m}\left(\frac{\partial u_2}{\partial z} + \frac{\partial w_2}{\partial x}\right) = f_{12} \text{ of } N_2(\zeta_1, \zeta_1),$$

$$P\left(\frac{\partial v_1}{\partial z} + \frac{\partial w_1}{\partial y}\right) - \frac{P}{m}\left(\frac{\partial v_2}{\partial z} + \frac{\partial w_2}{\partial y}\right) = f_{13} \text{ of } N_2(\zeta_1, \zeta_1),$$

the normal stress condition involves the f_{14} of $N_2(\zeta_1, \zeta_1)$,

$$[\![\tilde{\theta}]\!] - h[\![A]\!] = f_{15} \text{ of } N_2(\zeta_1, \zeta_1),$$

$$\frac{\partial \tilde{\theta}_1}{\partial z} - \frac{1}{\zeta}\frac{\partial \tilde{\theta}_2}{\partial z} = f_{16} \text{ of } N_2(\zeta_1, \zeta_1),$$

$$-w_1 + 2\sigma h = f_{17} \text{ of } N_2(\zeta_1, \zeta_1),$$

$$u = v = \tilde{\theta} = 0 \quad \text{at} \quad z = 0, 1.$$

Similarly, we find χ_{11} from $A\chi_{11} = N_2(\zeta_1, \bar{\zeta}_1)$. Here, $N_2(\zeta_1, \bar{\zeta}_1)$ is independent of x and y and the comments in the paragraph of (7b.4) - (7b.12) and thereafter are relevant for the calculation of χ_{11}: for example, set $\tilde{p}_1$ equal to 0 at $z = l_1$, and set average interface height equal to 0.

Let $\tilde{\theta}$, u, v, w, $\tilde{p}$ and h represent the components of χ_{11}. The divergence free condition on the velocity yields $w = 0$. The Laplacian Δ is $\partial^2/\partial z^2$. The components satisfy:

$$-\tilde{\theta}'' = f_1 \text{ of } N_2(\zeta_1, \bar{\zeta}_1),$$

$$-Pu'' = f_2 \text{ of } N_2(\zeta_1, \bar{\zeta}_1),$$

$$-Pv'' = f_3 \text{ of } N_2(\zeta_1, \bar{\zeta}_1),$$

$$\frac{\partial \tilde{p}}{\partial z} - RP\tilde{\theta} = f_4 \text{ of } N_2(\zeta_1, \bar{\zeta}_1),$$

$$-\frac{1}{\gamma}\tilde{\theta}'' = f_5 \text{ of } N_2(\zeta_1, \bar{\zeta}_1),$$

$$-\frac{r}{m}Pu'' = f_6 \text{ of } N_2(\zeta_1, \bar{\zeta}_1),$$

$$-\frac{r}{m}Pv'' = f_7 \text{ of } N_2(\zeta_1, \bar{\zeta}_1),$$

$$r\frac{\partial \tilde{p}}{\partial z} - \frac{RP}{\beta}\tilde{\theta} = f_8 \text{ of } N_2(\zeta_1, \bar{\zeta}_1),$$

$$u = v = \tilde{\theta} = 0 \quad \text{at} \quad z = 0, 1,$$

$$[\![u]\!] = f_9 \text{ of } N_2(\zeta_1, \bar{\zeta}_1),$$

$$[\![v]\!] = f_{10} \text{ of } N_2(\zeta_1, \bar{\zeta}_1),$$

$$P\frac{\partial u_1}{\partial z} - \frac{P}{m}\frac{\partial u_2}{\partial z} = f_{12} \text{ of } N_2(\zeta_1, \bar{\zeta}_1),$$

$$P\frac{\partial v_1}{\partial z} - \frac{P}{m}\frac{\partial v_2}{\partial z} = f_{13} \text{ of } N_2(\zeta_1, \bar{\zeta}_1),$$

$$-\tilde{p}_1 + \tilde{p}_2 + M_1 h = f_{14} \text{ of } N_2(\zeta_1, \bar{\zeta}_1),$$

$$[\![\tilde{\theta}]\!] - h(A_1 - A_2) = f_{15} \text{ of } N_2(\zeta_1, \bar{\zeta}_1),$$

$$\frac{\partial \tilde{\theta}_1}{\partial z} - \frac{1}{\zeta}\frac{\partial \tilde{\theta}_2}{\partial z} = f_{16} \text{ of } N_2(\zeta_1, \bar{\zeta}_1).$$

In the case of χ_{11}, there is an indeterminate solution; namely, the mode which represents the trivial solution $u = v = w = 0$, but with the interface displaced up or down and the temperature and pressure fields adjusted appropriately: for example, $h = 1$, $[\![\tilde{p}]\!] = M_1$, and $[\![\tilde{\theta}]\!] = A_1 - A_2$ would make the f_2, f_3, f_6, f_7, f_{14} and f_{15} of $N_2(\zeta_1, \bar{\zeta}_1)$ zero (here, the f_i denotes the quadratic part of H_i). Any eigenfunction plus this solution is another eigenfunction. In order to rule this out, we specify no movement of the interface: $h = 0$.

Finally, we note that our goal is to compute the coefficients given by (7c.5) below (at $\lambda = 0$) and only a subset of the ψ's and χ's is actually required for that purpose.

With the notations

$$\Phi_1 = 2 \operatorname{Re} \sum_{k=1}^{6} z_k \zeta_k, \qquad (7b.41)$$

$$\Psi_2 = 2 \operatorname{Re}\left(\sum_{i,j=1}^{6} z_i z_j \psi_{ij} + z_i \bar{z}_j \chi_{ij} \right), \qquad (7b.42)$$

we obtain the following reduced system, which is accurate to third order, from (7b.32)

$$\frac{dz_i}{dt} + \mu(\lambda)z_i = (b_i, N_2(\Phi_1, \Phi_1))$$

$$+2(b_i, N_2(\Phi_1, \Psi_2)) + (b_i, N_3(\Phi_1, \Phi_1, \Phi_1)). \qquad (7b.43)$$

Obviously this system is of the form (7a.2) and it satisfies the symmetry conditions (7a.3). To determine the form of the function F_1, we note that b_1 is of the form $\tilde{b}(z)\exp(i\alpha x)$ and hence we need to consider only those terms in $N_2(\Phi_1, \Phi_1) + 2N_2(\Phi_1, \Psi_2) + N_3(\Phi_1, \Phi_1, \Phi_1)$ which have an (x, y)-dependence which is also proportional to $\exp(i\alpha x)$. We obtain the following form for F_1:

$$F_1(z_1, z_2, z_3, w_1, w_2, w_3, \lambda) = \mu(\lambda)z_1$$

$$+\beta_1(\lambda)w_2 w_3 + \beta_2(\lambda)\bar{z}_2\bar{z}_3 + \beta_3(\lambda)(w_2\bar{z}_3 + w_3\bar{z}_2)$$

$$+\gamma_1(\lambda)|z_1|^2 z_1 + \gamma_2(\lambda)|w_1|^2 z_1 + \gamma_3(\lambda)z_1^2 w_1$$

$$+\gamma_4(\lambda)\bar{w}_1^2\bar{z}_1 + \gamma_5(\lambda)|z_1|^2\bar{w}_1 + \gamma_6(\lambda)|w_1|^2\bar{w}_1$$

$$+\gamma_7(\lambda)(|z_2|^2 + |z_3|^2)z_1 + \gamma_8(\lambda)(|z_2|^2 + |z_3|^2)\bar{w}_1$$

$$+\gamma_9(\lambda)(|w_2|^2 + |w_3|^2)z_1 + \gamma_{10}(\lambda)(|w_2|^2 + |w_3|^2)\bar{w}_1$$

$$+\gamma_{11}(\lambda)(z_2 w_2 + z_3 w_3)z_1 + \gamma_{12}(\lambda)(z_2 w_2 + z_3 w_3)\bar{w}_1$$

$$+\gamma_{13}(\lambda)(\bar{z}_2\bar{w}_2 + \bar{z}_3\bar{w}_3)z_1 + \gamma_{14}(\lambda)(\bar{z}_2\bar{w}_2 + \bar{z}_3\bar{w}_3)\bar{w}_1. \qquad (7b.44)$$

The coefficients β_i and γ_i are as follows:

$$\beta_1(\lambda) = -2(b_1, N_2(\zeta_5, \zeta_6)),$$

$$\beta_2(\lambda) = -2(b_1, N_2(\bar{\zeta}_2, \bar{\zeta}_3)),$$

$$\beta_3(\lambda) = -2(b_1, N_2(\zeta_5, \bar{\zeta}_3)),$$

$$\gamma_1(\lambda) = -2(b_1, N_2(\bar{\zeta}_1, \psi_{11})) - 4(b_1, N_2(\zeta_1, \chi_{11})) - 3(b_1, N_3(\zeta_1, \zeta_1, \bar{\zeta}_1)),$$

$$\gamma_2(\lambda) = -4(b_1, N_2(\bar{\zeta}_4, \psi_{14})) - 4(b_1, N_2(\zeta_4, \chi_{14})) - 4(b_1, N_2(\zeta_1, \chi_{44}))$$
$$-6(b_1, N_3(\zeta_1, \zeta_4, \bar{\zeta}_4)),$$

$$\gamma_3(\lambda) = -2(b_1, N_2(\zeta_4, \psi_{11})) - 4(b_1, N_2(\zeta_1, \psi_{14})) - 3(b_1, N_3(\zeta_1, \zeta_1, \zeta_4)),$$

$$\gamma_4(\lambda) = -2(b_1, N_2(\bar{\zeta}_1, \bar{\psi}_{44})) - 4(b_1, N_2(\bar{\zeta}_4, \bar{\psi}_{14})) - 3(b_1, N_3(\bar{\zeta}_1, \bar{\zeta}_4, \bar{\zeta}_4)),$$

$$\gamma_5(\lambda) = -4(b_1, N_2(\zeta_1, \bar{\psi}_{14})) - 4(b_1, N_2(\bar{\zeta}_1, \chi_{14})) - 4(b_1, N_2(\bar{\zeta}_4, \chi_{11}))$$
$$-6(b_1, N_3(\zeta_1, \bar{\zeta}_1, \bar{\zeta}_4)),$$

$$\gamma_6(\lambda) = -2(b_1, N_2(\zeta_4, \bar{\psi}_{44})) - 4(b_1, N_2(\bar{\zeta}_4, \chi_{44})) - 3(b_1, N_3(\zeta_4, \bar{\zeta}_4, \bar{\zeta}_4)),$$

$$\gamma_7(\lambda) = -4(b_1, N_2(\zeta_1, \chi_{22})) - 4(b_1, N_2(\zeta_2, \chi_{12})) - 4(b_1, N_2(\bar{\zeta}_2, \psi_{12}))$$
$$-6(b_1, N_3(\zeta_1, \zeta_2, \bar{\zeta}_2)),$$

$$\gamma_8(\lambda) = -4(b_1, N_2(\bar{\zeta}_4, \chi_{22})) - 4(b_1, N_2(\zeta_2, \bar{\psi}_{24})) - 4(b_1, N_2(\bar{\zeta}_2, \chi_{24}))$$
$$-6(b_1, N_3(\zeta_2, \bar{\zeta}_2, \bar{\zeta}_4)),$$

$$\gamma_9(\lambda) = -4(b_1, N_2(\zeta_1, \chi_{55})) - 4(b_1, N_2(\zeta_5, \chi_{15})) - 4(b_1, N_2(\bar{\zeta}_5, \psi_{15}))$$
$$-6(b_1, N_3(\zeta_1, \zeta_5, \bar{\zeta}_5)),$$

$$\gamma_{10}(\lambda) = -4(b_1, N_2(\bar{\zeta}_4, \chi_{55})) - 4(b_1, N_2(\zeta_5, \bar{\psi}_{45})) - 4(b_1, N_2(\bar{\zeta}_5, \chi_{54}))$$
$$-6(b_1, N_3(\bar{\zeta}_4, \zeta_5, \bar{\zeta}_5)),$$

$$\gamma_{11}(\lambda) = -4(b_1, N_2(\zeta_1, \psi_{25})) - 4(b_1, N_2(\zeta_2, \psi_{15})) - 4(b_1, N_2(\zeta_5, \psi_{12}))$$
$$-6(b_1, N_3(\zeta_1, \zeta_2, \zeta_5)),$$

$$\gamma_{12}(\lambda) = -4(b_1, N_2(\bar{\zeta}_4, \psi_{25})) - 4(b_1, N_2(\zeta_2, \chi_{54})) - 4(b_1, N_2(\zeta_5, \chi_{24}))$$
$$-6(b_1, N_3(\zeta_2, \zeta_5, \bar{\zeta}_4)),$$

$$\gamma_{13}(\lambda) = -4(b_1, N_2(\zeta_1, \bar{\psi}_{25})) - 4(b_1, N_2(\bar{\zeta}_2, \chi_{15})) - 4(b_1, N_2(\bar{\zeta}_5, \chi_{12}))$$
$$-6(b_1, N_3(\zeta_1, \bar{\zeta}_2, \bar{\zeta}_5)),$$

$$\gamma_{14}(\lambda) = -4(b_1, N_2(\bar{\zeta}_4, \bar{\psi}_{25})) - 4(b_1, N_2(\bar{\zeta}_2, \bar{\psi}_{45})) - 4(b_1, N_2(\bar{\zeta}_5, \bar{\psi}_{24}))$$
$$-6(b_1, N_3(\bar{\zeta}_2, \bar{\zeta}_4, \bar{\zeta}_5)). \tag{7b.45}$$

III.7(c) Transformation to Birkhoff Normal Form

The form (7a.4) represents a normal form (i.e., as many terms as possible are transformed out of the differential equation at leading order) into which (7b.43) can be put after a suitable coordinate transformation. We refer to Chow and Hale [1982] and Iooss and Joseph [1989] for a general discussion of the theory of normal forms and to Golubitsky and Stewart [1985] for applications to Hopf bifurcation with symmetry. The "phase shift" is an extra symmetry present in the normal form but not in the original problem. The original problem yields equation (7b.44). Here, if z_i is phase-shifted by $e^{i\phi}$, then $w_2 w_3$ is an $e^{2i\phi}$-term and $\bar{z}_2 \bar{z}_3$ is an $e^{-2i\phi}$-term so these have, in a way, the "wrong" shifts. However, these get transformed away in the process of transformations to the Birkhoff normal form. The terms that stay are the ones given in equation (7a.4). In Golubitsky and Stewart [1985], it is stated and proved that the terms in equation (7b.44) that *can* be transformed away are the ones that have the "wrong" phase shifts.

We first transform away the quadratic terms in (7b.44) by setting

$$\tilde{z}_1 = z_1 - \frac{\beta_1(\lambda)}{\mu(\lambda)} w_2 w_3 + \frac{\beta_2(\lambda)}{\mu(\lambda) - 2\bar{\mu}(\lambda)} \bar{z}_2 \bar{z}_3 - \frac{\beta_3(\lambda)}{\bar{\mu}(\lambda)}(w_2 \bar{z}_3 + w_3 \bar{z}_2). \quad (7c.1)$$

The variables $\tilde{z}_2$, $\tilde{z}_3$, $\tilde{w}_1$, $\tilde{w}_2$ and $\tilde{w}_3$ are defined in such a way that the hexagonal symmetry is preserved, i.e. using the same permutations of the arguments that appear in (7a.3). We obtain a new transformed system, which is again of the form (7a.2) and satisfies

$$\tilde{F}_1(\tilde{z}_1, \tilde{z}_2, \tilde{z}_3, \tilde{w}_1, \tilde{w}_2, \tilde{w}_3, \lambda) = \mu(\lambda)\tilde{z}_1$$

$$+\tilde{\gamma}_1(\lambda)|\tilde{z}_1|^2 \tilde{z}_1 + \tilde{\gamma}_2(\lambda)|\tilde{w}_1|^2 \tilde{z}_1 + \tilde{\gamma}_3(\lambda)\tilde{z}_1^2 \tilde{w}_1$$

$$+\tilde{\gamma}_4(\lambda)\tilde{w}_1^2 \bar{\tilde{z}}_1 + \tilde{\gamma}_5(\lambda)|\tilde{z}_1|^2 \bar{\tilde{w}}_1 + \tilde{\gamma}_6(\lambda)|\tilde{w}_1|^2 \bar{\tilde{w}}_1$$

$$+\tilde{\gamma}_7(\lambda)(|\tilde{z}_2|^2 + |\tilde{z}_3|^2)\tilde{z}_1 + \tilde{\gamma}_8(\lambda)(|\tilde{z}_2|^2 + |\tilde{z}_3|^2)\tilde{w}_1$$

$$+\tilde{\gamma}_9(\lambda)(|\tilde{w}_2|^2 + |\tilde{w}_3|^2)\tilde{z}_1 + \tilde{\gamma}_{10}(\lambda)(|\tilde{w}_2|^2 + |\tilde{w}_3|^2)\tilde{w}_1$$

$$+\tilde{\gamma}_{11}(\lambda)(\tilde{z}_2 \tilde{w}_2 + \tilde{z}_3 \tilde{w}_3)\tilde{z}_1 + \tilde{\gamma}_{12}(\lambda)(\tilde{z}_2 \tilde{w}_2 + \tilde{z}_3 \tilde{w}_3)\bar{\tilde{w}}_1$$

$$+\tilde{\gamma}_{13}(\lambda)(\bar{\tilde{z}}_2 \bar{\tilde{w}}_2 + \bar{\tilde{z}}_3 \bar{\tilde{w}}_3)\tilde{z}_1 + \tilde{\gamma}_{14}(\lambda)(\bar{\tilde{z}}_2 \bar{\tilde{w}}_2 + \bar{\tilde{z}}_3 \bar{\tilde{w}}_3)\tilde{w}_1 + ..., \quad (7c.2)$$

Here the $\tilde{\gamma}_i$ are as follows:

$$\tilde{\gamma}_i = \gamma_i, \ i = 1, 2, 3, 4, 5, 6,$$

$$\tilde{\gamma}_7 = \gamma_7 + \frac{|\beta_2|^2}{\mu - 2\bar{\mu}} - \frac{\beta_1 \beta_3}{\bar{\mu}},$$

$$\tilde{\gamma}_8 = \gamma_8 + \frac{\beta_2 \bar{\beta}_3}{\mu - 2\bar{\mu}} - \frac{\beta_3^2}{\bar{\mu}},$$

$$\tilde{\gamma}_9 = \gamma_9 - \frac{\beta_1 \beta_3}{\mu} - \frac{|\beta_3|^2}{\bar{\mu}},$$

$$\tilde{\gamma}_{10} = \gamma_{10} - \frac{\beta_1 \beta_2}{\mu} - \frac{\bar{\beta}_1 \beta_3}{\bar{\mu}},$$

$$\tilde{\gamma}_{11} = \gamma_{11} - \frac{\beta_1^2}{\mu} - \frac{\bar{\beta}_2 \beta_3}{\bar{\mu}},$$

$$\tilde{\gamma}_{12} = \gamma_{12} - \frac{\beta_1\beta_3}{\mu} - \frac{|\beta_3|^2}{\bar{\mu}},$$

$$\tilde{\gamma}_{13} = \gamma_{13} + \frac{\beta_2\bar{\beta}_3}{\mu - 2\bar{\mu}} - \frac{\beta_3^2}{\bar{\mu}},$$

$$\tilde{\gamma}_{14} = \gamma_{14} + \frac{\bar{\beta}_1\beta_2}{\mu - 2\bar{\mu}} - \frac{\beta_2\beta_3}{\bar{\mu}}. \tag{7c.3}$$

Finally, we can achieve the form (7a.4) by using the further transformation

$$\hat{z}_1 = \tilde{z}_1 - \frac{\tilde{\gamma}_3(\lambda)}{2\mu(\lambda)}\tilde{z}_1^2\tilde{w}_1 + \frac{\tilde{\gamma}_4(\lambda)}{\mu(\lambda) - 3\bar{\mu}(\lambda)}\bar{\tilde{w}}_1^2\tilde{z}_1$$

$$- \frac{\tilde{\gamma}_5(\lambda)}{2\bar{\mu}(\lambda)}|\tilde{z}_1|^2\bar{\tilde{w}}_1 - \frac{\tilde{\gamma}_6(\lambda)}{2\bar{\mu}(\lambda)}|\tilde{w}_1|^2\bar{\tilde{w}}_1$$

$$- \frac{\tilde{\gamma}_8(\lambda)}{2\bar{\mu}(\lambda)}(|\tilde{z}_2|^2 + |\tilde{z}_3|^2)\bar{\tilde{w}}_1 - \frac{\tilde{\gamma}_{10}(\lambda)}{2\bar{\mu}(\lambda)}(|\tilde{w}_2|^2 + |\tilde{w}_3|^3)\bar{\tilde{w}}_1$$

$$- \frac{\tilde{\gamma}_{11}(\lambda)}{2\mu(\lambda)}(\tilde{z}_2\tilde{w}_2 + \tilde{z}_3\tilde{w}_3)\tilde{z}_1 - \frac{\tilde{\gamma}_{13}(\lambda)}{2\bar{\mu}(\lambda)}(\bar{\tilde{z}}_2\tilde{w}_2 + \bar{\tilde{z}}_3\tilde{w}_3)\tilde{z}_1$$

$$+ \frac{\tilde{\gamma}_{14}(\lambda)}{\mu(\lambda) - 3\bar{\mu}(\lambda)}(\bar{\tilde{z}}_2\bar{\tilde{w}}_2 + \bar{\tilde{z}}_3\bar{\tilde{w}}_3)\bar{\tilde{w}}_1, \tag{7c.4}$$

with the other $\hat{z}_i$ defined using the permutations that appear in (7a.3).

The coefficients in the new system are

$$\alpha_1(\lambda) = \tilde{\gamma}_1(\lambda), \quad \alpha_2(\lambda) = \tilde{\gamma}_2(\lambda), \quad \alpha_3(\lambda) = \tilde{\gamma}_7(\lambda),$$

$$\alpha_4(\lambda) = \tilde{\gamma}_9(\lambda), \quad \alpha_5(\lambda) = \tilde{\gamma}_{12}(\lambda). \tag{7c.5}$$

III.7(d) Results and Discussion

Some guidance for choosing parameters for the numerical calculations is obtained from limiting situations where closed-form expressions for the interfacial eigenvalue are available. One such situation is the short-wave limit. In this case, the asymptotics of the interfacial eigenvalue can be obtained by focussing on a mode that decays rapidly away from the interface; the boundary conditions at the walls become irrelevant. It is found in section 5 that as $\alpha \to \infty$ the interfacial eigenvalue is, at leading order, given by equations (5.12) and (5.14). The dominant term is due to surface tension. Differences in density and the coefficient of cubical expansion enter at order α^{-1}. If the density and coefficient of cubical expansion are equal and there is no surface tension, then the stratification in thermal conductivity is important. Thermal conductivity stratification results in short-wave instability.

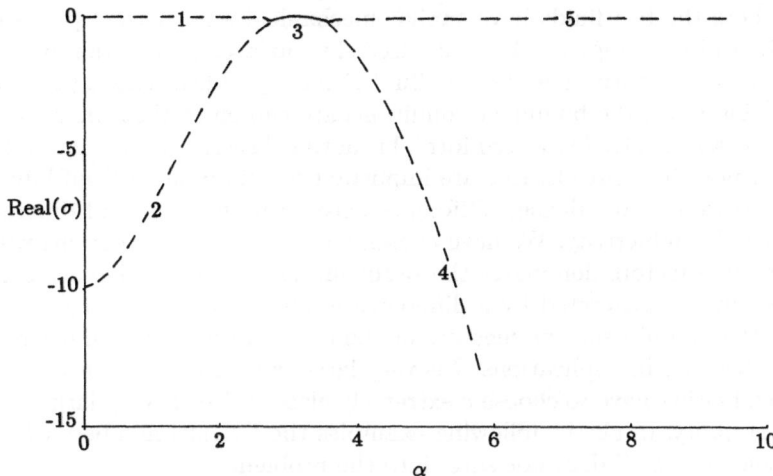

Fig. 7.5. [Renardy and Renardy, 1988, Elsevier] Surface tension is $S=0.01$; $R=1707.94031$, $l_1=0.5$, $\beta=\gamma=r=\zeta=m=P=1$. Branches 1,3 and 5 are associated with the interfacial mode. Branches 2,3 and 4 are associated with a mode that has an analogue in the one-fluid Bénard problem. The eigenvalues are real except on branch 3 (dark line), where they are complex conjugates; at $\alpha=3.11612$, they are at criticality. Branch 5 has a local maximum at about $\alpha=7$ but this is below criticality (see figure 7.6).

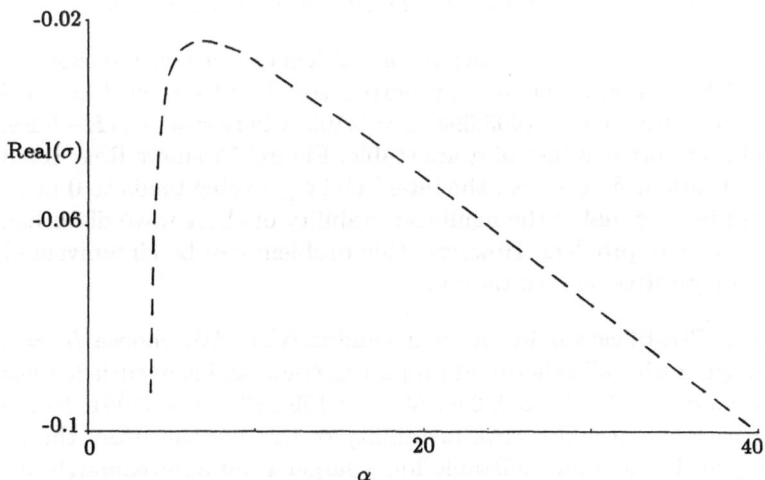

Fig. 7.6. [Renardy and Renardy, 1988, Elsevier] This is a magnification of branch 5 in figure 7.5, showing that the branch is below criticality.

A second situation where closed-form expressions can be obtained is the case where the two fluids have similar mechanical and thermal properties and the walls are replaced by stress-free slip surfaces (see section 6). This problem is a perturbation of a one-fluid Bénard problem with a (neutrally stable) interface; the boundary conditions are chosen so that the one-fluid problem can be solved in closed form. From this discussion of similar liquids we conclude that three factors are important for obtaining a Hopf bifurcation: surface tension, density difference across the interface and difference in thermal conductivity. We have chosen to look at three situations where one of these factors dominates the problem. The density difference across the interface is generated by a difference in cubical expansion coefficients rather than a difference in densities at the temperature of the upper plate. This is because in applications G is very large, and if r were larger than 1, we would either have to choose r extremely close to 1 or R very large to get any instability. In all the following examples the Prandtl number is $P = 1$ and, since $r = 1$, G does not enter into the problem.

Case (i): All fluid properties are equal but surface tension is non-zero. We choose $l_1 = 0.5$, $S = 0.01$. In this case, we find criticality at $R = 1707.94031$, $\alpha = 3.11612$ with a computed $\sigma = -.18E - 7 \pm .19i$. Eigenvalues for other values of α are stable. The least stable one-fluid eigenvalue at $\alpha = 0$ is $\sigma = -9.87$. It forms a complex conjugate pair with the interfacial eigenvalue for α between 2.8 and 3.7. As $\alpha \to \infty$, the one-fluid eigenvalue is proportional to $-\alpha^2$ and the interfacial eigenvalue is proportional to $-\alpha$. Figure 7.5 presents Re α versus α for α between 0 and 10. Figure 7.6 presents a magnification for $\alpha = 3.8$ to 40 showing a local maximum around $\alpha = 7$ and the asymptotic behavior of the interfacial mode as $\alpha \to \infty$.

Case (ii): Stratification in coefficients of cubical expansion. We choose $l_1 = 0.4$, $\beta = 0.8$ with all other fluid properties equal and zero surface tension. We find criticality at $R = 1903.694$, $\alpha = 3.105$, where $\sigma = -.17E - 5 \pm 9.19$. Eigenvalues at other values of α are stable. Figure 7.7 shows Re σ versus α for this situation. As $\alpha \to \infty$, the interfacial eigenvalue tends to 0 at order α^{-1}. This feature makes the nonlinear stability of short wave disturbances a rather delicate problem. However, this problem can be circumvented by adding any positive surface tension.

Case (iii): Stratification in thermal conductivity. We choose $l_1 = 0.3$, $\zeta = 0.8$, again with all other fluid properties equal and zero surface tension. This yields $\sigma = -.64E - 6 \pm 2.85$ at $R = 1692.881$, $\alpha = 3.081$. However, due to the inherent short-wave instability of this stratification, the interfacial eigenvalue becomes unstable for α larger than approximately 9. We use a small amount of surface tension to stabilize this. With a judicious amount of surface tension, it is possible to have two critical wavenumbers simultaneously. We add more surface tension so that the only criticality is the one around $\alpha = 3.08$. With $S = 0.03$, $l_1 = 0.3$, $\zeta = 0.8$, criticality

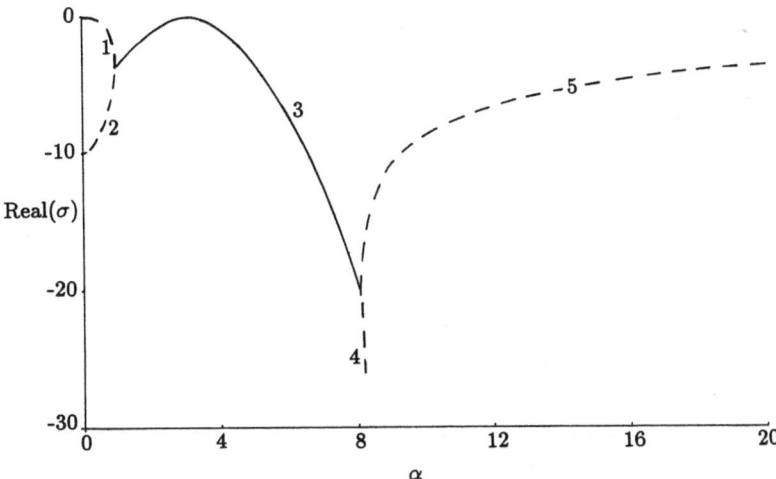

Fig. 7.7. [Renardy and Renardy, 1988, Elsevier] Stratification in coefficients of cubical expansion is $\beta=0.8$, $R=1903.694$, $l_1=0.4$, zero surface tension, $P=1$, other fluid properties are equal. On branch 3 (dark line), the eigenvalues are complex conjugates; at $\alpha=3.105$, they are at criticality.

occurs at $R = 1693.33478$, $\alpha = 3.08$ with $\sigma = -.1E - 7 \pm 2.86$. The addition of $S = 0.03$ raised the critical R by 0.03% over the case of $S = 0$ and the critical α remains about the same. Figure 7.8 shows Re σ versus α for this situation. Figure 7.9 is a magnification of figure 7.8 showing a local maximum of Re σ near $\alpha = 10.7$.

We note that the critical Rayleigh number and α in case (i) are very close to those of the one-fluid problem. In case (ii) the critical Rayleigh number is higher and in case (iii) it is lower than in the one-fluid case. Although the results of section 6 concern a different set of boundary conditions, and a direct comparison is therefore not possible, we remark that the results of section 6 (ii) would indeed predict a decrease in the critical Rayleigh number for case (iii), while case (ii) is close to marginal. Since the terms involving β are multiplied by the (large) factor R in the equations, a perturbation analysis based on nearly equal fluids is probably not applicable in case (ii).

We investigated the three situations discussed above. For the first case (non-zero surface tension), all bifurcating branches turned out to be subcritical. In the second case (density difference across the interface), standing rolls, travelling patchwork quilt (1) and oscillating triangles are supercritical but unstable; the remaining branches are subcritical. For the third case (thermal conductivity stratification), only the travelling rolls, travelling patchwork quilt (2) and wavy rolls (1) are subcritical; the remaining branches are supercritical, but unstable. Hence, in all three cases,

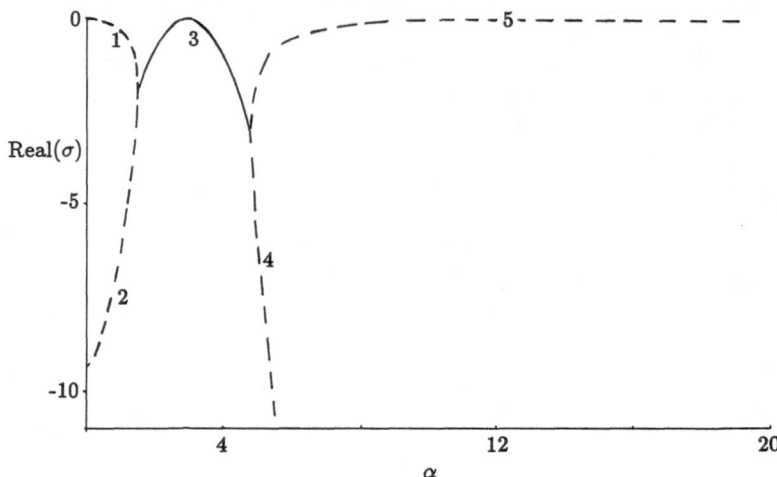

Fig. 7.8. [Renardy and Renardy, 1988, Elsevier] Stratification in thermal conductivity is $\zeta = 0.8$; $R = 1692.881$, $l_1 = 0.3$, $S = 0.03$, $P = 1$, other fluid properties are equal. On branch 3 (dark line), the eigenvalues are complex conjugates; they are at criticality at $\alpha = 3.08$. On branch 5, there is a local maximum at about $\alpha = 10.7$ (see figure 7.9).

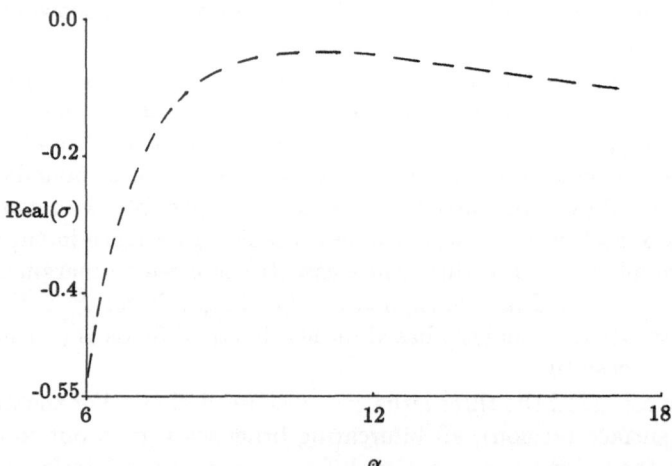

Fig. 7.9. [Renardy and Renardy, 1988, Elsevier] This is a magnification of branch 5 in figure 7.8 showing a local maximum. This branch is below criticality.

all branches are unstable near the bifurcation point. An open question is whether any of the branches are stable at a larger amplitude. The numbers α_i, $i = 1, ..., 5$ for the three cases are listed below. It should be noted that these values are not independent of the normalization of the eigenfunctions ζ_i and are thus determined only up to a common positive factor.

	Case (i)	Case (ii)	Case (iii)
α_1	-0.106E-7-0.198E-5i	-0.398E-5-0.373E-4i	-0.544E-5-0.153E-4i
α_2	-0.769E-8-0.190E-5i	0.105E-4-0.451E-5i	0.127E-4-0.183E-4i
α_3	-0.530E-8-0.194E-5i	0.876E-5-0.241E-4i	0.947E-5-0.291E-4i
α_4	-0.145E-7-0.209E-5i	-0.463E-4+0.328E-4i	-0.427E-5-0.510E-4i
α_5	-0.598E-8-0.144E-5i	0.873E-5-0.420E-5i	0.131E-4-0.248E-4i

These results might at first not appear believable for the following reason. The bifurcation problem for convective instability in the one-fluid Bénard problem has been analyzed by Buzano and Golubitsky [1983] and Golubitsky, Swift and Knobloch [1984]. In the presence of hexagonal symmetry, they arrive at an ordinary differential equation for the amplitudes in $\mathbb{R}^6$. The process of pattern selection depends on whether there is midplane symmetry or not. This is different from the case of Hopf bifurcation in the two-layer problem, where the selection mechanism does not depend on whether there is midplane symmetry or not. In the one-fluid problem with midplane symmetry, Golubitsky, Swift and Knobloch [1984] show that all bifurcating branches are supercritical. Our cases (i) - (iii) concern fluids that differ only slightly, so that at first glance, one might expect results that are a slight perturbation of the one-fluid results. In particular, case (i) is close to the one-fluid problem with midplane symmetry, so that one may expect the bifurcation to be supercritical.

In the following, we show a simple model problem which demonstrates that such an intuitive expectation is not justified.

The equation

$$\dot{x} = \lambda x - x^3, \qquad (7d.1)$$

with λ considered the bifurcation parameter, undergoes steady supercritical bifurcation at $\lambda = 0$ (set $\dot{x} = 0$. Then $\lambda x - x^3 = 0$ yielding $x = \pm\sqrt{\lambda}$ and 0, a pitchfork bifurcation). The variable x may be thought of as being analogous to the velocity field.

In going from the one-fluid to the two-fluid Bénard problem, a new variable, the interface position, is introduced. In analogy, let us add a new variable y to equation (7d.1). A small parameter ϵ is analogous to the difference in the physical properties of the fluids. Thus, for $\epsilon = 0$ (the one-fluid case), the dynamics of x is not influenced by y. We arrive at the following full system for our simple model problem:

$$\dot{x} = \lambda x - x^3 - \epsilon y,$$

$$\dot{y} = x + K\epsilon y^3. \tag{7d.2}$$

Now turn to the case of $\epsilon \neq 0$. By linearizing (7d.2), we can see that there is a Hopf bifurcation from the trivial solution $x = y = 0$ at $\lambda = 0$: the linearized system is

$$\begin{pmatrix} \dot{x} \\ \dot{y} \end{pmatrix} = \begin{pmatrix} \lambda & -\epsilon \\ 1 & 0 \end{pmatrix} \begin{pmatrix} x \\ y \end{pmatrix}.$$

Since the linearized solutions are assumed to depend on $e^{\sigma t}$, the eigenvalues σ satisfy

$$det \begin{pmatrix} \lambda - \sigma & -\epsilon \\ 1 & -\sigma \end{pmatrix} = 0,$$

so that

$$\sigma = (\lambda \pm \sqrt{\lambda^2 - 4\epsilon})/2.$$

At $\lambda = 0$, $\sigma = \pm i\sqrt{\epsilon}$, yielding a Hopf bifurcation. Is this subcritical or supercritical? To answer this, we introduce the scaling $x = \epsilon^{1/4}x'$, $y = \epsilon^{-1/4}y'$, $\lambda = \epsilon^{1/2}\lambda'$, $t = \epsilon^{-1/2}t'$ in equation (7d.2) $(d/dt \rightarrow \sqrt{\epsilon}d/dt')$. We obtain the new system

$$\dot{x}' = \lambda' x' - y' - x'^3,$$

$$\dot{y}' = x' + Ky'^3. \tag{7d.3}$$

This system no longer involves ϵ. Hence whether the Hopf bifurcation in (7d.2) is supercritical or subcritical depends only on K but not on ϵ. The cubic terms $-x'^3$ and Ky'^3 compete; the former for supercritical and the latter for subcritical. One would intuitively expect that if $K > 1$, the Ky'^3 term would win.

For further analysis, drop the primes in equation (7d.3) and seek solutions $x(wt)$, $y(wt)$, both 2π-periodic, and expand in powers of a small parameter δ:

$$x(wt) = \sum_{n=1}^{\infty} \delta^n x_n(wt), \quad y(wt) = \sum_{n=1}^{\infty} \delta^n y_n(wt), \tag{7d.4}$$

$$w = \sum_{n=0}^{\infty} \delta^n w_n, \quad \lambda = \sum_{n=1}^{\infty} \delta^n \lambda_n(wt).$$

By this means, we look at solutions bifurcating from the trivial solution $x = 0$ at $\lambda = 0$. The term w_0 would be 1 in this case because the eigenvalues are $\pm i$ after having scaled out the ϵ. The x_1 and y_1 are eigenfunctions. Equating the coefficients of δ^n, the odd-power terms in w and λ are zero. Hence, at leading order, $\lambda = \lambda_2\delta^2$, and λ_2 must be calculated. There are two possibilities for the Hopf bifurcation for non-zero λ_2. If $\lambda_2 > 0$, then the

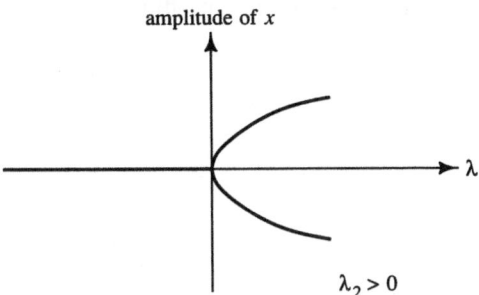

Fig. 7.10(a). $\lambda_2 > 0$.

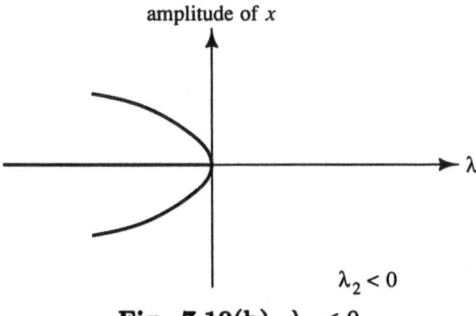

Fig. 7.10(b). $\lambda_2 < 0$.

picture is as in figure 7.10 (a) and the bifurcation is supercritical. If $\lambda_2 < 0$, then it is subcritical. Details of this analysis yield that the bifurcation is subcritical if $K > 1$: the bifurcation is then subcritical no matter how small ϵ is, even though the steady bifurcation at $\epsilon = 0$ is supercritical.

We conclude from this example that the supercriticality of the bifurcation in the one-fluid problem need not rule out subcritical bifurcation in the two-fluid problem, even if the two fluids are similar.

We emphasize that the steady solutions of (7d.1) do not correspond to steady solutions of (7d.2), even for $\epsilon = 0$. Instead, the steady solutions of (7d.2) for $\epsilon = 0$ are given by $x = 0$ and y arbitrary. This mimics the features of the two-fluid Bénard problem. Steady flows of the one-fluid problem do not remain steady solutions when the interface position is introduced as an extra variable. Instead, the "steady" solutions of the two-fluid problem with equal fluids are given by arbitrary interface deformations and no flow. Thus, even when the two fluids have identical properties, it is not possible to link up the bifurcation results to the one-fluid case.

Chapter IV
Plane Channel Flows

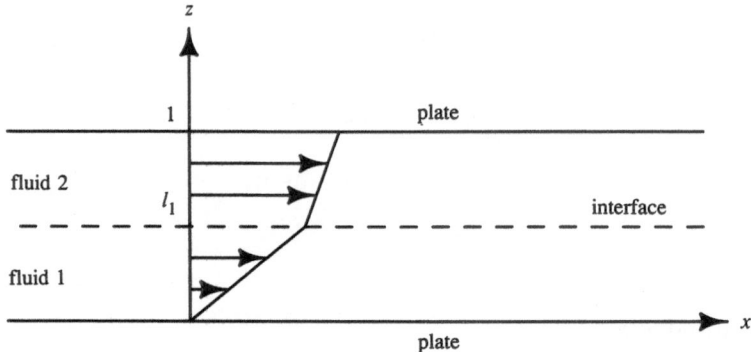

Fig. 1.1. Undisturbed two-layer Couette flow. Fluid 1 occupies $0 \leq z \leq l_1$. Fluid 2 occupies $l_1 \leq z \leq 1$. The fluids have viscosities μ_i $(i = 1, 2)$ and densities ρ_i. Upper plate speed is 1.

IV.1 Introduction

The problem of stability of plane Couette flow of two fluids, treated here (figure 1.1), is the simplest of all the shearing flows of two fluids we might consider. This is a good problem in its own right and is well suited for teaching the physical concepts and analytical methods which are appropriate in the study of shearing flows of two fluids. Squire's transformation for layered flows, discussed in section IV.3, is well known in the problem of stability of shearing flows of one fluid, where it guarantees that the smallest critical Reynolds number will occur when the disturbances are two-dimensional in the plane of the flow. The transformation leading to the theorem requires that planes parallel to the bottom and top of the channel should be unbounded in all directions. In practice, channels are nearly always bounded by side walls so that the application of Squire's Theorem to practical situations is still open to questions.

Yih [1967] did the first significant study of stability of shear flows of two fluids. He showed that a viscosity difference could lead to instability even for vanishingly small Reynolds numbers which was a surprise at the time. He introduced the notion of an *interfacial mode* which, in the one-fluid problem, can coincide with any streamline and is neutrally stable. The addition of an interface to the one-fluid flow sets up a mode that allows for the interface to be wavy while not affecting the velocity field. When the two fluids have different properties, the interfacial mode can become unstable at any Reynolds number.

Two-layer Couette flow has three scales of length: the plate separation, the thickness of one of the fluids, and a viscous diffusion length which can be defined by a kinematic viscosity and shear rate. The asymptotic analysis for short and long waves below and in sections IV.4 and IV.5 are to be interpreted with respect to all three scales of length.

Yih [1967] did an asymptotic analysis of the interfacial mode for long waves using a formal power series in the wavenumber $\alpha \ll 1$ for $\alpha \Re \ll 1$ where $\Re$ is a Reynolds number. This technique has now been applied to a variety of multi-layer shearing flows [Li 1969ab, 1970; Hickox 1971; Wang, Seaborg and Lin 1978; Waters 1983; Waters and Keeley 1987; Than, Rosso and Joseph 1987; Smith 1989]. The analysis of long waves can be done in closed form but the final expressions are often complicated enough that computing is necessary. Yih showed that a jump in viscosity across the interface can lead to long wave instability.

In section IV.4, asymptotic analyses for long waves in three situations are presented. Part (a) treats two-layer Couette flow. The method of calculation adopted here is different from Yih's, and is based on the standard Fredholm solvability condition for the adjoint systems. The viscosity, density and layer thickness ratios are essential factors for instability: for long waves, the arrangement with a thin layer of the less viscous fluid is stable. Part (b) treats combined Couette-Poiseuille flow. Part (c) treats a semi-infinite Couette flow [Hooper 1985] in which one of the walls in the two-layer Couette flow is placed at infinity.

Hooper and Boyd [1983] solve the linear stability problem for two-dimensional plane Couette flow when each fluid occupies a half-plane. They show that this flow is unstable to short waves, and there is no instability in the long-wave limit. They also introduce a perturbation scheme for short waves ($\alpha \gg 1$, where α is the wavenumber of the disturbance) of the interfacial eigenvalue for layered flows in general, and justify their scheme rigorously for their particular flow. Their scheme applies to more complicated layered parallel shearing flows because the eigenfunction for a short-wave disturbance dies out exponentially with distance away from the interface, so that these flows resemble that of unbounded Couette flow locally. There is a certain robustness, because the local analysis does not depend strongly on the detailed nature of the flow or on distant boundaries. This also explains why the algebraic calculations are much simpler than for long waves and the final expressions are obtained in closed form. This perturbation scheme has been applied to other layered flows: two-layer Taylor flow [Y. Renardy and Joseph 1985a], two-layer Benard problem [Y. Renardy and Joseph 1985b; Renardy and Renardy 1988], two-layer Poiseuille flow [Yiantsios and Higgins 1988].

The asymptotic expression for the growth rate of the interfacial mode for short waves ($\alpha \gg 1$) is essentially an expression of three terms. The dominant $O(\alpha)$ term is associated with surface tension and is always stabilizing. The next term is $O(1/\alpha)$ and it depends on gravity and the ratio of the densities of the fluids. The third term is $O(1/\alpha^2)$ and depends on the ratio of viscosities and densities: this term is always destabilizing if the densities are equal. It is surprising that in the absence of surface tension or a density difference, the interfacial mode is unstable for sufficiently short waves at any Reynolds number! Intuitively, one might have expected short waves to be damped by viscosity, but instead they are made unstable by the viscosity difference. On the other hand, the conditions under which instability to short waves may be expected to occur are unlikely to be realizable: surface tension will stabilize the shortest waves.

Asymptotic and numerical investigations indicate that there is a *rule of thumb* for finding a linearly stable arrangement, at least for the case of equal densities: stabilize the long waves by putting the less viscous fluid in a thin layer, and stabilize the short waves by including enough surface tension. A striking result is that in the case of a density difference, the mechanism just described can stabilize the arrangement in which the upper fluid is the heavier [Renardy 1985]. This has been referred to as the *thin-layer* effect, or *lubrication stabilization* [Hooper 1985; Y. Renardy and Joseph 1985a] and is consistent with experimental observations in more complicated shearing flows where the less viscous fluid is found in a thin layer next to walls. A particular example is presented in section IV.6. A posteriori, one may argue that when the density difference is small, gravity is a relatively small force, and although it tends to destabilize an arrangement with the heavier fluid on top, this destabilizing effect can be offset by choosing the viscosities so that stresses at the interface oppose the effect of gravity. Numerical results indicate that the *rule of thumb* is useful at low Reynolds numbers.

At higher Reynolds numbers, numerical calculations in Renardy [1985] reveal instabilities not described by the asymptotic results for short or long waves. Mechanisms for these instabilities can be associated with boundary layers at the walls and the interface in certain limiting cases. These are studied by Hooper and Boyd [1987] numerically and analytically using asymptotic methods, in the context of semi-infinite two-layer Couette flow. They find an instability which is caused neither by short nor long waves, but is relevant for high Reynolds numbers (see section IV.6b). A different branch of the neutral stability curve is analyzed by Blennerhassett and Smith [1987] for high Reynolds numbers (see end of section IV.6b). This analysis is related to the analysis for short waves of Hooper and Boyd [1983]. Linear stability calculations at high Reynolds numbers are also found in Blennerhassett [1980] (cf. Renardy [1989]), who gives data modelling air-water systems (see Valenzuela [1976] for a different basic flow state), and Yiantsios and Higgins [1988] for two-layer plane Poiseuille flow.

Section IV.8 (a) contains some analysis of nonlinear amplitude equations which are supposed to hold asymptotically for long weakly nonlinear waves [Hooper and Grimshaw 1985; Shlang, Sivashinsky, Babchin and Frenkel 1985]. This analysis leads to the Kuramoto-Sivashinsky equation. This model equation has arisen in a variety of contexts, including long waves on thin viscous films [Benney 1966; Homsy 1974; Atherton and Homsy 1976; Topper and Kawahara 1978; Nakoryakov and Alekseyenko 1981], unstable drift waves in plasmas [LaQuey, Mahajan, Rutherford and

Tang 1975; Cohen, Krommes, Tang and Rosenbluth 1976], and thermod-
iffusive flame instability [Sivashinsky 1983]. In this nonlinear model, the
energy in long-wave instabilities is transferred to short-wave modes, which
are then damped by surface tension. This provides a mechanism for main-
taining stable nonlinear states (that is, achieving 'saturation'). A critical
look on such long wave theories will be given in chapter VIII. Ooms, Segal,
Cheung and Oliemans [1985] did a numerical study of the propagation of
long waves of finite amplitude. They derive an equation for the evolution
of the interface height and computed periodic sawtooth shaped waves that
persist as steady solutions from an initial condition of symmetric sinusoidal
waves. Smith [1990a,b] describes a physical mechanism for the long-wave
instability that appears in thin liquid films.

In section IV.8 (b), the stability of plane Couette-Poiseuille flow of two
fluids is analyzed using bifurcation theory [Blennerhassett 1980; Renardy
1989]. The fluids have different viscosities and densities, and there is surface
tension at the interface. The nonlinear calculations are carried out for two
different cases. In one case, the combined volume flux is fixed, and in the
other the pressure gradient in the horizontal direction is fixed. Numerical
results are presented for some Couette flow profiles and a Poiseuille flow
profile at low speeds, showing that travelling waves are supported at the
interface. A computation for a fast flow is also presented. The derivation and
numerical results are compared with those of a formal approach, employing
multiple scales, which has been used on related problems.

In section IV.9, we study viscoelastic effects on the two-layer Couette
flow of an upper-convected Maxwell model. For long-wave disturbances,
the stratification of the relaxation times can stabilize or destabilize the
interfacial mode. Renardy [1988b] showed that for short waves, the effect of
elasticity stratification on growth rates is an order of magnitude larger than
density and viscosity stratifications, and an order of magnitude smaller than
the stabilizing effect of surface tension. Thus, if surface tension is small, a
suitable choice of relaxation times can give rise to a stronger short-wave
instability than in the Newtonian case. At the other extreme, a suitable
choice can be used to stabilize a short-wave instability caused by density
or viscosity stratifications.

Analysis of the energy of an unstable disturbance can be of enormous
value in understanding the mechanisms of instability. The energy equation was used by Hooper and Boyd [1983] to discuss instability to short
waves: the viscosity difference at the interface drives the instability, which
is created in the immediate neighborhood of the interface and in the less
viscous fluid. The case of long waves is discussed by Hooper [1988], and the
interface term can be either stabilizing or destabilizing. The basic mechanism for instability is similar to that of film flow down an inclined plane
[Kelly, Goussis, Lin and Hsu 1989; Smith 1989]. The key idea in using the
energy equation for diagnosing instability is to follow the evolution of the
energy of the most unstable mode. This method has been widely exploited

for core-annular flows by Joseph and co-workers and will be discussed in later chapters. Energy methods may be used to discuss the nonlinear stability of flows of one fluid (see Joseph [1976]) but this type of application is frustrated in two-fluid dynamics because terms on the interface cannot be estimated *a priori* in terms of the usual norms.

We mention here works which have contributed to the understanding of layered flows, but will not be presented in detail. These include layered flow down an inclined plane [Li 1970; Kao 1965ab, 1968; Wang *et al.* 1978; Lin 1983ab; Lin and Wang 1983; Weinstein and Kurz 1991]. Two-layer flow down an incline for very viscous fluids of the same density has been analyzed by Loewenherz and Lawrence [1989] with the zero Reynolds number approximation. In this case, the fast growing mode is associated with finite wavelength. Rayleigh-Taylor instability with viscous effects has been studied by Yiantsios and Higgins [1989]; Tryggvason and Unverdi [1990, 1991] have used a front-tracking scheme for the numerical study of this and related multi-fluid problems. Flows with two or more interfaces can have a 'resonating' instability due to an interaction of the interfaces [Li 1969b; Anturkar, Papanastasiou and Wilkes 1990a; Weinstein and Kurz 1991].

IV.2 Governing Equations for Two-Layer Couette Flow

Much of the analysis of linear stability in this chapter will be discussed in the context of two-layer Couette flow. (The nonlinear bifurcation analysis of section IV.8 will be presented for the more general case of two-layer Couette-Poiseuille flow.)

Two fluids of densities ρ_i (i=1,2) and viscosities μ_i lie in the (x^*, y^*, z^*) plane in layers between infinite parallel plates located at $z^* = 0, l^*$ (see figure 1.1). Asterisks denote dimensional variables. The location of the interface is given by $z^* = h^*(x^*, y^*, t^*)$. The average value of h^* is denoted by l_1^*. The average height of the upper fluid (fluid 2) is $l^* - l_1^* = l_2^*$. The velocity is denoted by $\mathbf{v}^* = (u^*, v^*, w^*)$ and the pressure by p^*. In each fluid, the governing equations are the Navier-Stokes equations and incompressibility.

The equations of motion in fluid i read as follows:

$$\rho_i\big(\dot{\mathbf{v}}^* + (\mathbf{v}^* \cdot \nabla)\mathbf{v}^*\big) = \mu_i \Delta \mathbf{v}^* - \nabla p^* - \rho_i g \mathbf{e}_z, \qquad (2.1)$$

$$\text{div } \mathbf{v}^* = 0.$$

At the walls, we have the no-slip boundary condition:

$$\mathbf{v}^* = \mathbf{0} \quad \text{at } z^* = 0,$$

$$\mathbf{v}^* = (U^*, 0, 0) \quad \text{at } z^* = l^*. \qquad (2.2)$$

At the interface, the conditions are the continuity of velocity and shear stress, the jump in the normal stress is balanced by surface tension, and the kinematic free surface condition holds. The formulation of these conditions is explained fully in section III.2 and the jump conditions are contained in equations (2.4) - (2.5) in that section.

Dimensionless variables (without asterisks) are as follows:

$$(x, y, z) = (x^*, y^*, z^*)/l^*,$$

$$t = t^* U^*/l^*,$$

$$\mathbf{v} = \mathbf{v}^*/U^*,$$

$$p = p^*/(\rho_1 U^{*2}). \tag{2.3}$$

There are six dimensionless parameters:

$$m = \mu_1/\mu_2,$$

$$r = \rho_1/\rho_2,$$

$$l_1 = l_1^*/l^*,$$

a Froude number F given by

$$F^2 = U^{*2}/gl^* \tag{2.4}$$

where g denotes the gravitational acceleration constant, a surface tension parameter

$$S = \frac{S^*}{\mu_1 U^*}, \tag{2.5}$$

where S^* is the dimensional surface tension coefficient, and a Reynolds number based on the lower fluid,

$$\Re = U^* l^*/\nu_1 \tag{2.6}$$

where $\nu_1 = \mu_1/\rho_1$ is the kinematic viscosity of fluid 1. We denote

$$l_2 = 1 - l_1. \tag{2.7}$$

Expressed in dimensionless form, the equations in fluid 1 read

$$\dot{\mathbf{v}} + (\mathbf{v} \cdot \nabla)\mathbf{v} = \frac{1}{\Re}\Delta\mathbf{v} - \nabla p - \frac{1}{F^2}\mathbf{e}_z, \tag{2.8}$$

$$\operatorname{div} \mathbf{v} = 0.$$

In fluid 2, the equations read:

$$\dot{\mathbf{v}} + (\mathbf{v} \cdot \nabla)\mathbf{v} = \frac{r}{m\Re}\Delta\mathbf{v} - r\nabla p - \frac{1}{F^2}\mathbf{e}_z, \tag{2.9}$$

$$\operatorname{div} \mathbf{v} = 0.$$

The boundary conditions at the plates are

$$\mathbf{v} = \mathbf{0}, \quad \text{at } z = 0,$$

$$\mathbf{v} = (1, 0, 0), \quad \text{at } z = 1. \tag{2.10}$$

We set

$$h(x, y, t) = h^*(x^*, y^*, t^*)/l^* - l_1. \tag{2.11}$$

The interface is then at $z = l_1 + h(x, y, t)$, and the conditions to be satisfied at the interface are, from equations (2.17) - (2.18) in section III.2,

$$[\![\mathbf{v}]\!] = 0, \quad [\![\mathbf{t}_i \cdot \mathbf{T} \cdot \mathbf{n}]\!] = 0, \quad i = 1, 2,$$

$$[\![\mathbf{n} \cdot \mathbf{T} \cdot \mathbf{n}]\!] = \frac{\frac{S}{\Re}\left\{\frac{\partial^2 h}{\partial x^2}\left(1 + \left(\frac{\partial h}{\partial y}\right)^2\right) + \frac{\partial^2 h}{\partial y^2}\left(1 + \left(\frac{\partial h}{\partial x}\right)^2\right) - 2\frac{\partial^2 h}{\partial x \partial y}\frac{\partial h}{\partial x}\frac{\partial h}{\partial y}\right\}}{(1 + (\nabla h)^2)^{3/2}},$$

$$\dot{h} + u\frac{\partial h}{\partial x} + v\frac{\partial h}{\partial y} = w. \tag{2.12}$$

Here we have set the dimensionless stress tensor

$$\mathbf{T} = \begin{cases} \frac{1}{\Re}[\nabla\mathbf{v} + (\nabla\mathbf{v})^T] - p\mathbf{1} & \text{in fluid 1,} \\ \frac{1}{m\Re}[\nabla\mathbf{v} + (\nabla\mathbf{v})^T] - p\mathbf{1} & \text{in fluid 2.} \end{cases} \tag{2.13}$$

A steady shearing flow solution to the two-layer problem is given by a velocity $(U_1(z), 0, 0)$, a pressure P, and a flat interface at $z = l_1$, where

$$U_1(z) = \frac{z}{l_1 + ml_2} \quad \text{for} \quad 0 \le z \le l_1, \tag{2.14}$$

$$= \frac{m(z - 1)}{l_1 + ml_2} + 1 \quad \text{for} \quad l_1 \le z \le 1, \tag{2.15}$$

$$P = \begin{cases} P_1 - z/F^2 & \text{for } 0 \le z \le l_1, \\ P_2 - z/F^2 r & \text{for } l_1 \le z \le 1. \end{cases} \tag{2.16}$$

and $P_1 - P_2$ must be chosen such that P is continuous at $z = l_1$.

We examine the linear stability of this solution by adding a small disturbance (u, v, w) to the velocity and h to the interface position, that are taken to be proportional to $\exp(i\alpha x + i\beta y + \sigma t)$. The resulting equations governing linear stability are: in fluid i,

$$\frac{1}{\Re_i}\triangle u - \frac{\rho_1}{\rho_i}\frac{\partial p}{\partial x} - wU_{1z}(z) - U_1(z)i\alpha u = \sigma u, \tag{2.17}$$

$$\frac{1}{\Re_i}\triangle v - \frac{\rho_1}{\rho_i}\frac{\partial p}{\partial y} - i\alpha vU_1(z) = \sigma v, \tag{2.18}$$

$$\frac{1}{\Re_i}\triangle w - \frac{\rho_1}{\rho_i}\frac{\partial p}{\partial z} - i\alpha wU_1(z) = \sigma w, \tag{2.19}$$

$$\nabla \cdot \underline{u} = 0, \tag{2.20}$$

where $\triangle = \partial^2/\partial z^2 - (\alpha^2 + \beta^2)$, and

$$\Re_i = U^* l^* / \nu_i. \tag{2.21}$$

The Froude number does not enter into these equations since the pressure gradient $\partial P / \partial z$ in the basic flow field cancels the effect of gravity.

The interface conditions, linearized at $z = l_1$, yield:

kinematic free surface condition: $w - hi\alpha U_1(l_1) = h\sigma,$ (2.22)

continuity of velocity: $[\![w]\!] = 0,$ (2.23)

$$[\![u]\!] + h[\![U_{1z}(l_1)]\!] = 0, \tag{2.24}$$

$$[\![v]\!] = 0, \tag{2.25}$$

continuity of shear stress: $[\![\mu(\dfrac{\partial u}{\partial z} + \dfrac{\partial w}{\partial x})]\!] = 0,$ (2.26)

$$[\![\mu(\dfrac{\partial v}{\partial z} + \dfrac{\partial w}{\partial y})]\!] = 0, \tag{2.27}$$

balance of normal stress:

$$[\![-p + \frac{\mu}{\mu_1}\frac{2}{\Re}\frac{\partial w}{\partial z}]\!] = h(-\frac{(\alpha^2 + \beta^2)S}{\Re} + \frac{1}{F^2}(\frac{1}{r} - 1)). \tag{2.28}$$

The boundary conditions are:

$$u = 0, \quad v = 0, \quad w = 0 \quad \text{at} \quad z = 0, 1. \tag{2.29}$$

The above equations are analogous to equations (2.28) - (2.39) of section III.2 with the Prandtl number P replaced by $1/\Re$, the Rayleigh number R set to zero, and the gravity parameter G replaced by $1/F^2$.

We state the linearized problem in the $x - z$ plane in terms of the streamfunction ψ where

$$u = \frac{\partial \psi}{\partial z}, \quad w = -\frac{\partial \psi}{\partial x} = -i\alpha\psi, \tag{2.30}$$

and we introduce the phase speed c given by

$$\sigma = -i\alpha c. \tag{2.31}$$

These will be used in later sections.

The equations of motion yield the Orr-Sommerfeld equation in each fluid:

$$\psi^{iv} - 2\alpha^2\psi'' + \alpha^4\psi = i\alpha\Re_i(U_1 - c)(\psi'' - \alpha^2\psi), \tag{2.32}$$

where $\Re_i = U l^* / \nu_i$, and prime denotes d/dz. The interface conditions (2.22) - (2.24), (2.26) and (2.28) become

$$\psi = h(c - U_1(l_1)) \tag{2.33}$$

$$[\![\psi]\!] = 0 \tag{2.34}$$

$$[\![\psi_z]\!] + h[\![U_{1z}]\!] = 0 \tag{2.35}$$

$$[\mu \psi_{zz}] + \alpha^2 \psi[\mu] = 0 \qquad (2.36)$$

$$i\alpha(-c + U_1(l_1))[\frac{\rho}{\rho_1} \psi_z] - i\alpha\psi[\frac{\rho}{\rho_1} U_{1z}] - [\frac{\rho}{\rho_1 \Re_i} \triangle \psi_z] + \frac{1}{\Re}[\frac{2\mu}{\mu_1} \alpha^2 \psi_z]$$

$$= i\alpha h(\frac{-\alpha^2 S}{\Re} + \frac{1}{F^2}(\frac{1}{r} - 1)), \qquad (2.37)$$

where $\triangle = \partial^2/\partial z^2 - \alpha^2$. The boundary conditions (2.29) become

$$\psi = \frac{\partial\psi}{\partial z} = 0 \quad \text{at } z = 0, 1. \qquad (2.38)$$

IV.3 Squire's Theorem

Squire [1933] showed that for a single fluid, the linear stability of a parallel flow can be determined by considering only two-dimensional perturbations. 'Squire's theorem' states that

there is a transformation ("Squire's transformation") that establishes the equivalence of a three-dimensional disturbance to a two-dimensional one. This transformation is such that if the flow is unstable at a particular Reynolds number, then it will be unstable at a lower value of the Reynolds number for two-dimensional disturbances. Further, there are some instances where Squire's transformation degenerates. These must be investigated separately.

Pearlstein [1985] proved Squire's theorem for the one-fluid case with continuously stratified viscosity, where the temperature and the concentration of a solute are allowed to vary. Hesla, Pranckh and Preziosi [1986] proved Squire's theorem for the incompressible parallel flow of two immiscible fluids, with constant surface tension, bounded by two parallel plates. Their result can be generalized to flows in unbounded domains, and to flows with more than two strata. They do not include the effect of gravity, and in fact, Squire's theorem does not always hold in the presence of gravity. We will discuss their results in more detail below.

Squire's transformation for the two-layer Couette flow [Gumerman and Homsy 1974; Blennerhassett 1980; Pearlstein 1985, 1987; Hesla, Pranckh and Preziosi 1986; Yiantsios and Higgins 1988] is in the following form:

$$\tilde{\alpha} = \alpha(1 + \frac{\beta^2}{\alpha^2})^{1/2}, \qquad \tilde{\alpha}\tilde{\Re} = \alpha\Re, \qquad \tilde{\sigma}/\tilde{\alpha} = \sigma/\alpha,$$

$$\tilde{\alpha}\tilde{u} = \alpha u + \beta v, \qquad \tilde{w} = w, \qquad \tilde{p}/\tilde{\alpha} = p/\alpha, \qquad \tilde{h}\tilde{\alpha} = h\alpha,$$

$$\tilde{S}/\tilde{\alpha} = S/\alpha, \qquad \tilde{F}^2\tilde{\alpha}^2 = F^2\alpha^2, \qquad (3.1)$$

where two-dimensional quantities have tildes.

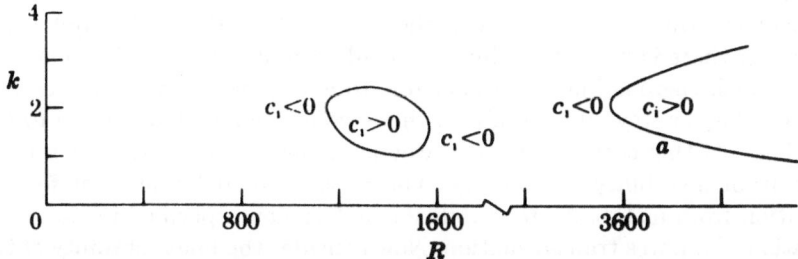

Fig. 3.1. [Figure 5 of Blennerhassett, 1980, On the generation of waves by wind, Phil. Trans. R. Soc. Lond. A Vol. 298, Royal Society of London] Neutral stability curve for surface waves in 'boundary layer 1' profile. This models an air-water system with equal depths for the fluids.

The transformed set of equations are the same ones you would get in a two-dimensional problem with disturbances proportional to $\exp(i\tilde{\alpha}x + \tilde{\sigma}t)$. Since $\tilde{\alpha} \geq \alpha$, it follows that $\operatorname{Re}\tilde{\sigma} \geq \operatorname{Re}\sigma$ and $\tilde{\Re} \leq \Re$. This shows that the growth rate is larger and the Reynolds number is smaller for the equivalent two-dimensional problem. Two-dimensional disturbances are the most dangerous.

In the two-dimensional problem, dropping the tildes, we get equations (2.17) - (2.29) with $v = 0$ in the $x - z$ plane. For the one-fluid problem ($r = 1$, $m = 1$, $S = 0$, with no interface at $z = l_1$), Romanov [1973] has proved that the real parts of all eigenvalues σ are negative at any Reynolds number and wavenumber α. The addition of an interface at $z = l_1$ with two identical fluids gives rise to the oscillating neutrally stable *interfacial mode*. If the two fluids have identical viscosity and density, and if there is no surface tension, then

$$u = 0, \quad w = 0, \quad h = \exp(i\alpha x + \sigma t), \quad \sigma = -i\alpha U_1(l_1) \qquad (3.2)$$

satisfies the equations. Thus, the eigenvalues of the two-layer problem are associated with an interfacial mode and an infinite number of the other modes that are analogous to the one-fluid modes. At higher Reynolds numbers, some of these modes can cross over. In this case we can diagnose the instability by analysis of the energy of the most rapidly growing disturbance.

The first part of Squire's theorem holds in the presence of gravity, even in the case of an adverse density stratification ($r < 1$). This needs some elaboration. (The second part concerns the degenerate cases, where $\alpha u + \beta v = 0$ or when $\alpha = 0$, and is discussed for the two-layer problem in the absence of gravity by Hesla, Pranckh and Preziosi [1986]). Consider the reduction of the three-dimensional problem to a two-dimensional problem in the $x-z$ plane by Squire's transformation. The shearing in the basic flow is in the x direction. One can find linearly stable arrangements with adverse density stratification, provided that the viscosity ratio and surface tension are such as to be stabilizing: for example, figure 6.1 displays a thin-layer arrangement with $r < 1$. Typically, there is a critical Reynolds number below which there is instability due to the adverse density stratification, and above which there is linear stability due to a stabilizing effect of shearing. (At even higher Reynolds numbers, there would be instabilities due to the shearing.) In this case, the critical Reynolds number is located at a transition from instability to stability. The critical Reynolds number for the transition from instability to stability is not the one specified in the interpretation of Squires transformation. Now consider the linear stability of the fully three-dimensional problem and look at the cross-stream direction (the $y-z$ plane). There is no shearing in the basic flow in that direction, so that the adverse density stratification leads to a long-wave instability ($\alpha = 0$, $h = h_0\exp(i\beta y + \sigma t)$) at any Reynolds number. This same instability holds also for small non-zero α. Thus, the three-dimensional problem has an instability for any Reynolds number, while its two-dimensional counterpart is unstable from zero up to a "critical" Reynolds number. The first part of Squire's theorem merely states that if the three-dimensional problem has an instability at a certain Reynolds number, then the corresponding two-dimensional situation given by Squire's transformation has an instability at a lower Reynolds number. This is certainly true if the onset of instability in both problems is at Reynolds number zero. Thus, the first part of Squire's theorem holds even with gravity. (But the second part of Squire's original theorem, concerning the degenerate cases, may not hold for an adverse density stratification.)

The usual interpretation of Squire's theorem is a stronger statement, which is not stated in Squire's paper [1933] but is construed from it, namely:

Squire's transformation exists, and if the two-dimensional problem is stable at some Reynolds number $\breve{\mathfrak{R}}$, then the three-dimensional problem is stable at the same Reynolds number $\breve{\mathfrak{R}}$.

Obviously, the usual interpretation does not hold for the case of adverse density stratification because the three-dimensional problem has an instability at any Reynolds number, but the two-dimensional problem given by Squire's transformation does not. A related situation, in terms of a two-layer Couette-Bénard problem is discussed by Gumerman and Homsy [Part I, 1974], and in terms of the two-layer Poiseuille flow, by Yiantsios and Higgins [1988]. A similar difficulty in deriving the usual interpretation of Squire's theorem from Squire's transformation arises when there are islands of instability as in figure 3.1.

IV.4 Asymptotic Analysis for Long Waves

The asymptotic analysis of the interfacial eigenvalue for long waves was first done by Yih [1967] for the two-layer Couette flow and the two-layer Poiseuille flow, in the case of equal densities. In part (a), we present the full calculation for the two-layer Couette flow, including the effect of density stratification. This is followed in part (b) by the results for the combined Couette-Poiseuille flow, which are stated without details (see also Blennerhassett [1980]). For two-layer Poiseuille flow, see Yiantsios and Higgins [1988]. In part (c), results of Hooper [1985] on a semi-infinite Couette flow are presented. This problem is equivalent to Yih's problem when one of the walls is allowed to go to infinity: however, in this limit, she has shown that Yih's perturbation expansion must be modified.

Li [1969b] investigated the three-layer plane Couette flow for long-wave disturbances and found a resonant instability. This instability appears in the zeroth-order approximation in the wavenumber α, in contrast to the instability found by Yih (cf. section IV.4(a)) which apears at $O(\alpha)$. It is due to a resonance between the two interfacial modes and may be related to a phenomenon in three-layer inviscid flow described by Taylor [1931]. Flows with three or more layers have the potential for this instability. Weinstein and Kurz [1991] have analyzed the long-wavelength instabilities in three-layer flow down an incline and also identified a resonant instability analogous to that of Li. They note that "neither inertial nor finite wavelength effects are necessary for this instability. It is found that destabilization occurs in cases where the middle-layer viscosity (for equal densities in each layer) or density (for equal viscosities in each layer) is smaller than those of the adjacent layers."

IV.4(a) Two-Layer Couette Flow

Yih [1967] formulated the problem in terms of a streamfunction ψ and a phase speed c, where $\sigma = -i\alpha c$. The formal expansions for long waves, for this particular flow, are:

$$\psi = \psi_0 + \alpha\psi_1 + \alpha\psi_2 + ..., \tag{4a.1}$$

$$c = c_0 + \alpha c_1 + \alpha^2 c_2 + ... \text{ for } \alpha << 1. \tag{4a.2}$$

Yih's variables α, τ, x, p are our $l_2\alpha$, t/l_2, x/l_2 and rp, respectively. Our velocities, c, c'_0, m, and r are identical to Yih's. Yih's lengths d_1, d_2 are our dimensional lengths l_2^*, l_1^* (our indices are opposite Yih's). Yih's parameters n, b, a_1, a_2, R, F^2 are our l_1/l_2, $U_1(l_1)$, $l_2U'_{12}(l_1)$, $l_2U'_{11}(l_1)$, $\Re l_2 m/r$, $(r-1)l_2/F^2$, respectively. Here, $U'_{1i}(l_1)$ denotes $U'_1(l_1)$ in fluid i.

We denote

$$c'_0 = c_0 - U_1(l_1). \tag{4a.3}$$

At the zeroth order, we have, from equations (2.32) - (2.38),

$$\psi_0^{iv} = 0 \quad \text{in both fluids,} \tag{4a.4}$$

$$[\![\psi_0]\!] = 0,$$

$$c_0'[\![\psi_{0z}]\!] + \psi_0[\![U_{1z}]\!] = 0, \tag{4a.5}$$

$$[\![\mu\psi_{0zz}]\!] = 0, \quad [\![\mu\psi_{0zzz}]\!] = 0,$$

$$\psi_0 = \psi_{0z} = 0 \text{ at } z = 0, 1. \tag{4a.6}$$

The linear eigenvalue problem is determined only up to a scale, which must be fixed by a normalizing condition chosen more or less arbitrarily, say

$$\psi(l_1) = 1. \tag{4a.7}$$

This implies that in the expansion (4a.1), $\psi_0(l_1) = 1$, and $\psi_i(l_1) = 0$ for $i = 1, 2, \ldots$. The solution is

$$\psi_0(z) = \begin{cases} 1 + B_1(l_1 - z) + B_2(l_1 - z)^2 + B_3(l_1 - z)^3 & \text{for fluid 1} \\ 1 + A_1(z - l_1) + A_2(z - l_1)^2 + A_3(z - l_1)^3 & \text{for fluid 2,} \end{cases} \tag{4a.8}$$

where

$$B_1 = -\frac{4ml_2^3 + l_1^3 + 3ml_1l_2^2}{2ml_1l_2^2}, \tag{4a.9}$$

$$A_1 = -\frac{4l_1^3 + ml_2^3 + 3l_2l_1^2}{2l_1^2l_2}, \tag{4a.10}$$

$$B_2 = \frac{A_2}{m}, \quad A_2 = \frac{ml_2^3 + l_1^3}{l_1^2l_2^2}, \tag{4a.11}$$

$$B_3 = -\frac{A_3}{m}, \quad A_3 = \frac{l_1^2 - ml_2^2}{2l_1^2l_2^2}. \tag{4a.12}$$

Equation (4a.5) yields

$$c_0' = -\frac{[\![U_{1z}]\!]}{[\![\psi_{0z}]\!]} = -\frac{[\![U_{1z}]\!]2ml_1^2l_2^2}{l_1^4 + m^2l_2^4 + 2ml_1l_2(2l_1^2 + 3l_1l_2 + 2l_2^2)}. \tag{4a.13}$$

This agrees with Yih's equation (34) for his c_0'. Since c_0' is a real quantity, we must calculate c_1 in order to find the contribution to the real part of $\sigma = -i\alpha c$, which determines stability.

The equations governing the first-order correction ψ_1 are, from equations (2.32) - (2.38),

$$\psi_1^{iv} = i\Re_i(U_1 - c_0)\psi_0''. \tag{4a.14}$$

At $z = l_1$,

$$\psi_1 = h_0c_1 + h_1c_0', \quad \text{where } h_0 = \frac{\psi_0}{c_0'}, \tag{4a.15}$$

$$[\![\psi_1]\!] = 0 \tag{4a.16}$$

$$c_1[\![\psi_{0z}]\!] + c_0'[\![\psi_{1z}]\!] + \psi_1[\![U_{1z}]\!] = 0 \tag{4a.17}$$

$$[\![\mu\psi_{1zz}]\!] = 0 \tag{4a.18}$$

and

$$[\![\mu\psi_{1zzz}]\!] = \frac{i\mu_1\Re}{\rho_1}\left(-c_0'[\![\rho\psi_{0z}]\!] - \psi_0[\![\rho U_{1z}]\!] + \frac{\psi_0[\![\rho]\!]}{c_0'F^2}\right). \tag{4a.19}$$

$$\psi_1 = \psi_{1z} = 0 \quad \text{at } z = 0, 1. \tag{4a.20}$$

Yih computed ψ_1 and c_1 from these equations. We will use an alternative method, based on the adjoint eigenfunction, for the calculation of the higher-order corrections c_1, c_2, ... The amount of messy algebra involved in either method is the same.

We next calculate the adjoint eigenfunction. Let ψ and ϕ be arbitrary functions satisfying

$$\psi = \psi_z = 0 \quad \text{at } z = 0, 1, \tag{4a.21}$$

$$\phi = \phi_z = 0 \quad \text{at } z = 0, 1, \tag{4a.22}$$

and regular enough so that integration by parts yields

$$\int_0^{l_1} \mu\psi^{iv}\phi\,dz + \int_{l_1}^1 \mu\psi^{iv}\phi\,dz = [\![\mu(\phi\psi''' - \phi'\psi'' + \phi''\psi' - \phi'''\psi)]\!]$$

$$+ \int_0^{l_1} \mu\psi\phi^{iv}\,dz + \int_{l_1}^1 \mu\psi\phi^{iv}\,dz. \tag{4a.23}$$

Here, let ψ be ψ_0. Then the left hand side vanishes because of equation (4a.4). Let ϕ be the adjoint eigenfunction at the zeroth order, denoted by ϕ_0. We choose

$$\phi_0^{iv} = 0 \quad \text{in fluids 1 and 2,} \tag{4a.24}$$

so that the integrals on the right hand side of equation (4a.23) vanish. The interface terms in (4a.23) are then

$$0 = \mu\psi_0'''[\![\phi_0]\!] - \mu\psi_0''[\![\phi_0']\!] + [\![\phi_0''\mu\psi_0']\!] - \psi_0[\![\mu\phi_0''']\!]. \tag{4a.25}$$

We choose

$$[\![\phi_0]\!] = 0, \quad [\![\phi_0']\!] = 0. \tag{4a.26}$$

Equation (4a.25) becomes

$$0 = [\![\phi_0''\mu\psi_0']\!] + \frac{c_0'[\![\psi_0']\!]}{[\![U_{1z}]\!]}[\![\mu\phi_0''']\!]. \tag{4a.27}$$

Here, we set

$$[\![\mu\phi_0'']\!]_{\text{fluid 1}} + \frac{c_0'}{[\![U_{1z}]\!]}[\![\mu\phi_0''']\!] = 0. \tag{4a.28}$$

The effect of this condition on (4a.27) is equivalent to setting

$$[\![\mu\phi_0'']\!] = 0. \tag{4a.29}$$

For the same reasons as that leading to equation (4a.7), we require a normalizing condition for the adjoint, which is chosen to be

$$\phi(l_1) = 1. \tag{4a.30}$$

This implies that in the long wave expansion for the adjoint in the manner of (4a.1), $\phi_0(l_1) = 1$, and $\phi_i(l_1) = 0$ for $i = 1, 2, ...$ Therefore, the adjoint equations are (4a.30), (4a.22), (4a.24), (4a.26), (4a.28) and (4a.29), with one of the latter two conditions being an "extra" condition, which can be used to check the final answer.

The solution for the adjoint equations is

$$\phi_0(z) = \begin{cases} 1 + B_1^*(l_1 - z) + B_2^*(l_1 - z)^2 + B_3^*(l_1 - z)^3 & \text{for fluid 1} \\ 1 + A_1^*(z - l_1) + A_2^*(z - l_1)^2 + A_3^*(z - l_1)^3 & \text{for fluid 2,} \end{cases} \tag{4a.31}$$

where

$$B_1^* = \frac{3(l_1^2 - ml_2^2)}{2l_1 l_2(l_1 + ml_2)}, \quad A_1^* = -B_1^*, \tag{4a.32}$$

$$B_2^* = \frac{-3}{l_1 l_2(l_1 + ml_2)}, \quad A_2^* = mB_2^*, \tag{4a.33}$$

$$B_3^* = \frac{4l_1 l_2 + 3l_1^2 + ml_2^2}{2l_1^3 l_2(l_1 + ml_2)}, \quad A_3^* = \frac{4ml_1 l_2 + 3ml_2^2 + l_1^2}{2l_2^3 l_1(l_1 + ml_2)}. \tag{4a.34}$$

Since c_0 is an algebraically simple eigenvalue, we have a unique eigenfunction ψ_0 and adjoint ϕ_0. There is a condition of solvability, derived below, which yields the correction term c_1. We let $\phi = \phi_0$ and $\psi = \psi_1$ in equation (4a.23). The conditions (4a.21) - (4a.22) are satisfied. The last two integrals in (4a.23) vanish, and we have

$$\int_0^{l_1} \mu_1 i\Re(U_1 - c_0)\psi_0'' \phi_0 \, dz + \int_{l_1}^1 \mu_2 i\Re_2(U_1 - c_0)\psi_0'' \phi_0 \, dz \tag{4a.35}$$

$$= [\mu(\phi_0 \psi_1''' - \phi_0' \psi_1'' + \phi_0'' \psi_1' - \phi_0''' \psi_1)]$$
$$= \phi_0 [\mu \psi_1'''] - \phi_0' [\mu \psi_1''] + \mu \phi_0'' [\psi_1'] - \psi_1 [\mu \phi_0'''] \tag{4a.36}$$

where (4a.16), (4a.26) and (4a.29) have been used. The second term vanishes by (4a.18). Making use of (4a.19), (4a.17) and (4a.28), equation (4a.36) becomes

$$\frac{\phi_0 i \mu_1 \Re}{\rho_1} \left(-c_0' [\rho \psi_0'] - \psi_0 [\rho U_{1z}] - \frac{[\psi_0'][\rho]}{[U_{1z}]F^2} \right)$$

$$+ \frac{[\mu \phi_0'']_{\text{fluid 1}}}{c_0'} \left(-c_1 [\psi_0'] - \psi_1 [U_{1z}] \right) + \psi_1 \left(\frac{[\mu \phi_0'']_{\text{fluid 1}} [U_{1z}]}{c_0'} \right).$$

The latter two terms cancel. Also, $m\Re = r\Re_2$ in (4a.35). We use (4a.5) in the gravity term to replace $[\psi_0']/[U_1']$ by $-\psi_0/c_0'$. This yields

$$c_1 = \frac{c_0'^2 i\Re}{\phi_{01}'' [U_{1z}]} \left(\int_0^{l_1} (U_1 - c_0)\psi_0'' \phi_0 \, dz + \frac{1}{r} \int_{l_1}^1 (U_1 - c_0)\psi_0'' \phi_0 \, dz \right.$$

$$+\frac{1}{\rho_1}(c_0'[\rho\psi_0'] + [\rho U_{1z}] - \frac{[\rho]}{c_0'F^2})),\tag{4a.37}$$

where ϕ_{01} denotes ϕ_0 in fluid 1 and

$$\phi_{01}'' = 2B_2^* = \frac{-6}{l_1 l_2 (l_1 + m l_2)}.\tag{4a.38}$$

Consider the evaluation of the integral over fluid 1 in equation (4a.37). In the integrand,

$$U_1(z) - c_0 = U_1'(z - l_1) - c_0', \quad \psi_0'' = 2B_2 + 6B_3(l_1 - z),\tag{4a.39}$$

so that $(U_1 - c_0)\psi_0''$ is a quadratic polynomial in $z - l_1$. As in the work of Yih [1967], we define

$$h_1(z) = a_1(z - l_1)^4 + b_1(z - l_1)^5 + d_1(z - l_1)^6.\tag{4a.40}$$

Note that h_1^{iv} is a quadratic polynomial. Its coefficients a_1, b_1 and d_1 will be determined in order to set

$$h_1^{iv}(z) = (U_1 - c_0)\psi_0''.\tag{4a.41}$$

This yields

$$a_1 = \frac{-c_0' B_2}{12}, \quad b_1 = \frac{B_2 U_{11}' + 3c_0' B_3}{60}, \quad d_1 = \frac{-U_{11}' B_3}{60},\tag{4a.42}$$

where U_{11} denotes the basic velocity $U_1(z)$ in fluid 1.

In equation (33) of Yih [1967], his C_2 is our $l_2^2 B_2$, his D_2 is our $-l_2^3 B_3$, his y is our $(z - l_1)/l_2$; thus, his $h_2(y)$ is our $h_1(z)/l_2^4$. Similarly, his $h_1(y)$ is our $h_2(z)/l_2^2$, defined below.

Therefore, the integral over fluid 1 in equation (4a.37) is

$$\int_0^{l_1} (U_1 - c_0)\psi_0''\phi_0 \, dz = \int_0^{l_1} h_1^{iv}(z)\phi_0 \, dz$$

$$= [h_1'''\phi_0 - h_1''\phi_0' + h_1'\phi_0'' - h_1\phi_0''']_0^{l_1} + \int_0^{l_1} h_1(z)\phi_0^{iv} \, dz.\tag{4a.43}$$

The interface terms vanish since $h_1 = h_1' = h_1'' = h_1''' = 0$ at $z = l_1$. At $z = 0$, $\phi_0 = \phi_0' = 0$. Hence, equation (4a.43) reduces to

$$-h_1'(0)\phi_{01}''(0) + h_1(0)\phi_{01}'''(0),\tag{4a.44}$$

where ϕ_{0i} denotes ϕ_0 in fluid i, and

$$\phi_{01}''(0) = 2B_2^* + 6B_3^* l_1, \quad \phi_{01}'''(0) = -6B_3^*.\tag{4a.45}$$

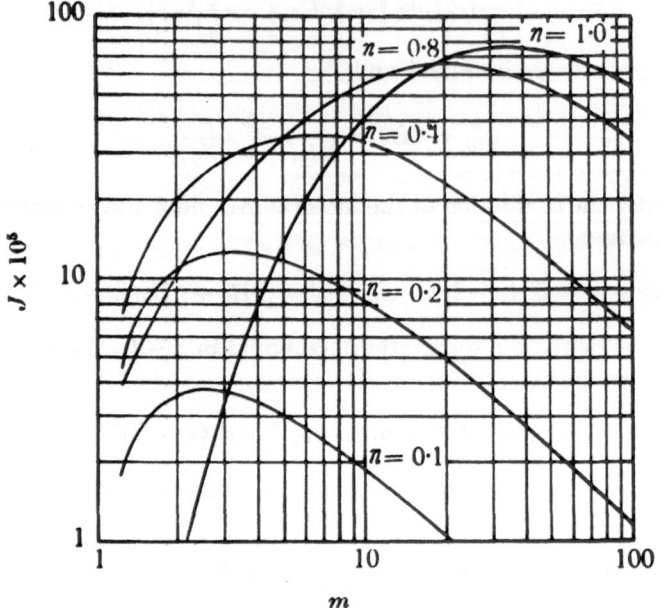

Fig. 4.1. [Figure 2(a) of Yih, 1967] The variation of J with the viscosity ratio m for various values of the depth ratio $n \leq 1$ for plane Couette flow with uniform density, showing instability.

Similarly, for fluid 2, we define

$$h_2(z) = a_2(z - l_1)^4 + b_2(z - l_1)^5 + d_2(z - l_1)^6, \qquad (4a.46)$$

where the coefficients a_2, b_2 and d_2 are determined in order to satisfy

$$h_2^{iv}(z) = (U_1 - c_0)\psi_0'' \quad \text{in fluid 2.} \qquad (4a.47)$$

This yields

$$a_2 = \frac{-A_2 c_0'}{12}, \quad b_2 = \frac{U_{12}' A_2 - 3c_0' A_3}{60}, \quad d_2 = \frac{A_3 U_{12}'}{60} \qquad (4a.48)$$

where U_{12} denotes the basic velocity U_1 in fluid 2. Therefore, the integral in equation (4a.37) over fluid 2 is

$$\int_{l_1}^{1} (U_1 - c_0)\psi_0'' \phi_0 \, dz = \int_{l_1}^{1} h_2^{iv}(z)\phi_0 \, dz$$

$$= h_2'(1)\phi_{02}''(1) - h_2(1)\phi_{02}'''(1) \qquad (4a.49)$$

where ϕ_{02} denotes ϕ_0 in fluid 2, and

$$\phi_{02}''(1) = 2A_2^* + 6A_3^* l_2, \quad \phi_{02}'''(1) = 6A_3^*. \qquad (4a.50)$$

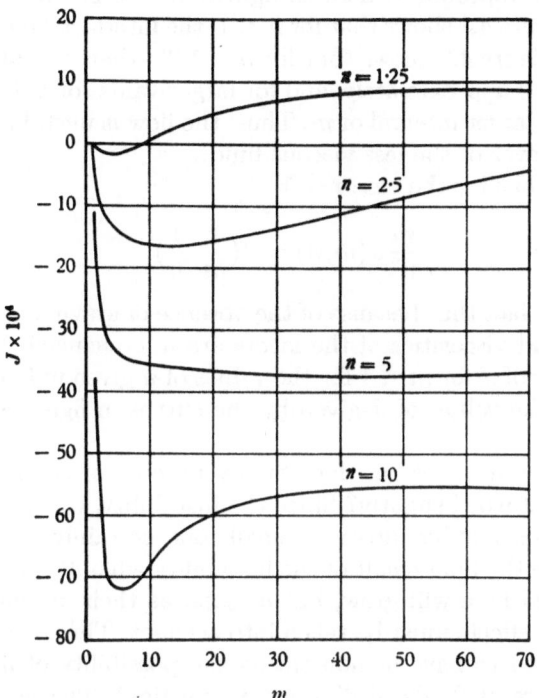

Fig. 4.2. [Figure 2(b) of Yih, 1967] The variation of J with the viscosity ratio m for various values of the depth ratio $n > 1$ for plane Couette flow with uniform density, showing extensive regions of instability.

Finally,

$$c_1 = \frac{c_0'^2 i\Re}{\phi_{01}''[U_{1z}]}\left(-h_1'(0)\phi_{01}''(0) + h_1(0)\phi_{01}'''(0) + \frac{1}{r}[h_2'(1)\phi_{02}''(1) - h_2(1)\phi_{02}'''(1)]\right.$$

$$\left. + \frac{1}{\rho_1}(c_0'[\rho\psi_0'] + [\rho U_{1z}] - \frac{[\rho]}{c_0'F^2})\right), \tag{4a.51}$$

and in terms of the function $J(m, n, r)$, $n = l_1/l_2$, defined in equation (42) of Yih [1967],

$$c_1 = i\Re l_2^2 \frac{m}{r} J(m, n, r). \tag{4a.52}$$

Note that αc_1 is Yih's Δc, defined in his equation (41). To establish the connection between the expression in equation (4a.51) and the quantity J defined by Yih, it is helpful to write

$$c_0'[\rho\psi_0'] + [\rho U_1'] = c_0'(\rho_1 - \rho_2)\psi_{01}' + c_0'\rho_2[\psi_0'] + \rho_2(U_{11}' - U_{12}') + [\rho]U_{11}'$$

$$= c_0'[\rho]\psi_{01}' + \rho_2(c_0'[\psi_0'] + [U_1']) + [\rho]U_{11}'$$

$$= c_0'[\rho]\psi_{01}' + [\rho]U_{11}'$$

by (4a.5).

Figures 2(a) and 2(b) of Yih [1967] show the variation of $J(m, n, r)$ for $r = 1$: these are reproduced here as figures 4.1 - 4.2. There is instability when $J > 0$. Figure 4.1 shows that for $n \leq 1$, the interfacial mode is unstable for all $m > 1$. Figure 4.2 shows that for $n = 1.25$, there is stability between $m = 1$ and $m = 9$ approximately, and for larger values of n displayed, there is stability for a larger interval of m. Thus, the flow is unstable except when there is a thin layer of the less viscous fluid.

Yih notes that for the case $r = 1$,

$$\frac{m}{n^2} J(m, n) = J(\frac{1}{m}, \frac{1}{n}). \tag{4a.53}$$

This reflects the fact that because of the absence of gravity, c_1 is unchanged if the depths and viscosities of the layers are interchanged. With this formula, the values of J for $m < 1$ for the values of n given in figure 4.2 can be obtained from the values of J given by the curves in figure 4.1 for $m > 1$, and vice versa.

Yih [1967] writes: "What does finally become of the flow when it is unstable and disturbed slightly? Since the instability, when it exists, exists for any Reynolds number however small, one certainly does not expect turbulence to be the final result of the instability when R is small. The long waves considered here will grow, but as soon as their amplitude becomes finite, nonlinear effects must be taken into account. Thus, the present work can be considered to have demonstrated the possibility of finite waves in superposed layers of fluids of different viscosities." The nonlinear effects will be addressed in section IV.8.

The perturbation problem at any order $n \geq 1$ is

$$\mu \psi_n^{iv} = \begin{cases} k_1(z) & \text{for fluid 1} \\ k_2(z) & \text{for fluid 2} \end{cases} \tag{4a.54}$$

$$\psi_n = \psi_n' = 0 \quad \text{at} \quad z = 0, 1, \tag{4a.55}$$

and at $z = l_1$,

$$c_0'[\![\psi_n']\!] + \psi_n[\![U_1']\!] = -c_n[\![\psi_0']\!] + H_1^{(n)}, \tag{4a.56}$$

$$[\![\mu \psi_n'']\!] = H_2^{(n)},$$

$$[\![\mu \psi_n''']\!] = H_3^{(n)},$$

$$[\![\psi_n]\!] = 0,$$

and the normalization condition (4a.7), where the inhomogeneous terms $k_i(z)$, $H_1^{(n)}$, $H_2^{(n)}$ and $H_3^{(n)}$ are known from lower-order terms: i.e., ψ_0 up to ψ_{n-1}.

This problem cannot be solved when these inhomogeneous terms are arbitrary. There is a condition of solvability, derived from equation (4a.23), with ψ replaced by ψ_n. At the first order, we have equation (4a.35)- (4a.36). More generally, at the nth order, setting $\phi = \phi_0$ in both fluids, and

$$[\![\phi_0]\!] = [\![\phi_0']\!] = [\![\mu\phi_0'']\!] = 0, \tag{4a.57}$$

we have

$$\int_0^{l_1} k_1(z)\phi_0 \, dz + \int_{l_1}^1 k_2(z)\phi_0 \, dz = \phi_0[\![\mu\psi_n''']\!] - \phi_0'[\![\mu\psi_n'']\!] + \mu\phi_0''[\![\psi_n']\!] - \psi_n[\![\mu\phi_0''']\!]$$

$$= \phi_0 H_3^{(n)} - \phi_0' H_2^{(n)} + \frac{\mu\phi_0''}{c_0'}\left(-c_n[\![\psi_0']\!] + H_1^{(n)} - \psi_n[\![U_1']\!]\right) - \psi_n[\![\mu\phi_0''']\!]. \tag{4a.58}$$

In order to carry out this calculation, the adjoint eigenfunction is only needed at the zeroth order (ϕ_0). In general, $c_n[\![\psi_0']\!] = H_1^{(n)} +$ terms depending on orders up to $n-1$. We then solve this for c_n. The streamfunction ψ_n is calculated before solving for c_{n+1}. In Yih's scheme, the streamfunction ψ_{n+1} is calculated before solving for c_{n+1}, but ϕ_0 need not be calculated: thus, at the nth step, the amount of labor involved in either approach is the same. The c_n and ψ_n are real when n is even and pure imaginary when n is odd.

IV.4(b) Two-Layer Couette-Poiseuille Flow

Our analysis for long waves for combined Couette - Poiseuille flow will be done with the perturbation method of Yih. The following results are implicitly contained in §2.1.1 of Blennerhassett [1980]. Kao and Park [1972] have performed experiments to investigate the stability of plane Poiseuille flow of oil and water in a two-dimensional channel. No long wave interfacial instability was recorded. For a comparison of the linear stability results with these experimental results, we refer to Hooper and Grimshaw [1985] and Yiantsios and Higgins [1988].

The velocity of the interface in the basic flow is $(U^*(l_1^*), 0)$ and for brevity, we denote $U^*(l_1^*)$ by U_i. The velocity, distance, time and pressure are made dimensionless with respect to U_i, l^*, l^*/U_i, and $\rho_1 U_i^2$. In Couette-Poiseuille flow, the basic flow has a pressure gradient $-G^*$ in the x-direction. Reynolds numbers for each fluid are denoted by $R_1 = U_i l^* \rho_1/\mu_1$ and $R_2 = U_i l^* \rho_2/\mu_2$. There are seven dimensionless parameters: a Reynolds number, say R_1, the undisturbed depth l_1 of fluid 1, a surface tension parameter $T = $ (surface tension coefficient) $/(\mu_2 U_i)$, a Froude number F given by $F^2 = U_i^2/gl^*$ where g is the gravitational acceleration constant, a dimensionless pressure gradient $G = G^* l^*/(\rho_1 U_i^2)$, the viscosity ratio $m = \mu_1/\mu_2$, and a density ratio $r = \rho_1/\rho_2$.

The dimensionless basic flow $(U(z), 0)$ is

$$U(z) = \begin{cases} -GR_1 z^2/2 + c_1 z, & 0 \leq z \leq l_1, \\ -rGR_2(z-1)^2/2 + c_2(z-1) + U_p, & l_1 \leq z \leq 1 \end{cases} \tag{4b.1}$$

where

$$c_1 = (1 + GR_1 l_1^2/2)/l_1, \quad l_2 = 1 - l_1,$$

$$c_2 = m(-GR_1 l_1 + c_1) - rGR_2 l_2,$$

and the upper plate speed U_p is

$$U_p = 1 + rGR_2 l_2^2/2 + c_2 l_2.$$

The basic pressure field P satisfies $dP/dx = -G$ and

$$\frac{dP}{dz} = \begin{cases} -1/F^2, & 0 \le z \le l_1, \\ -1/(rF^2), & l_1 \le z \le 1. \end{cases} \tag{4b.2}$$

Let $\psi(z)\exp(i\alpha x + \sigma t)$ be a streamfunction. The streamfunction and eigenvalue are expanded for small α and αR_1: $\sigma \sim \sigma_0 + \sigma_1 + \dots$. At the leading order, $\psi(z) = C_1 z^2 + D_1 z^3$ in fluid 1 and $\psi(z) = C_2(z-1)^2 + D_2(z-1)^3$ in fluid 2, where

$$D_1 = \frac{n^2 - m}{3ml_1 + l_2(2m - n^3)}, \tag{4b.3}$$

$$n = \frac{l_1}{l_2}, \quad l_2 = 1 - l_1, \tag{4b.4}$$

$$D_2 = mD_1, \tag{4b.5}$$

$$C_2 = n^2 + D_1 l_2(n^3 + m), \tag{4b.6}$$

and we may choose

$$C_1 = 1. \tag{4b.7}$$

At this order the eigenvalue is imaginary:

$$\sigma_0 + i\alpha U(l_1) \sim i\alpha \frac{(l_1^2 + D_1 l_1^3)[U']}{2l_1 + 3D_1(l_1^2 - ml_2^2) + 2C_2 l_2}. \tag{4b.8}$$

At the next order, $\psi^{iv} = i\alpha R_i((U + \sigma/i\alpha)\psi'' - U''\psi)$. Thus, in fluid 1,

$$\psi(z) = E_1 z^2 + F_1 z^3 + i\alpha R_1 h_1(z),$$

where

$$h_1(z) = a_4 z^4 + a_5 z^5 + a_6 z^6 + a_7 z^7, \tag{4b.9}$$

$$a_4 = (GR_1 l_1^2/2 - c_1 l_1 + \sigma_0'/i\alpha)/12, \tag{4b.10}$$

$$a_5 = (c_1 + 3D_1(GR_1 l_1^2/2 - c_1 l_1 + \sigma_0'/i\alpha))/60, \tag{4b.11}$$

$$a_6 = c_1 D_1/60, \quad a_7 = -GR_1 D_1/420, \tag{4b.12}$$

$$\sigma_0' = \sigma_0 + i\alpha U(l_1). \tag{4b.13}$$

Here, σ_0' is found from equation (4b.8) and the small script c_1 is defined in equation (4b.1). Similarly, in fluid 2,

$$\psi(z) = E_2(z-1)^2 + F_2(z-1)^3 + i\alpha R_2 h_2(z), \tag{4b.14}$$

where

$$h_2(z) = b_4(z-1)^4 + b_5(z-1)^5 + b_6(z-1)^6 + b_7(z-1)^7, \qquad (4b.15)$$

and b_i are defined analogously to a_i. We may choose

$$E_2 = i\alpha R_2. \qquad (4b.16)$$

Continuity of shear stress yields

$$2E_1 + 6F_1 l_1 + i\alpha R_1 h_1''(l_1) = \frac{1}{m}\left(2E_2 - 6F_2 l_2 + i\alpha R_2 h_2''(l_1)\right). \qquad (4b.17)$$

The normal stress condition yields

$$6F_1/R_1 = 6F_2/rR_2 + i\alpha f, \qquad (4b.18)$$

where

$$f = (1 - \frac{1}{r})(U + \sigma_0/i\alpha)(2l_1 + 3D_1 l_1^2)$$

$$+(l_1^2 + D_1 l_1^3)(\frac{1}{r} - 1)(U_1' + \frac{1}{F^2(U + \sigma_0/i\alpha)})$$

$$-h_1'''(l_1) + h_2'''(l_1)/r. \qquad (4b.19)$$

Here, σ_0 is given by equation (4b.8) and is imaginary. Continuity of horizontal velocity yields

$$(U + \sigma_0/i\alpha)\left(2E_1 l_1 + 3F_1 l_1^2 + i\alpha R_1 h_1'(l_1) + 2E_2 l_2 - 3F_2 l_2^2 - i\alpha R_2 h_2'(l_1)\right)$$

$$-\left(E_1 l_1^2 + F_1 l_1^3 + i\alpha R_1 h_1(l_1)\right)[\![U']\!]$$

$$= \frac{\sigma_1}{i\alpha}(2l_1 + 3D_1(l_1^2 - ml_2^2) + 2C_2 l_2) \qquad (4b.20)$$

where σ_1 is the correction to the eigenvalue and is real. Continuity of vertical velocity yields

$$E_1 l_1^2 + F_1 l_1^3 + i\alpha R_1 h_1(l_1) = E_2 l_2^2 - F_2 l_2^3 + i\alpha R_2 h_2(l_1). \qquad (4b.21)$$

We substitute for F_1 using (4b.18) in (4b.17) and (4b.21). We eliminate E_1 from (4b.17) and (4b.21) and use (4b.16) to find F_2. Equation(4b.17) then yields E_1. Finally, (4b.20) yields the real-valued term σ_1.

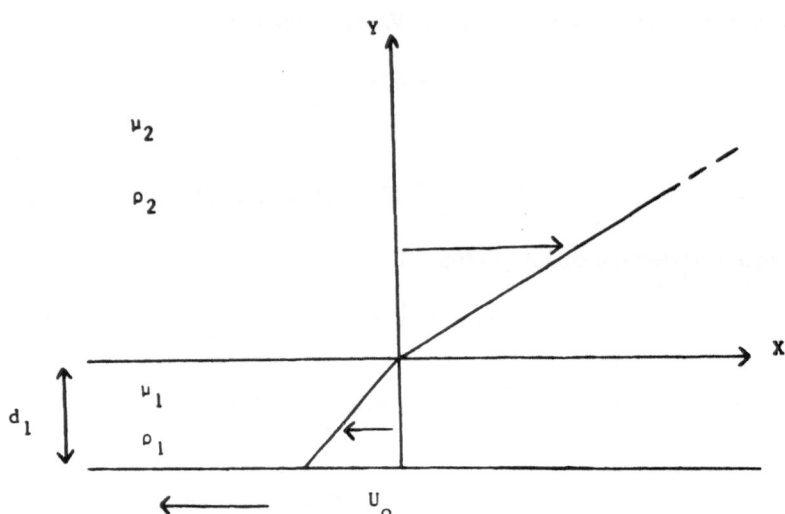

Fig. 4.3. [Figure 1 of Hooper, 1985, American Institute of Physics] The flow configuration for semi-infinite Couette flow.

IV.4(c) Two-Layer Semi-Infinite Couette Flow

This section is based on the paper by Hooper [1985] on long-wave instability at the interface between two viscous fluids when the upper fluid occupies a half-plane (see figure 4.3). This problem is intermediate between the bounded Couette flow of Yih and the unbounded flow studied by Hooper and Boyd [1983]. They found that the flow is always linearly unstable when the lower fluid is the more viscous. The lower fluid has finite depth d_1, viscosity μ_1, and density ρ_1. The lower boundary has velocity $(-U_0, 0)$ in the negative x-direction. The upper fluid has viscosity μ_2 and density ρ_2. The following are dimensionless parameters:

$$m = \mu_2/\mu_1, \tag{4c.1}$$

$$r = \rho_2/\rho_1, \tag{4c.2}$$

$$R = \rho_1 U_0 d_1/\mu_1, \tag{4c.3}$$

$$S = T/\rho_1 d_1 U_0^2, \tag{4c.4}$$

where T is the surface tension,

$$F = (1 - r)(gd_1/U_0^2). \tag{4c.5}$$

The dimensionless coordinates of the system are

$$(x, y) = (X, Y)/d_1, \tag{4c.6}$$

and the (X, Y)-system is used in figure 4.3. The dimensionless wavenumber is

$$\alpha = kd_1, \tag{4c.7}$$

where k is the dimensional wavenumber. The velocity is made dimensionless with respect to U_0. Thus, the basic flow $(u(y), 0)$ is given by

$$u(y) = \begin{cases} u_1(y) = y, & -1 < y < 0, \\ u_2(y) = y/m, & y > 0. \end{cases} \tag{4c.8}$$

The streamfunction of a small disturbance to the basic flow satisfies the linearized Navier-Stokes equation. The streamfunction is assumed to be of the form

$$\phi(y)e^{i\alpha(x-ct)}, \tag{4c.9}$$

where $2\pi/\alpha$ is the wavelength of the disturbance in the x-direction and c is complex. When this form of the disturbance streamfunction is substituted into the linearized Navier-Stokes equation, the Orr-Sommerfeld equation is derived (see, for example, equations (2.30) - (2.37)). Thus,

$$\left(\frac{d^2}{dy^2} - \alpha^2\right)^2 \phi_1(y) = i\alpha R(y - c)\left(\frac{d^2}{dy^2} - \alpha^2\right)\phi_1(y) \tag{4c.10}$$

for $-1 < y < 0$, and

$$\left(\frac{d^2}{dy^2} - \alpha^2\right)^2 \phi_2(y) = i\alpha \frac{Rr}{m^2}(y - mc)\left(\frac{d^2}{dy^2} - \alpha^2\right)\phi_2(y) \tag{4c.11}$$

for $y > 0$. There is no slip at the solid boundary at $y = -1$, so

$$\phi_1(-1) = \phi_1'(-1) = 0. \tag{4c.12}$$

The disturbance is bounded as $y \to \infty$, which implies that

$$\phi_2(y) \to 0 \quad \text{as} \quad y \to \infty. \tag{4c.13}$$

There are four conditions at the interface: continuity of normal and tangential velocity, and continuity of normal and tangential stress. Thus, at $y = 0$,

$$\phi_1 = \phi_2 = \phi(0), \tag{4c.14}$$

$$\frac{d\phi_1}{dy} = \frac{d\phi_2}{dy} + \frac{\phi(0)}{c}\left(\frac{1 - m}{m}\right), \tag{4c.15}$$

$$\left(\frac{d^2}{dy^2} + \alpha^2\right)\phi_1 = m\left(\frac{d^2}{dy^2} + \alpha^2\right)\phi_2, \tag{4c.16}$$

$$\left(\frac{d^2}{dy^2} - 3\alpha^2\right)\frac{d\phi_1}{dy}$$

$$= m\left(\frac{d^2}{dy^2} - 3\alpha^2\right)\frac{d\phi_2}{dy} + i\alpha R(\alpha^2 S + F)\frac{\phi(0)}{c}$$

$$+ i\alpha R\left(c(r\frac{d\phi_2}{dy} - \frac{d\phi_1}{dy}) + \phi(0)(\frac{r-m}{m})\right). \tag{4c.17}$$

Equations (4c.10) - (4c.17) define an eigenvalue problem where $\phi_1(y)$ and $\phi_2(y)$ are the eigenfunctions and c, the eigenvalue, is a function of α, R, m, r, S and F. The flow is said to be linearly unstable if the imaginary part of c is positive.

The vorticity of the disturbance in fluid j is denoted by $w_j(y)$ and is defined by

$$w_j(y) = -\left(\frac{d^2}{dy^2} - \alpha^2\right)\phi_j(y), \quad j = 1, 2. \tag{4c.18}$$

Inspection of equations (4c.10) - (4c.11) reveals that the vorticity $w_j(y)$ satisfies the Airy equation. In particular, because the disturbance is bounded as $y \to \infty$, we find that

$$w_2(y) = b_2 A_2(\alpha^{1/3}y), \tag{4c.19}$$

where b_2 is a constant and

$$A_2(y) = Ai\left(e^{i\pi/6}(\frac{Rr}{m^2})^{1/3}[y - (mc + \frac{i\alpha m^2}{rR})\alpha^{1/3}]\right), \tag{4c.20}$$

where Ai denotes the Airy function.

The streamfunction $\phi_2(y)$ is found from equations (4c.18) - (4c.19) to have the following exact form:

$$\phi_2(y) = b_1 e^{-\alpha y} + \frac{b_2}{2\alpha^{4/3}}\left(e^{-\alpha y}\int_0^{\alpha^{1/3}y} e^{\alpha^{2/3}z}A_2(z)dz\right.$$

$$\left. + e^{\alpha y}\int_{\alpha^{1/3}y}^{\alpha} e^{-\alpha^{2/3}z}A_2(z)dz\right). \tag{4c.21}$$

Equation (4c.21) is valid for all values of y and α. Near the interface, y is small and $\phi_2(y)$ is approximated by its Taylor series:

$$\phi_2(y) = \beta_{00} + \beta_{01}y + \beta_{02}y^2 + \beta_{03}y^3 + O(y^4), \tag{4c.22}$$

where

$$\beta_{00} = b_1 + b_2(J_2/2)\alpha^{-4/3}, \tag{4c.23}$$

$$\beta_{01} = -\alpha b_1 + b_2(J_2/2)\alpha^{-1/3}, \tag{4c.24}$$

$$\beta_{02} = \alpha^2\beta_{00} - b_2 A_2(0), \tag{4c.25}$$

$$\beta_{03} = \alpha^2\beta_{01} - e^{i\pi/6}(Rr/m^2)^{1/3}A_2'(0)b_2\alpha^{1/3}, \tag{4c.26}$$

$$J_2 = \int_0^\infty e^{-\alpha^{2/3}z}A_2(z)dz. \tag{4c.27}$$

This series is valid near the interface for all α.

The lower fluid is bounded between the surfaces $y = -1$ and $y = \eta(x, t)$, where η represents the disturbed interfacial position. Yih's perturbation expansion is valid in this region. Therefore, when $\alpha << 1$ and R is $O(1)$, the eigenfunction $\phi_1(y)$ is expanded as

$$\phi_1(y) = \phi_{10}(y) + O(\alpha) \tag{4c.28}$$

such that

$$\frac{d^4\phi_{10}}{dy^4} = 0, \tag{4c.29}$$

subject to

$$\phi_{10}(-1) = 0, \quad \frac{d\phi_{10}}{dy}(-1) = 0. \tag{4c.30}$$

Hence,

$$\phi_{10} = a_{10}(1 + y)^2 + a_{20}(1 + y)^3, \tag{4c.31}$$

where a_{10} and a_{20} are arbitrary constants to be determined via the four interfacial conditions. The equations are linear and homogeneous and, therefore, we let $\phi(0) = 1$ in each fluid (as in equations (4a.7), (4a.30)). However, since the constants β_{0j} defined in equations (4c.22) - (4c.27) depend not only on the constants b_1 and b_2 but also on the functions $A_2(0)$, $A_2'(0)$, and J_2, which are functions of α and c, it is clear that progress cannot continue unless some a priori assumption about the asymptotic behavior of c when $\alpha << 1$ is made. Thus, we assume that $(Rrm)^{1/3}c$ is $O(1)$ or less, which implies that the terms J_2, $A_2(0)$, and $A_2'(0)$ which appear in β_{0j}, are all $O(1)$. Then when α is small, the interfacial conditions equations (4c.14) - (4c.17) reduce to

$$a_{10} + a_{20} + O(\alpha) = b_1 + b_2(J_2/2)\alpha^{-4/3} = 1, \tag{4c.32}$$

$$2a_{10} + 3a_{20} + O(\alpha) = -\alpha b_1 + (b_2/2)J_2\alpha^{-1/3} + (1 - m)/mc, \tag{4c.33}$$

$$2a_{10} + 6a_{20} + O(\alpha) = -b_2 m A_2(0) + O(\alpha^2), \tag{4c.34}$$

$$6a_{20} + O(\alpha) = -b_2 m e^{i\pi/6}(Rr/m^2)^{1/3}A_2'(0)\alpha^{1/3} + O(\alpha R, \alpha^2\beta_{01}). \tag{4c.35}$$

The right hand side of equations (4c.32) - (4c.35) are functions of α and so the constants a_{10} and a_{20} must also be functions of α but are less than $O(1)$ as $\alpha \to 0$. Equation (4c.34) shows that b_2 is $O(1)$. Hence, β_{01} is $O(\alpha^{-1/3})$. Then, equation (4c.35) implies that for moderate Reynolds number R, a_{20} is $O(\alpha^{1/3})$. Therefore,

$$a_{10} = 1 + O(\alpha^{1/3}), \tag{4c.36}$$

$$b_2 = -2/mA_2(0) + O(\alpha^{1/3}). \tag{4c.37}$$

Equation (4c.33) can only be satisfied if c is $O(\alpha^{1/3})$. This is consistent with the original hypothesis that c is $O(1)$ or less. At $O(\alpha^{1/3})$, equation (4c.33) reduces to

$$-b_2 J_2\alpha^{-1/3} = (1 - m)/mc, \tag{4c.38}$$

which implies that to leading order,

$$c = \alpha^{1/3} c_0 + O(\alpha^{2/3}) \qquad (4c.39)$$

where

$$c_0 = (1 - m) A_2(0)/2J_2 \qquad (4c.40)$$

$$= (1 - m)(rR/m^2)^{1/3}(0.53254)e^{i\pi/6} + O(\alpha^{2/3}). \qquad (4c.41)$$

Therefore, the flow is linearly unstable when $m < 1$, that is, when the lower film of fluid is the more viscous fluid.

The iterations for finding corrections to the above expansion are given in Hooper [1985]. Gravity enters at order α in c in equation (4c.39): as expected, when the lower fluid is the more dense, gravity has a stabilizing effect.

This analysis is only valid when $\alpha << 1$ if certain other terms also remain small. In particular,

$$\alpha R << 1 \qquad (4c.42)$$

is required, so that the expansion given by equation (4c.28) - (4c.31) is valid. The parameters R, S, and F are restricted to be $O(1)$, so that the perturbation scheme holds. Further, c has been shown to be $O(\alpha^{1/3})$. In this case, equations (4c.39) - (4c.41) show that the term

$$(1 - m)(r/m^2)^{1/3} \qquad (4c.43)$$

must be at most $O(1)$. This means that if m is large ($m > 1$ is the stable arrangement) then r must be small to keep this term $O(1)$.

The equation governing the nonlinear development of the unstable mode is discussed in Hooper's paper [1985]. Also, the exact dispersion relation is derived for the case $S = 0$ and $r = 1$. It is demonstrated that there are an infinite number of modes which govern the stability of the flow in the configuration shown in figure 4.3. All these modes are stable in the long-wave limit except one, which is unstable only when the lower fluid is the more viscous fluid. The nature of the modes, which are stable when $\alpha << 1$ irrespective of the value of m, is identical to the nature of those modes found for semi-infinite Couette flow bounded by a wall at $y = 0$ (i.e. these are 'one-fluid modes'). The one-fluid semi-bounded Couette flow, with one solid boundary, has been shown to be stable [Zondek and Thomas 1953].

The unbounded two-layer Couette flow analyzed by Hooper and Boyd [1983] has no instability in the long-wave limit. The existence of the unstable mode in the semi-infinite problem for $\alpha << 1$ shows that the presence of one wall is sufficient to cause the instability in the long-wave limit. The growth rate of this instability is $O(\alpha^{4/3})$. On the other hand, the growth rate for Yih's problem of bounded two-layer Couette flow (see section IV.4 (a), (b)) is $O(\alpha^2)$. Therefore, it seems that the presence of the second wall has a stabilizing effect, probably caused by the fact that the disturbance is now constrained between two solid boundaries. A similar effect has been predicted by Davis, Herbolzheimer and Acrivos [1983], who studied sedimentation in narrow tilted channels. Their analysis suggests that the flow during sedimentation can be stabilized by reducing the spacing between the two walls of the channel.

How is the long-wave analysis for the semi-infinite problem relevant to that of the bounded channel problem? The instability discussed here is expected to be apparent in stratified viscous flow in a channel when

$$\alpha^{1/3}n > 1, \quad n = d_2/d_1, \tag{4c.44}$$

where d_2 is the depth of the upper fluid. This is evident from the Orr-Sommerfeld equation equation (4c.11) that governs the motion of the disturbance in the upper fluid and is defined on the region $0 < y < n$. When $\alpha << 1$ and $c \sim O(1)$, this equation becomes, at leading order,

$$\frac{d^4\hat{\phi}}{dY^4} = i(\frac{Rr}{m^2})Y\frac{d^2\hat{\phi}}{dY^2}, \quad 0 < Y < \alpha^{1/3}n, \tag{4c.45}$$

at leading order, where

$$\phi(y) = \hat{\phi}(Y) \quad \text{and} \quad Y = \alpha^{1/3}y. \tag{4c.46}$$

Therefore, the term on the right-hand side of equation (4c.45) is $O(\alpha^{1/3}n)$ and it can only be neglected when $\alpha^{1/3}n << 1$, giving Yih's solution. When $\alpha^{1/3}n$ is not small, this term cannot be ignored and the analysis discussed here applies. This is reminiscent of the Stokes paradox for flow problems involving an unbounded domain: if one considers flow past a body, then even at low Reynolds numbers, the Stokes equation does not yield the right decay at infinity and convective terms must be taken into account. At leading order, if one solves the Stokes equation here, one would be solving the left-hand side of equation (4c.45) set equal to zero, but this yields a cubic polynomial as the solution, and this does not have the decay property required of the solution for the unbounded problem, so that the term on the right-hand side of equation (4c.45) must be included.

The condition $\alpha^{1/3}n > 1$ is equivalent to

$$d_1 < (\frac{2\pi d^3}{\lambda})^{1/2}, \tag{4c.47}$$

where d is the total channel depth $d_1 + d_2$, λ is the wavelength of the disturbance, and the depth of the lower fluid $d_1 << d$. Therefore, we see that for a fixed depth of the lower fluid in a channel such that $d_1 << d$, the growth rate of an unstable disturbance will be $O(\alpha^{4/3})$ for moderately long waves and $O(\alpha^2)$ for very long waves.

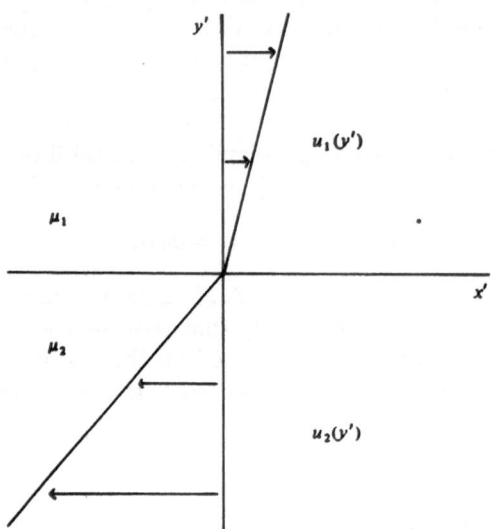

Fig. 5.1. [Figure 1 of Hooper and Boyd, 1983] The flow configuration for un-bounded Couette flow.

IV.5 Asymptotic Analysis for Short Waves

Hooper and Boyd [1983] consider the linear stability of the cocurrent flow of two fluids of different viscosity and density, when each fluid occupies a half-plane (see figure 5.1). The basic velocity field in dimensional coordinates (x', y') is

$$\mathbf{u} = \begin{cases} (a_1 y', 0), & y' > 0 \\ (a_2 y', 0), & y' < 0 \end{cases} \tag{5.1}$$

where, by continuity of tangential stress,

$$\mu_1 a_1 = \mu_2 a_2. \tag{5.2}$$

They find the exact form of the eigenfunctions and find a mode that is unstable (for zero surface tension and equal density). They show that this is the only unstable mode and that it leads to instability only for short waves. They evaluate this eigenfunction and corresponding eigenvalue in closed form in the short-wave limit.

Their analysis is relevant in a local sense to any situation in which a shear flow acts in the neighborhood of a viscosity jump [Yiantsios and Higgins 1988; Hooper and Boyd 1983; Y. Renardy and Joseph 1985 a,b]. In addition, they introduce a perturbation procedure which can be used for the asymptotic analysis for short waves (wavenumber $\alpha \gg 1$) of the interfacial mode in layered flows, and provide a rigorous proof of the procedure for the situation in figure 5.1. They find that if surface tension is absent and the densities are matched, the interface is unstable to short-wave disturbances. This is like the Kelvin-Helmholtz problem, with the velocity discontinuity replaced by a discontinuity in the velocity gradient. The Kelvin-Helmholtz problem is Hadamard unstable with growth rates that tend to infinity with α. Hooper and Boyd's problem is the viscous analog of the Kelvin-Helmholtz problem (see Joseph and Saut [1990]): it is also unstable to short waves under the stated conditions but the growth rates are uniformly bounded in α. The conditions under which short-wave instability may be expected to occur are unlikely to be realizable. In particular, surface tension will stabilize the shortest waves. For a discussion of the practical difficulties in observing the short-wave instability due to viscosity stratification, see Hinch [1984]. It is also possible for diffusion at the interface to stabilize the short waves in the absence of surface tension. In the Hele-Shaw problem (cf. section X.14 of volume 2), diffusion is stabilizing and enters at an order of α larger than the surface tension term here.

We shall now carry out the asymptotic analysis for short waves for the two-layer Couette flow. The governing equations for the streamfunction are (2.32) - (2.38) given in section IV.2. The procedure is reminiscent of the boundary-layer analysis, where the vertical variable is rescaled, some boundary conditions are lost, and the asymptotic solution is valid locally. We rescale the vertical variable z to

$$\eta = \alpha(z - l_1), \quad \eta = O(1), \quad \alpha \gg 1. \tag{5.3}$$

The differential equation for the streamfunction (2.32) becomes

$$(D^2 - 1)^2 \psi(\eta) = \frac{i \Re_i}{\alpha}(U_1(z) - c)(D^2 - 1)\psi(\eta), \tag{5.4}$$

where $\sigma = -i\alpha c$ as in (2.31) and

$$D = d/d\eta. \tag{5.5}$$

The interface conditions (2.33) - (2.37) become: at $\eta = 0$,

$$\psi = h(c - U_1(l_1)), \tag{5.6}$$

$$[\![\psi]\!] = 0, \tag{5.7}$$

$$\alpha[\![D\psi]\!] + h[\![U_1']\!] = 0, \tag{5.8}$$

$$[\![\mu D^2 \psi]\!] + \psi[\![\mu]\!] = 0, \tag{5.9}$$

$$-i\alpha^2(c - U_1(l_1))[\![\frac{\rho_i}{\rho_1}D\psi]\!] - i\alpha\psi[\![\frac{\rho_i}{\rho_1}U_1']\!] - \frac{\alpha^3}{\mu_1 \Re}[\![\mu(D^3\psi - 3D\psi)]\!]$$

$$= i\alpha h(\frac{-\alpha^2 S}{\Re} + \frac{1}{F^2}(\frac{1}{r} - 1)). \tag{5.10}$$

The boundary conditions (2.38) at $z = 0, 1$ become conditions on the decay of ψ:

$$\psi, \ D\psi \to 0 \quad \text{as} \quad \eta \to \pm\infty. \tag{5.11}$$

The interface conditions (5.6) and (5.8) combined yield

$$\alpha[D\psi](c - U_1(l_1)) + \psi[U_1'] = 0. \tag{5.12}$$

These conditions suggest the following expansions for $\alpha \gg 1$ in powers of $1/\alpha$:

$$\psi = \psi_0 + \frac{\psi_1}{\alpha} + \frac{\psi_2}{\alpha^2} + \dots \tag{5.13}$$

$$c - U_1(l_1) = \frac{c_0}{\alpha} + \frac{c_1}{\alpha^2} + \frac{c_2}{\alpha^3} + \dots \tag{5.14}$$

$$h = \alpha h_0 + h_1 + \frac{h_2}{\alpha} + \dots \ . \tag{5.15}$$

Equation (5.4) yields

$$(D^2 - 1)^2 \psi_0 = 0, \tag{5.16}$$

so that the solution satisfying the boundary conditions (5.11) is

$$\psi_0 = \begin{cases} (a_1 + a_2\eta)e^\eta, & \eta < 0 \ \text{(fluid 1)} \\ (b_1 + b_2\eta)e^{-\eta}, & \eta > 0 \ \text{(fluid 2)} \end{cases}. \tag{5.17}$$

The interface conditions are

$$\psi_0 = h_0 c_0, \quad [\psi_0] = 0, \quad [D\psi_0] + h_0[U_1'] = 0,$$

$$[\mu D^2 \psi_0] + \psi_0[\mu] = 0, \quad [\mu(D^3\psi_0 - 3D\psi_0)] = 0. \tag{5.18}$$

The second and last conditions together yield

$$a_1 = b_1 = 0,$$

and the rest of the conditions yield

$$b_2 = -ma_2,$$

$$h_0 = \frac{-a_2(1 + m)}{[U_1']}. \tag{5.19}$$

Finally,

$$\psi_0 = \begin{cases} a_2\eta e^\eta, & \eta < 0, \\ -ma_2\eta e^{-\eta}, & \eta > 0, \end{cases} \tag{5.20}$$

and

$$c_0 = 0. \tag{5.21}$$

In order to evaluate c_1, we need the equations satisfied by ψ_1. The differential equation for ψ_1 is the same as that for ψ_0. The interface conditions are that at $\eta = 0$,

$$\psi_1 = h_0 c_1, \quad [\![\psi_1]\!] = 0,$$
$$c_1[\![D\psi_0]\!] + \psi_1[\![U_1']\!] = 0, \tag{5.22}$$
$$[\![\mu D^2 \psi_1]\!] + \psi_1[\![\mu]\!] = 0, \quad [\![\mu(D^3\psi_1 - 3D\psi_1)]\!] = 0.$$

Rather than to carry out the asymptotic analysis exactly in the manner of Hooper and Boyd [1983], we will use the method based on the use of the adjoint eigenfunction and a solvability condition. As noted in section IV.4 (a), the amount of algebra required by either method is the same. For the calculation of c_1, we require the adjoint eigenfunction. Let ϕ denote the adjoint eigenfunction. We find the adjoint equations by integrating the following integral by parts:

$$\int_{-\infty}^{0} \mu\phi(D^4 - 2D^2 + 1)\psi_1 \, d\eta + \int_{0}^{\infty} \mu\phi(D^4 - 2D^2 + 1)\psi_1 \, d\eta \tag{5.23}$$

$$= [\![\mu\phi(D^3\psi_1 - 2D\psi_1) + \mu D\phi(-D^2\psi_1 + 2\psi_1) + \mu D^2\phi D\psi_1 - \mu D^3\phi\psi_1]\!]$$

$$+ \int_{-\infty}^{0} \mu\psi_1(D^4 - 2D^2 + 1)\phi \, d\eta + \int_{0}^{\infty} \mu\psi_1(D^4 - 2D^2 + 1)\phi \, d\eta, \tag{5.24}$$

provided ϕ satisfies

$$\phi, \phi_\eta \to 0 \quad \text{as} \quad \eta \to \pm\infty. \tag{5.25}$$

After some rearrangement, (5.24) is equal to

$$[\![\mu\phi(D^3\psi_1 - 3D\psi_1)]\!] + [\![\mu D\phi(-D^2\psi_1 - \psi_1)]\!]$$

$$+ [\![(\mu D^2\phi + \mu\phi)D\psi_1]\!] + [\![(3\mu D\phi - \mu D^3\phi)\psi_1]\!]$$

$$+ \int_{-\infty}^{0} \mu\psi_1(D^4 - 2D^2 + 1)\phi \, d\eta + \int_{0}^{\infty} \mu\psi_1(D^4 - 2D^2 + 1)\phi \, d\eta \tag{5.26}$$

We choose

$$[\![\phi]\!] = 0, \quad [\![D\phi]\!] = 0, \quad [\![\mu\phi + \mu D^2\phi]\!] = 0, \tag{5.27}$$

and

$$(D^2 - 1)^2\phi(\eta) = 0 \text{ for } \eta < 0 \text{ and } \eta > 0. \tag{5.28}$$

Since (5.23) is zero, (5.26) yields

$$[\![D\psi_1]\!][\mu\phi + \mu D^2\phi]_{fluid\ 1} + [\psi_1]_{fluid\ 1}[-\mu D^3\phi + 3\mu D\phi] = 0. \tag{5.29}$$

We next solve for the adjoint eigenfunction. The solution to (5.28) satisfying (5.25) is

$$\phi(\eta) = \begin{cases} (d_1 + d_2\eta)e^\eta, & \eta < 0, \\ (f_1 + f_2\eta)e^{-\eta}, & \eta > 0. \end{cases} \tag{5.30}$$

The interface conditions (5.27) yield

$$d_1 = -d_2 = f_1 = f_2,$$

so that

$$\phi(\eta) = \begin{cases} d_1(1-\eta)e^\eta, & \eta < 0, \\ d_1(1+\eta)e^{-\eta}, & \eta > 0. \end{cases} \tag{5.31}$$

This yields that

$$[\mu\phi + \mu D^2\phi]_{fluid\ 1} = 0, \tag{5.32}$$

and

$$[-\mu D^3\phi + 3\mu D\phi] = 2d_1\mu_2(m+1). \tag{5.33}$$

Equations (5.29), (5.32) and (5.33) imply that

$$[\psi_1]_{fluid\ 1} = 0. \tag{5.34}$$

This is substituted into (5.22) to yield

$$c_1 = 0. \tag{5.35}$$

In order to evaluate c_2, we must calculate ψ_1. The differential equation and the conditions (5.22) yield the same form of solution as at the zeroth order:

$$\psi_1(\eta) = \begin{cases} a_3\eta e^\eta, & \eta < 0, \text{ (fluid 1)}, \\ -ma_3\eta e^{-\eta}, & \eta > 0, \text{ (fluid 2)}, \end{cases} \tag{5.36}$$

$$h_1 = -\frac{a_3(1+m)}{[U_1']}. \tag{5.37}$$

The equations for ψ_2 are required. These are, from (5.4) and substituting the expansion for c,

$$(D^2 - 1)^2\psi_2 = i\Re_i(U_1(z) - U_1(l_1))(D^2 - 1)\psi_0. \tag{5.38}$$

Substitution for ψ_0 from equation (5.20) yields

$$(D^2 - 1)^2\psi_2 = i\Re_i \begin{cases} 2a_2\eta e^\eta/(l_1 + ml_2), & \eta < 0, \\ 2a_2m^2\eta e^{-\eta}/(l_1 + ml_2), & \eta > 0. \end{cases} \tag{5.39}$$

In the interface conditions, we assume

$$\frac{\alpha^3 S}{\Re} = O(1), \quad \frac{\alpha}{F^2}(\frac{1}{r} - 1) = O(1). \tag{5.40}$$

This is a distinguished limit taken by Hooper and Boyd [1983]. It is also possible to carry out the perturbation scheme for other assumptions on the magnitudes of S and $1/F^2$ (see Yiantsios and Higgins [1988] and the end of this section). Then at $\eta = 0$,

$$\psi_2 = h_0 c_2, \quad [\psi_2] = 0,$$

$$[D\psi_0]c_2 + \psi_2[U_1'] = 0, \quad [\mu D^2\psi_2] + \psi_2[\mu] = 0,$$

$$[\mu(D^3\psi_2 - 3D\psi_2)] = -i\mu_1\Re\left(\psi_0[\frac{\rho_i}{\rho_1}U_1'] + h_0\left(\frac{-\alpha^3 S}{\Re} + \frac{\alpha}{F^2}(\frac{1}{r} - 1)\right)\right). \tag{5.41}$$

In equations (5.23) - (5.26), replace ψ_1 by ψ_2, and use the differential equation (5.39) in (5.23). At $\eta = 0$, $[\![\psi_0]\!]$(fluid 1) $= 0$ and $\psi_2 = -[\![D\psi_0]\!]\, c_2/[\![U_1']\!]$. Then

$$\int_{-\infty}^{0} \mu_1 \phi i \Re \frac{2a_2 \eta e^{\eta}}{(l_1 + m l_2)}\, d\eta + \int_{0}^{\infty} \mu_2 \phi i \Re_2 \frac{2a_2 m^2 \eta e^{-\eta}}{(l_1 + m l_2)}\, d\eta \tag{5.42}$$

$$= -[\phi]_{fluid\ 1} i \mu_1 \Re h_0 \left(\frac{-\alpha^3 S}{\Re} + \frac{\alpha}{F^2}\left(\frac{1}{r} - 1\right) \right) - \frac{[\![D\psi_0]\!] c_2}{[\![U_1']\!]}[3\mu D\phi - \mu D^3 \phi]. \tag{5.43}$$

We substitute for ψ_0 and h_0 from (5.19) - (5.20), for ϕ from (5.32) and use (5.34). This simplifies (5.42) - (5.43) to

$$c_2 = \frac{i\Re m}{2(m+1)}\left(\frac{-\alpha^3 S}{\Re} + \frac{\alpha}{F^2}\left(\frac{1}{r} - 1\right) + \frac{(1-m)}{(l_1 + m l_2)^2}\frac{(1 - \frac{m^2}{r})}{(1+m)} \right). \tag{5.44}$$

This agrees with the results obtained by Hooper and Boyd [1983]. Thus, the growth rate $\mathrm{Re}\,\sigma$ of the interfacial mode, in the limit of short disturbance wavelength α is an expression consisting of three terms. In the absence of surface tension or density difference, there is always instability.

In order to calculate c_n, the functions $\psi_0, ... \psi_{n-1}$ must be known, and the equations satisfied by ψ_n must be known. In general,

$$\frac{1}{\Re_i}(D^2 - 1)^2 \psi_n = \begin{cases} k_1(\eta), & \eta < 0, \\ k_2(\eta), & \eta > 0, \end{cases} \tag{5.45}$$

where

$$\psi_n,\ D\psi_n \to 0 \quad \text{as} \quad \eta \to \pm\infty. \tag{5.46}$$

The interface conditions (5.6) - (5.10) are

$$\psi_n[\![U_1']\!] = -c_n[\![D\psi_0]\!] + H_1^{(n)}, \tag{5.47}$$

$$[\![\psi_n]\!] = 0, \tag{5.48}$$

$$[\![\mu D^2 \psi_n]\!] + \psi_n[\![\mu]\!] = 0, \tag{5.49}$$

$$[\![\mu(D^3 \psi_n - 3D\psi_n)]\!] = H_2^{(n)}, \tag{5.50}$$

where the inhomogeneous terms $k_i(\eta)$, $H_1^{(n)}$ and $H_2^{(n)}$ depend on lower-order terms (i.e., on the functions $\psi_0, ... \psi_{n-1}$). The solvability condition to find c_n is, from equations (5.23) and (5.43),

$$\int_{-\infty}^{0} \mu\phi(D^4 - 2D^2 + 1)\psi_n\, d\eta + \int_{0}^{\infty} \mu\phi(D^4 - 2D^2 + 1)\psi_n\, d\eta$$

$$= [\phi]_{fluid\ 1}[\![\mu(D^3 \psi_n - 3D\psi_n)]\!] + [\psi_n]_{fluid\ 1}[3\mu D\phi - \mu D^3 \phi]$$

$$= [\phi]_{fluid\ 1} H_2^{(n)} + \frac{(-c_n[\![D\psi_0]\!] + H_1^{(n)})}{[\![U_1']\!]}[3\mu D\phi - \mu D^3 \phi]. \tag{5.51}$$

Therefore,

$$c_n = \frac{1}{[D\psi_0]}\Big(\frac{[U_1']}{2d_1(\mu_1 + \mu_2)}(d_1 H_2^{(n)}) - \int_{-\infty}^0 \mu_1 \Re\phi k_1(\eta)d\eta$$

$$- \int_0^\infty \mu_2 \Re_2 \phi k_2(\eta)d\eta) - H_1^{(n)}\Big). \tag{5.52}$$

Hooper and Boyd [1983] note that for sufficiently short waves, any parallel shear flow with velocity components $(U_j(z), 0)$, $U_j'(l_1) \neq 0$ is locally Couette near the interface (i.e. has the appearance of figure 5.1). Thus, very similar results are expected to hold for any such flow. A related result is that of Blennerhassett and Smith [1987] for high Reynolds numbers, mentioned in section IV.6b. Although a simple physical explanation of the instability appears to be unavailable, some insight into the nature of the instability is provided by energy considerations. Hooper and Boyd obtain the energy equation for the system, expressing the balance between the material rate of change of kinetic energy of the perturbed flow, the rate at which the Reynolds stress is transferring energy between the basic flow and the perturbed flow, the rate of viscous dissipation of the perturbed flow and an interface term (the rate at which the disturbed flow is being supplied with energy at the interface). By substituting their expressions for the interfacial eigenfunction and eigenvalue in the short-wave limit, they expand each term for large α. At the leading order, they find that the Reynolds stress term is zero, and the interface term is balanced by viscous dissipation. At the next order, the Reynolds stress term can be of either sign, and the energy transfer is stabilizing in the more viscous fluid. In the less viscous fluid, the energy transfer is destabilizing near the interface, but is stabilizing far from the interface. Thus, one may say that the interface term drives the short-wave instability, which is essentially created in the immediate neighborhood of the interface and in the less viscous fluid.

The method of expansion for short waves introduced by Hooper and Boyd [1983] is for the case $\alpha^3 S = O(1)$, but schemes may also be constructed for cases when S or $\alpha^2 S$ is $O(1)$ [Yiantsios 1988]. The following is for $S = O(1)$, $\alpha \to \infty$. Instead of (5.14), put

$$c - U_1(l_1) \sim c_0 + \frac{c_1}{\alpha} + \frac{c_2}{\alpha^2} + \dots. \tag{5.53}$$

The calculation is analogous to the above and we find

$$c_0 = \frac{-imS}{2(1+m)}. \tag{5.54}$$

This is the same as the first term of (5.44). Of course, the values of c_1, c_2, ..., differ from those of the case $\alpha^3 S = O(1)$, but the leading contribution to c is given by the same surface-tension-term as in the leading term of Hooper and Boyd. The formula was compared with numerical results on the exact interfacial eigenvalue for two-layer Couette flow with $l_1 = 0.5$, $m = 0.5$, $S = .0006$, $\Re = .03$ and $\alpha = 10$. For these data, we expect (5.53) - (5.54) to be a good approximation. The parameters were also chosen to compare with the results in figure 5 of Hooper and Boyd [1987] where, in their notation, $m = 2$, $R = .01$, $\beta = 13.6$, $\alpha = 5$, $S = .1$, Im $c/R \sim -.016$. We find good agreement between the exact value of the interfacial eigenvalue and the short-wave asymptotic formulas. Thus, the surface-tension term in the leading-order term of Hooper and Boyd is the leading-order term for $\alpha \rightarrow \infty$ when $S \neq 0$.

IV.6(a) Asymptotic Analysis For The Thin-Layer Effect

The analyses for long waves of section IV.4 have demonstrated that in the presence of viscosity stratification, the interfacial mode is stable if the less viscous fluid is placed in a sufficiently thin layer. It is therefore conceivable that a linearly stable arrangement can be achieved by a sufficiently thin lubricating layer of less viscous fluid, provided that there is enough surface tension at the interface to damp the short-wave instability due to viscosity stratification. In fact, such an arrangement may be linearly stable even with an adverse density stratification. We refer to this stabilizing effect of viscosity stratification as the "thin-layer effect". The stability of a thin layer of less viscous fluid is relevant to the lubrication industry and to coating technology.

A particular example is figure 6.1 for the two-layer Couette flow. The notation for the physical parameters for this graph is defined in section IV.2. The figure shows the growth rate Re σ against the wavenumber α, $0 \leq \alpha \leq 1$, for various depth ratios l_1. The eigenvalues were computed numerically with the Chebyshev-tau method [Orszag 1971]. The upper more viscous fluid is slightly more dense than the lower fluid. A sufficient amount of surface tension is included to stabilize short waves. The Froude number is chosen not too small: otherwise, gravity would destabilize long waves because of the adverse density stratification. We choose $F^2 = 0.1$. The Reynolds number in each fluid is low. We choose m small and r close to 1. The parameter l_1 ($0 \leq l_1 \leq 1$) measures the depth of the lower fluid. We note that for small l_1 (e.g. $l_1 = 0.05, 0.1$), the growth rate Re σ is negative for all wavelengths of the disturbance α, showing stability. For larger l_1, instabilities arise from long waves, due to gravity. The flow is linearly stable at sufficiently large α for each l_1, as can be deduced from equation (5.44).

Another example of the thin-layer effect occurs in the two-layer Taylor flow (see [Renardy and Joseph 1985]), where the fluids are assumed to lie concentrically between concentric cylinders. The outer cylinder is fixed and the inner one rotates. When the linear stability of the cylindrical interface is examined at low speeds, a thin-layer effect is found. The configuration with a thin layer of the less viscous but heavier fluid lying next to the inner

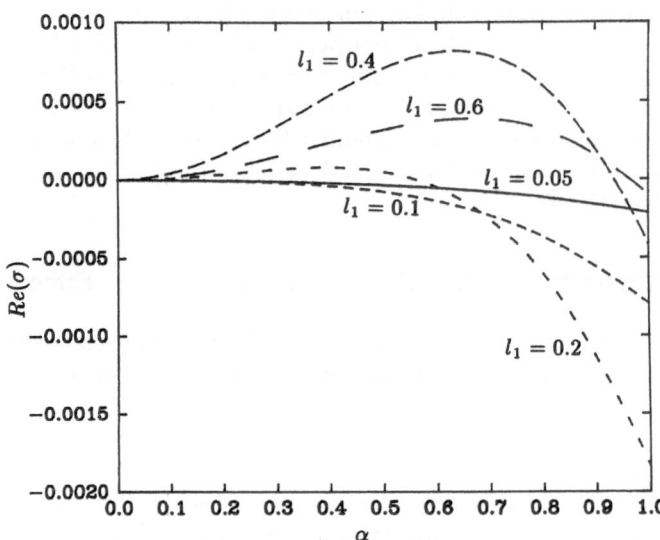

Fig. 6.1. [Figure 2 of Renardy, 1985, American Institute of Physics] Growth rate versus α. $\Re$ for lower fluid $= 10$, viscosity ratio $m = \mu_1/\mu_2 = 0.01$, density ratio $r = \rho_1/\rho_2 = 0.95$, dimensional surface tension coefficient$/\mu_2 U^* = 0.1$, U^*=dimensional plate speed, $U^{*2}/gl^* = 0.1$, l^*= dimensional plate separation.

cylinder may be linearly stable provided the rotation rate is sufficiently slow and there is an appropriate amount of surface tension. This stabilizing mechanism of viscosity stratification is absent if the basic flowfield has no shearing. For example, if the basic flowfield is rigid-body rotation, the linear stability criterion (see chapter II) can be found in closed form and the critical condition is independent of the viscosity ratio. Close to criticality in the two-layer Taylor flow, what happens can be counter to intuition: that is, one may expect the flow to become more stable as the amount of the more viscous or stabilizing fluid is increased, and to become more unstable as the amount of the less viscous fluid is increased. However, the onset of Taylor instability in one-fluid flow is promoted by the addition of a thin layer of a more viscous fluid next to the inner cylinder and delayed by the addition of a thin layer of a less viscous fluid there. A 'thin-layer effect' occurs in water-lubricated pipelines discussed in chapter V.

Why does the viscosity stratification play such a crucial role in the linear stability of thin layers? Renardy [1987a] answered this question for two-layer Couette flow of similar liquids by obtaining the interfacial eigenvalue in closed form and investigating the thin-layer asymptotics of that eigenvalue. The analysis is rather lengthy and will not be presented here. The method is like the one used in chapter III for similar liquids. The two fluids are assumed to have mechanical properties which differ only by a small amount, say of $O(\epsilon)$, and the surface tension is also of $O(\epsilon)$. It is shown that the dominant term involves the viscosity difference: it is stabilizing if the thin layer is the less viscous and destabilizing otherwise. Terms involving surface tension and density difference enter at next order in the thickness of the layer. This indicates the importance of viscosity stratification in the stability of thin layers to disturbances that are not too short compared with the thickness of the thin layer.

IV.6(b) Asymptotic Analysis for High Reynolds Numbers

Consider the semi-bounded two-layer Couette flow of figure 4.3(c). The lower fluid has viscosity μ_1 and density ρ_1, and has depth d_1. The surface tension is T. In this section, d_1 will be denoted by d. The upper fluid is unbounded and has viscosity μ_2 and density ρ_2. The kinematic viscosities of the fluids are ν_i. The shear rate in fluid i is a_i. We may think of the shear rate in the lower fluid a_1 as U_0/d, where U_0 is defined in section IV.4(c) as the speed of the lower wall in the negative X-direction. The linear stability of this flow was studied for all values of the Reynolds number and wavelength by Hooper and Boyd [1987], using asymptotic and numerical methods. We reproduce their section 3.1 concerning high Reynolds numbers here, using their notation.

The equations of motion are written in dimensionless form using the scales d, the distance from the wall to the interface, and a_1, the shear rate of the lower fluid. The dimensionless numbers are the Reynolds number of the lower fluid,

$$R = \frac{a_1 d^2}{\nu_1},$$ (6b.1)

the viscosity ratio,

$$m = \frac{\mu_2}{\mu_1},$$ (6b.2)

the density ratio,

$$r = \frac{\rho_2}{\rho_1},$$ (6b.3)

the dimensionless surface tension parameter,

$$S = \frac{T}{\rho_1 a_1^2 d^3},$$ (6b.4)

and

$$F = \frac{(1-r)g}{a_1^2 d}.$$ (6b.5)

The dimensionless surface tension parameter of the unbounded problem of Hooper and Boyd [1983] is related to S by

$$S_{HB} = S R^{3/2}.$$ (6b.6)

Similarly,

$$F_{HB} = FR^{1/2}. \tag{6b.7}$$

The parameters S_{HB} and F_{HB} are independent of the depth of the lower fluid.

These parameters are like those used in equations (4c.1) - (4c.5). As in equation (4c.6), dimensionless coordinates are defined by

$$(x, y) = (X, Y)/d, \tag{6b.8}$$

The basic flow is given by equation (4c.8).

Disturbances are assumed to be proportional to $\exp(i\alpha(x - ct))$. The growth rate of the disturbance is given by $Im\ \alpha c$. The streamfunction satisfies equations (4c.9) - (4c.17).

There are three important length scales: d, the distance from the wall to the interface; λ, the wavelength of the disturbance; and l a viscous length-scale of the disturbance. The viscous lengthscale l equals $(\lambda\nu_1/a_1)^{1/3}$. This lengthscale arises naturally in the linearized equations for the disturbance.

Corresponding to these three lengthscales are three dimensionless ratios:

$$\alpha = 2\pi\frac{d}{\lambda}, \tag{6b.9}$$

$$(\alpha R)^{1/3} = (2\pi)^{1/3}\frac{d}{l}, \tag{6b.10}$$

and

$$\beta = (2\pi)^{2/3}\frac{l}{\lambda} = \frac{\alpha}{(\alpha R)^{1/3}}. \tag{6b.11}$$

Here, R is the Reynolds number of the lower fluid and $\beta^{3/2}$ is the wavenumber used in the shortwave analysis of Hooper and Boyd [1983] (see section IV. 5), and is denoted by α in their paper. It is a dimensionless wavenumber measured on a viscous lengthscale. The first two of the above parameters are of course widely used in hydrodynamic stability calculations (see, for example, Drazin and Reid [1981], chapter 4) but the importance of β seems not to have been widely appreciated.

The nature of the instability for the flow configuration of figure 4.3 is determined primarily by the magnitude of the parameters $(\alpha R)^{1/3}$ and β. $\beta >> 1$ corresponds to the shortwave instability of section IV.5. The asymptotic analysis for large Reynolds number of Blennerhassett and Smith [1987] is relevant for this case. Small β and small $(\alpha R)^{1/3}$ corresponds to the long-wave instability of section IV.4(c). When $(\alpha R)^{1/3} >> 1$ and $\beta << 1$, there is a new instability due to a viscous boundary layer at the wall. When the fluids have the same density, the growth rate of this instability is determined by the ratio of the viscosities. If the unbounded fluid is also more viscous, then the flow is unstable and vice versa.

A Singular Perturbation Scheme for $(\alpha R)^{1/3} >> 1$ and $\beta << 1$. We follow section 3.1 of Hooper and Boyd [1987] for the case $(\alpha R)^{1/3} >> 1$ and

$\beta << 1$, for equal density and zero surface tension. Their paper also contains results for unequal density and non-zero surface tension, the derivation and numerical solution of the exact secular equation, the asymptotic analysis of the secular equation, the extension to channel flow configurations, and a study of the energy equation.

When $(\alpha R)^{1/3} >> 1$ and $\beta << 1$, the viscous length l is much less than the distance between the wall and the interface d and the wavelength of the disturbance λ. We can therefore argue that viscous forces are much less important than inertial forces and may be neglected except in small regions where viscous and inertial forces are of the same order. These regions are the viscous boundary layer at the wall, the viscous boundary layer at the interface and the critical layer where the velocity of the disturbance equals the basic flow velocity. The singular perturbation scheme for the case $r = 1$ and $S = 0$ is described below.

The viscous part of the Orr-Sommerfeld equation (left-hand sides of equations (4c.10)- (4c.11)) may be neglected except within the viscous boundary layers at the wall and interface and the critical layer. Thus, outside these layers, the eigenfunctions ϕ_1 and ϕ_2 both satisfy

$$\left(\frac{d^2}{dy^2} - \alpha^2\right)\phi = 0 \qquad (6b.12)$$

Hence

$$\phi_1^{(O)} = b_{00}e^{-\alpha y} + d_{00}e^{\alpha y}, \qquad (6b.13)$$

and, since ϕ_2 is bounded as $y \to \infty$,

$$\phi_2^{(O)} = a_{00}e^{-\alpha y}. \qquad (6b.14)$$

The superscript (O) denotes the 'outer' solution.

Fluid 1 is bounded by a wall at $y = -1$ and hence ϕ_1 contains an inner solution $\phi_1^{(W)}$, valid within the viscous boundary layer near the wall. The boundary-layer thickness is determined by balancing the highest derivative in the viscous term on the left of equation (4c.10) with the highest derivative on the right hand side, and is $O(\alpha R)^{-1/2}$ for large Reynolds numbers. To find this inner solution, we transform equation (4c.10) by changing the variable to

$$z = (\alpha R)^{1/2}(1 + y). \qquad (6b.15)$$

Then (4c.10) becomes

$$\frac{d^2 w_1}{dz^2} = \left(-i(1 + c) + i\frac{z}{(\alpha R)^{1/2}} + \frac{\alpha^2}{\alpha R}\right)w_1, \qquad (6b.16)$$

where

$$w_1 = \left(\frac{d^2}{dz^2} - \frac{\alpha^2}{\alpha R}\right)\phi_1^{(W)}. \qquad (6b.17)$$

The form of equations (6b.16)- (6b.17) suggests an expansion for w_1, $\phi_1^{(W)}$ and c of the form

$$w_1 = \sum_{n=0}^{\infty} \left(\frac{1}{(\alpha R)^{1/2}}\right)^n w_{1n}, \tag{6b.18}$$

$$\phi_1^{(W)} = \sum_{n=0}^{\infty} \left(\frac{1}{(\alpha R)^{1/2}}\right)^n \phi_{1n}^{(W)}, \tag{6b.19}$$

$$c = \sum_{n=0}^{\infty} \left(\frac{1}{(\alpha R)^{1/2}}\right)^n c_n \tag{6b.20}$$

w_{1n} satisfies

$$\left(\frac{d^2}{dz^2} + i(1 + c_0)\right) w_{1n} = iz w_{1(n-1)} - i \sum_{j=1}^{n} c_j w_{1(n-j)} + \alpha^2 w_{1(n-2)}, \tag{6b.21}$$

w_{1j} is identically zero when j is negative, and $\phi_{1n}^{(W)}$ satisfies

$$\left(\frac{d^2}{dz^2} - \frac{\alpha^2}{\alpha R}\right) \phi_{1n}^{(W)} = w_{1n}. \tag{6b.22}$$

We are thus able to find a series expansion in $(\alpha R)^{-1/2}$ for $\phi_1^{(W)}$ which is valid within the viscous boundary layer at the wall. We match this solution to the inviscid solution (outer solution) for ϕ_1 (equations (6b.13)-(6b.14)) and in this way generate a series expansion in powers of $(\alpha R)^{-1/2}$ for the outer solution $\phi_1^{(O)}$ (see equation (6b.51)). We assume that the critical layer (where the fluid speed is equal to the wave speed) is not close to either the viscous boundary layer at the wall or the viscous boundary layer at the interface. This assumption is verified _a posteriori_ in equation (6b.83). It can then be shown that the eigenfunction ϕ has the same form inside and outside the critical layer since $U'' = 0$. Therefore, the presence of the critical layer, whether it occurs within fluid 1 or fluid 2, can be ignored.

We find next a series expansion in $(\alpha R)^{-1/2}$ for the eigenfunctions ϕ_j, $j = 1, 2$, within the viscous boundary layer at the interface and match the inner interfacial viscous solution $\phi_j^{(i)}$ to the corresponding outer inviscid solution $\phi_j^{(O)}$, $j = 1, 2$. The eigenfunctions $\phi_j^{(i)}$ are substituted into the four interfacial conditions at $y = 0$ (equations (4c.14) - (4c. 17)). We thus find a series expansion in $(\alpha R)^{-1/2}$ for the eigenvalue c (see equations (6b.83) - (6b.98)).

We first solve the boundary layer equations at the wall, equations (6b.21) - (6b.22), at leading order. This gives a series expansion in $(\alpha R)^{-1/2}$ for the outer inviscid solution $\phi_1^{(O)}$ which is valid up to $O(\alpha R)^{-1/2}$.

The boundary conditions at the interface show that the viscous boundary layer at the interface does not affect the disturbance until $O(\alpha R)^{-1}$. Therefore, we can determine c up to $O(\alpha R)^{-1/2}$ by substituting the inviscid eigenfunctions $\phi_j^{(O)}$, $j = 1, 2$ into continuity of normal velocity (4c.14) and continuity of normal stress (4c.17) at the interface.

The first two terms in the expansion for the eigenvalue c are given in equations (6b.83) - (6b.89). To determine the higher-order terms of c, however, we must determine the eigenfunctions ϕ_j valid within the interfacial viscous boundary layer. The eigenvalue c then satisfies equation (6b.98).

The leading-order term for $\phi_1^{(W)}$ is found from equations (6b.21) - (6b.22) and satisfies

$$\frac{d^2 w_{10}}{dz^2} + i(1 + c_0)w_{10} = 0, \quad \left(\frac{d^2}{dz^2} - \frac{\alpha^2}{\alpha R}\right)\phi_{10}^{(W)} = w_{10}.$$

Hence

$$\phi_{10}^{(W)} = d_{10}\exp[\alpha z/(\alpha R)^{1/2}] + b_{10}\exp[-\alpha z/(\alpha R)^{1/2}] + d_{1i}e^{pz} + b_{1i}e^{-pz},$$
$$(6b.23)$$

where

$$p = e^{-i\pi/4}(1 + c_0)^{1/2}, \tag{6b.24}$$

with Re p taken to be non-negative. We require that this leading-order solution match the outer solution in the overlap region when $1 + y << 1$ but $(\alpha R)^{1/2}(1 + y) >> 1$. In this region both the inner wall solution of equation (6b.23) and the outer inviscid solution of equation (6b.13) are valid. The outer solution may be expressed in terms of $1 + y$ by

$$\phi_1^{(O)} = b_{00}e^{\alpha}e^{-\alpha(1+y)} + d_{00}e^{-\alpha}e^{\alpha(1+y)},$$

$$b_{00}e^{\alpha} = \sum_{n=0}^{\infty}\left(\frac{1}{(\alpha R)^{1/2}}\right)^n b_{00n}, \quad d_{00}e^{-\alpha} = \sum_{n=0}^{\infty}\left(\frac{1}{(\alpha R)^{1/2}}\right)^n d_{00n}, \quad (6b.25)$$

and the boundary-layer solution by

$$\phi_{10}^{(W)} = d_{10}e^{\alpha(1+y)} + b_{10}e^{-\alpha(1+y)} + d_{1i}e^{p\sqrt{\alpha R}(1+y)} + b_{1i}e^{-p\sqrt{\alpha R}(1+y)}.$$

In the overlap region, $1 + y << 1$ and $(\alpha R)^{1/2}(1 + y) >> 1$. We may put

$$d_{1i} = 0, \tag{6b.26}$$

because there is nothing in $\phi_1^{(O)}$ to match the exponentially growing term. Also,

$$b_{000} = b_{10}, \tag{6b.27}$$

and

$$d_{000} = d_{10}. \tag{6b.28}$$

The term b_{1i} is indeterminate at this stage because it is exponentially small in the overlap region.

The no-slip and no-flux boundary conditions at the wall, where $z = 0$, are

$$\phi_{10}^{(W)}(0) = 0, \quad \frac{d\phi_{10}^{(W)}(0)}{dz} = 0, \tag{6b.29}$$

where $\phi_{10}^{(W)}$ is given by (6b.23). Hence

$$d_{10} = \frac{1}{2}b_{1i}\frac{(\alpha R)^{1/2}p}{\alpha}\left(1 - \frac{\alpha}{(\alpha R)^{1/2}p}\right) \tag{6b.30}$$

and

$$b_{10} = -\frac{1}{2}b_{1i}\frac{(\alpha R)^{1/2}p}{\alpha}\left(1 + \frac{\alpha}{(\alpha R)^{1/2}p}\right). \tag{6b.31}$$

After substituting equations (6b.26) - (6b.31) into (6b.13) we find that the outer solution for ϕ_1 may be expressed in terms of only one unknown constant and that at leading order $\phi_1^{(O)}$ satisfies

$$\phi_1^{(O)} = \bar{b}_0\left(\sinh\alpha(1+y) - \frac{\alpha}{(\alpha R)^{1/2}p}\cosh\alpha(1+y)\right) \tag{6b.32}$$

where

$$\bar{b}_0 = b_{1i}\frac{(\alpha R)^{1/2}p}{\alpha}. \tag{6b.33}$$

The next-order of approximation for $\phi_1^{(O)}$ is found from $\phi_1^{(W)}$ at order $O(\alpha R)^{-1/2}$. The equations at this order are (6b.22) with $n = 1$

$$\left(\frac{d^2}{dz^2} + i(1+c_0)\right)w_{11} = i(z - c_1)b_{1i}\left(p^2 - \frac{\alpha^2}{\alpha R}\right)e^{-pz}, \tag{6b.34}$$

where (6b.34) is obtained from (6b.21)-(6b.23). A particular solution of (6b.34) is in the form $(Cz + Dz^2)\exp(-pz)$ so that

$$w_{11} = f_1e^{-pz} + f_2e^{pz} + (Cz + Dz^2)e^{-pz}, \tag{6b.35}$$

where

$$C = \frac{(-i)}{2p}b_{1i}(p^2 - \frac{\alpha^2}{\alpha R})(\frac{1}{2p} - c_1), \quad D = \frac{(-i)}{4p}b_{1i}(p^2 - \frac{\alpha^2}{\alpha R}), \tag{6b.36}$$

and f_1, f_2 and b_{1i} are undetermined constants.

Matching (6b.35) to an outer solution as in (6b.25) is possible provided that

$$f_2 = 0, \tag{6b.37}$$

Equation (6b.22) then reduces to

$$\left(\frac{d^2}{dz^2} - \frac{\alpha^2}{\alpha R}\right)\phi_{11}^{(W)} = (f_1 + Cz + Dz^2)e^{-pz}, \tag{6b.38}$$

so that a particular solution for $\phi_{11}^{(W)}$ has the form

$$(Ez^2 + Fz + G)e^{-pz}, \tag{6b.39}$$

where

$$E = \frac{(-i)}{4p}b_{1i}, \quad F = \frac{(-i)}{2p}b_{1i}(\frac{1}{2p} - c_1) + \frac{(-i)}{(p^2 - \frac{\alpha^2}{\alpha R})}b_{1i},$$

$$G = \frac{(-i)}{(p^2 - \frac{\alpha^2}{\alpha R})}b_{1i}(-c_1) + \frac{(-i)2p}{(p^2 - \frac{\alpha^2}{\alpha R})^2}b_{1i} + \frac{f_1}{(p^2 - \frac{\alpha^2}{\alpha R})}. \tag{6b.40}$$

When αR is large,

$$F = \frac{(-i)}{2p}b_{1i}(\frac{1}{2p} - c_1) + \frac{(-i)}{p^2}b_{1i}(1 + O(\alpha R)^{-1})$$

$$\sim \frac{(-i)b_{1i}}{p^2}(\frac{5}{4} - \frac{c_1 p}{2}), \tag{6b.41}$$

$$G \sim \frac{(-i)}{p^2}b_{1i}(-c_1) + \frac{(-i)2p}{p^4}b_{1i} + \frac{f_1}{p^2}$$

$$= \frac{(-i)b_{1i}2}{p^3}(1 - \frac{c_1 p}{2}) + \frac{f_1}{p^2}. \tag{6b.42}$$

Finally,

$$\phi_{11}^{(W)} = d_{11}\exp[\alpha z/(\alpha R)^{1/2}] + b_{11}\exp[-\alpha z/(\alpha R)^{1/2}] + e^{-pz}\frac{f_1}{p^2}$$

$$+ b_{1i}e^{-pz}(-i)\left(\frac{1}{4p}z^2 + \frac{1}{p^2}\left(\frac{5}{4} - \frac{c_1 p}{2}\right)z + \frac{2}{p^3}\left(1 - \frac{c_1 p}{2}\right)\right), \tag{6b.43}$$

where b_{1i} is the undetermined constant in $\phi_{10}^{(W)}$ and f_1 is an undetermined constant from $\phi_{11}^{(W)}$. The constants d_{11} and b_{11} are determined in terms of b_{1i} and f_1 from the boundary conditions (6b.29) which hold at $y = -1$. Thus

$$b_{11} = \frac{1}{2}\left(b_{1i}(-i)\left[\frac{-2}{p^3}(1 - \frac{c_1 p}{2}) + \frac{\sqrt{\alpha R}}{\alpha p^2}(\frac{-3}{4} + \frac{c_1 p}{2})\right] - \frac{f_1}{p}(\frac{\sqrt{\alpha R}}{\alpha} + \frac{1}{p})\right), \tag{6b.44}$$

$$d_{11} = \frac{1}{2}\left(-b_{1i}(-i)\left[\frac{\sqrt{\alpha R}}{\alpha p^2}(\frac{-3}{4} + \frac{c_1 p}{2}) + \frac{2}{p^3}(1 - \frac{c_1 p}{2})\right] + \frac{f_1}{p}(\frac{\sqrt{\alpha R}}{\alpha} - \frac{1}{p})\right). \tag{6b.45}$$

In the overlap region, the terms in equation (6b.43) proportional to $\exp(-pz)$ are exponentially small. The other terms are proportional to $\exp(\pm\alpha(y+1))$ and these are matched to the outer solution, given by equation (6b.25):

$$\phi_1^{(O)} \sim (b_{000} + \frac{b_{001}}{\sqrt{\alpha R}})e^{-\alpha(y+1)} + (d_{000} + \frac{d_{001}}{\sqrt{\alpha R}})e^{\alpha(y+1)}. \tag{6b.46}$$

The matching yields

$$b_{001} = b_{11}, \quad d_{001} = d_{11}, \tag{6b.47}$$

given by (6b.44) - (6b.45). Therefore,

$$\phi_1^{(O)} \sim \sinh\alpha(y+1)\left(\left(\bar{b}_0 + \frac{f_1}{\alpha p}\right) - \frac{\bar{b}_0 i}{\sqrt{\alpha R}p^3}\left(\frac{3}{4} - \frac{c_1 p}{2}\right)\right)$$

$$-\alpha\cosh\alpha(y+1)\left(\frac{1}{\sqrt{\alpha R}p}\left(\bar{b}_0 + \frac{f_1}{\alpha p}\right) - \frac{i\bar{b}_0}{\alpha R p^4}(2 - c_1 p)\right), \qquad (6b.48)$$

where $\bar{b}_0$ is defined in equation (6b.33). The undetermined constants b_{1i}, f_1,... and others arise at each stage of approximation, from a solution of the homogeneous part of the differential equation (6b.21). These constants appear in a multiplicative constant in $\phi_1^{(O)}$ which could be determined by a normalization condition as in (4a.7). One implicit normalization condition is implied by putting $f_1,\ldots$ to zero. Then

$$\bar{b}_0 + \frac{f_1}{\alpha p} \rightarrow \bar{b}_0 \qquad (6b.49)$$

gives rise to

$$\phi_1^{(O)} = \bar{b}_0\left(\sinh\alpha(1+y)\left(1 - \frac{i}{(\alpha R)^{1/2}p^3}\left(\frac{3}{4} - \frac{1}{2}c_1 p\right)\right)\right.$$

$$\left. - \frac{\alpha}{(\alpha R)^{1/2}p}\cosh\alpha(1+y)\left(1 - \frac{i}{(\alpha R)^{1/2}p^3}(2 - c_1 p)\right)\right). \qquad (6b.50)$$

This can be more conveniently written in the form

$$\phi_1^{(O)} = b_0\left(\sinh\alpha(1+y)\right.$$

$$\left. +\alpha\cosh\alpha(1+y)\left(-\frac{1}{(\alpha R)^{1/2}p} + \frac{i}{\alpha R p^4}\left(\frac{5}{4} - \frac{1}{2}c_1 p\right)\right) + O(\alpha R)^{-3/2}\right), \qquad (6b.51)$$

where

$$b_0 = \bar{b}_0\left(1 - \frac{i}{(\alpha R)^{1/2}p^3}\left(\frac{3}{4} - \frac{1}{2}c_1 p\right) + O(\alpha R)^{-1}\right).$$

Thus, the expansion in powers of $(\alpha R)^{-1/2}$ for ϕ_1 within the viscous boundary layer at the wall leads to another expansion in powers of $(\alpha R)^{-1/2}$ for ϕ_1 in the outer inviscid region. Similarly, we see that in fluid 2, the outer inviscid solution for ϕ_2 has the form

$$\phi_2^{(O)} = \sum_{j=0}^{\infty} a_{0j}(\alpha R)^{-j/2}e^{-\alpha y}. \qquad (6b.52)$$

The value of c is found from the interface conditions at $y = 0$. However, there is a viscous boundary layer at the interface, so that in order to implement the interface conditions, we must find the boundary-layer solution there.

To find a solution for ϕ_1 and ϕ_2 valid at the interface, we change variables in the Orr - Sommerfeld equations (4c.10) - (4c.11), with r set to 1,

$$u = \sqrt{\alpha R}y, \tag{6b.53}$$

$$\frac{d^2w_1}{du^2} = \left[-ic + \frac{iu}{\sqrt{\alpha R}} + \frac{\alpha^2}{\alpha R}\right]w_1, \tag{6b.54}$$

and

$$\frac{d^2w_2}{du^2} = \left[\frac{-ic}{m} + \frac{iu}{\sqrt{\alpha R m^2}} + \frac{\alpha^2}{\alpha R}\right]w_2. \tag{6b.55}$$

The boundary layer solutions $\phi_j^{(i)}$ then satisfy

$$\left(\frac{d^2}{du^2} - \frac{\alpha^2}{\alpha R}\right)\phi_j^{(i)} = w_j \quad (j = 1, 2). \tag{6b.56}$$

We look for solutions to equations (6b.54) - (6b.56) of the form

$$\phi_j^{(i)} = \sum_{n=0}^{\infty} (\alpha R)^{-n/2}\phi_{jn}^{(i)} \quad (j = 1, 2) \tag{6b.57}$$

and

$$w_j = \sum_{n=0}^{\infty} (\alpha R)^{-n/2}w_{jn} \quad (j = 1, 2), \tag{6b.58}$$

where c is expanded as in (6b.20). The w_{1n} satisfy

$$\left(\frac{d^2}{du^2} + ic_0\right)w_{1n} = (-ic_1 + iu)w_{1(n-1)} + (-ic_2 + \alpha^2)w_{1(n-2)} - i\sum_{j=0}^{n-3} c_{n-j}w_{1j}, \tag{6b.59}$$

and the w_{2n} satisfy

$$\left(\frac{d^2}{du^2} + \frac{ic_0}{m}\right)w_{2n} = \left(\frac{-ic_1}{m} + \frac{iu}{m^2}\right)w_{2(n-1)}$$

$$+\left(\frac{-ic_2}{m} + \alpha^2\right)w_{2(n-2)} - i\sum_{j=0}^{n-3} \frac{c_{n-j}w_{2j}}{m}. \tag{6b.60}$$

We define

$$q^2 = -ic_0 \tag{6b.61}$$

with Re q taken to be non-negative, so that the solutions to the homogeneous part of equations (6b.59) - (6b.60) are $\exp(\pm qu)$ and $\exp(\pm qu/\sqrt{m})$, respectively. For w_1, we have $u < 0$, so that the solution $\exp(-qu)$ is discarded, since it is exponentially large in the overlap region and does not match to an outer solution. Similarly, the solution $\exp(qu/\sqrt{m})$ is discarded in w_2.

The solution in fluid 1 at zeroth order is

$$\phi_{10}^{(i)} = g_{i0}e^{qu} + A_0 \cosh(\frac{\alpha u}{\sqrt{\alpha R}} + \alpha) + B_0 \sinh(\frac{\alpha u}{\sqrt{\alpha R}} + \alpha) \qquad (6b.62)$$

where g_{i0}, A_0 and B_0 are undetermined constants. The latter two are determined by matching (6b.62) to the outer solution (6b.51) in the overlap region; they correspond to the coefficients of $\cosh \alpha(y+1)$ and $\sinh \alpha(y+1)$ in (6b.51). Combining the zeroth and first order solutions, we find that

$$\phi_1^{(i)} \sim \phi_1^{(O)} + (g_{i0} + \frac{g_{i1}}{\sqrt{\alpha R}})e^{qu}$$

$$+ig_{i0}\frac{e^{qu}}{\sqrt{\alpha R}}\left(\frac{2+c_1q}{q^3} - \frac{u}{q^2}(\frac{5}{4} + \frac{c_1q}{2}) + \frac{u^2}{4q}\right). \qquad (6b.63)$$

The solution for fluid 2 at zeroth order is

$$\phi_{20}^{(i)} = h_{i0}e^{-q\sqrt{\alpha R}y/\sqrt{m}} + D_0e^{\alpha y} + E_0e^{-\alpha y}, \qquad (6b.64)$$

where h_{i0} is an undetermined constant, and D_0 and E_0 are determined by matching to the outer solution $\phi_2^{(O)}$ of equation (6b.52) in the overlap region. Thus,

$$E_0 = a_{00}, \quad D_0 = 0. \qquad (6b.65)$$

Higher order solutions can be found in the same way. The combination of zeroth and first order solutions given by (6b.63) and

$$\phi_2^{(i)} \sim \phi_2^{(O)} + e^{-qu/\sqrt{m}}(h_{i0} + \frac{h_{i1}}{\sqrt{\alpha R}})$$

$$+\frac{ih_{i0}}{mq\sqrt{\alpha R}}e^{-qu/\sqrt{m}}\left(\frac{\sqrt{m}}{q^2}(\sqrt{m}qc_1-2)+\frac{u}{q}(\frac{-5}{4}+\frac{\sqrt{m}qc_1}{2})-\frac{u^2}{4\sqrt{m}}\right) \quad (6b.66)$$

are sufficient to determine c_0.

Next, the interface conditions (4c.14) - (4c.17) of section IV.4(c), with $r = 1$, $S = 0$ and $F = 0$, must be written in terms of the boundary layer variable u. The conditions at $u = 0$ (or $y = 0$) are

$$\phi_1 = \phi_2 = \phi(u = 0), \qquad (6b.67)$$

$$c(\frac{d\phi_1}{du} - \frac{d\phi_2}{du}) = \frac{1}{\sqrt{\alpha R}}\phi(u = 0)\frac{(1-m)}{m}, \qquad (6b.68)$$

$$\frac{d^2\phi_1}{du^2} - m\frac{d^2\phi_2}{du^2} = \frac{\alpha^2}{\alpha R}\phi_1(m-1), \qquad (6b.69)$$

and after making use of (6b.68)

$$(\frac{d^2}{du^2} - \frac{3\alpha^2}{\alpha R})\frac{d\phi_1}{du} = m(\frac{d^2}{du^2} - \frac{3\alpha^2}{\alpha R})\frac{d\phi_2}{du}. \qquad (6b.70)$$

The outer solution does not enter into equations (6b.68) - (6b.70). Equations (6b.67) - (6b.70) yield

$$b_0 \sinh \alpha + g_{i0} = a_{00} + h_{i0}, \tag{6b.71}$$

$$q g_{i0} = \frac{-q}{\sqrt{m}} h_{i0}, \tag{6b.72}$$

$$q^2 g_{i0} = q^2 h_{i0}, \tag{6b.73}$$

$$q^3 g_{i0} = \frac{-q^3}{\sqrt{m}} h_{i0}. \tag{6b.74}$$

Equations (6b.72) - (6b.74) combine to give

$$g_{i0} = h_{i0} = 0, \tag{6b.75}$$

and (6b.71) is

$$b_0 \sinh \alpha = a_{00}. \tag{6b.76}$$

At the next order, equations (6b.68) - (6b.70) give rise to

$$b_0 \alpha \cosh \alpha + q g_{i1} = -\alpha a_{00} - \frac{q}{\sqrt{m}} h_{i1} + \frac{a_{00}}{c_0} \frac{(1 - m)}{m}, \tag{6b.77}$$

$$q^2 g_{i1} = q^2 h_{i1}, \tag{6b.78}$$

and

$$q^3 g_{i1} = \frac{-q^3}{\sqrt{m}} h_{i1}. \tag{6b.79}$$

Hence

$$g_{i1} = h_{i1} = 0. \tag{6b.80}$$

It now follows that the inner solutions (6b.63) and (6b.66) reduce to

$$\phi_1^{(i)} = \phi_1^{(O)} + O(\alpha R)^{-1}, \tag{6b.81}$$

and

$$\phi_2^{(i)} = \phi_2^{(O)} + O(\alpha R)^{-1}. \tag{6b.82}$$

Equations (6b.76), (6b.77) and (6b.80) yield

$$c_0 = \frac{(1 - m)}{m} \left(\frac{1 - e^{-2\alpha}}{2\alpha} \right). \tag{6b.83}$$

Thus at leading order, the viscous portion of the disturbance at the interface is negligible and the first term in the series expansion for c, equation (6b.20), is found from the inviscid boundary conditions at the interface: continuity of normal velocity and stress. Since c_0 is real, we need to go to higher orders to find the growth rate, $Im \, \alpha c$.

At the next order of approximation, the calculations mirror those leading to equations (6b.63) and (6b.66). Thus,

$$\phi_1^{(i)} \sim \phi_1^{(O)} + \left(\frac{g_{i2}}{\alpha R} + \frac{g_{i3}}{(\alpha R)^{3/2}} \right) e^{qu}$$

$$+ig_{i2}\frac{e^{qu}}{(\alpha R)^{3/2}}\left(\frac{2+c_1 q}{q^3}-\frac{u}{q^2}(\frac{5}{4}+\frac{c_1 q}{2})+\frac{u^2}{4q}\right), \qquad (6b.84)$$

and

$$\phi_2^{(i)} \sim \phi_2^{(O)} + e^{-qu/\sqrt{m}}(\frac{h_{i2}}{\alpha R}+\frac{h_{i3}}{(\alpha R)^{3/2}})$$

$$+\frac{ih_{i2}}{mq(\alpha R)^{3/2}}e^{-qu/\sqrt{m}}\left(\frac{\sqrt{m}}{q^2}(\sqrt{m}qc_1-2)+\frac{u}{q}(\frac{-5}{4}+\frac{\sqrt{m}qc_1}{2})-\frac{u^2}{4\sqrt{m}}\right),$$
$$(6b.85)$$

where $\phi_1^{(O)}$ and $\phi_2^{(O)}$ are given by equations (6b.50) and (6b.52).

The interface condition (6b.67) yields

$$a_{01} = \frac{-b_0 \alpha}{p}\cosh \alpha. \qquad (6b.86)$$

The normal stress condition (6b.70) yields

$$g_{i2} = \frac{-h_{i2}}{\sqrt{m}} \qquad (6b.87)$$

and (6b.69) then implies that

$$h_{i2} = \frac{2b_0 \alpha^2 \sqrt{m}(1-m)\sinh \alpha}{q^2(1+\sqrt{m})}. \qquad (6b.88)$$

Using the results obtained so far, we evaluate from (6b.68)

$$c_1 = \frac{-(1-m)}{mp}e^{-2\alpha}$$

with p given by (6b.24). Thus,

$$c_1 = \frac{-(1-m)}{m}e^{-2\alpha}\frac{e^{i\pi/4}}{\sqrt{1+c_0}}, \qquad (6b.89)$$

where c_0 is given by (6b.83).

If $m < 1$, then $c_0 > 0$. Therefore, $Im\ c_1 < 0$ giving stability. If $m > 1$, then $(1+c_0) > 0$ for the following reason. The function $(1-e^{-x})/x$ for $x > 0$ has a maximum when $e^{-x} = 1/(1+x)$, and the maximum is $1/(1+x) = e^{-x} < 1$. Therefore, $(1-e^{-2\alpha})/(2\alpha) < 1$. Also, since $(1-m)/m > -1$ when $m > 1$, we have that $c_0 > -1$. Thus, if $m > 1$, then $Im\ c_1 > 0$ giving instability.

One may proceed in a similar fashion to calculate higher-order approximations to c. The interface conditions (6b.67) - (6b.70) yield

$$a_{02} = b_0\alpha\frac{i}{p^4}(\frac{5}{4}-\frac{c_1 p}{2})\cosh \alpha + g_{i2} - h_{i2}, \qquad (6b.90)$$

$$c_0 \left(b_0 \alpha^2 \frac{i}{p^4} \sinh \alpha (\frac{5}{4} - \frac{c_1 p}{2}) + q g_{i3} + i q g_{i2} \frac{(2 + c_1 q)}{q^3} - \frac{i g_{i2}}{q^2}(\frac{5}{4} + \frac{c_1 q}{2}) \right.$$

$$\left. + \alpha a_{02} + \frac{q}{\sqrt{m}} h_{i3} + \frac{i h_{i2}}{m q^2}(c_1 q \sqrt{m} - 2) - \frac{i h_{i2}}{m q^2}(\frac{-5}{4} + \frac{c_1 q \sqrt{m}}{2}) \right)$$

$$+ c_1 \left(-b_0 \frac{\alpha^2}{p} \sinh \alpha + q g_{i2} + \alpha a_{01} + \frac{q}{\sqrt{m}} h_{i2} \right) + c_2 \left(b_0 \alpha \cosh \alpha + \alpha a_{00} \right)$$

$$= \frac{(1 - m)}{m} (a_{02} + h_{i2}), \qquad (6b.91)$$

$$-b_0 \frac{\alpha^3}{p} \cosh \alpha + q^2 g_{i3} + i g_{i2} \frac{(2 + c_1 q)}{q} - 2i \frac{g_{i2}}{q}(\frac{5}{4} + \frac{c_1 q}{2}) + \frac{i g_{i2}}{2q}$$

$$-m \left(\alpha^2 a_{01} + \frac{q^2}{m} h_{i3} + \frac{i h_{i2} \sqrt{m}}{m^2 q}(c_1 q \sqrt{m} - 2) \right.$$

$$\left. - \frac{2 i h_{i2}}{m q \sqrt{m}}(\frac{-5}{4} + \frac{c_1 q \sqrt{m}}{2}) - \frac{i h_{i2}}{2 m q \sqrt{m}} \right)$$

$$= \alpha^2 (m - 1) a_{01}, \qquad (6b.92)$$

and

$$b_0 \alpha^3 \cosh \alpha + q^3 g_{i3} + i g_{i2}(2 + c_1 q) - 3 i g_{i2}(\frac{5}{4} + \frac{c_1 q}{2}) + \frac{3 i g_{i2}}{2}$$

$$-m \left(-\alpha^3 a_{00} - \frac{q^3}{m \sqrt{m}} h_{i3} - \frac{i h_{i2}}{m^2}(c_1 q \sqrt{m} - 2) + \frac{3 i h_{i2}}{m^2}(\frac{-5}{4} + \frac{c_1 q \sqrt{m}}{2}) + \frac{3 i h_{i2}}{2 m^2} \right)$$

$$= 3 \alpha^2 (b_0 \alpha \cosh \alpha + m \alpha a_{00}). \qquad (6b.93)$$

Equation (6b.92) yields

$$g_{i3} - h_{i3} = (1 - m) \frac{2 b_0 \alpha^3 \cosh \alpha}{p q^2}. \qquad (6b.94)$$

Equation (6b.93) yields

$$g_{i3} + \frac{h_{i3}}{\sqrt{m}} = \frac{\alpha^3 b_0}{q^3} \left(e^\alpha (1 + m + \frac{\sqrt{m}}{2}) + e^{-\alpha}(1 - m) \right). \qquad (6b.95)$$

In the above equations, terms involving g_{i2}, g_{i3}, h_{i2} and h_{i3} arise from the viscous solution, and terms involving cosh and sinh arise from the outer inviscid solution. We denote by c_{2i} the contribution to c_2 from the the inviscid solution, and by c_{2v} the contribution from the viscous solution. Then,

$$c_{2i} = -i \frac{(1 - m)}{m} \frac{e^{-2\alpha}}{(1 + c_0)^2} \left(\alpha(1 + c_0) + \frac{5}{4} + \frac{(1 - m)}{2m} e^{-2\alpha} \right), \qquad (6b.96)$$

and

$$c_{2v} = -2i\alpha^2 \Big((1+m) + e^{-2\alpha}(1-m)\Big). \tag{6b.97}$$

We thus have found an expansion for c of the form

$$c = c_0 + \frac{1}{(\alpha R)^{1/2}}c_1 + \frac{1}{\alpha R}(c_{2i} + c_{2v}) + ..., \tag{6b.98}$$

where c_0 is given in (6b.83), c_1 in (6b.89), and c_{2i} and c_{2v} in (6b.96) - (6b.97).

This series expansion for c has been checked numerically by Hooper and Boyd [1987]. Exact secular equations for c for the semi-bounded problem are derived in their section 4 and solved using both numerical and asymptotic techniques. The same asymptotic expansion for c is found when $\beta << 1$ and $(\alpha R)^{1/3} >> 1$ and the expansion is shown to agree with numerical results (see their figure 4).

The terms c_0, c_1 and c_{2i} all arise from the expansion for ϕ in the inviscid region and hence are due to the disturbance vorticity created at the wall. The term c_0 is real and therefore does not affect the stability of the flow. The imaginary part of both c_1 and c_{2i} is positive when $m > 1$ which implies that the viscous boundary layer at the wall can be a destabilizing influence on the flow.

The term c_{2v} arises from the expansion for ϕ in the viscous boundary layer near the interface. This term is purely imaginary and always negative. Therefore, when $\beta << 1$ and $(\alpha R)^{1/3} >> 1$, the viscous boundary layer at the interface has a stabilizing effect on the flow.

As the separation between the solid boundary and the interface increases with other parameters fixed, the destabilizing effect of the viscous boundary layer at the wall is reduced and the stability of the flow is determined by the viscous boundary layer at the interface. Neutral stability results when the destabilizing effect of the wall (c_1 being the dominant term) balances the stabilizing effect of the interface (expressed by c_{2v}). Therefore, when $(\alpha R)^{1/3} >> 1$ and $m > 1$, we get neutral stability whenever $Im\,(c_1/\sqrt{\alpha R} + c_{2v}/\alpha R) = 0$; that is, when

$$\Big(\frac{m-1}{m}\Big)\frac{e^{-2\alpha}}{(2(1+c_0)\alpha R)^{1/2}} = \frac{2\alpha^2}{\alpha R}\Big((1+m) + (1-m)e^{-2\alpha}\Big). \tag{6b.99}$$

Since $(\alpha R)^{1/3} >> 1$, $\alpha^2/\sqrt{\alpha R}$ is small. This implies that $\alpha << R^{1/3}$ consistent with the assumption that $\beta << 1$, where β is defined in equation (6b.11). The magnitude of the left hand side of (6b.99), $e^{-2\alpha}$, must also be small. This implies that α is large. Thus, the balance in equation (6b.99) occurs when α is large, but still smaller than R, and R is large. The log of equation (6b.99) when $m > 1$ shows that there is a branch of the neutral stability curve for which

$$\alpha \sim \frac{1}{4}lnR \qquad \text{as } (\alpha R)^{1/3} \to \infty. \tag{6b.100}$$

Numerical results by Hooper and Boyd [1987] show that the appropriate branch of the neutral stability curve approaches this asymptotic behavior when $(\alpha R)^{1/3} \geq 20$ approximately. Their figure 3 shows that when $R = 900$, $(\alpha R)^{1/3} = 7.86$ and this is not yet in the asymptotic regime. They do not show figures that extend into the asymptotic regime.

Blennerhassett and Smith [1987] have done an asymptotic analysis of the upper branch of the neutral case shown in figure 7.1, where α tends to infinity with R. They seek an explanation of water waves generated by wind as a linear instability of a two-layer flow. The velocity profiles can be of a general form: their analysis is localized at the interface. They fix the fluid properties and seek the form of the neutral stability curve for large Reynolds number. They find a range of parameters for which the neutral stability curve is in the short-wave regime. They suggest that their results describe the very short waves seen when a puff of wind blows over an otherwise smooth water surface. The wind produces a patch of small waves called 'cat's paws', which appear to move at the surface speed. They consider a regime in which the surface tension term balances the viscous term, and the gravity term is negligible in (5.44). Here, the viscous term must be positive, so that

$$(1 - m)(1 - \frac{m^2}{r}) > 0.$$

The balance required for this in (5.44) implies that

$$\alpha^3 S/\Re = O(1),$$

for large $\Re$. However, the way we have made S dimensionless in equation (2.5) means that S depends on the Reynolds number as $1/\Re$ (keeping S^* fixed). Therefore, $\alpha^3/\Re^2 = O(1)$, or

$$\alpha = O(\Re^{2/3}), \quad \Re \to \infty.$$

This is equation (1.1) of their paper. They find a self-consistent set of equations for this scaling. For neutral stability, c_2 in equation (5.44) is thus set to zero and they analyze a perturbation around this situation. Their short waves are thus driven by surface tension effects combined with the local basic-flow shear at the interface and with viscous diffusion. They also develop perturbation expansions for other regimes.

IV.7 Analysis of Stability in the General Case: Energy Equation

In the case of Couette flow of two fluids, there are sets of prescribed data for which the basic flow with a flat interface is stable to waves of all lengths at small Reynolds numbers. In these cases we expect to find a neutral curve of the usual kind (cf figure 7.1). In other cases, we have no stable flows; there are some unstable waves at every Reynolds number. In either case we may find valuable information about the instability from the linear theory using Rayleigh's idea that the disturbance with the maximum growth rate will determine the size of structures arising from instability.

Rayleigh's idea may be beautifully implemented using the energy equation for the most dangerous disturbances. This idea is used for core-annular flows studied in chapters VI - VIII and it has been done for the problem at hand by Nelson, Mohr and Joseph [1992]. They carried out this type of analysis for the problems first treated by Renardy [1985] in which the viscosity ratio $m = 0.5$ and volume ratio $l_1 = 1/3$ were chosen so that long waves would be stable. The suitable range for these parameters are found in figures 2(a) and 2(b) of Yih [1967], reproduced in this book in figures 4.1 and 4.2. The surface tension parameter $S^* = 0.75$ dyne/cm ($SR = 0.476$) was chosen so as to compare our figure 7.2 with figure 3(a) of Hooper and Boyd [1983] with $S_{HB} = 0.1$. Note that $R = U^* l^* / \nu_1$.

In figure 7.1 we have plotted neutral curves $\operatorname{Re} \sigma = \alpha c_i = 0$ for conditions specified above and in the caption. We propose to diagnose the instabilities in the unstable regions with the energy equation using the most dangerous mode.

The most dangerous mode is the eigenfunction belonging to the wave number $\alpha = \tilde{\alpha}$ which maximizes the growth rate $\alpha c_i = \operatorname{Re} \sigma(R, \alpha)$:

$$\operatorname{Re} \sigma(R, \tilde{\alpha}) = \frac{\max}{\alpha \geq 0} \operatorname{Re} \sigma(R, \alpha). \tag{7.1}$$

Squire's transformation shows that the most dangerous disturbance is two-dimensional; $v = 0$, $\partial/\partial y = 0$. The equations satisfied by $u(z)$ and $w(z)$ are (2.17), (2.19) and (2.20) together with interface conditions (2.22)-(2.28) and boundary conditions (2.29).

In figure 7.2, we have plotted the growth rate $\operatorname{Re} \sigma(R, \alpha)$ for the conditions in the caption of figure 7.1 and three different representative values of R. There are two relative maxima in all cases but the one on the right for larger α is the larger of the two. We can say that there are two unstable modes, the more unstable one for short waves on the right and the unstable one for longer waves on the left. There is a different cause for instability for each unstable mode as we shall see now.

Suppose u, w, p are eigenfunctions of the most dangerous mode, belonging to the eigenvalue σ satisfying (7.1). We may form an energy equation for this most dangerous disturbance in the usual way (cf chapter VI) by

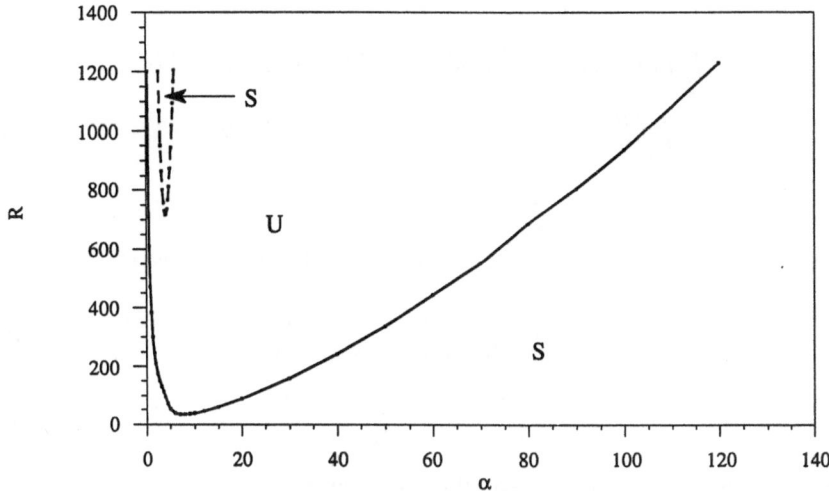

Fig. 7.1. Neutral stability curves for plane Couette flow of two fluids: $\mu_1 = 2P$, $\rho_1 = 1$ gm/cm^3, $S^* = 0.75$ dyn/cm, $m = 0.5$, $r = 1$ and $l_1 = 1/3$. $R = U^* l^* / \nu_1$, $R_c = 34.3$. There are two branches of neutral curves when R is above about 750.

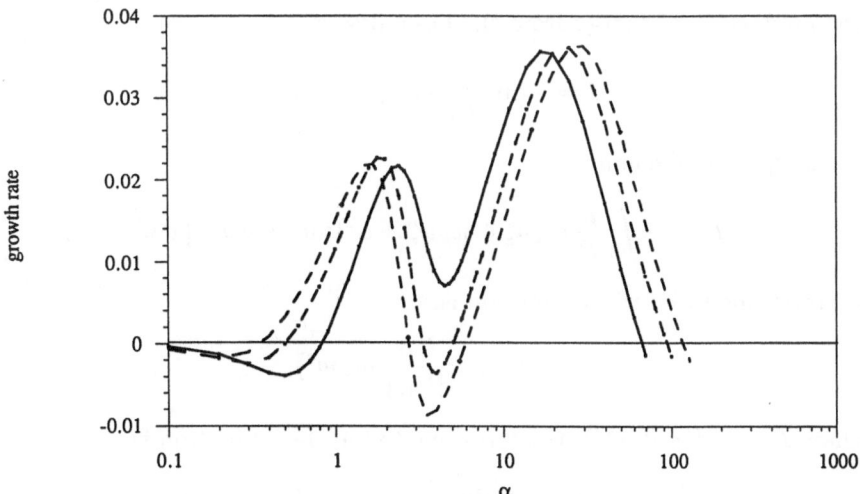

Fig. 7.2. Maximum growth rate versus wavenumber. —— $R = 514$, $- \blacksquare - R = 857$, $- \diamond - R = 1200$. There are two peaks at each R. The bigger one on the right is for short waves and it is caused by interfacial friction. The smaller one on the left is for longer waves and is produced by the Reynolds stress in the low viscosity fluid.

multiplying (2.17) by u^* and (2.19) by w^*, complex conjugates of u and w, and then integrating from $z = 0$ to $z = 1$. After integrating by parts we get some terms on the boundary. The resulting equation is (7.2) below

$$\int_\Omega \rho[\alpha \, (U_1(z) - c) \, (uu^* + ww^*) - iU_{1z}(z)wu^*]dz$$

$$= \int_\Omega \frac{i\mu}{R}[u_z u_z^* + w_z w_z^* + \alpha^2(uu^* + ww^*)]dz \qquad (7.2)$$

$$+ [ipw^*] - [\frac{i\mu}{R}(u^* u_z + w^* w_z)]$$

where [ ] represents the jump in the quantity at the interface of the fluids, and

$$\rho = \begin{cases} 1 & \text{for} \quad 0 < z < l_1 \\ 1/r & \text{for} \quad l_1 < z < 1 \end{cases} \qquad (7.3a)$$

$$\mu = \begin{cases} 1 & \text{for} \quad 0 < z < l_1 \\ 1/m & \text{for} \quad l_1 < z < 1 \end{cases} . \qquad (7.3b)$$

The imaginary part of (7.2) describes the energy:

$$\dot{E} = I + D + B_1 + B_2 + B_3 \qquad (7.4)$$

where

$$\dot{E} = \int_\Omega \rho\alpha \, \text{Im} \, c \, (uu^* + ww^*)dz.$$

The production of energy from the basic flow is

$$I = -\text{Im} \int_\Omega \rho U_{1z}(z)wu^* dz,$$

the dissipation of energy is

$$D = -\int_\Omega \frac{\mu}{R} \left[u_z u_z^* + w_z w_z^* + \alpha^2(uu^* + ww^*) \right] dz,$$

the term due to interfacial tension is

$$B_1 = -\text{Im} \, \frac{T\alpha}{[U_{1z}]}[w_z w^*],$$

where $T = S^*/(\rho_1 l^* U^{*2})$, the term due to interfacial friction is

$$B_2 = -\text{Im} \, [\frac{i\mu}{R}(w_z w^* - \frac{w_z^* w_{zz}}{\alpha^2})],$$

and the term due to interfacial buoyancy is

$$B_3 = -\text{Im} \, \frac{[\rho]}{\alpha F^2 [U_{1z}]}[w_z w^*].$$

We calculated the terms of the energy equation (7.4) for three Reynolds numbers under the same conditions as in figures 7.1 and 7.2. The results of the calculations are displayed in tables 7.1 - 7.3. In the second column of the tables, the value of $\dot{E}$ has been normalized to be the same as the growth rate. The density difference was put to zero so that $B_3 = 0$. The reader can identify the two relative maxima by looking for the two largest values in the columns $\alpha Im\ c$. Inspection of the different terms in the energy balance shows that the most unstable (larger α) mode draws its energy from interfacial friction B_2 and the second most unstable (long wave) mode draws its energy from the production I. We may write that $I = I_1 + I_2$ where I_1 is the production in fluid one and I_2 the production in fluid two. The fluid in region one is less viscous and I_1 is much larger than I_2. For example, when $R = 857$ and $\alpha = 2.0$, $I_1/I_2 = 7.66$. We may conclude that the second most unstable mode draws its energy from the Reynolds stress in the less viscous fluid.

We cannot predict what hydrodynamics will be generated by these instabilities with certainty, because nonlinear effects are neglected. We might guess, based on our experiences with core-annular flows, that we will see a wave driven by interfacial friction with a wavelength and celerity like those associated with the maximum growth rate together with turbulent-like structure arising from the weaker long-wave instability.

Table 7.1. Energy analysis for $R = 36$.

α	$\alpha Im\ c$	$\dot{E}$	I	D	B_1	B_2
6.0	-0.0033598	-0.0033598	0.024289	-28.562	0.0035664	28.530
7.0	0.0030863	0.0030863	-0.013793	-33.741	-0.0044518	33.762
8.0	0.0042719	0.0042719	-0.051120	-38.570	-0.0079932	38.634
9.0	0.00075032	0.00075032	-0.088099	-43.755	-0.0017846	43.845
10.0	-0.0061803	-0.0061803	-0.12368	-49.623	0.018484	49.722

Table 7.2. Energy analysis for $R = 514$.

α	$\alpha Im\ c$	$\dot{E}$	I	D	B_1	B_2
0.7	-0.0063750	-0.0063750	0.070988	-0.14190	9.8E-08	0.064536
1.0	0.010471	0.010471	0.10714	-0.15712	-3.1E-07	0.060450
2.5	0.063070	0.063070	0.22343	-0.24956	-1.26E-05	0.25317
5.0	0.022563	0.022563	0.27099	-0.75777	-3.9E-05	1.4642
20.0	0.10391	0.10391	0.036805	-16.983	-0.015748	31.115
60.0	0.0094330	0.0094330	-0.14622	-102.10	-0.024314	102.27
70.0	-0.0039586	-0.0039586	-0.13642	-137.0	0.016298	137.12

Table 7.3. Energy analysis for $R = 857$.

α	$\alpha Im\, c$	$\dot{E}$	I	D	B_1	B_2
0.1	-0.0016423	-0.0016423	0.0043119	-0.077709	2.5E-10	0.071755
0.5	1.4420E-05	1.4420E-05	0.054342	-0.086870	-3.8E-11	0.032543
2.0	0.066021	0.066021	0.18966	-0.14808	-2.32E-06	0.024444
3.0	0.015605	0.015605	0.15671	-0.18792	-1.52E-06	0.046812
3.5	-0.0055739	-0.0055739	0.13991	-0.22143	8.9E-07	0.075945
4.5	-0.0069565	-0.0069565	0.15587	-0.33399	2.6E-06	0.17116
5.0	-0.00068366	-0.00068366	0.17674	-0.41429	3.7E-07	0.23687
5.5	0.0062351	0.0062351	0.20128	-0.51182	-4.7E-06	0.31678
25.0	0.10616	0.10616	0.061501	-16.180	-0.011531	16.235
90.0	0.0034690	0.0034690	-0.12317	-135.98	-0.010674	136.13
100.0	-0.0042309	-0.0042309	-0.11574	-166.76	0.017760	166.85

IV.8 Nonlinear Analysis

IV.8(a) Weakly Nonlinear Amplitude Equations for Long Waves

Most of this section is based on the papers of Hooper and Grimshaw [1985] and Shlang, Sivashinsky, Babchin and Frenkel [1985]. They examine the nonlinear instability of two-layer Couette or Poiseuille flow with the long wave assumption. Multiple scaling is used to find an amplitude evolution equation for long wavelength weakly nonlinear waves. This technique was developed by Benney and Newell [1967] and has been used successfully to find amplitude evolution equations for interfacial waves [Homsy 1974; Atherton and Homsy 1976; Nakoryakov and Alekseyenko 1981; Krantz and Goren 1970; Sivashinsky and Michelson 1980; Shlang and Sivashinsky 1982]. Other works relevant in the area of nonlinear long waves include Lin [1969], Krishna and Lin [1977] and Lin and Krishna [1977]. Surface tension is assumed to be strong.

Consider the configuration in figure 8.1 with either Couette or Poiseuille flow. The Navier-Stokes equations (2.1) hold in each fluid. Lengths are made dimensionless with the depth of the lower fluid d_1, velocity with u_0 (the horizontal speed of the interface in the basic velocity), time with d_1/u_0, and pressure with $\rho_1 u_0^2$ where ρ_1 is the density of the lower fluid. A Reynolds number

$$R = \rho_1 u_0 d_1 / \mu_1$$

where μ_1 is the viscosity of the lower fluid, and a gravity parameter

$$F = (1 - r)g d_1 / u_0^2$$

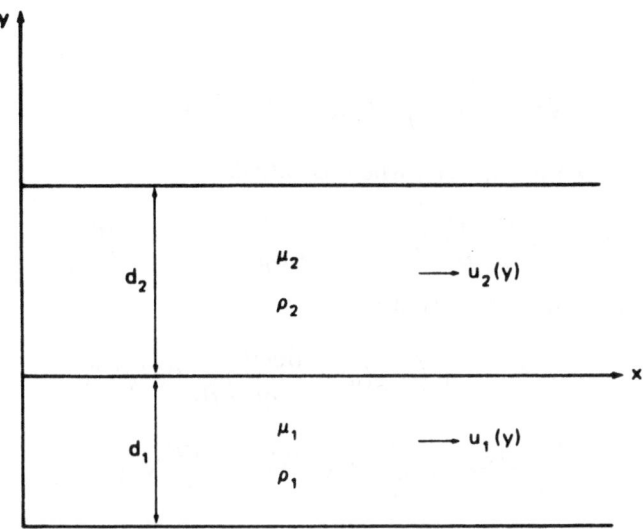

Fig. 8.1. [Figure 1 of Hooper and Grimshaw, 1985, American Institute of Physics] Shear flow of two fluids in a channel. The upper and lower fluids have viscosity μ_2 and μ_1, and density ρ_2 and ρ_1, respectively. The interface is located at $y = 0$.

emerge after making the equations dimensionless. Therefore, in dimensionless coordinates (x,y) (see equation (4c.6)), the basic flow is equal given by

$$u_1(y) = A_1 y^2 + a_1 y + 1, \qquad y < 0, \qquad (8a.1)$$
$$u_2(y) = A_2 y^2 + a_2 y + 1, \qquad y > 0,$$

where $A_i = G/2\mu_i$, $i = 1, 2$, and G is the pressure gradient. We define the the ratios

$$m = \frac{\mu_2}{\mu_1}, \quad r = \frac{\rho_2}{\rho_1}, \quad n = \frac{d_2}{d_1}. \qquad (8a.2)$$

and find that

$$A_1 = mA_2. \qquad (8a.3)$$

The continuity of shear stress at the interface $y = 0$ yields

$$a_1 = ma_2. \qquad (8a.4)$$

For plane Couette flow, $a_1 = 1$ and $A_1 = A_2 = 0$. For plane Poiseuille flow, the no-slip conditions at $y = -1$ and $y = n$ yield

$$a_1 = (n^2 - m)/n(1 + n), \quad a_2 = (n^2 - m)/mn(1 + n),$$

$$A_1 = -(m + n)/n(1 + n), \quad A_2 = -(m + n)/mn(1 + n). \qquad (8a.5)$$

Suppose that the system is disturbed so that the streamfunction (see equation (2.30)) is

$$\psi(x, y, t) = \int^y u_1(y)dy + \psi^{(1)}(x, y, t) \tag{8a.6}$$

below the interface, and

$$\psi(x, y, t) = \int^y u_2(y)dy + \psi^{(2)}(x, y, t) \tag{8a.7}$$

above the interface $y = \eta(x, t)$, where ψ satisfies

$$\frac{\partial}{\partial t}(\nabla^2 \psi) + \frac{\partial \psi}{\partial y}\frac{\partial}{\partial x}(\nabla^2 \psi) - \frac{\partial \psi}{\partial x}\frac{\partial}{\partial y}(\nabla^2 \psi) = \frac{m(y)}{R}\nabla^4 \psi$$

and the perturbations of ψ satisfy

$$\frac{\partial}{\partial t}(\nabla^2 \psi^{(i)}) + \left(u_{(i)}(y) + \frac{\partial \psi^{(i)}}{\partial y}\right)\frac{\partial}{\partial x}(\nabla^2 \psi^{(i)})$$

$$-\frac{\partial \psi^{(i)}}{\partial x}\left(2A_i + \frac{\partial}{\partial y}(\nabla^2 \psi^{(i)})\right) = \frac{m(y)}{R}\nabla^4 \psi^{(i)} \tag{8a.8}$$

where

$$m(y) = \begin{cases} 1 & \text{below the interface,} \\ m/r & \text{above the interface.} \end{cases} \tag{8a.9}$$

The no-slip conditions at each solid boundary require that

$$\frac{\partial \psi}{\partial x} = 0, \quad \frac{\partial \psi}{\partial y} = u_1 \quad \text{at} \quad y = -1, n,$$

so that the perturbations satisfy

$$\frac{\partial \psi^{(i)}}{\partial x} = 0, \quad \text{at} \quad y = -1, n, \quad \frac{\partial \psi^{(1)}}{\partial y} = u_1(-1), \quad \frac{\partial \psi^{(2)}}{\partial y} = u_1(n). \tag{8a.10}$$

There are four continuity conditions at the disturbed interface $y = \eta(x, t)$:

(1) continuity of tangential velocity

$$u_1(\eta) + \frac{\partial \psi^{(1)}}{\partial y}(\eta) = u_2(\eta) + \frac{\partial \psi^{(2)}}{\partial y}(\eta); \tag{8a.11}$$

(2) continuity of normal velocity

$$\frac{\partial \psi^{(1)}}{\partial x}(\eta) = \frac{\partial \psi^{(2)}}{\partial x}(\eta); \tag{8a.12}$$

(3) continuity of tangential stress

$$e_{xx}^{(1)}t_x n_x + e_{yy}^{(1)}t_y n_y + e_{xy}^{(1)}(t_x n_y + t_y n_x) = m\left(e_{xx}^{(2)}t_x n_x\right.$$

$$\left. + e_{yy}^{(2)}t_y n_y + e_{xy}^{(2)}(t_x n_y + t_y n_x)\right), \tag{8a.13}$$

where

$$(t_x, t_y) = \frac{(1, \eta_x)}{(1 + \eta_x^2)^{1/2}}, \quad (n_x, n_y) = \frac{(-\eta_x, 1)}{(1 + \eta_x^2)^{1/2}}, \tag{8a.14}$$

and

$$e_{xx}^{(i)} = \left(\frac{\partial^2 \psi^{(i)}}{\partial x \partial y}\right)_{y=\eta(x,t)}, \quad e_{yy}^{(i)} = \left(\frac{-\partial^2 \psi^{(i)}}{\partial x \partial y}\right)_{y=\eta(x,t)},$$

$$e_{xy}^{(i)} = \frac{1}{2}\left(\frac{\partial^2 \psi^{(i)}}{\partial y^2} - \frac{\partial^2 \psi^{(i)}}{\partial x^2}\right)_{y=\eta(x,t)}, \quad i = 1, 2; \tag{8a.15}$$

(4) continuity of normal stress

$$-p^{(1)} + 2(e_{xx}^{(1)} n_x n_x + 2e_{xy}^{(1)} n_x n_y + e_{yy}^{(1)} n_y n_y)$$

$$= -p^{(2)} + 2m(e_{xx}^{(2)} n_x n_x + 2e_{xy}^{(2)} n_x n_y + e_{yy}^{(2)} n_y n_y) + \frac{S\eta_{xx}}{(1 + \eta_x^2)^{3/2}} - F\eta, \tag{8a.16}$$

where $p^{(i)}$ is the pressure at the interface of fluid i, $i = 1, 2$. The dimensionless surface tension parameter S is defined by

$$S = \frac{T}{\rho_1 d_1 u_0^2}, \tag{8a.17}$$

where T is the surface tension.

Finally, the kinematic condition at the interface is

$$\frac{D}{Dt}\left(y - \eta(x, t)\right) = 0. \tag{8a.18}$$

The position of the interface $\eta(x, t)$ is unknown.

In the weakly nonlinear long-wave analyses under consideration it is assumed that

$$\eta = \epsilon A(\xi, \tau) \tag{8a.19}$$

where

$$\xi = \epsilon(x - c_0 t), \quad \tau = \epsilon^2 t, \tag{8a.20}$$

ϵ is a small and sometimes unspecified parameter, and $c \sim c_0 + O(\alpha)$ (see equation (8a.43), and section IV.4(a)). The streamfunction is then expanded

$$\psi^{(i)}(x, y, t) = \epsilon \psi_0^{(i)}(\xi, y, \tau) + \epsilon^2 \psi_1^{(i)}(\xi, y, \tau) + \epsilon^3 \psi_2^{(i)}(\xi, y, \tau) + O(\epsilon^4), \quad i = 1, 2. \tag{8a.20}$$

The Navier-Stokes equations (8a.8), the boundary conditions (8a.10) and the interfacial conditions (8a.11)- (8a.18) are rewritten in terms of the new variables ξ, τ and η ($\partial/\partial x = \epsilon \partial/\partial \xi$, $\partial/\partial t = -\epsilon c_0 \partial/\partial \xi + \epsilon^2 \partial/\partial \tau$) and the expansion (8a.20). The interface conditions can be expanded in a Taylor series about $y = 0$.

At O(ϵ), equation (8a.8) yields:

$$\frac{m(y)}{R}\frac{\partial^4\psi_0^{(i)}}{\partial y^4} = 0. \tag{8a.21}$$

The boundary conditions (8a.10) yield

$$\frac{\partial\psi_0^{(1)}}{\partial x} = \frac{\partial\psi_0^{(1)}}{\partial y} = 0, \quad y = -1, \tag{8a.22}$$

$$\frac{\partial\psi_0^{(2)}}{\partial x} = \frac{\partial\psi_0^{(2)}}{\partial y} = 0, \quad y = n. \tag{8a.23}$$

The interface conditions (8a.11) - (8a.18) at $O(\epsilon)$, at $y = 0$, are

$$\frac{\partial\psi_0^{(1)}}{\partial y} + Au_1'(0) = \frac{\partial\psi_0^{(2)}}{\partial y} + Au_2'(0), \tag{8a.24}$$

$$\frac{\partial\psi_0^{(1)}}{\partial \xi} = \frac{\partial\psi_0^{(2)}}{\partial \xi}, \tag{8a.25}$$

$$\frac{\partial^2\psi_0^{(1)}}{\partial y^2} = m\frac{\partial^2\psi_0^{(2)}}{\partial y^2}, \tag{8a.26}$$

$$p_0^{(1)} = p_0^{(2)}. \tag{8a.27}$$

Here, $p^{(i)}$ is expanded in the form $p_0^{(i)} + \epsilon p_1^{(i)} + O(\epsilon^2)$, $i = 1, 2$. The value of the pressure $p^{(i)}$ is found from the ξ-component of the Navier-Stokes equations

$$\epsilon\frac{\rho_1}{\rho_i}\frac{\partial p^{(i)}}{\partial \xi} = \frac{m(y)}{R}\left(\frac{\partial^2 u^{(i)}}{\partial y^2} + \epsilon^2\frac{\partial^2 u^{(i)}}{\partial \xi^2}\right) - \epsilon(u^{(i)} - c_0)\frac{\partial u^{(i)}}{\partial \xi}$$

$$-\left(\epsilon^2\frac{\partial u^{(i)}}{\partial \tau} + v^{(i)}\frac{\partial u^{(i)}}{\partial y}\right), \tag{8a.28}$$

where $u^{(i)} = u_i(y)$ (the basic flow) $+\partial\psi^{(i)}/\partial y$ in fluid i and $v^{(i)} = -\epsilon\partial\psi^{(i)}/\partial\xi$. The interface equation (8a.18) reduces to

$$(u_1(0) - c_0)\frac{\partial A}{\partial \xi} = -\frac{\partial\psi_0^{(i)}}{\partial \xi}, \tag{8a.29}$$

at $y = 0$, where $u_1(0) = u_2(0) = 1$.

The above equations can be solved for $\psi_0^{(i)}$ in the following separated form

$$\psi_0^{(1)}(\xi, y, \tau) = A(\xi, \tau)\phi_0(y), \tag{8a.30}$$

$$\psi_0^{(2)}(\xi, y, \tau) = A(\xi, \tau)\chi_0(y). \tag{8a.31}$$

After substituting this into equation (8a.28) we find that at $O(\epsilon)$,

$$\frac{\partial p_0^{(1)}}{\partial \xi} = \frac{1}{R} A(\xi, \tau) \phi_0'''(y), \tag{8a.32}$$

$$\frac{\partial p_0^{(2)}}{\partial \xi} = \frac{m}{R} A(\xi, \tau) \chi_0'''(y). \tag{8a.33}$$

The separated forms of equations (8a.30) - (8a.33) are put into equations (8a.21) -(8a.27). This yields the following system of equations

$$\phi_0^{iv}(y) = 0, \quad -1 \le y \le 0; \qquad \chi_0^{iv}(y) = 0, \quad 0 \le y \le n, \tag{8a.34}$$

$$\phi_0(-1) = 0, \quad \phi_0'(-1) = 0, \quad \chi_0(n) = 0, \quad \chi_0'(n) = 0, \tag{8a.35}$$

$$\phi_0'(0) - \chi_0'(0) = a_2 - a_1 = (1 - m)\frac{a_1}{m}, \tag{8a.36}$$

$$\phi_0(0) = \chi_0(0), \tag{8a.37}$$

$$\phi_0''(0) = m\chi_0''(0), \tag{8a.38}$$

$$\phi_0'''(0) = m\chi_0'''(0), \tag{8a.39}$$

$$c_0 - 1 = \phi_0(0). \tag{8a.40}$$

Thus,

$$\phi_0(y) = \alpha_{00} + \alpha_{01}y + \alpha_{02}y^2 + \alpha_{03}y^3, \tag{8a.41}$$

$$\chi_0(y) = \alpha_{00} + \beta_{01}y + \beta_{02}y^2 + \beta_{03}y^3, \tag{8a.42}$$

$$\alpha_{00} = c_0 - 1 = \frac{2n^2(1+n)(1-m)a_1}{m^2 + 2mn(2n^2 + 3n + 2) + n^4}, \tag{8a.43}$$

$$\alpha_{01} = \frac{(m + 4n^3 + 3n^2)m\alpha_{00}}{2mn^2(1+n)}, \tag{8a.44}$$

$$\beta_{01} = -\frac{(4mn + 3mn^2 + n^4)\alpha_{00}}{2mn^2(1+n)}, \tag{8a.45}$$

$$\beta_{02} = \frac{(m + n^3)\alpha_{00}}{mn^2(1+n)}; \qquad \alpha_{02} = m\beta_{02}, \tag{8a.46}$$

$$\beta_{03} = \frac{(m - n^2)\alpha_{00}}{2mn^2(1+n)}; \qquad \alpha_{03} = m\beta_{03}. \tag{8a.47}$$

This procedure therefore recovers Yih's [1967] results for the perturbed flow at $O(\alpha^0)$ (see section IV.4 (a)). The term α_{00} is equivalent to Yih's value of c at this order and determines c_0. Throughout the nonlinear perturbation scheme we expect to recover Yih's results from his linear analysis as well as identifying the nonlinear terms.

At the next order, the equation of motion (8a.8) at $O(\epsilon^2)$ becomes

$$\frac{m(y)}{R} \frac{\partial^4 \psi_1^{(i)}}{\partial y^4} = (u_1(y) - c_0)\frac{\partial^3 \psi_0^{(i)}}{\partial \xi \partial y^2} - 2A_1 \frac{\partial \psi_0^{(i)}}{\partial \xi}. \tag{8a.48}$$

The boundary conditions become

$$\frac{\partial \psi_1^{(1)}}{\partial y} = \frac{\partial \psi_1^{(1)}}{\partial x} = 0 \quad \text{at} \quad y = -1, \tag{8a.49}$$

$$\frac{\partial \psi_1^{(2)}}{\partial y} = \frac{\partial \psi_1^{(2)}}{\partial x} = 0 \quad \text{at} \quad y = n. \tag{8a.50}$$

The interfacial conditions, calculated at $y = 0$, at $O(\epsilon^2)$, are

$$\frac{1}{2} A^2 u_1''(0) + A \frac{\partial^2 \psi_0^{(1)}}{\partial y^2} + \frac{\partial \psi_1^{(1)}}{\partial y} = \frac{1}{2} A^2 u_2''(0) + A \frac{\partial^2 \psi_0^{(2)}}{\partial y^2} + \frac{\partial \psi_1^{(2)}}{\partial y}, \tag{8a.51}$$

$$\frac{\partial \psi_1^{(1)}}{\partial \xi} = \frac{\partial \psi_1^{(2)}}{\partial \xi}, \tag{8a.52}$$

$$\frac{\partial^2 \psi_1^{(1)}}{\partial y^2} = m \frac{\partial^2 \psi_1^{(2)}}{\partial y^2}, \tag{8a.53}$$

$$p_1^{(1)} = p_1^{(2)} - S\epsilon^2 A_{\xi\xi} + FA, \tag{8a.54}$$

where $S\epsilon^2$ is assumed to be $O(1)$.[1] When terms of $O(\epsilon^2)$ are equated in equation (8a.28), we see that the pressure must satisfy

$$\frac{\rho_1}{\rho_i} \frac{\partial p_1^{(i)}}{\partial \xi} = \frac{m(y)}{R} \frac{\partial^3 \psi_1^{(i)}}{\partial y^3} - (u_i(y) - c_0) \frac{\partial^2 \psi_0^{(i)}}{\partial y \partial \xi} + \frac{\partial \psi_0^{(i)}}{\partial \xi} u_i'(y), \quad i = 1, 2. \tag{8a.55}$$

Therefore, $p_1^{(i)}$ in equation (8a.54) is known in terms of $\psi_1^{(i)}$ and other functions already calculated.

Equation (8a.48) shows that at $O(\epsilon^2)$, $\psi_1^{(i)}$ satisfies an inhomogeneous equation with a forcing term which involves A_ξ. The interface conditions (8a.51)-(8a.55) contain A_ξ, $A_{\xi\xi\xi}$ and A^2. Hence, we may seek solutions of the form

$$\psi_1^{(1)} = A_\xi \phi_{11}(y) + A^2 \phi_{12}(y) + A_{\xi\xi\xi} \phi_{13}(y), \tag{8a.56}$$

$$\psi_1^{(2)} = A_\xi \chi_{11}(y) + A^2 \chi_{12}(y) + A_{\xi\xi\xi} \chi_{13}(y), \tag{8a.57}$$

$$p_1^{(1)} = A f_{11}(y) + C(\xi, \tau) f_{12}(y) + A_{\xi\xi} f_{13}(y), \tag{8a.58}$$

$$p_1^{(2)} = A g_{11}(y) + C(\xi, \tau) g_{12}(y) + A_{\xi\xi} g_{13}(y), \tag{8a.59}$$

where

$$\frac{\partial C}{\partial \xi}(\xi, \tau) = A^2,$$

$$f_{11}(y) = \frac{1}{R} \phi_{11}'''(y) - (u_1(y) - c_0)\phi_0'(y) + u_1'(y)\phi_0,$$

---

[1]   This restrictive assumption is required for several reasons not the least of which is that without it the system can undergo a Hadamard instability. These questions are discussed by Atherton and Homsy [1976, equation (40)], and after equation (8a.68), and again in the latter half of chapter VIII)

$$g_{11}(y) = \frac{m}{R}\chi_{11}'''(y) - r(u_2(y) - c_0)\chi_0'(y) + ru_2'(y)\chi_0,$$

$$f_{12}(y) = \frac{1}{R}\phi_{12}'''(y), \qquad g_{12}(y) = \frac{m}{R}\chi_{12}'''(y),$$

$$f_{13}(y) = \frac{1}{R}\phi_{13}'''(y), \qquad g_{13}(y) = \frac{m}{R}\chi_{13}'''(y).$$

Substitution of equations (8a.56) - (8a.59) into equations (8a. 48), (8a.51) - (8a.55) yields systems of equations to be solved for ϕ_{11}, χ_{11}, ϕ_{12}, χ_{12}, ϕ_{13} and χ_{13}.

The functions ϕ_{12}, χ_{12}, ϕ_{13}, and χ_{13} satisfy a homogeneous fourth order differential equation and are therefore cubic polynomials in y. The functions ϕ_{11} and χ_{11} satisfy an inhomogeneous fourth-order equation and are found to have the following form:

$$\phi_{11}(y) = \alpha_{10} + \alpha_{11}y + \alpha_{12}y^2 + \alpha_{13}y^3 + Rh_1(y), \tag{8a.60}$$

$$\chi_{11}(y) = \beta_{10} + \beta_{11}y + \beta_{12}y^2 + \beta_{13}y^3 + \frac{Rr}{m}h_2(y), \tag{8a.61}$$

where $h_1(y)$ and $h_2(y)$ are defined in equations (15) of the paper by Hooper and Grimshaw [1985].

Finally, the kinematic condition at $O(\epsilon^2)$ becomes

$$\frac{\partial A}{\partial \tau} + \frac{\partial \psi_0}{\partial y}(0)\frac{\partial A}{\partial \xi} + u_i'(0)AA_\xi = -\frac{\partial \psi_1}{\partial \xi}(0) - A\frac{\partial^2 \psi_0}{\partial y\partial \xi},$$

which reduces to

$$\frac{\partial A}{\partial \tau} + l(m,n)AA_\xi + RJ(m,n,r,A_1)A_{\xi\xi} + \epsilon^2 RS(m,n)A_{\xi\xi\xi\xi} = 0, \tag{8a.62}$$

where

$$l(m,n) = 2\phi_0'(0) + u_1'(0) + 2\phi_{12}(0), \tag{8a.63}$$

$$RJ(m,n,r,A_1) = \phi_{11}(0), \tag{8a.64}$$

$$RS(m,n) = \phi_{13}(0), \tag{8a.65}$$

where

$$S(m,n) = \frac{2n^3(1+n)}{K(m,n)}S, \quad K(m,n) = m^2 + 2mn(2n^2 + 3n + 2) + n^4,$$

and the expressions for $J(m,n,r,A_1)$ and $l(m,n)$ are given in equations (17) of the paper by Hooper and Grimshaw.

The term $\phi_{11}(0)$ can be identified with the growth rate of the linear stability theory, $\phi_{13}(0)$ is proportional to surface tension which was neglected in the analysis of Yih [1967]. Analysis of the linear stability from (8a.62) leads to the formula

$$c = c_0 + i\alpha R\Big(J(m,n,r,A_1) - \alpha^2 S(m,n)\Big) + O(\alpha^2). \tag{8a.66}$$

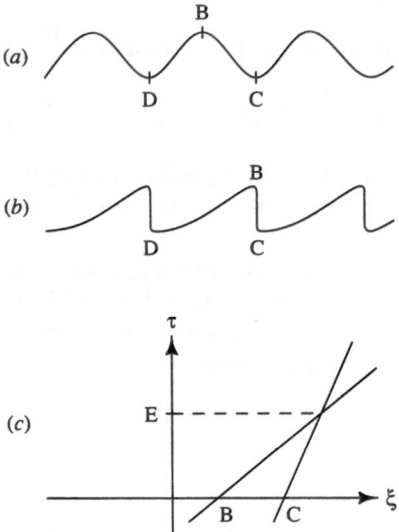

Fig. 8.2(a-c). (a) The sinusoidal initial condition. (b) The solution at a later time $\tau = E$. (c) The $\tau - \xi$ plane, with characteristics.

The function J is related to that found by Yih who showed that it was positive for certain values of m and n. Thus, when S is O(1), the flow can be linearly unstable. Note that when S is O(1), its contribution is smaller than the error term in (8a.66). To retain this term we must assume that $\alpha^2 S$ (or $\epsilon^2 S$) is O(1).

The evolution equation (8a.62) may be simplified by replacing lA with A:

$$A_\tau + AA_\xi + dA_{\xi\xi} + sA_{\xi\xi\xi\xi} = 0, \qquad (8a.67)$$

where

$$d = RJ, \quad s = R\epsilon^2 S. \qquad (8a.68)$$

When $s = 0$ and $d = 0$, we have the well-known equation (see Whitham [1974])

$$A_\tau + AA_\xi = 0.$$

The characteristics are defined by

$$\frac{d\xi}{d\tau} = A(\xi, \tau).$$

Along characteristics,

$$\frac{d}{d\tau} A(\xi(\tau), \tau) = A_\xi \frac{d\xi}{d\tau} + A_\tau = A_\xi A + A_\tau = 0.$$

Hence, the solution $A(\xi(\tau), \tau)$ is constant along characteristics, and the characteristics are lines in the $\xi - \tau$ plane with slope A.

Figures 8.2 illustrate what will happen to a sinusoidal initial condition when $(s, d) = (0, 0)$. If the solution $A(\xi, \tau)$ is a decreasing function of ξ, as along the points B to C in (a), then the slope of the characteristics in the $\xi - \tau$ plane is steeper at point B than at point C. Hence, characteristics cross and the waves steepen to form shocks at time $\tau = E$, as along B to C in (b). On the other hand, if $A(\xi, \tau)$ is an increasing function of ξ, as along D to B in (a), then the characteristics flatten out in (c) and the solution flattens out, as along D to B in (b). Equation (8a.67) also arises in the study of stability of liquid films with matched viscosities with the heavy film above, which in the stagnant case would give rise to a Rayleigh-Taylor instability. This instability may evidently be suppressed in a certain range of shear rates by nonlinear terms [Babchin, Frenkel, Levich and Sivashinsky 1983]. This type of nonlinear flow-induced and surface-tension-assisted mechanism for nonlinear saturation of instabilities is a possibility for a restricted class of film flows, including the shearing flows with nonequal viscosities [Frenkel, Babchin, Levich, Shlang and Sivashinsky 1987; Frenkel 1988], which are discussed in the latter sections of chapter VIII.

When $s = 0$, the surface tension term is removed and (8a.67) reduces to Burger's equation. The Cole-Hopf transformation further reduces it to the heat equation. When $d < 0$, the initial value problem may then be solved and it can be shown that an initial disturbance will evolve into a sequence of decaying shock waves (see, for example, Whitham [1974]). However, when $d > 0$ the Cole-Hopf transformation produces a heat equation backward in time. An initial disturbance will then grow without limit. Hence, we shall include the surface tension term and discuss equation (8a.67) when $d > 0$ and $s > 0$.

Equation (8a.67) is the Kuramoto-Sivashinsky equation. It is capable of generating solutions in the form of irregularly fluctuating quasi-periodic waves [LaQuey, Mahajan, Rutherford and Tang 1975; Kuramoto and Tsuzuki 1976; Sivashinsky 1977; Sivashinsky 1983; Manneville 1981; Babchin, Frenkel, Levich and Sivashinsky 1983; Pumir, Manneville and Pomeau 1983]. This model provides a mechanism for the saturation of an instability, in which the energy in long-wave instabilities is transferred to shortwave modes which are then damped by surface tension. In (8a.67), the terms $A_\tau + AA_\xi$ lead to steepening and wave breaking in the absence of stabilizing terms. The term $dA_{\xi\xi}$ destabilizes shorter wavelength modes preferentially and therefore aggravates wave steepening (this follows if one substitutes $A \sim \exp{ik\xi}$; then $A_\tau = - dA_{\xi\xi} + ... = dk^2 A + ...$ Hence, the $A_{\xi\xi}$ term makes a positive contribution to A_τ). The term $sA_{\xi\xi\xi\xi}$ is required for saturation.

Explicit analytic solutions are not available, but approximate analytic solutions and numerical solutions are discussed in Hooper and Grimshaw [1985]. Their analysis is closely related to that of Cohen, Krommes, Tang and Rosenbluth [1976] and like these authors, they consider solutions which are periodic, of period 2L. The linearized version of equation (8a.67) has solutions of the form $\exp(\sigma\tau + ik\xi)$ where $k = \pi/L$, and

$$\sigma = dk^2 - sk^4. \tag{8a.69}$$

This is just the linear dispersion relation (8a.66), now expressed in different variables. Equation (8a.69) shows that short waves are stable, and long waves are unstable; the critical wavenumber is

$$k_c = \left(\frac{d}{s}\right)^{1/2} \tag{8a.70}$$

which ought to be small for the analysis of long waves to make sense. The maximum growth rate is $d^2/(4s)$ and occurs at $k_c/\sqrt{2}$. It is anticipated that the effect of the nonlinear term in (8a.67) will be to allow energy exchange between a wave with wavenumber k and its harmonics with the end result being nonlinear saturation. The final state may be either chaotic oscillatory motion [Kawahara 1983; Sivashinsky 1983] or a steady state involving only a few harmonics [Cohen, Krommes, Tang and Rosenbluth 1976].

The energy equation corresponding to (8a.67) is obtained by multiplying (8a.67) by A and integrating by parts, assuming A is periodic with period $2L$:

$$\frac{\partial}{\partial t}\int_0^{2L} \frac{1}{2}A^2 d\xi = \int_0^{2L}(dA_\xi^2 - sA_{\xi\xi}^2)d\xi. \tag{8a.71}$$

The right hand side may be estimated by writing it as

$$\int_0^{2L} dA_\xi^2 d\xi\left(1 - \frac{\int_0^{2L} sA_{\xi\xi}^2 d\xi}{\int_0^{2L} dA_\xi^2 d\xi}\right),$$

and using the result that

$$\frac{s\pi^2}{dL^2} = \min\frac{\int_0^{2L} sA_{\xi\xi}^2 d\xi}{\int_0^{2L} dA_\xi^2 d\xi}, \tag{8a.72}$$

the minimum being taken over all periodic functions. This is established by noting that the right hand side of equation (8a.72) is equivalent to the minimization of

$$\int_0^{2L} sA_{\xi\xi}^2 d\xi \quad \text{subject to} \quad \int_0^{2L} dA_\xi^2 d\xi = 1.$$

The Euler-Lagrange equation for this is obtained by substituting $(A + \phi)$ for A above, where ϕ is a test function, also periodic with period $2L$, and linearizing, to obtain

$$2\int_0^{2L} sA_{\xi\xi}\phi_{\xi\xi}d\xi - \lambda 2\int_0^{2L} dA_\xi\phi_\xi d\xi = 0,$$

where λ is the Lagrange multiplier. This is integrated by parts to yield

$$sA_{\xi\xi\xi\xi} + \lambda dA_{\xi\xi} = 0.$$

The periodic eigenfunctions are $A = \exp i2\pi N\xi/2L$, $N = 1, 2, ...$, so that the eigenvalues are $\lambda = s\pi^2 N^2/dL^2$, and the smallest is $s\pi^2/dL^2$.

Thus, the right hand side of (8a.71) will be negative for $\pi/L > k_c$, and therefore the nonlinear equation (8a.67) is globally stable for an initial condition with a wavenumber satisfying the linear stability criterion. In other words, if you put in an initial disturbance (e.g. $\sin k\xi$) with a wavenumber k greater than k_c, then the nonlinear term in (8a.67) creates higher harmonics, but it will not create waves with wavenumbers smaller than k, so there will be stability. If you want to generate a component with a wavenumber in the unstable region, you have to put in an initial condition with a wavenumber less than k_c. Of course, we do not know what kind of wave-shortening will result from the action of inertial terms which have been discarded in the analysis of long waves.

Hence, we need to consider only the case $k < k_c$. The periodic boundary conditions allow A to be written as the Fourier series

$$A = \sum_{-\infty}^{\infty} A_n(\tau)\exp(ink\xi), \qquad (8a.73)$$

where $A_{-n} = A_n^*$, and without loss of generality we may put $A_0 = 0$ (it may be readily shown from (8a. 67) that A_0 is a constant). Substitution of (8a.73) into (8a. 67) gives

$$\frac{\partial A_n}{\partial \tau} - \sigma_n A_n + inkB_n = 0, \qquad (8a.74)$$

where

$$B_n = \sum_{r=1}^{\infty} A_r^* A_{r+n} + \frac{1}{2} \sum_{r=1}^{n-1} A_r A_{n-r},$$

and

$$\sigma_n = d(nk)^2 - s(nk)^4.$$

The significant feature of (8a.74) is that, for any given k, only a finite number of Fourier modes, $A_1, ..., A_n$ say, are unstable ($\sigma_n > 0$), and all higher modes are stable. Note that the nth mode has a critical wavenumber of k_c/n, and a maximum growth rate of $d^2/4s$ at $k_c/(n\sqrt{2})$; in particular, its maximum growth rate is independent of n. This implies that unstable modes will be stabilized by energy transfer to higher harmonics. This view is particularly clear when (8a. 73) is substituted into (8a.72) with the result that

$$\frac{\partial}{\partial \tau} \sum_{n=1}^{\infty} |A_n|^2 = 2 \sum_{n=1}^{\infty} \sigma_n |A_n|^2. \qquad (8a.75)$$

From the work of Cohen et al. [1976] it is deduced that the stabilization to bounded states takes one of two forms. Either a time-independent 'two-mode equilibrium' is achieved by which the mode with the greatest growth rate is stabilized by its first harmonic, or a time-varying 'bouncy' state is achieved in which energy is continually exchanged between several two-mode states. Note that whenever k is such that A_n has the most rapid growth, A_{2n} is stable. As k is decreased, more and more Fourier modes participate and stabilization takes longer to achieve. It is suspected that the 'bouncy' state then becomes indistinguishable from the chaotic behavior found by Kawahara [1983].

The simplest case amenable to some analysis is when $k_c/2 < k < k_c$. Only $n = 1$ is unstable and in the following, it is assumed that it is sufficient to consider just the interaction between $n = 1$ and $n = 2$. The approximate version of (8a.74) is then (see section 3.1 of Cohen et $al.$)

$$\frac{\partial A_1}{\partial \tau} - \sigma_1 A_1 + ik A_2 A_1^* = 0, \tag{8a.76}$$

$$\frac{\partial A_2}{\partial \tau} - \sigma_2 A_2 + ik A_1^2 = 0.$$

Note that since A_1 is unstable and A_2 is stable, $\sigma_1 > 0$ and $\sigma_2 < 0$, equation (8a.76) has the steady solution

$$|A_1| = \left(\frac{-\sigma_1 \sigma_2}{k^2}\right)^{1/2}, \quad A_2 = \frac{ik A_1^2}{\sigma_2}. \tag{8a.77}$$

Here, A_1 is growing and A_2 is stabilizing and the pair form a steady solution (see equation (12) in Cohen et $al.$ [1976]).

The phase of A_1 is arbitrary, reflecting an indeterminancy in the choice of the origin of time. Note that $\sigma_1|A_1|^2 + \sigma_2|A_2|^2 = 0$, reflecting the required energy balance in the approximate version of (8a.75). This solution is stable to perturbations of A_1 and A_2 as can be seen from a linear stability analysis of (8a.76) (see equation (13) of Cohen et $al.$) when $k_c/\sqrt{3} < k < k_c$. Indeed, as k is decreased through k_c, there is a bifurcation from the stable zero solution for $k > k_c$, to the stable solution (8a.77); as k continues to decrease there is a further bifurcation at $k = k_c/\sqrt{3}$ (where $2\sigma_1 + \sigma_2 = 0$) to an oscillatory solution of arbitrary frequency and amplitude which exists only for $k = k_c/\sqrt{3}$. However, as k is decreased, the hypothesis that only two modes are involved becomes more suspect. Indeed Cohen et $al.$ have shown that as k is decreased the steady solution of (8a.74) given approximately by (8a.77) is at first modfied by the presence of a small correction due to A_3, and then when $k_c/3 < k < k_c/2$ (i.e. $\sigma_2 > 0$ but $\sigma_3 < 0$) is replaced by another 'two-mode equilibrium' in which A_2 and A_4 are the dominant components. Further decrease in k then leads to a succession of states, alternating between 'two-mode equilibria' and 'bouncy states'.

Hooper and Grimshaw note that the various truncations of (8a.74) are presumably equivalent, in the long-wave limit, to a direct Fourier expansion of the original equations of the kind carried out for thin viscous films by Gjevik [1970] and Lin [1969]. In particular, if the steady solution for A_2 in

(8a.77) is substituted into the first equation in (8a.76) a Landau equation is obtained for A_1; it is valid for k close to k_c.

Numerical simulations for a sinusoidal initial condition are given by Hooper and Grimshaw [1985]. Their results are consistent with the Fourier analysis described above (their numerical method does not use a Fourier decomposition) and with the numerical results of Cohen et al [1976]. They also consider the possibility that equation (8a.67) possesses travelling wave solutions i.e. solutions which are functions of $\xi - V\tau$. See also Hooper and Grimshaw [1988], Grimshaw and Hooper [1991], Chang [1986], Papageorgiou and Smyrlis [1991], Smyrlis and Papageorgiou [1991].

The preceding analysis indicates that surface tension is required to stabilize a short-wave instability of co-current plane Couette or plane Poiseuille flow of two viscous fluids. Weakly nonlinear effects do not alter this conclusion. The crucial parameter is $k^2 s/d$, where d is the growth rate of the linear instability, s represents the stabilizing effect of surface tension, and k is the wavenumber of the disturbance. For $k^2 s/d > 1$, all waves decay; for $k^2 s/d < 1$, the linearly unstable wave will steepen because of the convective nonlinearity and then evolve, either to a steady state containing just a few harmonics, or to a quasiperiodic state. If $l(m,n) > 0$, then the convective nonlinearity will cause a sinusoidal disturbance to deform into a wave with a steep front face and a long gradual tail; for $l(m,n) < 0$, the reverse occurs. The values for $J(m,n,r,A_1)$ are given by Yih [1967] for $r = 1$ and $A_1 = 0$. From these results and figure 2 of Hooper and Grimshaw [1985], it may be deduced that when plane Couette flow of two fluids of equal density is linearly unstable, the constant $l(m,n) > 0$.

If the depth of the lower layer of fluid is small, so that $n >> 1$, the term $J(m,n,r,A_1)$ simplifies; one can see that for $m > 1$ (lower fluid is less viscous) and $r < 1$, $J < 0$, so the flow is linearly stable to long waves (see 'thin-layer effect' in this chapter). For instance, this implies that the co-current flow of oil above a thin film of water is stable to long-wave perturbations. However, although the system may have a stable density stratification ($r < 1$), it is possible to find values of $m < 1$ (lower fluid is more viscous) such that $J > 0$ when the depth of the lower layer is small. When the upper fluid is air and the lower fluid is water, a long-wave instability occurs whenever the depth of the water is decreased below a critical depth d_c. In the limit $n >> 1$, $l(m,n) \sim 1$, and hence this instability will manifest itself as long interfacial waves with steep fronts and gradual tails. The long-wave analysis should not be valid for waves with steep fronts.

Craik [1966] has observed interfacial waves of this kind when air is blown over a thin film of water, the waves appearing when the depth of the water layer is sufficiently small. These observations are in an apparent but not tested agreement with the preceding analysis. Craik observes that the speed of this long-wave disturbance is less than the interfacial velocity. In contrast, Saric and Marshall [1971] found, in their experiments on supersonic flow over thin films, that long interfacial waves at low Reynolds number travel faster than the interfacial velocity. In a linear stability analysis of thin films in uniform shearing motion, Miles [1960] and Smith and Davis [1982] found that the wave speed of a disturbance is less than the interfacial velocity. However, in both these studies, the effect of the upper fluid is ignored. Miles [1960] infers that if the upper fluid is taken into account, then a new class of modes will be introduced with wave speeds in excess of the interfacial velocity. Yih's [1967] analysis and equation (8a.43) show that this is indeed the case. Miles [1960] also derives a criterion to justify the hypothesis that the air flow in the upper layer can be taken as laminar in terms of a parameter α_g. When α_g is $O(1)$ or larger, we can assume a laminar velocity profile in the upper layer. For the experiments of Saric and Marshall [1971], α_g is approximately 0.5; this suggests that the flow in the upper layer can be regarded as laminar and their observation that the wave speed is greater than the interfacial velocity does not disagree with the preceding theory. However, in the experiments of Craik [1966], $\alpha_g \ll 1$ and the upper layer flow must be regarded as turbulent.

The propagation of long interfacial waves of finite amplitude for two-layer plane Couette-Poiseuille flow has also been examined by Ooms, Segal, Cheung and Oliemans [1985]. They do not assume that the amplitude of the interfacial disturbance is small compared with the plate separation. Their long-wave assumption is that the disturbance wavelength is long compared with the plate separation, and there is also a condition on the Reynolds number. Surface tension is neglected. They obtain a scalar equation (their equation (86)) describing the evolution of the interfacial wave, and this was used for numerical simulations. For example, they display a case where the initial wave has the shape of a sine function, which eventually develops into a sawtoothlike shape. The equilibrium shape of the wave is a function of the viscosity ratio of the fluids, the density difference between the fluids, and the applied pressure gradient. For very small wave amplitudes, the shape of the wave does not change with time and its velocity agrees with that found by Yih [1967] in his first approximation. There is also some qualitative agreement with observations on oil-water core-annular flow experiments reported by Ooms, Segal, Van der Wees, Meerhoff and Oliemans [1984], and with the aforementioned simulations of Hooper and Grimshaw, in that waves with steep fronts and gradual tails are simulated (see figures 36, 37 of Oliemans [1986]). It is not yet known if these casual qualitative agreements can be taken as serious experimental validations of any of the long-wave theories so far proposed. It would be interesting to see if the solutions of the long-wave equations agree with solutions of the full equations in the limit of parameters in which the long-wave theories are said to arise.

IV.8(b) Bifurcation Analysis

In this chapter we shall study plane Couette-Poiseuille flow following the work of Y. Renardy [1989]. The center manifold theorem (cf. chapter III) and the method of reduction to normal forms are used to obtain the Landau equation.

The two fluids are contained between infinite parallel plates and have different viscosities and densities, with surface tension at the interface. Gravity and a horizontal pressure gradient are incorporated. The focus is on the calculation of the Stuart-Landau coefficient κ. Situations of practical interest cover a wide range of flow parameters.

Since the earliest experimental results [Yu and Sparrow 1969], waves have been reported to travel on the interface of vertically stratified flows. In the linear stability analysis, the spectrum at low speeds consists of an interfacial mode and an infinite number of 'internal' modes, which have a counterpart in the one-fluid flow. The internal modes are typically stable at sufficiently low speeds and it is then that the interfacial mode can cause instability. Yih [1967] suggested that the instability of the interfacial mode may lead to the establishment of waves on the interface.

In this section, the bifurcation from the arrangement with a flat interface is analyzed for the case of a single critical wavelength, with all other wavelengths linearly stable. The procedure used in this section is like that of section III.7, where the two-layer Benard problem is considered, except that there the critical eigenvalue has a sixfold degeneracy. There is no such degeneracy in the present problem and the algebra is correspondingly simpler. The method of normal forms is used to reduce the nonlinear partial differential equations governing the perturbed motion to one ordinary differential equation. This procedure is standard in bifurcation theory but, as discussed later in section (ii), this is not the only approach that can be made for a nonlinear analysis. An alternative approach in arriving at a Ginzburg-Landau equation is that of Blennerhassett [1980] and is mentioned in section (ii).

The present study is restricted to uniform wavetrains, but a comparison with the modulated wave solutions in Blennerhassett [1980] yields some unexpected results. In addition, the results indicate that the Landau constant κ depends strongly on whether it is the pressure gradient that is kept fixed throughout the nonlinear analysis, or whether the combined volume flux is kept fixed.

Numerical results and neutral stability curves for low speeds are presented in section (iii). Different situations are analyzed to illustrate the effects of viscosity and density stratifications, and surface tension. In particular, a situation with the heavier fluid on top is shown to support supercritical travelling interfacial waves. In section (iv) we consider more extreme cases in which the bifurcating interfacial waves are subcritical.

IV.8(b)(i) Governing Equations

Two fluids of densities ρ_i (i=1,2), and viscosities μ_i lie in the (x^*, z^*) plane in layers between infinite parallel plates located at $z^* = 0, l^*$. The upper plate moves with velocity $(U_p^*, 0)$ and the bottom plate is at rest. In the basic flow, fluid 1 occupies $0 \leq z^* \leq l_1^*$ and fluid 2 occupies $l_1^* \leq z^* \leq l^*$. The velocity of the interface in the basic flow is $(U^*(l_1^*), 0)$ and we denote $U^*(l_1^*)$ by U_i. The velocity, distance, time and pressure are made dimensionless with U_i, l^*, l^*/U_i, and $\rho_1 U_i^2$. The dimensionless parameters are given at the beginning of section IV.4 (b), and the basic flow is defined by equations (4b.1) - (4b.2).

Perturbations of the velocity, pressure and interface are denoted by (u, w), p and h respectively. The Navier-Stokes equations in each fluid are

$$\frac{\partial u}{\partial t} + U\frac{\partial u}{\partial x} + wU' - \frac{1}{R_i}\triangle u + \frac{\rho_1}{\rho_i}\frac{\partial p}{\partial x} = -u\frac{\partial u}{\partial x} - w\frac{\partial u}{\partial z}, \qquad (8b.1)$$

$$\frac{\partial w}{\partial t} + U\frac{\partial w}{\partial x} - \frac{1}{R_i}\triangle w + \frac{\rho_1}{\rho_i}\frac{\partial p}{\partial z} = -u\frac{\partial w}{\partial x} - w\frac{\partial w}{\partial z}. \qquad (8b.2)$$

where $i = 1, 2$ and

$$\frac{\partial u}{\partial x} + \frac{\partial w}{\partial z} = 0. \qquad (8b.3)$$

The boundary conditions are $u = w = 0$ at $z = 0, 1$. The conditions at the interface are posed at the unknown position $z = l_1 + h(x, t)$, where $h(x, t)$ is small. The interfacial conditions are expanded as Taylor series about $z = l_1$ and truncated after the cubic term. Continuity of velocity then yields

$$h[U'] + [u] = -h[\frac{\partial u}{\partial z}] - \frac{h^2}{2}[U''] - \frac{h^2}{2}[\frac{\partial^2 u}{\partial z^2}], \qquad (8b.4)$$

and

$$[w] = -h[\frac{\partial w}{\partial z}] - \frac{h^2}{2}[\frac{\partial^2 w}{\partial z^2}], \qquad (8b.5)$$

where $[x]$ denotes x(fluid 1) - x(fluid 2). Continuity of shear stress can be expressed as

$$\frac{1}{R_1}[\frac{\mu}{\mu_1}(\frac{\partial u}{\partial z} + \frac{\partial w}{\partial x})] = \frac{1}{R_1}\left(2\frac{\partial h}{\partial x}[\frac{\mu}{\mu_1}(\frac{\partial u}{\partial x} - \frac{\partial w}{\partial z})] - h[\frac{\mu}{\mu_1}(\frac{\partial^2 u}{\partial z^2} + \frac{\partial^2 w}{\partial z\partial x})]\right.$$

$$+ 2\frac{\partial h}{\partial x}h[\frac{\mu}{\mu_1}(\frac{\partial^2 u}{\partial z\partial x} - \frac{\partial^2 w}{\partial z^2})] + (\frac{\partial h}{\partial x})^2[\frac{\mu}{\mu_1}(\frac{\partial u}{\partial z} + \frac{\partial w}{\partial x})]$$

$$\left. - \frac{h^2}{2}[\frac{\mu}{\mu_1}(\frac{\partial^3 u}{\partial z^3} + \frac{\partial^3 w}{\partial z^2\partial x})]\right). \qquad (8b.6)$$

In forming this condition we note that the normal stress condition for the basic flow implies that $[\mu U''] = 0$. The balance of normal stress for the perturbation implies that

$$[\frac{2\mu}{R_1\mu_1}\frac{\partial w}{\partial z} - p] - \frac{T}{mR_1}\frac{\partial^2 h}{\partial x^2} - h[\frac{\partial P}{\partial z}]$$

$$= \frac{\partial h}{\partial x} \frac{2}{R_1} [\frac{\mu}{\mu_1}(\frac{\partial u}{\partial z} + \frac{\partial w}{\partial x})] - h[\frac{2\mu}{R_1\mu_1} \frac{\partial^2 w}{\partial z^2} - \frac{\partial p}{\partial z}]$$

$$-(\frac{\partial h}{\partial x})^2 [\frac{2}{R_1} \frac{\mu}{\mu_1} \frac{\partial u}{\partial x} - p] + (\frac{\partial h}{\partial x})h\frac{2}{R_1}[\frac{\mu}{\mu_1}(\frac{\partial^2 u}{\partial z^2} + \frac{\partial^2 w}{\partial z\partial x})]$$

$$-\frac{h^2}{2}[\frac{2\mu}{R_1\mu_1} \frac{\partial^3 w}{\partial z^3} - \frac{\partial^2 p}{\partial z^2}] - \frac{T}{2mR_1} \frac{\partial^2 h}{\partial x^2}(\frac{\partial h}{\partial x})^2 + h(\frac{\partial h}{\partial x})^2[\frac{\partial P}{\partial z}]. \qquad (8b.7)$$

The kinematic condition for the perturbed interface can be written as

$$\frac{\partial h}{\partial t} + U(l_1)\frac{\partial h}{\partial x} - w_1 = -\frac{\partial h}{\partial x}(u_1 + hU_1')$$

$$+h\frac{\partial w_1}{\partial z} - h\frac{\partial h}{\partial x}\frac{\partial u_1}{\partial z} + \frac{h^2}{2}\frac{\partial^2 w_1}{\partial z^2} - \frac{\partial h}{\partial x}\frac{h^2}{2}U_1''. \qquad (8b.8)$$

In the linearized stability analysis u, w, p and h are proportional to $\exp(i\alpha x + \sigma t)$, where σ is an eigenvalue. Asymptotic formulas for $\alpha \gg 1$ and $\alpha \ll 1$, as well as numerical results for linear stability are discussed in section (iii).

IV.8(b)(ii) The Nonlinear Problem

We now ask whether the travelling wave solutions bifurcating from the basic flow at criticality are stable or unstable. Equations (8b.1)-(8b.8) are reduced to an ordinary differential equation by the method of normal forms.

Let Φ represent the set of unknowns (u, w, p, h). The nonlinear terms in the equations of motion are retained. Terms up to and including the third order are retained in the interfacial conditions. Thus, (8b.1)- (8b.8) have the form of equation (7b.1) of chapter III, where N_2 contains quadratic terms and N_3 contains cubic terms from the right hand sides of equations (8b.1) -(8b.8).

We let λ be the bifurcation parameter: this could be any of the physical parameters, e.g., m, l_1. At $\lambda = 0$, there is one eigenvalue, the interfacial mode, at $\sigma = -ic, c > 0$, for $\alpha = \alpha_c > 0$ and a corresponding eigenvalue $\sigma = ic$ for $\alpha = -\alpha_c$, and the rest of the eigenvalues are stable ($\mathrm{Re}\,\sigma < 0$). The eigenfunction with wavenumber α is denoted by $\zeta(\lambda)$ and that with wavenumber $-\alpha$ by $\overline{\zeta}(\lambda)$, where the overbar denotes the complex conjugate.

For $\lambda > 0$, $-ic$ becomes $-s(\lambda)$ and ic becomes the complex conjugate of $-s(\lambda)$ ($s(\lambda)$ corresponds to the $\mu(\lambda)$ of equation (7b.13) in chapter III). The eigenfunction satisfies $A\zeta(\lambda) = s(\lambda)B\zeta(\lambda)$.

The solution Φ can be decomposed (as in equation (7b.22) of chapter III) as follows:

$$\Phi = Z\zeta + \overline{Z}\overline{\zeta} + \Psi. \qquad (8b.9)$$

Here, the $Z(t)$ is the complex-valued amplitude function, and Ψ denotes a linear combination of the other eigenfunctions, and possibly generalized eigenfunctions. The adjoint eigenfunction with wavenumber α is denoted by $b(\lambda)$ and satisfies

$$(b, L(-s(\lambda))\Phi) = 0, \qquad (8b.10)$$

for every Φ where the brackets denote an inner product.

For computations, the Euclidean inner product is used, defined in analogy with equations (7b.39) - (7b.40) of chapter III as follows. In the linear calculations, the equations are discretized in the z-direction using a spectral method, namely, the Chebyshev -tau method [Orszag 1971]. This method approximates discrete eigenvalues belonging to C^∞ eigenfunctions with infinite-order accuracy. Thus, the functions that arise in the nonlinear calculations are readily available as series in the Chebyshev polynomials $T_i(z)$. Hence,

$$(g_1, g_2) = \frac{\int_A \exp i(-\alpha_1 + \alpha_2)x\,dx}{\int_A dx} \times \sum_{i=0}^{N} \overline{g_{1i}}g_{2i}, \qquad (8b.11)$$

$$g_1 = \sum_{i=0}^{N} g_{1i}T_i(z_j)\exp i\alpha_1 x,$$

$$g_2 = \sum_{i=0}^{N} g_{2i}T_i(z_j)\exp i\alpha_2 x,$$

$$z_1 = 2z/l_1 - 1, \ z_2 = 2(z-1)/l_2 + 1,$$

where A denotes one wavelength $[0, 2\pi/\alpha_c]$. Note that the complex conjugate is placed on the first member of the inner product. In the actual computations, $\lambda=0$, and the interaction equations involve the eigenfunction and adjoint eigenfunction of the linearized problem at criticality. Since the $-\alpha_1+\alpha_2$ then appear as an integral multiple of the critical wavenumber α_c, the inner product vanishes unless $-\alpha_1 + \alpha_2 = 0$.

Since $L(-s) = A - sB$, the equation satisfied by the adjoint eigenfunction is

$$A^*b(\lambda) = \overline{s(\lambda)}B^*b(\lambda). \qquad (8b.12)$$

The adjoint eigenfunction is normalized by

$$(b, B\zeta) = 1, \qquad (8b.13)$$

so that we also have $(\bar{b}, B\bar{\zeta}) = 1$. Since the eigenfunction satisfies $(A - s(\lambda)B)\zeta = 0$ and A and B are real operators, $(A - \overline{s(\lambda)}B)\bar{\zeta} = 0$. Hence , $0 = (b, (A - \overline{s(\lambda)}B)\bar{\zeta}) = (A^*b, \bar{\zeta}) - \overline{s(\lambda)}(b, B\bar{\zeta})$ and using (8b.12) for the adjoint, this yields $0 = s(\lambda)(b, B\bar{\zeta}) - \overline{s(\lambda)}(b, B\bar{\zeta})$. Therefore,

$$(b, B\bar{\zeta}) = 0. \qquad (8b.14)$$

Similar manipulations yield $(\bar{b}, B\zeta) = 0$. Further, Ψ is a linear combination of eigenvectors belonging to eigenvalues σ that are not equal to $-s(\lambda)$.

If ψ is such an eigenvector and $L(\sigma)\psi = 0$, then $0 = (b, L(\sigma)\psi)$ where $L(\sigma) = A + \sigma B$, so that $0 = (b, A\psi) + \sigma(b, B\psi)$ and using (8b.12), $0 = (s(\lambda)B^*b, \psi) + \sigma(b, B\psi) = (s(\lambda) + \sigma)(b, B\psi)$. Hence, $(b, B\psi) = 0$. Since Ψ is a linear combination of such ψ's,

$$(b, B\Psi) = 0. \tag{8b.15}$$

Also, $0 = ((A^* - \overline{s(\lambda)}B^*)b, \Psi) = (b, A\Psi) - s(\lambda)(b, B\Psi)$ and together with (8b.15), we have

$$(b, A\Psi) = 0. \tag{8b.16}$$

The inner product of the adjoint with $B\Phi$ where the expression for Φ in equation (8b.9) is used, together with (8b.15) yields $(b, B\Phi) = \overline{Z}(b, B\overline{\zeta}) + Z(b, B\zeta)$. Using (8b.14) and the orthonormality condition (8b.13),

$$Z = (b, B\Phi). \tag{8b.17}$$

The inner product of the adjoint with $L\Phi$ is $(b, (A + Bd/dt)\Phi) = (b, A\Phi) + (b, Bd\Phi/dt)$. By (8b.17), the last term is dZ/dt. The second last term is (A^*b, Φ) which, on using the equation satisfied by the adjoint becomes $(s(\lambda)b, B\Phi) = s(\lambda)Z$. This, together with equation (7b.1) of chapter III for $L\Phi$ yields

$$\frac{dZ}{dt} + s(\lambda)Z = (b, N_2) + (b, N_3). \tag{8b.18}$$

The right hand side of this equation will be shown to be of the form $\kappa|Z|^2 Z$. The purpose of this analysis is to calculate κ. If the real part of κ is negative, then the bifurcating solution is supercritical and the travelling wave solution would be stable for small amplitudes. If the real part of κ is positive, then the bifurcating solution would be unstable (see, for instance, Chow and Hale [1982]; Iooss and Joseph [1990]). Discussion about the equilibrium amplitude is found in Craik [1985]. (Fujimura [1991a] gives a comparison of alternative derivations for the Landau equation for single-fluid motions.)

The center manifold theorem is discussed after equation (7b.34) of chapter III. The center manifold is of the form $\Psi = \tau(Z, \overline{Z})$. The asymptotic expansion of the center manifold for small solutions yields $\Psi = \Psi_1 +$ higher order terms where

$$\Psi_1 = Z^2\eta + Z\overline{Z}\chi + \overline{Z}Z\chi + \overline{Z}^2\overline{\eta}. \tag{8b.19}$$

Here, χ represents the distortion to the mean flow and η is the second harmonic.

In order to evaluate the functions η and χ, an equation that yields the leading terms in Ψ in terms of the leading terms in Φ is required. To this end, note that

$$\Phi = \overline{(b, B\Phi)\zeta} + (b, B\Phi)\zeta + \Psi.$$

Analogously, any real-valued f representing the right hand side of equation (7b.1) of chapter III can be written as

$$f = (b, f)B\zeta + (\bar{b}, f)B\bar{\zeta} + g, \tag{8b.20}$$

where $(b, g) = 0$ by equation (8b.13).

A projection operator Π is defined so that $\Pi\Phi$ picks out the portion of Φ that depends on the critical mode:

$$\Psi = \Phi - \Pi\Phi. \tag{8b.21}$$

Analogously, for any real-valued function f, a projection $\tilde{\Pi}$ is defined so that $\tilde{\Pi}f$ equals the first two terms in the right hand side of (8b.20):

$$g = (I - \tilde{\Pi})f. \tag{8b.22}$$

Since

$$L\Phi = (s + d/dt)ZB\zeta + (\bar{s} + d/dt)\bar{Z}B\bar{\zeta} + L\Psi, \tag{8b.23}$$

the application of $(I - \tilde{\Pi})$ to (8b.23) yields $L\Psi = (I - \tilde{\Pi})(N_2 + N_3)$, or

$$(A + Bd/dt)\Psi = N_2 + N_3 - (b, N_2 + N_3)B\zeta - (\bar{b}, N_2 + N_3)B\bar{\zeta}. \tag{8b.24}$$

The leading terms from (8b.19) are substituted into the left hand side of above and the leading terms in Φ are substituted into the nonlinear terms in the right hand side. The coefficient of Z^2 in the resulting expression yields $(A - 2s(\lambda)B)\eta = $ coefficient of Z^2 in $N_2 - (b, N_2)B\zeta - (\bar{b}, N_2)B\bar{\zeta}$, and the coefficient of $Z\bar{Z}$ yields $A\chi - (s(\lambda) + \bar{s}(\lambda))\chi = $ coefficient of $Z\bar{Z}$ in $N_2(\zeta, \bar{\zeta}) - (b, N_2)B\zeta - (\bar{b}, N_2)B\bar{\zeta}$. For the definition of N_2 in the case of unequal arguments, see equations (7b.2) - (7b.3) of chapter III.

Many of the inner products vanish. For example, since ζ and b are proportional to $\exp i\alpha x$, $N_2(\zeta, \zeta)$ is proportional to $\exp 2i\alpha x$ so that $(b, N_2(\zeta, \zeta))$ and $(\bar{b}, N_2(\zeta, \zeta))$ both vanish. Therefore,

$$(A - 2s(\lambda)B)\eta = N_2(\zeta, \zeta), \tag{8b.25}$$

where, in actual computations, η is proportional to $\exp 2i\alpha_c x$ and λ is set to zero. Similarly, the equation for χ simplifies to

$$(A - (s(\lambda) + \bar{s}(\lambda))\chi = N_2(\zeta, \bar{\zeta}). \tag{8b.26}$$

Since ζ depends on $\exp i\alpha x$ and $\bar{\zeta}$ on $\exp - i\alpha x$, χ has no sinusoidal dependence on x. At criticality, $s(0) + \bar{s}(0) = 0$, so that $A\chi = N_2(\zeta, \bar{\zeta})$. The component w of χ satisfies $dw/dz = 0$ by incompressibility. Since $w = 0$ at $z = 0, 1$, $w = 0$ in the entire domain. Denote the quadratic terms in the right hand sides of equations (8b.1) and (8b.2) by f_1 and f_2 in fluid 1, and by f_3 and f_4 in fluid 2 respectively. Denote the quadratic terms on the right hand sides of equations (8b.4)-(8b.8) by f_5, f_6, f_7, f_8, and f_9, respectively. We note that f_9 in $N_2(\zeta, \bar{\zeta})$ vanishes. Putting $d/dx = 0$, the velocity component u in χ satisfies:

$$\frac{-1}{R_1}\frac{d^2u}{dz^2} \text{ in fluid } 1 = f_1, \tag{8b.27}$$

$$\frac{-1}{R_2}\frac{d^2u}{dz^2} \text{ in fluid } 2 = f_3, \tag{8b.28}$$

$$u = 0 \text{ at } z = 0, 1,$$

$$h[dU/dz] + [u] = f_5, \tag{8b.29}$$

and

$$\frac{1}{R_1}\left[\frac{\mu}{\mu_1}\frac{\partial u}{\partial z}\right] = f_7. \tag{8b.30}$$

There is no condition on h and it could be set to anything. However, there is a nullspace for the linear operator L with eigenvalue 0 which affects the component in Φ that does not depend sinusoidally on x. One can solve the above problem for u with zero right hand sides, and find an eigenfunction with $h =$ any constant. This is a mode that simply shifts the interface up or down. Because this violates the condition that the volumes of the fluids are given, we must put

$$h = 0 \tag{8b.31}$$

for χ. Therefore the problem for p becomes

$$\frac{\partial p}{\partial z} \text{ in fluid } 1 = f_2, \tag{8b.32}$$

$$r\frac{\partial p}{\partial z} \text{ in fluid } 2 = f_4, \tag{8b.33}$$

$$-[p] = f_8. \tag{8b.34}$$

The problem for p requires an additional condition. This corresponds to the fact that the operator L has a nullspace consisting of $p=$constant, $u = w = h = 0$, and the addition of an arbitrary constant to the pressure does not affect the problem. To rule out this eigenfunction, p is made unique by setting

$$p \text{ for fluid } 1 = 0 \text{ at } z = l_1, \tag{8b.35}$$

so that equation (8b.34) becomes

$$p \text{ for fluid } 2 = f_8 \text{ at } z = l_1. \tag{8b.36}$$

Finally, the functions involved in (8b.19) have been calculated so that the solution Φ, known up to quadratic terms in the neighborhood of criticality, can be substituted into the nonlinear terms in the right hand side of (8b.18). Again, many of the inner products vanish, and the final equation for the amplitude function is

$$\frac{dZ}{dt} + s(\lambda)Z = \kappa|Z|^2Z,$$

$$\kappa = (b, 4N_2(\zeta, \chi) + 2N_2(\overline{\zeta}, \eta) + 3N_3(\zeta, \zeta, \overline{\zeta})). \tag{8b.37}$$

This is the Landau equation and κ is the Stuart-Landau coefficient (see §8.3 (iii) of Craik [1985]).

The rest of this section concerns comparisons made with known results for single-fluid flow [Pekeris and Shkoller 1967; Reynolds and Potter 1967; Davey, Hocking and Stewartson 1974] and with the results of Blennerhassett [1980] for a two-layer system. Agreement for the single-fluid case is found to be good. Section (iv) is devoted to the comparison with the two-layer system.

The computer code to carry out the calculation of κ was checked against the known values for the one-fluid Poiseuille flow. Two entries in table 1 in Pekeris and Shkoller [1967] were considered. The first is their $R = 4000, \alpha = 1.0$. Converting to the present notation, $R_1 = 8000, \alpha = 2.0$. Our other parameters were chosen to be $l_1 = 0.5, m = r = 1, T = 1/F^2 = 0$. With 25 Chebyshev modes in each fluid, the resulting least stable shear mode eigenvalue is $\sigma = (-.9892E - 2, -.55708)$. The complex parameter 2β, listed in their table 1, is related to our κ by $\overline{\kappa} = -16\beta|w(l_1 = 0.5)|^2/\alpha^2$, where the present code yields $|w(l_1 = 0.5)| = 0.7402$. This would imply κ should be (21.149,-110.346). However, Pekeris and Shkoller carry out the nonlinear part of the calculations by putting the growth rate $\mathrm{Re}\,\sigma$ equal to 0 even when it is not. This is an approximation they use for regions close to criticality. When this adjustment is made in the present computations, $\kappa = (21.135, -110.180)$. Without this adjustment, the computed value is (18.57,-113.30). A second check with their table 1 is at their $R = 6000, \alpha = 1.0$. Our parameters are $R_1 = 12000, \alpha = 2.0, l_1 = 0.5, \sigma = (.646E - 3, .51963)$ and $\kappa = (23.59, -147.69)$ with the adjustment that $\mathrm{Re}\,\sigma = 0$, and (23.66,-147.31) without. Their β_i/β_r is 6.24 while our κ_i/κ_r is 6.26.

The next discussion concerns the results of Blennerhassett [1980] for a two-layer system. Whilst the present analysis is restricted to uniform wavetrains, comparison with his modulated wave solutions provides some unexpected results. In particular, the results below (see also section (iv)) indicate that the Landau constant κ is strongly dependent on the normalization chosen for the distortion to the mean flow in the case of uniform waves, and further that it is different from the value of κ obtained when modulation is allowed.

Section (iv) shows the results for some of the data corresponding to tables 2 - 5 and 7 - 10 of Blennerhassett [1980]. A comparison reveals disagreement in the nonlinear calculations. There are two reasons for the difference, apart from the possibility of bugs in the computer code. The first reason is the same one that explains the difference between the one-fluid results of Pekeris and Shkoller [1967] and those of Reynolds and Potter [1967] and Davey, Hocking and Stewartson [1974], and lies in the calculation of χ, the distortion to the mean flow. These works represent two different approaches. In our case, equations (8b.27) - (8b.36) show that the pressure gradient in the x-direction in the entire nonlinear analysis is fixed, while the combined volume flux is not fixed. In Blennerhassett [1980], Reynolds and Potter [1967] and Davey, Hocking and Stewartson [1974], the streamfunction is used, and is set to zero at the walls. This fixes the combined volume flux in the entire analysis, while the pressure gradient in the x-direction is not fixed. In our notation, the latter approach is implemented by changing the calculation of the function χ in equations (8b.27) - (8b.30), for which the unknowns would then be the streamfunction and a constant dp/dx. Let ψ denote the streamfunction for χ. Then (8b.27)-(8b.28) are replaced by

$$\frac{-1}{R_1}\frac{d^3\psi}{dz^3} + \frac{\partial p}{\partial x} \text{ in fluid } 1 = f_1, \tag{8b.38}$$

$$\frac{-1}{R_2}\frac{d^3\psi}{dz^3} + \frac{\partial p}{\partial x} \text{ in fluid } 2 = f_3. \tag{8b.39}$$

Equations (8b.29)-(8b.30) are replaced by

$$\frac{\partial\psi}{\partial z} = 0 \text{ at } z = 0, 1,$$

$$[\frac{\partial\psi}{\partial z}] = f_5, \tag{8b.40}$$

$$\frac{1}{R_1}[\frac{\mu}{\mu_1}\frac{\partial^2\psi}{\partial z^2}] = f_7. \tag{8b.41}$$

In addition,

$$\psi = 0 \text{ at } z = 0, 1,$$

and the Taylor expansion around $z = l_1$ for the continuity of the streamfunction across $z = l_1 + h(x, t)$ reads

$$[\![\psi]\!] = f_{10}, \tag{8b.42}$$

where f_{10} is the quadratic term from $N_2(\zeta, \bar{\zeta})$ denoted by

$$f_{10} = -h[\![u]\!] - \frac{h^2}{2}[\![dU/dz]\!].$$

Eqns. (8b.31) - (8b.36) stay as is. This approach was also programmed in order to check our code with the one-fluid results of Reynolds and Potter [1967] and Davey, Hocking and Stewartson [1974] at the first criticality of plane Poiseuille flow: $R_1 = 11544.44, \alpha_c = 2.04112, l_1 = 0.5, m = r = 1, T = 1/F^2 = 0$. Our κ is related to the $k = (30.8, -173)$ in equation (2.35) of Davey, Hocking and Stewartson [1974] by $\kappa = 2k|w(l_1 = 0.5)|^2/\alpha_c^2$, where we find $|w(l_1 = 0.5)| = .7457$. This conversion yields $\kappa = (32.9, -184.93)$, compared with the present calculation, with 35 Chebyshev modes in each fluid, of (33.0, -184.6). On the other hand, when the pressure gradient is fixed, $\kappa = (31.7, -153.53)$, so that either approach yields a similar result for the real part of κ, which is of interest. These nonlinear calculations were also checked with $l_1 = 0.3$.

The one-fluid combined Couette-Poiseuille flow of table 4 in Reynolds and Potter [1967] was checked for agreement. The value of $b^{(2)}/a^{(2)}$ for $u_w = 0.3$ in their table 4 is 2.006, and this corresponds to our κ_i/κ_r. The same ratio from table A5 of Blennerhassett [1980] is 2.06. Our computation with fixed volume flux, using 35 Chebyshev modes in each fluid at $R_1 = 38929.371, G = .0001712, l_1 = 0.5, \alpha_c = .958, m = r = 1, R = 1/F^2 = 0$ yields $\sigma = (.43E - 5, -.2222), \kappa = (193.56, -396.77), \kappa_i/\kappa_r = 2.05$. When the pressure gradient is fixed, $\kappa = (192.91, -384.18)$. Thus, there is not much difference in $Re \, \kappa$ between the two approaches. This is not always true for two-fluid problems as evidenced in section (iv); for example, tables 8.8 and 8.10 contain situations where the $Re \, \kappa$ from the two approaches have opposing signs!

The second reason for the difference between Blennerhassett's approach and the present one is the fact that he has a dispersion relation (his equation (2.37)), whereas there is none in the present work. A standard idea in bifurcation theory used here is that the wavenumber is considered to be fixed and the harmonics are taken into the nonlinear solution. This gives rise to the function χ, which has no dependence on x. This leads to equation (8b.31). This approach is different from his, which uses a formal method, employing multiple scales for describing amplitude modulations. Our χ corresponds to his ϕ_{02} in his equation (2.66). His ϕ_{02} has an x-dependence, where the wavenumber is assumed to be small but non-zero, and then the limit as the wavenumber approaches zero is taken. For every non-zero wavenumber, his interface has a non-zero amplitude h, in contrast with our equation (8b.31). This implies that for small wavenumbers, the volume ratio of the two fluids is not equal to the given original value over large lengths of the channel, but on average, the volume ratio is preserved. His kinematic condition for ϕ_{02} is, from equation (8b.8), essentially

$$\sigma h + i\alpha U(l_1)h = -i\alpha\phi + \text{nonlinear terms}, \qquad (8b.43)$$

where ϕ is the streamfunction. At small α, his σ is therefore $O(\alpha)$ and h is $O(\phi)$ so that the limit $\alpha \to 0$ yields a non-zero h. A simple analogy of this type of limiting process is to consider the function $\cos \alpha x$, and to think of this as the interface shape. The average height of this over one wavelength is then $\int_0^{2\pi/\alpha} \cos \alpha x \, dx = 0$ for any finite non-zero α. Now let $\alpha \to 0$ in $\cos \alpha x$. Since $\lim_{\alpha \to 0} \cos \alpha x = 1$, the average height is no longer zero.

This may be one of the reasons for the numerical difference between his results and the present one. (However, in the single-fluid case, this issue does not arise and the comparisons made above with Davey, Hocking and Stewartson [1974], where amplitude modulation in space is allowed too, gives satisfactory agreement.)

The computations presented in the tables and figures that follow have been tested for convergence. A consistency test has been performed for the case where the layer densities are the same (or gravity is absent) by interchanging the layers as follows. In the original data, let ρ_1, ρ_2, l_1 and l^* be fixed. Values for μ_1 and μ_2 are chosen, and this yields the dimensionless upper plate speed U_p. A choice for R_1 yields the value of the interfacial speed U_i since $U_i = R_1\mu_1/(\rho_1 l^*)$. A choice for T yields the value of the dimensional surface tension coefficient as $T\mu_2 U_i$. The dimensional speed of the upper plate is $U_p^* = U_p U_i$. In order to interchange the layers, the values for μ_1 and μ_2 are interchanged. The plate speed U_p^* remains as before and this yields a new value for U_i, and consequently for R_1 and T. Since $\sigma t = (\sigma U_i)t^*/l^*$, the quantity σU_i is compared for the two sets of data. It is found that in the linear calculations, $\text{Re}\,\sigma U_i$ stays the same. However, $\text{Im}\,\sigma U_i$ changes for the following reason. If the flowfield is simply turned upside down (e.g. if $l_1 = 0.5$, reflection across the centerline), then the lower plate would move and the results should remain the same. In the above transformation of 'interchanging layers', not only have the fluids been exchanged but the shear has been reversed and a uniform flow has been superposed (in order to keep the lower plate fixed). This accounts for the difference in the $\text{Im}\,\sigma U_i$, and, in the nonlinear calculations, in the reversal of the sign of $\text{Im}\,\kappa$. The values for κ for the two sets of data then agree to within a constant multiple which arises from the normalizations involved in the functions present in equation (8b.37).

In conclusion, it is noted that there is no 'unique' way of carrying out the nonlinear analysis, and it is difficult to say which approach is relevant to practical situations.

IV.8(b)(iii) Neutral Stability Curves and Numerical Results

The asymptotic analysis of the interfacial eigenvalue for short-wave disturbances was first carried out by Hooper and Boyd [1983] (see section IV.5). They showed that short-wave instability is possible if, for example, there is not enough surface tension, or the density stratification is adverse. In particular, the presence of viscosity stratification alone causes short-wave instability. Their analysis proceeds by rescaling the vertical variable z to η, such that $z - l_1 = \eta/\alpha$ and setting $\eta = O(1)$, $\alpha \gg 1$. The velocity is expanded in a series in $1/\alpha$ and

$$\sigma + i\alpha U(l_1) \sim c_0 + \frac{c_1}{\alpha} + \frac{c_2}{\alpha^2} + \dots \qquad (8b.44)$$

for $\alpha \gg 1$. The boundary conditions at $z = 0, 1$ become conditions on the decay of the velocity as $\eta \to \pm\infty$. In the normal stress condition, one assumes a distinguished limit: $\alpha^3 T = O(1)$, $\alpha[\![\partial P/\partial z]\!] = O(1)$, where P is the basic pressure field given by equation (4b.2) of section IV.4 (b). The results are $c_0 = c_1 = 0$ and

$$\sigma \sim -i\alpha U(l_1) - \frac{mR_1}{\alpha^2 2(1+m)}\left(\frac{\alpha^3 T}{mR_1} - \frac{\alpha}{F^2}\left(\frac{1}{r} - 1\right)\right.$$

$$\left. -\left(\frac{1}{l_1} - \frac{GR_1 l_1}{2}\right)^2\frac{(1-m)}{(1+m)}\left(1 - \frac{m^2}{r}\right)\right). \qquad (8b.45)$$

Stability to long-wave disturbances is necessary but not sufficient for linear stability. The asymptotic analysis of the interfacial mode for long-wave disturbances was first carried out by Yih [1967] for two-layer Couette flow and for the two-layer Poiseuille flow. The results for combined Couette-Poiseuille flow are implicitly contained in §2.1.1 of Blennerhassett [1980]; these results are made explicit in section IV.4 (b).

We next discuss the calculation of the interfacial eigenvalue for the general case without the asymptotic approximations. The neutral stability curves presented in the figures of this section are determined numerically. The full linearized equations are discretized with the Chebyshev-tau method [Orszag 1971], which has infinite-order accuracy for C^∞-eigenfunctions. The accuracy of the computer code in calculating the interfacial eigenvalue was checked against the long-wave formulas of Yih, the short-wave formula of Hooper and Boyd, and tables 2 - 5 of Blennerhassett (see section (iv)). The code was also checked in the case when the fluids are the same, to retrieve the one-fluid modes, at the criticality of plane Poiseuille flow and at the higher Reynolds number listed in table 5 of Orszag [1971].

For plane Poiseuille flow, neutral stability curves for the interfacial and shear modes have been presented by Yiantsios and Higgins [1988]. Their figure 2b gives the neutral stability diagram for long-wave disturbances plotted in the m-n plane where n is their thickness ratio. Other neutral stability curves for the interfacial mode are plotted in the $\alpha - n$ or the α-Reynolds number plane. The shear mode of instability takes place at relatively higher Reynolds numbers and will not be pursued here.

Figure 8.3 displays neutral stability curves for Couette flow for a typical situation at low Reynolds number, with equal density for $m \leq 1$. The dashed curve is derived from the asymptotic formula for long waves and is the neutral stability curve for situations, in which instabilities from order one and shorter wavelengths are suppressed by large amounts of surface tension. The bold curve is for T=0.01. As the surface tension increases, the curve moves towards that of the long waves. The region to the left of each curve is linearly stable and the region to the right is unstable. For small m and for m close to 1, the neutral stability curve for T=0.01 coincides with the long-wave curve. This is explicitly illustrated in figure 8.4 which shows the critical wavenumbers for the case of T=0.01 from figure 8.3.

Neutral stability curves for $m \geq 1$ are displayed in figures 8.5 and 8.6. In figure 8.5, the dashed curve is again obtained from the asymptotic formula for long waves. The bold curves are for T=0.01 and 0.1, showing how the curves tend to the one for long waves as the surface tension increases. Here, the region to the right of each curve is stable and that to the left is unstable. As m increases, the bold curves tend toward $l_1 = 1$, away from the long-wave curve. This exemplifies the tendency for thin-layer arrangements to be linearly stable. Figure 8.6 illustrates that the critical wavenumber tends to infinity as m increases. For such short waves, equation (8b.45) is

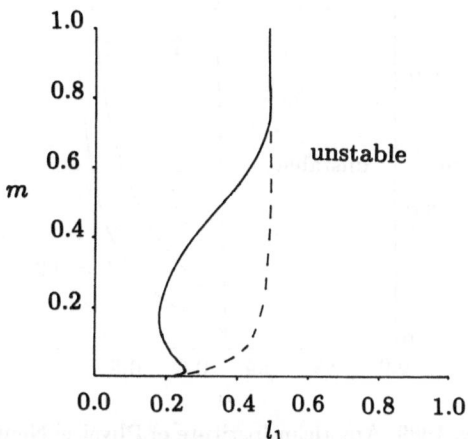

Fig. 8.3. [Renardy, 1989, American Institute of Physics] Neutral stability curves for Couette flow for $R_1 = 10, G = 0, r = 1, 1/F^2 = 0, m \le 1$ in the $l_1 - m$ plane. The bold curve is for T=0.01. The dashed curve is asymptotic for long waves: this also represents neutral stability for $T \ge 0.1$ when surface tension is large enough to suppress instability arising from waves which are not long. The region to the right of each curve represents instability and the region to the left represents linear stability.

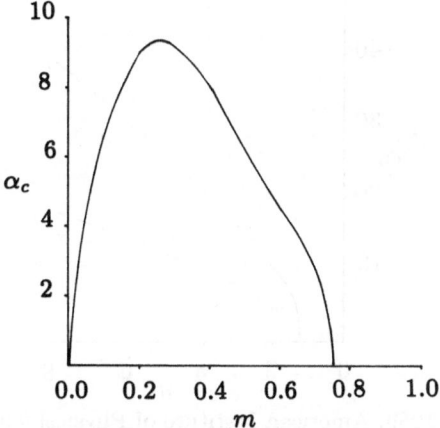

Fig. 8.4. [Renardy, 1989, American Institute of Physics] Neutral curves for the case T=0.01 of figure 8.3, against m for $m \le 1$.

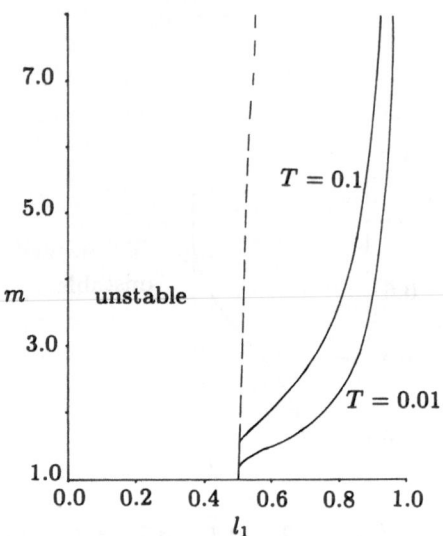

Fig. 8.5. [Renardy, 1989, American Institute of Physics] Neutral stability curves for $R_1 = 10, G = 0, r = 1, 1/F^2 = 0, m \geq 1$ in the $l_1 - m$ plane. The dashed curve is for long waves. The two bold curves are for T=0.01 and T=0.1. The region to the left of each curve represents instability.

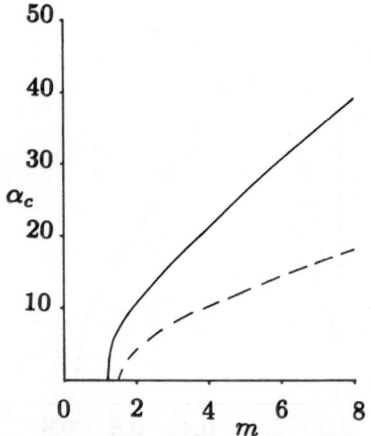

Fig. 8.6. [Renardy, 1989, American Institute of Physics] The critical wavenumbers α_c for the cases T = 0.01, 0.1 in figure 8.5, against m.

valid and the neutral curve is given explicitly by putting the real part of σ to zero (also see Blennerhassett and Smith [1987]).

Nonlinear calculations along the curves of figures 8.3 - 15 can be carried out provided there is a single finite critical wavelength (see the latter half of chapter VIII for further discussion). In the case of long waves, we are forced to calculations of the type given in section IV.8(a). Nonlinear calculations were carried out at several points in the range of viscosity ratios $0.005 \leq m \leq 0.727$ and $1.3 \leq m \leq 8$. These yield supercritical bifurcations. The values of κ for several situations are tabulated in table 8.1. In the column caption, F.P.G. stands for 'fixed pressure gradient' and refers to the approach where the pressure gradient in the x-direction is fixed in the nonlinear calculations and the equations for χ are (8b.27)-(8b.36). F.V.F. stands for 'fixed volume flux' and refers to the approach where the combined volume flux is fixed in the nonlinear calculations and the equations for χ are (8b.31)-(8b.36) and (8b.38) - (8b.42). The data shown have been checked for convergence. In table 8.1, both the fixed volume flux approach and the fixed pressure gradient approach are quantitatively similar. This need not be the case when the parameters are more extreme; for example, see the results for large Reynolds number and large m in section (iv).

The introduction of a small adverse density stratification on the arrangement in figure 8.5 at T=0.1 is shown in figure 8.7. The main difference here is that the long-wave curve approaches $l_1 = 1$ as m approaches 1. This is because at $m = 1$, the adverse density stratification yields long-wave instability for all arrangements with a heavier fluid above($l_1 < 1$). Comparing with figure 8.5, the stable region is decreased slightly, as expected, but the overall difference is slight. Nonlinear results are obtained at several points along the neutral stability curve for T=0.1 where $1.72 \leq m \leq 8.0$, i.e., the critical situation is not for long waves. Some of these results are presented in table 8.2 and they give rise to supercritical bifurcations so that waves are supported even in the presence of an adverse density stratification. A more pronounced density stratification is $r = 0.95$, $1/F^2 = 10.0$, with the other parameters as in figure 8.5, and T=0.1. This situation is shown in figure 8.8. Nonlinear computations at several points along the neutral stability curve for the range of viscosity ratio $1.5 \leq m \leq 7.0$ again yield supercritical bifurcations.

Table 8.1. Nonlinear results for Couette flow, $R_1 = 10$, $G = 0$, $r = 1$, $1/F^2 = 0$, $T = 0.01$

m	l_1	α	Re σ	κ(F.P.G.)	κ(F.V.F.)
.005	.23	1.0	.39E-4	-.487,32.4	-.443,50.2
.05	.215	4.5	.14E-5	-4.31,129.9	-4.31,127.3
.25	.191	9.4	-.19E-3	-68.7,1509	-67.4,1274
.5	.372	6.3	-.16E-4	-156,980	-157,963
.727	.495	2.3	-.32E-6	-3.98,115	-3.98,114
1.3	.523	4.0	.85E-4	-37.3,-531	-37.3,-528
2.0	.755	10.9	.52E-4	-1063,-4210	-1062,-4140
4.0	.915	21.0	-.39E-3	-199,-7395	-194,-6140
6.0	.951	30.0	-.66E-3	-198,-18403	-182,-13730
8.0	.966	39.0	-.38E-3	-284,-37960	-251,-27085

Table 8.2. Nonlinear results for Couette flow, $R_1 = 10$, $G = 0$, $r = .95$, $1/F^2 = 1$, $T = 0.1$

m	l_1	α	Re σ	κ(F.P.G.)	κ(F.V.F.)
1.72	.59	2.1	-.23E-3	-8.1,-90	-8.2,-86
2.0	.621	4.3	.47E-3	-43,-272	-43,-260
3.0	.767	7.9	.72E-3	-173,-935	-172,-837
4.0	.84	10.2	-.68E-3	-166,-1582	-162,-1283
5.0	.879	12.3	-.46E-3	-175,-2501	-166,-1867
8.0	.932	18.5	-.53E-3	-277,-7414	-245,-4807

Computations for a Poiseuille flow were performed for the equal density case as shown in figure 8.9: $\rho_1 = 1$, $\mu_1 = 1$, $l^* = 5$, $G^* = 2$, $\mu_2 = 1/m$, $m \geq 1$. It is more appropriate to keep these dimensional quantities fixed rather than to keep the dimensionless parameters fixed, for the calculation of neutral stability curves in the $m-l_1$ plane. This is because for a Poiseuille flow, the interfacial speed $U_i \to 0$ as $l_1 \to 1$ and the dimensionless parameters G and T contain U_i in their denominators. Thus, keeping G and T fixed in the $m - l_1$ plane would not reflect the parameter variations that would be analogous to an experiment; e.g., the stability of the thin-layer arrangement (l_1 close to 1) for large m would not show up. Therefore, with regard to experiments, it is more natural in this case to keep the above quantities fixed and to vary μ_2.

The neutral stability curves in figure 8.9 are similar to those of the

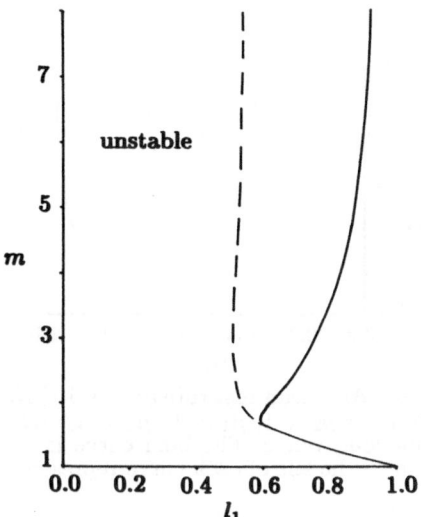

Fig. 8.7. [Renardy, 1989, American Institute of Physics] Neutral stability curves for $R_1 = 10, G = 0, r = 0.95, 1/F^2 = 1.0, m \geq 1$, in the $l_1 - m$ plane. The dashed curve is for the long wave case. The bold curve is the case T=0.1. The region to the left of each curve represents instability, and the region to the right represents linear stability.

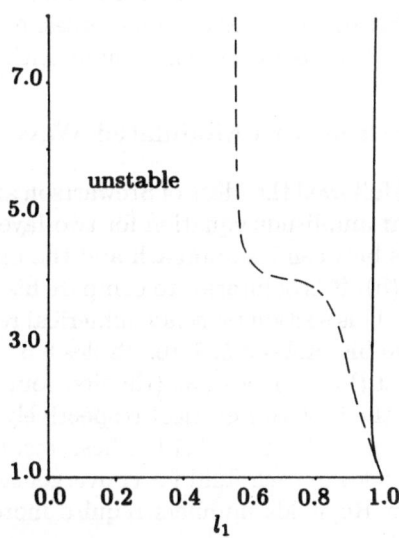

Fig. 8.8. [Renardy, 1989, American Institute of Physics] Neutral stability curves for $R_1 = 10, G = 0, r = 0.95, 1/F^2 = 10.0, m \geq 1$ in the $l_1 - m$ plane. The dashed curve is for the long wave case. The bold curve is for T=0.1. The region to the left of each curve represents instability.

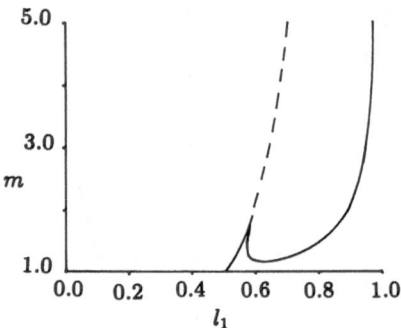

Fig. 8.9. [Renardy, 1989, American Institute of Physics] Neutral stability curves for Poiseuille flow with $\rho_1 = \rho_2 = 1$, $\mu_1 = 1$, $\mu_2 = 1/m$, $l^* = 5$, $G^* = 2$, $m \geq 1$. The dashed curve is for long waves. The bold curve is for dimensional surface tension 0.05. The region to the left of each curve represents instability.

Couette flow of figure 8.5. There is a difference in the angle that the curve for surface tension 0.05 makes at the junction with the long wave curve. In the Poiseuille flow, there are situations where there is linear stability for two disjoint intervals of l_1 at fixed m; e.g., at $m = 1.2$, stability for $0.53 \leq l_1 \leq 0.58$ and $l_1 \geq 0.69$. In particular, at the junction around $m = 1.7, l_1 = 0.58$ are two critical situations, one for long waves and one at a single wavenumber α. This junction represents a cross-over of the two neutral stability curves. For large m, there is the familiar stability of thin layers; e.g., at $m = 5$, stability for $l_1 \geq 0.97$. As in the Couette case of figure 8.5, the critical wavenumber α_c increases as m increases. Nonlinear calculations at several points along the bold curve show supercritical bifurcations.

IV.8(b)(iv) Comparison with Modulated Wave Solutions

Blennerhassett [1980] followed the ideas of Stewartson and Stuart [1971] and analyzed the nonlinear amplitude equation for two-layer Couette-Poiseuille flow. Some differences between his approach and the present one have been mentioned in section (ii). It is of interest to compare his results with those of the present approach. This section presents numerical results for some of the situations tabulated in his tables 2-5, 7-10. Tables 8.3 - 8.6 here correspond to his tables 2 (the first three entries), 3a (the first four entries),4a (the first three entries) and 5 (the first two entries) respectively. The corresponding nonlinear results are in tables 8.7 - 8.10. These results require less than seventy Chebyshev modes in each fluid for convergence. Those data sets in his tables with higher Reynolds numbers require more Chebyshev modes and are not pursued here.

Table 8.3. Critical conditions for P.P.F. profile.

R_1	α	T	$1/F^2$
2983.0065	3.6354	24379.048	6.8033538
3490.6171	5.0852	41667.624	39.748162
6190.3348	7.5008	46991.229	101.10745

Table 8.4. Critical conditions for B.L.1 profile.

G	R_1	α	T	$1/F^2$
.1504E-2	1837.6977	2.8322	39572.811	17.926
.6614E-3	4179.0355	3.5002	34803.66	27.731247
.7767E-3	3558.7535	3.9178	81739.704	305.92551
.3807E-3	7259.5709	4.5	80140.125	588.13848

Table 8.5. Critical conditions for B.L.2. profile.

G	R_1	α	T	$1/F^2$
.5455E-3	4124.2369	2.62392	26449.572	12.012078
.5008E-3	4491.3724	4.0956	48575.041	81.028383
.2182E-3	10312.35	4.6311	42312.101	122.96174

Table 8.6. Critical conditions for P.C.F. profile

R_1	α	T	$1/F^2$
1292.1093	3.0004	56282.285	36.260272
8594.5979	4.562	33845.847	52.451793

Table 8.7. Nonlinear results for P.P.F. profile

σ	κ (F.P.G.)	κ (F.V.F.)
.124E-4-8.4099i	-5.648-56.7256i	-5.4912-78.376i
.9095E-5-19.394i	-.4945E-1-6.51324i	-.1573E-2-16.132i
-.184E-5-35.282i	-.3489-18.988i	.19915-56.153i

Table 8.8. Nonlinear results for B.L.1. profile

σ	κ (F.P.G.)	κ (F.V.F.)
-.2E-2-9.332i	-.27-6.91i	.012-30.06i
-.1E-2-12.601i	-.49+4.41i	-.34-19.85i
-.2E-2-37.374i	-.37 +1.15i	-.20-12.6i
-.2E-2-54.720i	-.11-.94i	-.017-17.04i

Table 8.9. Nonlinear results for B.L.2 profile

σ	κ (F.P.G.)	κ (F.V.F.)
.3E-5-7.740i	-.395+.78i	-.25-37.51i
.5E-5-22.046i	-.231-9.06i	-.109-34.72i
-.8E-5-28.077i	-.075-12.88i	-.022-45.02i

Table 8.10. Nonlinear results for P.C.F. profile

σ	κ (F.P.G.)	κ (F.V.F.)
.44E-4-12.849i	-.14-.32i	.135-20.46i
.116E-4-19.128i	-.4974+6.402i	-.3852-20.955i

In all situations, fluid 1 is water and fluid 2 is air. The viscosity and density ratios are $m = 64.009897$, $r = 815.59184$. Table 8.3 is his P.P.F. (two-layer plane Poiseuille flow) profile with $l_1 = 0.5$, table 8.4 is his B.L.1 (boundary layer 1) profile with $l_1 = 0.5$, table 8.5 is his B.L.2 profile with $l_1 = 2/3$, and table 8.6 is his P.C. F. (two-layer plane Couette flow) profile with $l_1 = 0.5$. Tables 8.7 - 10 present the corresponding critical eigenvalues and the κ of (8b.37). In the column caption, F.P.G. stands for 'fixed pressure gradient' and refers to the approach where the pressure gradient in the x-direction is fixed in the nonlinear calculations and the equations for χ are (8b.27)-(8b.36). F.V.F. stands for 'fixed volume flux' and refers to the approach where the combined volume flux is fixed in the nonlinear calculations and the equations for χ are (8b.31)-(8b.36) and (8b.38) - (8b.42). The data shown have been checked for convergence.

The critical eigenvalues agree with those of Blennerhassett. For example, his first b.l.2 profile in table 4a lists $c_r = 1.180$, which, in our notation yields that the imaginary part of σ should be -7.7433. Using 35 Chebyshev modes in each fluid, the computed result with quadruple precision on a Vax 11/785 was $\sigma = 0.699E - 6 - 7.7403i$.

Comparing the fixed volume flux results for κ in tables 8.7 - 8.10 with those of Blennerhassett, both approaches predict the same type of bifurcation, supercritical ($Re\ \kappa < 0$) or subcritical ($Re\ \kappa > 0$), except for the first entries in B.L.1 and P.C.F. profiles. In Blennerhassett, these entries are supercritical. Both approaches yield supercritical bifurcations for most of the cases displayed, and both yield subcritical for the last entry in the P.P.F. profile. The fixed pressure gradient approach yields supercritical bifurcations for all the cases displayed. These comparisons therefore reveal some cases where the agreement is not even qualitative.

It is interesting that there are flow profiles where the bifurcation is supercritical for one formulation and subcritical for another (F.P.G. or F.V.F formulation). This is obviously not intended to imply that the flow might be both supercritical and subcritical 'at the same time': the two formulations are two quite distinct problems.

IV.9 Two-Layer Couette Flow of Upper-Convected Maxwell Liquids

This section is based on the paper of Y. Renardy [1988 b]. Stratified flows composed of liquids that exhibit non-Newtonian (or viscoelastic) properties occur in many industrial processes: examples (see chapter I.1) include the formation of bicomponent fibers. The modeling of these flows is complicated by the question of how to incorporate the elastic and viscous properties of the liquids.

The study of viscoelastic effects on the stability of layered flows has been conducted on, for example, the flow down an inclined plane [Li 1970], plane Couette flow [Waters 1983; Li 1969; Waters and Keeley 1987; Chen 1991b], plane Poiseuille flow [Khan and Han 1977; Anturkar et al. 1990b], and core-annular flow [Chen 1991a, 1992; Chen and Joseph 1992]. Many of these have utilized the method of asymptotic analysis for long waves of

Yih. These studies have shown that the elasticity in the liquid can stabilize or destabilize the flow depending on the values of the depth ratio, viscosity ratio, density ratio and other parameters.

For two-layer Couette flow, Li [1969a] used the constant-viscosity Ol-droyd model to do the asymptotic analysis for long waves of the interfacial eigenvalue. His long-wave analysis yields a complicated expression which requires some computations. Li presented results for the growth rate.

The viscosity of a non-Newtonian fluid usually decreases as shearing increases. For two-layer Couette flow, the effect of a shear-dependent viscosity was analyzed for long waves with the inelastic power-law liquids [Waters 1983]. It was found that the ratios of the power-law parameters for each layer have a significant role in determining the stability. Later, the four-constant Oldroyd model [Waters and Keeley 1987] was used to investigate both the shear-thinning effect of the power-law model, and the elasticity effect previously considered by Li.

The early investigations [Li 1969; Waters and Keeley 1987; Anturkar et al. 1990b] conclude that fluid elasticity has no effect on the stability for long waves when the viscosities of the two liquids are the same. On the other hand, Renardy [1988 b] found a new instability in the short-wave limit which is solely due to the difference in the elastic properties of two upper convected Maxwell fluids in plane Couette flow. This instability persists even when the viscosities of the two fluids are the same. Chen [1991 a] also found this type of instability in the long-wave limit for concentric co-extrusion flow of two viscoelastic fluids in a circular pipe.

Chen [1991b] re-examined the problem of plane Couette flow of two viscoelastic fluids in the long-wave limit considered by Li [1969a] and Waters and Keeley [1987] and found an error in their interfacial shear stress condition. This mistake is not uncommon. Chen noted that for the basic flow, the jump in the first normal stress difference across the unperturbed interface does not need to be balanced. When the interface is perturbed, it is this unbalanced jump (see equation (9a.12)) that causes the purely elastic instability found in section IV.9(b). The corrected result shows that *the instability resulting from the different elastic properties persists in the long-wave limit, even in the absence of viscosity stratification*, consistent with the short-wave calculation in section IV.9(b). It is then apparent that this type of elastic instability will occur for any disturbance wavenumber.

In this chapter, the linear stability of a two-layer Couette flow of upper-convected Maxwell liquids in the two-dimensional plane is considered. The fluids have different densities, viscosities, and relaxation times, with surface tension at the interface. At low speeds, the interfacial mode may become unstable, while other modes stay stable, and we concentrate on such a case where an analysis of the interfacial mode alone is sufficient for the study of linear stability.

In section IV.9 (b), an asymptotic analysis for short waves of the inter-facial mode is presented. As the viscosity difference increases, the range of relaxation times for which there is short-wave stability widens. We note that in the Newtonian case, short-wave instabilities are very weak and are easily stabilized by surface tension (see section IV.5). In contrast, the contribution from the stratification in relaxation times is found to be an order of magnitude larger than those due to either density or viscosity stratification.

In section IV.9 (c), a numerical study of instability for all values of α is presented. We focus on a particular situation that exemplifies the thin-layer effect (see section IV.6) in the presence of elasticity (relaxation time) stratification. This is a linearly stable situation with a thin layer of a much less viscous fluid below a heavier one. In the Newtonian case, surface tension is required to suppress short-wave instabilities, but here, these can be suppressed by the stratification of relaxation times.

Chen and Joseph [1992] do an analysis for the concentric core-annular flow of upper-convected Maxwell liquids. Their short-wave results are analogous to those here. They show that in the short-wave limit, elastic effects dominate over viscous effects, while in the long-wave limit, these effects are of the same order of magnitude. The short-wave instability is stabilized by surface tension, and they estimate the cut-off wavelength for stabilization by surface tension. They introduce a two-fluid model for the problem of slip at the wall and discuss some speculative ideas about the relationship of the elastic instability to sharkskin (section IV.9(d)).

IV.9(a) Governing Equations

Two immiscible fluids with different densities ρ_i $(i = 1, 2)$, viscosities μ_i and relaxation times λ_i, are situated in the (x^*, z^*) plane in layers between walls located at $z^* = 0$ and l^*. Fluid 1 occupies $0 \leq z^* \leq l_1^*$ and fluid 2 occupies $l_1^* \leq z^* \leq l^*$. The upper wall has velocity $(U^*, 0)$ and the lower wall is at rest. Asterisks denote dimensional variables and are dropped for dimensionless variables. The velocity, distance, time and stress components are made dimensionless with respect to U^*, l^*, l^*/U^*, and $\mu_1 U^*/l^*$. Reynolds numbers in fluid j (j=1,2) are defined as $R_j = U^* l^*/\nu_j$ and Weissenberg numbers are defined as $W_j = U^* \lambda_j/l^*$. A Froude number F is defined by $F^2 = U^{*2}/l^* g$ where g is the gravitational acceleration constant. The depth of fluid 1 is $l_1 = l_1^*/l^*$, a capillary number S is (surface tension coefficient)$/\mu_1 U^*$, and the ratios of fluid properties are $r = \rho_1/\rho_2$, $m = \mu_1/\mu_2$, $w = W_1/W_2 = \lambda_1/\lambda_2$. There are 8 dimensionless parameters: $R_1, W_1, F, l_1, S, r, m$ and w. The depth of fluid 2 is $l_2 = 1 - l_1$.

The dimensionless governing equations in each fluid are [Porteous and Denn 1972; Joseph 1990a] the constitutive law for the upper convected Maxwell fluid

$$\mathbf{T} + W_j \frac{D\mathbf{T}}{Dt} = \frac{\mu_j}{\mu_1}(\nabla \mathbf{u} + (\nabla \mathbf{u})^T), \tag{9a.1}$$

where $j = 1, 2$ and D/Dt is the upper convected time derivative

$$\frac{D\mathbf{T}}{Dt} = \frac{\partial \mathbf{T}}{\partial t} + (\mathbf{u} \cdot \nabla)\mathbf{T} - (\nabla \mathbf{u})\mathbf{T} - \mathbf{T}(\nabla \mathbf{u})^T \qquad (9a.2)$$

and the equations of motion

$$R_1\left(\frac{\partial \mathbf{u}}{\partial t} + \mathbf{u} \cdot \nabla \mathbf{u}\right) = \frac{\rho_1}{\rho_j}\nabla \cdot \underline{\tau} - R_1 F^{-2}\mathbf{e_z}, \qquad (9a.3)$$

and the stress tensor is

$$\tau_{ik} = -pg_{ik} + T_{ik},$$

g_{ik} is the 2×2 identity matrix and $\mathbf{T}$ is the extra stress tensor.

At the interface, the velocity and tangential stress are continuous, the jump in the normal stress is balanced by surface tension and curvature, and the kinematic free surface condition holds. At $z = 0$, $\mathbf{u} = (0,0)$ and at $z = 1$, $\mathbf{u} = (1,0)$.

The following two-layer solution with a flat interface at $z = l_1$ and velocity $(U(z), 0)$ satisfies the equations, boundary conditions and interfacial conditions:

$$\mathbf{T} = \begin{pmatrix} C_1 & C_2 \\ C_2 & 0 \end{pmatrix}, \qquad (9a.4)$$

$$C_1 = \frac{\mu}{\mu_1}2W[U'(z)]^2, \quad C_2 = \frac{\mu}{\mu_1}U'(z), \qquad (9a.5)$$

$$p_1 = \frac{R_1}{F^2}(-z + l_1(1 - 1/r)) + c,$$

$$p_2 = -\frac{R_1}{rF^2}z + c, \qquad (9a.6)$$

where c is an arbitrary constant,

$$U(z) = \frac{z}{(l_1 + ml_2)}, \quad 0 \le z \le l_1, \qquad (9a.7)$$

$$= \frac{m(z-1)}{(l_1 + ml_2)} + 1, \quad l_1 \le z \le 1.$$

The linear stability of this solution to perturbations proportional to $\exp(i\alpha x + \sigma t)$ will be examined. Perturbations of the velocity, stress and interface position are denoted by $(\partial\psi/\partial z, -i\alpha\psi)$, $\begin{pmatrix} T_{11} & T_{12} \\ T_{12} & T_{22} \end{pmatrix}$, and h respectively, where $\psi(z)$ denotes a streamfunction. At $z = 0$ and 1, ψ and $d\psi/dz$ vanish. The constitutive equations (9a.1) - (9a.2) yield the following coupled equations for the extra stress components:

$$T_{11}(1 + W[\sigma + i\alpha U(z)]) = 2W(i\alpha C_1\psi' + C_2\psi'' + U'T_{12}) + \frac{\mu}{\mu_1}2i\alpha\psi',$$

$$T_{12}(1 + W[\sigma + i\alpha U(z)]) = W(U'T_{22} + \alpha^2 C_1\psi) + \frac{\mu}{\mu_1}(\psi'' + \alpha^2\psi), \qquad (9a.8)$$

$$T_{22}(1 + W[\sigma + i\alpha U(z)]) = 2\alpha^2 C_2 W\psi - 2i\alpha\frac{\mu}{\mu_1}\psi'.$$

After eliminating the pressure in (9a.3) we get

$$R_1(\sigma + i\alpha U(z))(\psi'' - \alpha^2\psi) = \frac{\rho_1}{\rho}(\alpha^2 T_{12} - i\alpha\frac{\partial T_{22}}{\partial z} + i\alpha\frac{\partial T_{11}}{\partial z} + \frac{\partial^2 T_{12}}{\partial z^2}). \quad (9a.9)$$

The kinematic free surface condition gives rise to

$$h(\sigma + i\alpha U(l_1)) = -i\alpha\psi(l_1). \quad (9a.10)$$

The continuity of velocity implies that

$$h[\![U'(l_1)]\!] + [\![\psi']\!] = 0, \quad [\![\psi]\!] = 0, \quad (9a.11)$$

where $[\![x]\!]$ denotes x(fluid 1) - x(fluid 2). The continuity of shear stress requires that

$$[\![T_{12}]\!] = i\alpha h(2W_1[U_1'(l_1)]^2 - 2\frac{W_2}{m}[U_2'(l_1)]^2), \quad (9a.12)$$

where U_1 and U_2 are the basic flows given by (9a.7) in fluids 1 and 2 respectively. The normal stress condition is

$$\frac{hR_1}{F^2}(1 - 1/r) - [\![p]\!] + [\![T_{22}]\!] = -\alpha^2 Sh,$$

where p denotes the perturbation to the basic pressure. On eliminating the pressure, this yields

$$R_1\Big(\frac{\alpha h}{F^2}(1 - 1/r) - (i\sigma - \alpha U(l_1))[\frac{\rho}{\rho_1}\psi'] - \alpha\psi(l_1)[\frac{\rho}{\rho_1}U']\Big)$$

$$-\alpha[\![T_{11}]\!] + i[\![\frac{\partial T_{12}}{\partial z}]\!] + \alpha[\![T_{22}]\!] = -\alpha^3 Sh. \quad (9a.13)$$

There is a Squire's transformation for this problem, and the transformation is basically the same as equation (14) of Tlapa and Bernstein [1970]. In addition, F^2, h and $1/S$ transform like the Reynolds number.

In the one-fluid problem at low speeds [M. Renardy and Y. Renardy 1986], the spectrum consists of three types of modes. First, there is a continuous spectrum lying vertically in the complex plane, at

$$\sigma = -1/W - i\alpha z, \quad 0 \leq z \leq 1. \quad (9a.14)$$

Secondly, there is an infinite number of stable discrete eigenvalues that lie close to the line

$$\sigma = -1/2W, \quad (9a.15)$$

and which asymptote to this line for large imaginary parts. Thirdly, there is a pair of eigenvalues, which approaches the continuous spectrum as $W \to 0$, and which approach $\text{Im}\,\sigma = 0$ and $-\alpha$, $\text{Re}\,\sigma = -1/2W$ for large W. This

pair of eigenvalues was found in closed form at zero Reynolds number by
Gorodtsov and Leonov [1967].

In the two-fluid problem, the continuous spectrum is split into two
portions, given by the vanishing of the coefficient of the term with the
highest derivative in the differential equation for ψ:

$$\sigma = -1/W_1 - i\alpha y, \quad 0 \leq y \leq U(l_1), \tag{9a.16}$$

and

$$\sigma = -1/W_2 - i\alpha y, \quad U(l_1) \leq y \leq 1. \tag{9a.17}$$

The discrete eigenvalues consist of an interfacial mode, for which the asymptotic analysis for short waves is presented below, eigenvalues that have analogues in the one-fluid problem, and some others that are discussed in the next section.

IV.9(b) Asymptotic Analysis for Short Waves

The asymptotic analysis for short waves of the interfacial eigenvalue for unbounded two-layer Couette flow involving Newtonian fluids can be obtained by rescaling z,

$$\eta = \alpha(z - l_1), \quad \eta = O(1), \tag{9b.1}$$

for $\alpha \gg 1$ (see section IV.5). The interfacial eigenfunction thus decays rapidly away from the interface and the boundary conditions at the walls do not affect its short-wave behavior. The result at leading order is

$$\sigma = -i\alpha U(l_1) + \frac{m(1-m)(1-m^2/r)R_1}{2(1+m)^2\alpha^2(l_1+ml_2)^2}$$

$$-\frac{\alpha Sm}{2(1+m)} - \frac{m(1-1/r)R_1}{2(1+m)\alpha F^2}, \quad \alpha \to \infty. \tag{9b.2}$$

In what follows, we set

$$r = 1, \quad 1/F^2 = 0, \quad S = 0$$

because we are going to concentrate on the effect of the relaxation times.

The rescaling given by equation (9b.1) transforms equations (9a.8) - (9a.9) as follows:

$$R_1(\sigma + i\alpha U)(\psi_{\eta\eta} - \psi) =$$

$$(\frac{\rho_1}{\rho})(T_{12} - iT_{22\eta} + iT_{11\eta} + T_{12\eta\eta}), \tag{9b.3}$$

$$\frac{T_{11}}{\alpha^2}(1 + W[\sigma + i\alpha U]) = 2W(C_1 i\psi_\eta + C_2\psi_{\eta\eta} + U'\frac{T_{12}}{\alpha^2})$$

$$+2i(\frac{\mu}{\mu_1})\psi_\eta, \tag{9b.4}$$

$$\frac{T_{12}}{\alpha^2}(1 + W[\sigma + i\alpha U]) = W(U'\frac{T_{22}}{\alpha^2} + C_1\psi) + \frac{\mu}{\mu_1}(\psi_{\eta\eta} + \psi), \tag{9b.5}$$

$$\frac{T_{22}}{\alpha^2}(1 + W[\sigma + i\alpha U]) = 2C_2W\psi - 2i\frac{\mu}{\mu_1}\psi_\eta. \tag{9b.6}$$

The interfacial conditions with h eliminated are, at $\eta = 0$,

$$-i\psi[U_z(l_1)] + (\sigma + i\alpha U(l_1))[\psi_\eta] = 0, \tag{9b.7}$$

$$[\psi] = 0, \tag{9b.8}$$

$$[\frac{T_{12}}{\alpha^2}](\sigma + i\alpha U(l_1)) = [C_1]\psi(0), \tag{9b.9}$$

$$-[T_{11}] + i[T_{12\eta}] + [T_{22}] = 0. \tag{9b.10}$$

We assume that

$$\sigma = -i\alpha U(l_1) + \sigma_0 + \frac{\sigma_1}{\alpha^2} + \frac{\sigma_2}{\alpha^4} + \cdots$$

In order to have nontrivial solutions, it is found that the stress components are $O(\alpha^2\psi)$. Thus, at leading order, equation (9b.3) becomes

$$T_{12} - iT_{22\eta} + iT_{11\eta} + T_{12\eta\eta} = 0. \tag{9b.11}$$

This equation was analyzed by Gorodtsov and Leonov [1967] who studied the one-fluid problem in the limit of zero Reynolds number. We follow their analysis below.

A new variable:

$$\eta_* = \frac{U'(z)\eta}{\alpha} - i\frac{\sigma_0}{\alpha} - \frac{i}{\alpha W}$$

is introduced. Hence,

$$1 + W(\sigma + i\alpha U(z)) = i\alpha W\eta_*$$

for large α. This expression is substituted into the left hand sides of equations (9b.4) - (9b.6) for the stress components. The expression for T_{22}/α^2 given by (9b.7) is substituted into (9b.5) and the expression for T_{12}/α^2 is then substituted into (9b.4) so that all stress components are expressed in terms of the stream function. Finally, equation (9b.3) is expressed in terms of the stream function as follows:

$$\alpha^2\eta_*^2\psi^{iv} + \psi'''(-2\alpha\eta_*U' + 2iWU'\alpha^2\eta_*^2)$$

$$+\psi''(2U'^2 - 2\alpha^2\eta_*^2 - 2\alpha^2\eta_*^2W^2U'^2 - 4i\alpha\eta_*WU'^2)$$

$$+\psi'(2\alpha\eta_*U' + 4W^2U'^3\alpha\eta_* + 4iWU'^3 - 2iU'W\alpha^2\eta_*^2)$$

$$+\psi(-4W^2U'^4 - 2U'^2 + 2\alpha^2\eta_*^2W^2U'^2 + \alpha^2\eta_*^2) = 0,$$

where $U' = dU/dz$ at $z = l_1$. This equation can be compared with equation (1.4) of Gorodtsov and Leonov [1967] noting that our WU', η_*/U', α, z, $i(\sigma_0 - i\alpha U(l_1))/\alpha U'$ and $\eta_*/U' + i/\alpha U'W$ correspond to their Γ, y_*, α, y, c and $y - c$ respectively. Thus

$$\left(\eta_*^2\alpha^2 D^2/U'^2 - 2\eta_*\alpha D/U' + (2 - \alpha^2\eta_*^2/U'^2)\right)$$

$$\cdot\left(\alpha^2 D^2 + 2i\alpha^2 WU'D - (\alpha^2 + 2\alpha^2 W^2 U'^2)\right)\psi = 0, \qquad (9b.12)$$

where $D = d/d\eta$. The solutions are

$$\psi(\eta) = c_1\psi_1(\eta) + c_3\psi_3(\eta), \quad \eta \le 0, \qquad (9b.13)$$

$$\psi(\eta) = c_2\psi_2(\eta) + c_4\psi_4(\eta), \quad \eta \ge 0, \qquad (9b.14)$$

where

$$\psi_1(\eta) = (\eta - i\frac{\sigma_0}{U'})\exp(\eta),$$

$$\psi_2(\eta) = (\eta - i\frac{\sigma_0}{U'})\exp(-\eta),$$

$$\psi_3(\eta) = \exp(U'W(-i + \beta)\eta),$$

$$\psi_4(\eta) = \exp(-U'W(i + \beta)\eta), \qquad (9b.15)$$

$$\beta = (1 + (U'W)^{-2})^{1/2}.$$

Substituting (9b.13) - (9b.15) into the interfacial conditions (9b.7) - (9b.10), and letting

$$A_1 = 1 + \sigma_0 W_1, \quad A_2 = 1 + \sigma_0 W_2, \qquad (9b.16)$$

we find that

$$c_1\left(-i\psi_1(0)[U_z] + \sigma_0\psi_1'(0)\right) - c_2\sigma_0\psi_2'(0) + c_3\left(-i\psi_3(0)[U_z] + \sigma_0\psi_3'(0)\right)$$

$$-c_4\sigma_0\psi_4'(0) = 0, \qquad (9b.17)$$

$$c_1\psi_1(0) - c_2\psi_2(0) + c_3\psi_3(0) - c_4\psi_4(0) = 0, \qquad (9b.18)$$

$$c_1\left(\psi_1''(0)/A_1 - 2iU_1'W_1\psi_1'(0)/A_1^2 + \psi_1(0)(1/A_1 + 2W_1^2U_1'^2/A_1 + 2W_1^2U_1'^2/A_1^2)\right.$$

$$\left. - \frac{2W_1U_1'^2}{\sigma_0}(1 - m/w)\psi_1(0)\right)$$

$$-\frac{c_2}{m}\left(\psi_2''(0)/A_2 - 2iU_2'W_2\psi_2'(0)/A_2^2\right.$$

$$\left. +\psi_2(0)(1/A_2 + 2W_2^2U_2'^2/A_2 + 2W_2^2U_2'^2/A_2^2)\right)$$

$$+c_3\left(\psi_3''(0)/A_1 - 2iU_1'W_1\psi_3'(0)/A_1^2\right.$$

$$+\psi_3(0)(1/A_1 + 2W_1^2U_1'^2/A_1 + 2W_1^2U_1'^2/A_1^2) - \frac{2W_1U_1'^2}{\sigma_0}(1 - m/w)\psi_3(0)\Big)$$

$$-\frac{c_4}{m}\Big(\psi_4''(0)/A_2 - 2iU_2'W_2\psi_4'(0)/A_2^2$$

$$+\psi_4(0)(1/A_2 + 2W_2^2U_2'^2/A_2 + 2W_2^2U_2'^2/A_2^2)\Big) = 0, \tag{9b.19}$$

$$c_1\Big(i\psi_1'''(0)/A_1 + \psi_1''(0)(-2W_1U_1'/A_1 + W_1U_1'/A_1^2)$$

$$+\psi_1'(0)(-2iW_1^2U_1'^2/A_1 - 3i/A_1 + 2iW_1^2U_1'^2/A_1^2)$$

$$+\psi_1(0)(-U_1'W_1/A_1^2 - 2W_1^3U_1'^3/A_1^2 + 2U_1'W_1/A_1)\Big)$$

$$-\frac{c_2}{m}\Big(i\psi_2'''(0)/A_2 + \psi_2''(0)(-2W_2U_2'/A_2 + W_2U_2'/A_2^2)$$

$$+\psi_2'(0)(-2iW_2^2U_2'^2/A_2 - 3i/A_2 + 2iW_2^2U_2'^2/A_2^2)$$

$$+\psi_2(0)(-U_2'W_2/A_2^2 - 2W_2^3U_2'^3/A_2^2 + 2U_2'W_2/A_2)\Big)$$

$$+c_3\Big(i\psi_3'''(0)/A_1 + \psi_3''(0)(-2W_1U_1'/A_1 + W_1U_1'/A_1^2)$$

$$+\psi_3'(0)(-2iW_1^2U_1'^2/A_1 - 3i/A_1 + 2iW_1^2U_1'^2/A_1^2)$$

$$+\psi_3(0)(-U_1'W_1/A_1^2 - 2W_1^3U_1'^3/A_1^2 + 2U_1'W_1/A_1)\Big)$$

$$-\frac{c_4}{m}\Big(i\psi_4'''(0)/A_2 + \psi_4''(0)(-2W_2U_2'/A_2 + W_2U_2'/A_2^2)$$

$$+\psi_4'(0)(-2iW_2^2U_2'^2/A_2 - 3i/A_2 + 2iW_2^2U_2'^2/A_2^2)$$

$$+\psi_4(0)(-U_2'W_2/A_2^2 - 2W_2^3U_2'^3/A_2^2 + 2U_2'W_2/A_2)\Big) = 0. \tag{9b.20}$$

For nontrivial solutions, we have the 4 by 4 determinant equation

$$det\mathbf{A} = 0,$$

where the entries a_{ij} of $\mathbf{A}$ are the coefficients of c_j from (9b.17) - (9b.20). After some simplifications, these may be expressed

$$a_{11} = -i\sigma_0/U_1', \quad a_{12} = i\sigma_0/U_2', \quad a_{13} = 1, \quad a_{14} = -1,$$

$$a_{21} = \sigma_0(m - i\sigma_0/U_1'), \quad a_{22} = -\sigma_0(1 + i\sigma_0/U_2'),$$

$$a_{23} = -[U']i + \sigma_0U_1'W_1(-i + \beta_1), \quad a_{24} = \sigma_0U_2'W_2(i + \beta_2)$$

$$a_{31} = \frac{1}{A_1^2} - \frac{i\sigma_0}{U_1'A_1} - iW_1U_1'm/w, \quad a_{32} = \frac{i\sigma_0}{U_2'mA_2} + iU_1'W_2 + \frac{1}{mA_2^2},$$

$$a_{33} = (-i\beta_1U_1'^2W_1^2 + 1 + W_1^2U_1'^2)/A_1$$

$$-i\beta_1U_1'^2W_1^2/A_1^2 - W_1U_1'^2(1 - m/w)/\sigma_0,$$

$$a_{34} = -\frac{1}{m}((i\beta_2U_2'^2W_2^2 + 1 + W_2^2U_2'^2)/A_2 + i\beta_2U_2'^2W_2^2/A_2^2),$$

$$a_{41} = -\frac{\sigma_0}{A_1 U_1'} - W_1 U_1',$$

$$a_{42} = -\frac{\sigma_0}{m A_2 U_2'} - W_2 U_1',$$

$$a_{43} = -\frac{i\beta_1 U_1' W_1}{A_1}, \quad a_{44} = -\frac{i\beta_2 U_2' W_2}{m A_2}, \qquad (9b.21)$$

$$\beta_j = (1 + (U_j' W_j)^{-2})^{1/2},$$

and U_j' is dU/dz at $z = l_1$ for fluid j. To put the determinant equation into polynomial form, we multiply the third row by $A_1^2 A_2^2$. The factor σ_0 in the denominator of a_{33} is canceled by that factor in a_{21}, a_{22}, and a_{24}. The factors A_1 and A_2 appearing in the denominators of the fourth row also cancel because, for example, a_{41} and a_{43} appear in the combination $a_{33}a_{41} - a_{31}a_{43}$ which yields a cancellation. Therefore, the determinant equation is a fifth degree polynomial in σ_0. In the case of zero elasticities, it is evident that $A_1 = A_2 = 1$, and we retrieve $\sigma_0 = 0$. A direct factorization of the polynomial is not obvious, so that numerical methods are used to find the five roots. This is achieved by re-expressing the matrix $\mathbf{A}$ as $D_0 + D_1\sigma_0 + ...D_5\sigma_0^5$. We note that det $\mathbf{A} = $ det $\mathbf{B}$ where $\mathbf{B}$ is the following 20 by 20 matrix:

$$\mathbf{B} = \begin{pmatrix} \sigma_0 & 0 & 0 & 0 & D_0 \\ 1 & \sigma_0 & 0 & 0 & -D_1 \\ 0 & 1 & \sigma_0 & 0 & D_2 \\ 0 & 0 & 1 & \sigma_0 & -D_3 \\ 0 & 0 & 0 & 1 & \sigma_0 D_5 + D_4 \end{pmatrix} \qquad (9b.22)$$

so that $\mathbf{B}$ is of the form $\mathbf{B}_1 - \sigma_0 \mathbf{B}_2$ and the eigenvalues are readily computed.

The solutions of det $\mathbf{B} = 0$ are associated with the interfacial mode and four other modes. The latter arise from the splitting of the continuous spectrum of the one-fluid problem by the presence of the interface. Evidently, in the present case, the four other eigenfunctions possess the same type of short-wave behavior as the interfacial mode.

The 'neutral stability' curves (Re σ_0=0) are generated in a range of parameters such that $0 \le l_1 \le 1$, $0 \le m < \infty$, $0 \le W_1 < \infty$ and $0 \le W_2 < \infty$. We note that when the positions of the fluids are interchanged, Re σ_0 stays the same and Im σ_0 changes sign. This is because gravity is left out of the analysis. Thus, the transformation $l_1 \to 1 - l_1$, $m \to 1/m$, $W_1 \to W_2$, $W_2 \to W_1$ does not affect the location of the neutral stability curves. Therefore, we only need to study either $m \ge 1$ or $m \le 1$. We choose to study $0 < m \le 1$.

When the viscosities are equal (m=1), the shear rates in both fluids are identical, no matter what l_1. In addition, the wall conditions do not enter into the short-wave analysis so that the neutral stability curves in the $W_1 - W_2$ plane are identical for all l_1. Figure 9.1 gives the neutral stability curves for $m = 1$. There is a neutral line at $W_1 = W_2$ and there is symmetry across that line. For large W_2, the lower branch appears to become linear, with symmetric behavior for the upper branch.

When the viscosities of the fluids are different, the neutral stability curves in the $W_1 - W_2$ plane for any l_1 can be related to the curves obtained for one particular value of l_1. To show this, we first note that the parameters U_1', U_2', W_1, W_2, σ_0 occur in the determinant equation in the combinations $U_1'W_1$, $U_2'W_2$, $\sigma_0 W_1$, $\sigma_0 W_2$. Since l_1 enters only in U_1' and $U_2' = mU_1'$, changing l_1 to L_1 means changing $U_1' = 1/(l_1 + ml_2)$ to $1/(L_1 + mL_2)$, say aU_1'. The transformation

$$U_1' \to aU_1', \ U_2' \to aU_2', \ W_1 \to W_1/a, \ W_2 \to W_2/a \qquad (9b.23)$$

leaves products $U_i'W_i$ unchanged but changes σ_0 to $a\sigma_0$. The transformation (9b.23) and

$$l_1 \to L_1,$$

$$a = \frac{l_1 + ml_2}{L_1 + mL_2}, \qquad (9b.24)$$

relate the results for all values of l_1 to results at one value. We have chosen to work with $l_1 = 0.5$.

Figures 9.2 - 9.6 show how the neutral stability curves change as the viscosity ratio is decreased from 1. At $m = 0.9$ (figure 9.2), the middle branch of figure 9.1 has split into two. Thus, two branches emanate from the W_1-axis, one from the origin, and one from the W_2-axis. For small elasticities, the branch from the origin asymptotes to the line $w/m = 1$, which represents the condition that the normal stress ratio of the basic flow is 1. Thus, as m decreases, the initial slope of this line decreases. For large elasticities, $\mathrm{Re}\,\sigma$ decreases in magnitude.

When m decreases, the two branches emanating from the W_1-axis move closer together and eventually coalesce, while the branch emanating from the W_2-axis moves to the right. At $m = 0.5$ (figure 9.3), the upper two branches have already joined to form a loop. The computations were continued to Weissenberg numbers 50 and no new branches were found; the general picture for order 1 elasticities continues monotonically for the larger values. As m decreases further, the upper loop continues to move out to infinity. The junction at the W_2-axis of the branch emanating from that axis continues to move to the right. Its slope at the junction becomes negative as m decreases so that the branch develops a finger pointing to the neutral stability curve emanating from the origin. Eventually the two curves coalesce. The situations before and after they coalesce are illustrated in figures 9.4 ($m = 0.2$) and 9.5 ($m = 0.19$). Figure 9.5 shows two upper looped regions of instability that move out to infinity as m decreases and a lower closed region of instability that flattens out as m decreases. Figure 9.6 ($m = 0.1$) shows that the closed region of instability is already quite small, with the junction on the W_2-axis being around 21. The other looped

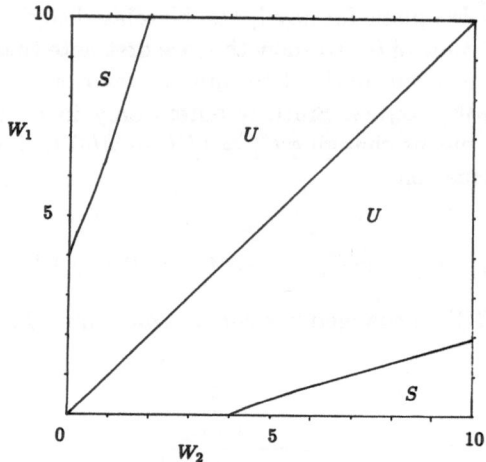

Fig. 9.1. [Renardy, 1988, J. Non-Newt. Fluid Mech. 28, 99, Elsevier] Neutral stability curves for short waves, in the $W_1 - W_2$ plane for equal viscosities (m=1). U denotes unstable, and S denotes stable regions.

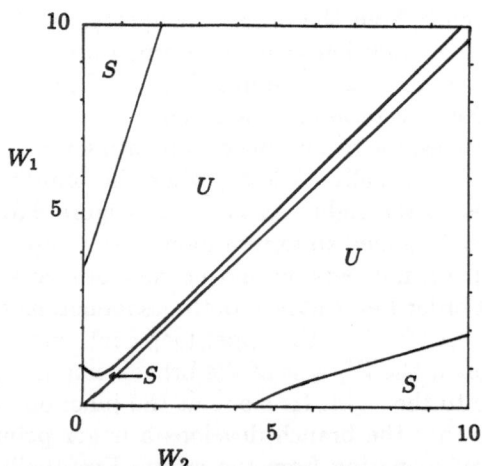

Fig. 9.2. [Renardy, 1988, J. Non-Newt. Fluid Mech. 28, 99, Elsevier] Neutral stability curves for short waves, in the $W_1 - W_2$ plane for $m = 0.9$.

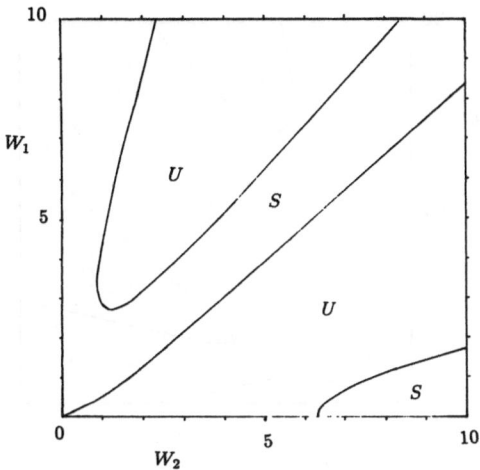

Fig. 9.3. [Renardy, 1988, J. Non-Newt. Fluid Mech. 28, 99, Elsevier] Neutral stability curves for short waves, in the $W_1 - W_2$ plane for $m = 0.5$.

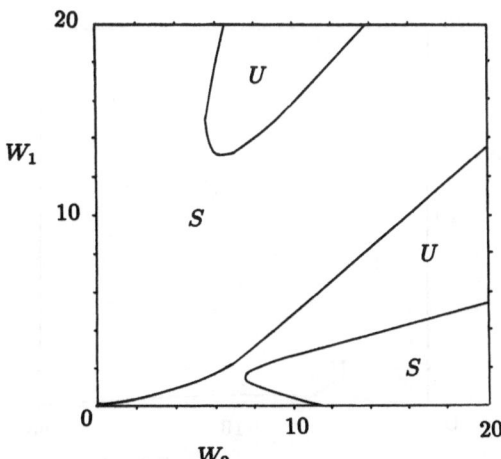

Fig. 9.4. [Renardy, 1988, J. Non-Newt. Fluid Mech. 28, 99, Elsevier] Neutral stability curves for short waves, in the $W_1 - W_2$ plane for $m = 0.2$.

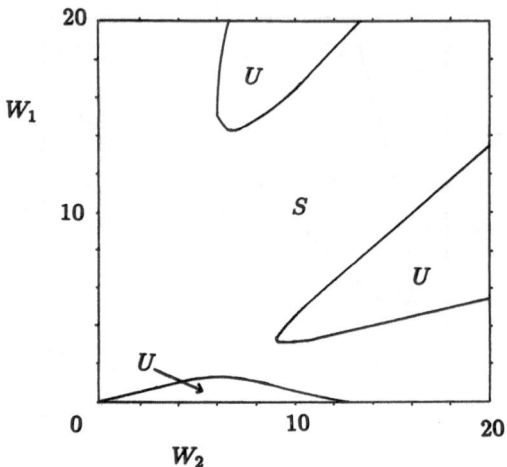

Fig. 9.5. [Renardy, 1988, J. Non-Newt. Fluid Mech. 28, 99, Elsevier] Neutral stability curves for short waves, in the $W_1 - W_2$ plane for $m = 0.19$.

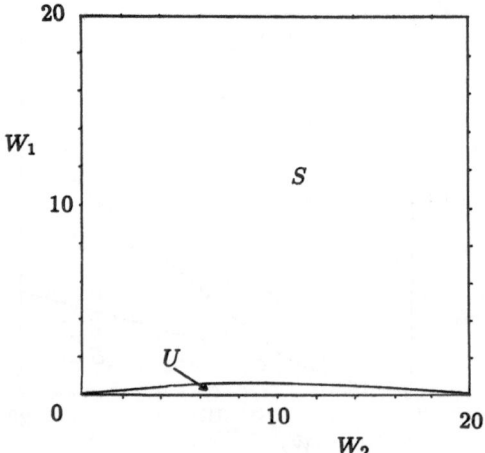

Fig. 9.6. [Renardy, 1988, J. Non-Newt. Fluid Mech. 28, 99, Elsevier] Neutral stability curves for short waves, in the $W_1 - W_2$ plane for $m = 0.1$.

regions of instability have moved out of the picture. As m approaches 0, the entire $W_1 - W_2$ plane would approach stability. In this limit, the lower fluid would sense a solid wall at the interface.

IV.9(c) Numerical Study of the Spectrum

The eigenvalues σ are computed numerically as follows. Equations (9a.8) - (9a.13) are discretized using the Chebyshev-tau method [Gottlieb and Orszag 1983]. The streamfunction, T_{11}, T_{12} and T_{22} are expressed in terms of Chebyshev polynomials up to degrees N, N-2, N-2 and N-1 respectively in each fluid as follows

$$\psi(z) = \sum_{j=0}^{N} p_j T_j(z_1), \quad z_1 = \frac{2}{l_1}z - 1, \quad -1 \leq z_1 \leq 1,$$

$$\psi(z) = \sum_{j=0}^{N} q_j T_j(z_2), \quad z_2 = \frac{2}{l_2}(z - 1) + 1, \quad -1 \leq z_2 \leq 1, \qquad (9c.1)$$

where T_j is the Chebyshev polynomial of jth degree. Together with the unknown h, this yields 8(N-1) unknown coefficients. As in the one-fluid case, the Chebyshev-tau method approximates the continuous spectrum poorly, and any discrete modes that are close are also poorly approximated.

In the Newtonian case, it is known (see section IV.6) that in the absence of density stratification, a linearly stable arrangement consists of a thin layer of a much less viscous liquid, provided there is a sufficient amount of surface tension at the interface. As evident from equation (9b.2), surface tension is necessary to stabilize short waves. In contrast, in the present situation, the stability of the thin-layer arrangement can be achieved for much of the range of elasticities without surface tension. When the viscosities of the fluids are very different (m either small or large), the flow is stable to short-waves for most values of W_1 and W_2 (cf figure 9.6). Figure 9.7 shows growth rates with no elasticity, equal densities and zero surface tension. This is a candidate for thin-layer stability: the viscosity stratification and volumes are such as to stabilize long waves. The lubrication effect introduced by the viscosity stratification is intensified by the addition of elasticities. This is a desirable effect in a number of applications. Figure 9.8 shows growth rates for the same situation with a small amount of elasticities added, for two values of density ratios. The density ratios are $r = 1$ and 0.8. At $r = 0.8$, the Froude number is 0.1, and the density stratification is adverse, but there is linear stability. Here, stability can also be achieved for many other values of the elasticities, as can be inferred from the analyses for long waves [Chen 1991ab].

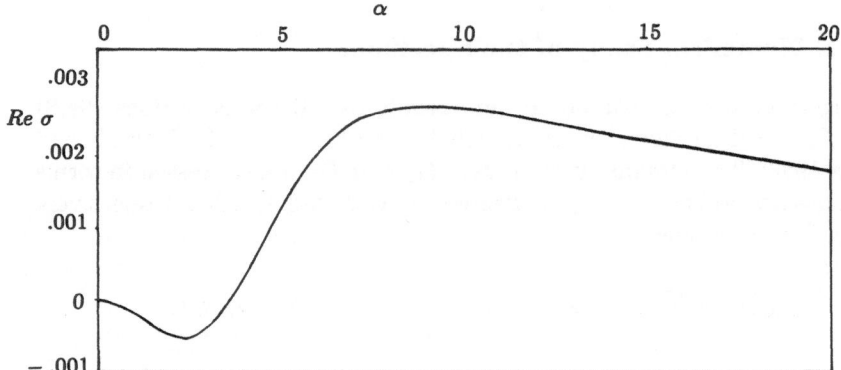

Fig. 9.7. [Renardy, 1988, J. Non-Newt. Fluid Mech. 28, 99, Elsevier] Growth rates $Re\,\sigma$ for the interfacial eigenvalue at $l_1 = 0.1, S = 0.0, m = 0.01, R_1 = 4, r = 1$, zero elasticities.

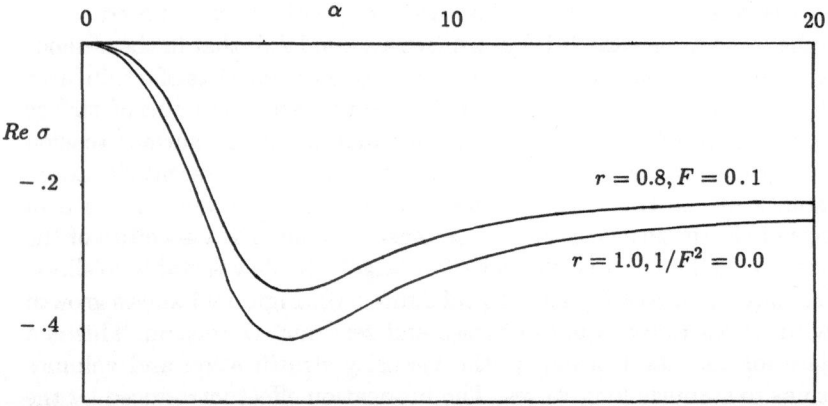

Fig. 9.8. [Renardy, 1988, J. Non-Newt. Fluid Mech. 28, 99, Elsevier] Growth rates $Re\,\sigma$ for the interfacial eigenvalue at $l_1 = 0.1, W_1 = 0.3, w = 10.0, S = 0.0, m = 0.01, R_1 = 4, r = 1, 1/F^2 = 0.0$ and for $r = 0.8, F = 0.1$. Linear stability is achieved without surface tension.

The asymptotic analysis for short waves for the case of no density difference in figure 9.8, based on the analysis of section IV.9 (b), yields $Re\,\sigma = -.24235$. The actual value at $\alpha = 30$ is $Re\,\sigma = -.243$. For the case of $r = 0.8$, $1/F^2 = 100$ at $\alpha = 30$, the actual value is $Re\,\sigma = -.227$. If the density difference were more adverse or the Froude number were smaller, gravitational effects eventually would destabilize long waves, as expected;

for example, at $r = 0.5$, $F = 0.1$, the flow is unstable to long waves. The results of this section show that the lubrication effect set up by a thin layer of a much less viscous liquid is intensified by the addition of elasticities, and even the presence of small elasticities is sufficient to stabilize short waves.

IV.9(d) Wet Slip and Extrudate Sharkskin Formation

This section is based on the paper of Chen and Joseph [1992]. Melt fracture is a problem that has puzzled the scientific community for many years. It refers to an instability observed in the extrusion of a polymeric liquid. Below a certain critical throughput, the surface of the extrudate is smooth. At a first critical stress, small-amplitude short-wavelength disturbances called sharkskin appear at the extrudate surface. At a second critical stress, the extrudate surface shows alternatively smooth and sharkskin regions; this is commonly called stick-slip or spurt flow. As the stress is increased further, the extrudate surface becomes rough and wavy [Denn 1990]. Melt fracture is a special phenomenon for elastic liquids because it does not occur in Newtonian fluids or in dilute to moderately concentrated solutions. Inertia is not relevant since melt fracture is observed at low Reynolds numbers, even as low as 10^{-15} [Tordella 1958]. Numerous attempts have been made to explain this instability with very limited success [Petrie and Denn 1976; Denn 1990].

Cohen and Metzner [1985] distinguish between *apparent slip* and *true slip*. Perhaps *true slip* is a dry slip, with one surface sliding along another without benefit of a wet lubricant. Using a slip boundary condition replacing the slip layer simplifies the analysis tremendously. However, we have seen already that the results obtained from this simplification are not satisfactory. In *apparent slip*, the polymer-deficient layer of wet solvent which develops at the wall could be and has been modeled as a lubricating layer of pure solvent. For this problem, we could imagine a core fluid in a core-annular flow, and it is natural to ask whether or not such a flow is stable. Since there is perhaps a polymer-depleted layer, rather than two spatially-segregated solutions, we are justified in thinking that surface tension effects are negligible. We are therefore led to an analysis of short waves of the type constructed in this section. We want to know if after a lubrication layer has formed, can there be a short-wave instability induced by the elasticity of the core?

In fact, it is probable that sharkskin can never develop in the rather mobile solutions studied by Cohen and Metzner [1985]. Sharkskin is a phenomenon which is associated with melts, amorphous polymers, and concentrated solutions in solvents. In one current line of thought, sharkskin arises in melts as intermittent adhesive failure, sticking and slipping [Denn 1990]. Many think that the melt must be prestressed at the inlet and in the die and that the slipping and sticking occur locally at the die exit [Moynihan, Baird and Ramanathan 1990].

For the purpose of our discussion, we shall assume that some form of slip occurs in the melt when the stress reaches a certain critical value at the wall. We want to see how far we can go with the argument that we are getting a sort of wet slip, with a segregation or fractionation of molecules on the wall, with large molecules inside. This kind of segregation occurs in additive-containing mixtures, say of PVC, where the additives migrate toward the wall forming a lubricating layer between the polymer and the wall. The PVC slips along this lubricating film [Funatsu et al. 1984; Knappe et al. 1984]. We should inquire whether such a lubricating layer is stable and what might be the nature of any instability. We are pursuing the thought that the mechanisms for creating slip and the stability of the slip or lubrication layer may be only weakly related or unrelated. For example, the high molecular weight part of a polydisperse polymer could be pulled off the wall at a critical stress, leaving the small molecules behind. This kind of arrangement could then be stable or unstable, depending on the conditions.

The segregation or fractionation which leaves small molecules on the wall is an old idea which has been advanced by polymer chemists from time to time for thirty years. The idea seems to have been put forward first by Busse [1964] who says that:

"The first conclusion, which is at least of some theoretical interest, is that a capillary viscometer should tend to fractionate polymer molecules with respect to molecular weight along the radius of the capillary. Near the wall, molecules of high molecular weight acquire relatively large amounts of free energy of elastic deformation, while very small molecules do not. Hence, there is a thermodynamic force that tends to increase the concentration of very small molecules at the wall, and of the larger molecules nearer the axis.

"No measurements of such separations have been reported, to the author's knowledge, but this factor might play a part in the action of die lubricants. It may also cause some of the change in apparent viscosity with the ratio of the capillary length to diameter."

Schreiber and Storey [1965] and Schreiber, Storey and Bagley [1966] have given indirect experimental support for Busse's idea, but more needs to be done. It seems not to be known whether additives can promote fractionation, but Moynihan et al. [1990] have suggested that 3M's Dynamar additive, which is known to promote slip in LLDPE, forms an LLDPE/fluoroelastomer blend at the surface of the melt.

Additives may enhance or suppress slip. De Smedt and Nam [1987] studied fluoroelastomer additives in the extrusion flow of polyethylene (PE) through capillaries of various dimensions. The effect of the additive is to reduce the apparent viscosity of the PE. Spectroscopic analysis indicated that the additive was concentrated at the free surface of the extrudate. This additive suppresses the surface defects which would appear after extrusion in additive-free PE.

However, these results contradict some of those obtained by Rama-murthy [1986], who used a fluoroelastomer which suppressed slip in a flat die, but not a round one. He says that suppressing slip eliminates sur-face defects in the melt. By using two different fluoroelastomers, both the suppression and enhancement of slip were achieved in the experiments of Hatzikiriankos and Dealy [1991]. They studied the slip of a polydisperse polyethylene in a sliding plane rheometer and they reduced slip by coating the steel plates with DFL (dry film lube) and increased the slip by coating it with Dynamar 9613.

It appears not to be known whether or not the various forms of slip are wet or dry. The evidence for wet slip is clear in certain solutions, like those studied by Cohen and Metzner [1985] or in additive-bearing mixtures like the PVC mentioned earlier.

Slip layers which form under adhesive fracture of the high molecular weight molecules from a wall are presently not well understood. We have been looking at the consequence of modeling such layers as a *wet slip* layer with an average thickness, density, viscosity, relaxation time and shear mod-ulus. Using this concept, we must allow that the results of adhesive fracture may lead to a stable or unstable layer, depending on the conditions. The analysis of Chen and Joseph [1992] was confined to short waves, so that an unstable layer could conceivably lead to sharkskin. Of course, we have only the vaguest idea of how to characterize these layers, but they obviously should be much less viscous and much less elastic than the polymer core.

Chen and Joseph [1992] look at a hypothetical example of core-annular flow in which the pipe radius is slightly larger than the nominal radius separating the high and low molecular weight polymers. The interfacial tension is zero because our polymers of different molecular weights can mix. The viscosity of the polymer core is, say, ten times greater than the average viscosity of the low molecular weight annulus and the core is more elastic. Their short-wave analysis yields stability in the lubrication layer when the molecular weight of the polymer is low and instability (sharkskin) when it is high. At least this doesn't disagree with experiments, but their comparison is much too casual to be anything more than suggestive.

Even in the case of cohesive fracture of the type leading to spurt, it is hard to imagine slipping without lubrication of some sort. We can expect a migration of low-viscosity constituents into regions of high shear. These high-shear regions may take form as a wet layer of small thickness or perhaps as a region deficient in high molecular weights defined by large gradients of molecular weight. The slip in such layers is apparent; there is no slip surface, rather there are large gradients across narrow layers which are perceived as slip. In any case, the study of the stability of such layers is of interest.

We conjecture that adhesive fracture in capillaries and possibly cohesive fractures could give rise to wet slip layers by leaching the polymers from the solution at the wall or by fractionation in amorphous polymers. The properties of such lubricating layers are not well understood, but they may be similar to the better understood lubricating layers in our two-fluid problem.

IV.10 Liquid-Vapor Films Between Heated Walls

This section is based on two papers of Huang and Joseph [1991a,b]. who treat a two-fluids problem with phase change between inclined walls. We heat one wall above the saturation temperature and cool the other below the saturation temperature. A water film is on the cold wall and a vapor film on the hot wall. We get an elementary solution depending only on the variable y normal to the walls with steady distributions of temperature and velocity driven by gravity. The basic flow is in thermodynamic equilibrium: the phases stay as they are and the phase boundary does not move. However, phase changes must be taken into account in the stability problem.

The problem we are considering is related to the problem of stability of laminar film condensation (Nusselt's solution) on an inclined cool plate (see Hwang and Weng [1987] for a review of this literature) and the problem of a falling film of liquid down an inclined plate [Yih 1963]. We generalize Yih's problem to two phases of a liquid between parallel horizontal or vertical plates.

A number of papers have appeared in the last 20 years on the problem of mantle convection with phase change and related problems. This literature is set in the frame of the Benard problem, horizontal layers without a mean flow, heated from below or above, but with change of phase included. The aim of these works is to determine the effect of phase change on convection and the pioneering work was done by Busse and Schubert [1971]. The literature of this topic is cited in the list of references of the paper by Sotin and Parmentier [1989]. All but one of these papers restricted their considerations to the case of small density differences in which the term $(\rho_1 - \rho_2)\mathbf{u}_\Sigma \cdot \mathbf{n}$ in our (10a.3) may be thought to be negligible. The one exception is the paper on geothermal systems, treated as a porous media problem, by Schubert and Strauss [1980] in which the density difference is large and (10a.3) is not simplified.

The problem we consider in this section differs from those mentioned in the foregoing paragraph. For instance, the problem of Yih has a fully-developed basic flow but no phase change. In the present study, we assume that the saturation temperature is a function of pressure. In the linear stability analysis, we allow a phase change but cannot accomodate applied pressure gradients. This is because if there is an axial pressure variation, the transition temperature would change as you go along the pipe, and you cannot set up a parallel flow with a flat interface. Thus, if the basic flow is driven by pressure gradients and involves phase change, it would not be unidirectional. Poiseuille flow cannot be treated in our framework. For a similar reason, we can not obtain a steady laminar flow for core-annular flow because there is a pressure jump across the cylindrical interface due to interfacial tension which prohibits the existence of a common saturation temperature there. On the other hand, we can treat the flow in free fall between heated inclined plates if the change of density with temperature is neglected. We follow this path, assuming in all that follows that the densities of the water and vapor are fixed constants independent of pressure and temperature. In fact, the density of water does not change much with temperature. But in the vapor, we have to allow only small temperature changes, say on the scale of $10°C$ above the boiling temperature, because the density of an ideal gas is inversely proportional to the temperature.

Our problem differs from that of laminar film condensation which is usually set in a semi-infinite region with phase change in the basic flow. In this section, the second wall and the vapor are active and the basic flow is fully-developed and not of boundary-layer type. The second wall allows the system to attain a steady fully-developed temperature profile in the basic state which cannot exist in a semi-infinite region. The basic flow which we study is amenable to an exact computational study of stability and the flow could conceivably be attained in experiments.

IV.10(a) Governing Equations and Interface Conditions for Two-Phase Flow of Vapor and Liquid

The most general balance equations we shall need are (2a.2, 2a.9, 2b.11, 2b.12, 2c.5 and 2c.6) of sections I.2 (a) - (c). We are going to assume that the density of the vapor and the density of the water are constants independent of variation of temperature or pressure across the channel. This assumption is usually made for condensation problems, and it means that the pressure will be a dynamic variable uncoupled from thermodynamics and that convective currents set up by gravity on thermally-induced variations of temperature will be ignored. It is conceivable that thermally-induced buoyancy could produce some important effects, but the work of Spindler [1982] indicates that they are not important.

The consequence of incompressibility is that div $\mathbf{u} = 0$ replaces I.(2a.2) and the stress in a Newtonian fluid

$$\mathbf{T} = -p\mathbf{1} + 2\mu\mathbf{D}[\mathbf{u}] \qquad (10a.1)$$

is determined by an unknown dynamic pressure p, the viscosity μ, and the rate of strain

$$\mathbf{D}[\mathbf{u}] = \frac{1}{2}(\mathbf{L} + \mathbf{L}^T)$$

where $\mathbf{L} = \nabla\mathbf{u}$ is the velocity gradient. Huang and Joseph [1991a,b] consider the limit of incompressibility after replacing the internal energy e per unit mass with the enthalpy h, using

$$h = e + p/\rho.$$

This change of variable is inserted into I.(2c.4). We will use the notation $\tilde{m}$ for the m used in section I.2(c) because m will denote the viscosity ratio later in this chapter. Thus,

$$\tilde{m} = \rho \mathbf{n_{12}} \cdot (\mathbf{u}_\Sigma - \mathbf{u}), \tag{10a.2}$$

and we also have

$$[\![p(\mathbf{u}_\Sigma - \mathbf{u}) \cdot \mathbf{n_{12}}]\!] + \tilde{m}[\![e]\!] = -[\![\rho h \mathbf{n_{12}} \cdot (\mathbf{u} - \mathbf{u}_\Sigma)]\!] = \tilde{m}[\![h]\!] \tag{10a.3}$$

and this is inserted into I.(2c.6b).

The equations to be used in all that follows are listed below. We have assumed that the viscosities μ_2 and μ_1 in the water and vapor are different constants. The heat flux is given by Fourier's law $\mathbf{q} = -k\nabla T$ with constants k_1 and k_2 and temperature T. The enthalpy is related to the temperature by

$$dh = C_p dT \tag{10a.4}$$

with constant specific heats C_{p1} and C_{p2}. The surface tension σ is constant and the dissipation, the last term of (10a.2), is negligible. Thus,

$$\operatorname{div} \mathbf{u} = 0, \quad \rho\frac{d\mathbf{u}}{dt} = -\nabla p + \mu\nabla^2\mathbf{u} + \rho\mathbf{g}, \tag{10a.5}$$

$$\rho C_p \frac{dT}{dt} = k\nabla^2 T. \tag{10a.6}$$

On the interface Σ we have

$$\tilde{m}[\![\mathbf{u}]\!] - [\![p]\!]\mathbf{n_{12}} + 2[\![\mu\mathbf{D}[\mathbf{u}] \cdot \mathbf{n_{12}}]\!] = 2H\sigma\mathbf{n_{12}}, \tag{10a.7}$$

$$-[\![k\nabla T]\!] \cdot \mathbf{n_{12}} = 2[\![\mu(\mathbf{u} - \mathbf{u}_\Sigma) \cdot \mathbf{D}[\mathbf{u}] \cdot \mathbf{n_{12}}]\!] + \tilde{m}[\![h + \frac{1}{2}|\mathbf{u} - \mathbf{u}_\Sigma|^2]\!] \tag{10a.8}$$

$$[\![\mathbf{u}]\!] \cdot \mathbf{t} = 0, \tag{10a.9}$$

and

$$\mathbf{u}_\Sigma \cdot \mathbf{n_{12}} = \frac{\partial I/\partial t}{|\nabla I|}, \tag{10a.10}$$

where $I(\mathbf{x}, t) = 0$ is the equation of the interface as in (10b.1) and H is defined in (10b.8). The saturation temperature T_s at equilibrium is determined as given by the Clapeyron equation as a function $\tilde{T}$ of pressure

$$T = T_s = \tilde{T}(p). \tag{10a.11}$$

The function $\tilde{T}(p)$ gives the equation of state for the phase transition. By equilibrium, we mean the condition in which there is no evaporation or condensation: $\tilde{m} = 0$. This means that pressure across a phase-change boundary must be continuous.

Obviously, if the pressure is different on the two sides of the interface, (10a.11) yields $T_{s1} = \tilde{T}(p_{s1}) \neq \tilde{T}(p_{s2}) = T_{s2}$: the temperature cannot be at the saturation value in both the water and water vapor. There will be a temperature discontinuity where

$$T_1 = \tilde{T}(p_v), \quad T_2 = \tilde{T}(p_w). \tag{10a.12}$$

This happens, for instance, in the core-annular flow where the interface is cylindrical and surface tension imposes a pressure jump there. A similar jump occurs in the linearly perturbed problem we consider here. This shows that thermodynamic equilibrium means that the water and its vapor are not in thermal equilibrium. The existence of a temperature discontinuity evidently cannot be eliminated by rigorous application of first principles. Schrage [1953] says "...There is no reason why the temperature of the gas phase should necessarily be the same as that of the liquid or solid surface in all cases." Indeed, classic kinetic theory calculations [Pao 1971; Sone and Onishi 1978; Aoki and Cercignani 1983; Onishi 1984; Cercignani, Fiszdon and Frezzotti 1985] indicate that for monatomic vapor, large temperature jumps exist at interfaces. Shankar and Deshpande [1990] have measured the temperature distribution in the vapor between an evaporating liquid surface and a cooler condensing surface in water, Freon 113 and mercury. The temperature profiles obtained in mercury showed large jumps at the interface, as large as almost 50% of the applied temperature difference.

The usual approximation made in the study of phase change of liquid and vapor is to require thermal equilibrium

$$[\![T]\!] = T_1 - T_2 = 0 \tag{10a.13a}$$

together with thermodynamic equilibrium in the vapor

$$T_1 = \tilde{T}(p_v) \quad \text{on the vapor side} \tag{10a.13b}$$

[Plesset and Zwick 1954; Gebhardt 1961; Ishii 1975]. In general, the pressure on the water side $p_w \neq p_v$, so that the water is not in thermodynamic equilibrium.

The choice of temperature conditions at the phase-change boundary appears to be an unresolved question of physics. Two of the possibilities are expressed as (10a.12) and (10a.13). There are many other possibilities.

The enthalpies h of water and water vapor are determined by the temperature and density, and the difference between them is called the heat of vaporization

$$h_{fg} = [\![h]\!]. \tag{10a.14}$$

At $T_s = 100°$ and atmospheric pressure, h_{fg} is 2.257×10^6 joules per kilogram.

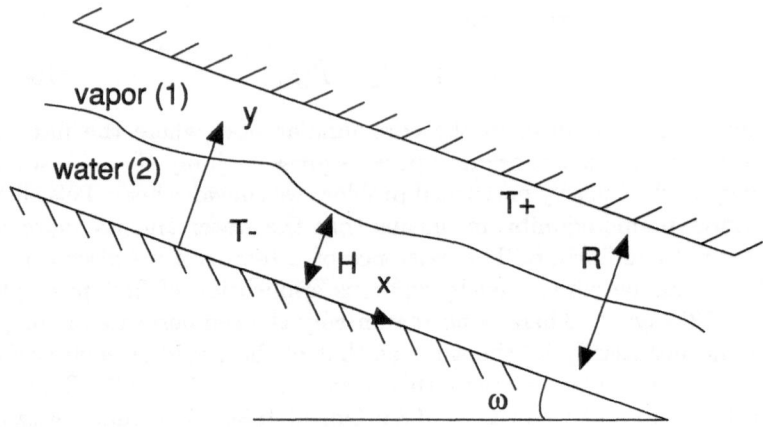

Fig. 10.1. [Huang and Joseph, 1992] Schematic of free fall of water 2 and vapor 1 down an inclined channel of height R. The interface is at $y = R - \delta(x,t)$. ω is the angle of inclination from the horizontal.

The values of the parameters which shall be taken as constants in the analysis to follow are their tabulated values at $T = T_s = 100°C$:

$$
\begin{aligned}
\rho_1 &= 0.585 \times 10^{-3} \text{ g/cm}^3 \\
\rho_2 &= 0.965 \text{ g/cm}^3 \\
\mu_1 &= 0.0125 \text{ cp} \\
\mu_2 &= 0.28 \text{ cp} \\
k_1 &= 2.5 \times 10^{-4} \text{ joule/cm sec}°\text{C} \\
k_2 &= 6.8 \times 10^{-3} \text{ joule/cm sec}°\text{C} \\
C_{p1} &= 1.96 \text{ joule/gm}°\text{C} \\
C_{p2} &= 4.18 \text{ joule/gm}°\text{C} \\
h_{fg} &= 2.257 \times 10^3 \text{ joule/gm} \\
\sigma &= 64.4 \text{ dynes/cm}.
\end{aligned}
\tag{10a.15}
$$

IV.10(b) Governing Equations for the Inclined Channel

We now write the governing equations for flow down an inclined channel (figure 10.1), using Squire's theorem as a justification for restricting the analysis to two space dimensions.

The interface is represented by

$$I(x, y, t) = y - (R - \delta(x, t)) = 0. \tag{10b.1}$$

Then

$$-\mathbf{n}_{12} = \mathbf{n}_{21} = \frac{\nabla I}{|\nabla I|} = \frac{\mathbf{e_y} + \mathbf{e_x}\delta_x}{\sqrt{1+\delta_x^2}}, \qquad (10b.2)$$

$$-\mathbf{t}_{12} = \mathbf{t}_{21} = \frac{-\mathbf{e_y}\delta_x + \mathbf{e_x}}{\sqrt{1+\delta_x^2}}. \qquad (10b.3)$$

The equations of motion and energy are resolved in the usual way after noting that

$$\mathbf{g} = -\mathbf{e_z}g = -g[(\mathbf{e_x} \cdot \mathbf{e_z})\mathbf{e_x} + (\mathbf{e_y} \cdot \mathbf{e_z})\mathbf{e_y}] = g[\sin\omega\, \mathbf{e_x} - \cos\omega\, \mathbf{e_y}]. \quad (10b.4)$$

Then

$$\rho\frac{du}{dt} = \rho g \sin\omega - \frac{\partial\phi}{\partial x} + \mu\nabla^2 u, \qquad (10b.5)$$

$$\rho\frac{dv}{dt} = -\frac{\partial\phi}{\partial y} + \mu\nabla^2 v, \qquad (10b.6)$$

where

$$\phi = p + \rho g y \cos\omega \qquad (10b.7)$$

and (10a.6) governs T. These equations hold both in the water and in the vapor with appropriate values for the constants.

The interface conditions (10a.7) - (10a.13) may be expressed in coordinate form using (10b.6) - (10b.7) and

$$2H = \frac{\delta_{xx}}{(1+\delta_x^2)^{3/2}}. \qquad (10b.8)$$

The continuity of the tangential component of velocity (10a.9) implies that on $y = R - \delta$

$$[\![-v\delta_x + u]\!] = 0. \qquad (10b.9)$$

The normal and tangential components of (10a.7) can be written, using (10b.7), as

$$-[\![\phi]\!] + [\![\rho]\!]g(R - \delta)\cos\omega - \tilde{m}\frac{[\![v + \delta_x u]\!]}{(1+\delta_x^2)^{1/2}}$$

$$+\frac{2}{1+\delta_x^2}[\![\mu(\frac{\partial v}{\partial y} + \delta_x(\frac{\partial v}{\partial x} + \frac{\partial u}{\partial y}) + \delta_x^2\frac{\partial u}{\partial x})]\!] = \frac{\sigma\delta_{xx}}{(1+\delta_x^2)^{3/2}} \qquad (10b.10)$$

and

$$(1 - \delta_x^2)[\![\mu(\frac{\partial u}{\partial y} + \frac{\partial v}{\partial x})]\!] - 2\delta_x[\![\mu(\frac{\partial v}{\partial y} - \frac{\partial u}{\partial x})]\!] = 0. \qquad (10b.11)$$

The condition (10a.10) may be written

$$v_\Sigma = -\delta_t - u_\Sigma\delta_x. \qquad (10b.12)$$

Then, since the tangential component of velocity is continuous across $y = R - \delta$

$$\mathbf{u} - \mathbf{u}_\Sigma = [(\mathbf{u} - \mathbf{u}_\Sigma) \cdot \mathbf{n}_{12}]\mathbf{n}_{12} = [(\mathbf{u} - \mathbf{u}_\Sigma) \cdot \mathbf{n}_{21}]\mathbf{n}_{21}$$

and
$$(\mathbf{u} - \mathbf{u}_{\Sigma}) \cdot \mathbf{n_{21}} = (v + \delta_x u + \delta_t)/(1 + \delta_x^2)^{1/2}. \tag{10b.13}$$

It follows that
$$\tilde{m} = \rho(v + \delta_x u + \delta_t)/(1 + \delta_x^2)^{1/2} \tag{10b.14}$$

and the energy balance (10a.8) reduces to

$$\frac{[\![k(-\delta_x \frac{\partial T}{\partial x} - \frac{\partial T}{\partial y})]\!]}{(1 + \delta_x^2)^{1/2}} = -\tilde{m}[\![h]\!] - \frac{\tilde{m}[\![(v + \delta_x u + \delta_t)^2]\!]}{2(1 + \delta_x^2)}$$

$$+ \frac{2}{(1 + \delta_x^2)^{3/2}} [\![\mu(v + \delta_x u + \delta_t)\left(\frac{\partial v}{\partial y} + \delta_x(\frac{\partial v}{\partial x} + \frac{\partial u}{\partial y}) + \delta_x^2 \frac{\partial u}{\partial x}\right)]\!]. \tag{10b.15}$$

We have already mentioned that we do not know what temperature condition to apply at the phase-change boundary. Since the pressure is not continuous in the perturbed flow, the temperature will not be continuous. We could demand that the temperature be at the saturation values in the water and water vapor. Or, we could require that either water or its vapor be at saturation and enforce the continuity of temperature. Then the second phase will not be at saturation.

IV.10(c) Basic Flow

There is a steady solenoidal solution

$$(u, v, \phi, T, \delta) = (U(y), 0, 0, T, \delta_0) \tag{10c.1}$$

where δ_0 is a constant equal to the mean value of δ, and

$$U_1(y) = -\frac{\rho_1 g \sin \omega}{2\mu_1}(R - y)^2 + C_1(R - y), \qquad \text{for all } y \in (R - \delta_0, R) \tag{10c.2}$$

$$U_2(y) = -\frac{\rho_2 g \sin \omega}{2\mu_2}(R - y)^2 + C_2(R - y) + C_3, \qquad \text{for all } y \in (0, R - \delta_0)$$

where the constants are determined by satisfying the continuity of velocity and stress $[\![U]\!] = 0$, $[\![\mu U]\!] = 0$, and the boundary conditions $U = 0$ at $y = 0, R$.

$$\delta_0 = \frac{(T_+ - T_s)k_1 R}{(T_+ - T_s)k_1 + (T_s - T_-)k_2} =: \beta R, \tag{10c.3}$$

defines β, the vapor volume fraction or relative thickness of the vapor layer, and

$$C_1 = \frac{g \sin \omega \ R[\beta^2(\rho_1\mu_2 - 2\rho_1\mu_1 + \rho_2\mu_1) + 2\beta\mu_1(\rho_1 - \rho_2) + \mu_1\rho_2]}{2\mu_1(\mu_1 + \beta(\mu_2 - \mu_1))} \tag{10c.4}$$

$$C_2 = \frac{g \sin \omega \ R[\beta^2(2\rho_2\mu_2 - \rho_2\mu_1 - \rho_1\mu_2) + \mu_1\rho_2]}{2\mu_2(\mu_1 + \beta(\mu_2 - \mu_1))}$$

$$C_3 = \frac{g \sin \omega \ R^2 [\rho_2 \beta(\mu_2 - \mu_1) - \beta^2 (2\rho_2 \mu_2 - \rho_2 \mu_1 - \rho_1 \mu_2)]}{2\mu_2(\mu_1 + \beta(\mu_2 - \mu_1))}.$$

The temperature is given by

$$T_1(y) = T_+ - (T_+ - T_s)\frac{R - y}{\delta_0} \qquad \text{for all } y \in (R - \delta_0, R) \qquad (10c.5)$$

$$T_2(y) = T_s - (T_s - T_-)\frac{R - \delta_0 - y}{R - \delta_0} \qquad \text{for all } y \in (0, R - \delta_0).$$

Note that at $y = R - \delta_0$, $T_1 = T_2 = T_s$. Since $\phi = $constant,

$$P_1 = -\rho_1 gy \cos \omega + A_1 \qquad (10c.6)$$

and

$$P_2 = -\rho_2 gy \cos \omega + A_2 \qquad (10c.7)$$

with A_1 and A_2 selected so that

$$[\![P]\!] = -[\![\rho]\!]g(R - \delta_0)\cos \omega + [\![A]\!] = 0. \qquad (10c.8)$$

Since the pressure at the vapor-water interface is continuous, the basic flow is in thermodynamic equilibrium with

$$T_s = T(P(R - \delta_0)), \qquad \tilde{m} = 0. \qquad (10c.9)$$

IV.10(d) Equations for Linear Stability Analysis

Let u, v, ϕ, θ, h be perturbations of $U, 0, 0, T, R - \delta_0$. The linearized equations for the perturbations are

$$u_x + v_y = 0, \qquad (10d.1)$$

$$u_t + Uu_x + vU' = -\frac{1}{\rho}\phi_x + \nu\nabla^2 u, \qquad (10d.2)$$

$$v_t + Uv_x = -\frac{1}{\rho}\phi_y + \nu\nabla^2 v, \qquad (10d.3)$$

$$\rho C_p(\theta_t + U\theta_x + vT') = k\nabla^2\theta. \qquad (10d.4)$$

u, v and θ vanish on $y = 0$ and $y = R$. The interface conditions are evaluated on $y = R - \delta_0$.

$$\tilde{m} = \rho_1(v_1 - U_1 h_x - h_t) = \rho_2(v_2 - U_2 h_x - h_t), \qquad (10d.5)$$

$$[\![u + U_y h]\!] = 0, \qquad (10d.6)$$

$$-[\![\phi]\!] + [\![\rho]\!]gh \cos \omega + 2[\![\mu v_y]\!] = -\sigma h_{xx}, \qquad (10d.7)$$

$$[\![\mu(u_y + v_x + U_{yy}h)]\!] = 0, \qquad (10d.8)$$

$$[\![k\theta_y]\!] - \tilde{m}h_{fg} = 0. \qquad (10d.9)$$

We have to choose a temperature condition at the phase change boundary. We require saturation on the vapor side 1, and the linearized condition using $T = \tilde{T}(p)$ is

$$\theta_1 = \tilde{T}'(p_s)p_1 + [-T_1' + \tilde{T}'(p_s)P_1']h \qquad (10d.10)$$

where p_i denotes the perturbation to the pressure. Then either

$$\theta_2 = \tilde{T}'(p_s)p_2 + [-T_2' + \tilde{T}'(p_s)P_2']h \qquad (10d.11a)$$

or

$$[\theta] + [\frac{dT}{dy}]h = 0. \qquad (10d.11b)$$

We do not know which of the two choices (10d.11a) or (10d.11b) better represents conditions to be described at a phase change boundary.

IV.10(e) Dimensionless Variables and Normal Modes

We introduce the following scales to make the variables dimensionless:

$$\text{length}: \quad H_0 = R - \delta_0$$
$$\text{velocity}: \quad V_0 = \frac{(\rho_2 + \beta(\rho_1 - \rho_2))g\sin\omega}{2(\mu_1 + \beta(\mu_2 - \mu_1))}\delta_0 H_0$$
$$\text{time}: \quad \frac{H_0}{V_0}$$
$$\text{pressure}: \quad \rho_i V_0^2$$
$$\text{temperature}: \quad T_s - T_-,$$

where V_0 is the velocity at the interface.

We shall use the same letters for dimensional and dimensionless variables. The dimensionless parameters listed below appear in the dimensionless equations and $i = (1,2)=$(vapor, water).

$$\Re_i = \frac{\rho_i V_0 H_0}{\mu_i} \qquad \text{Reynolds number}$$

$$Pr_i = \frac{\mu_i c_i}{k_i} \qquad \text{Prandtl number}$$

$$Pe_i = \Re_i Pr_i \qquad \text{Peclet number}$$

$$W = \frac{\sigma}{\rho_1 H_0 V_0^2} \qquad \text{Weber number}$$

$$\xi = \frac{k_2}{k_1} \qquad \text{Thermal conductivity ratio}$$

$$\Gamma = \frac{k_1(T_s - T_-)}{h_{fg}\mu_1} \qquad \text{Phase change number}$$

$$\Pi_i = \frac{V_0^2 T_s \rho_i}{h_{fg}(T_s - T_-)}\left(\frac{1}{\rho_1} - \frac{1}{\rho_2}\right) \qquad \text{Dynamic pressure over latent heat}$$

$$\tilde{G} = \frac{gH_0 \cos\omega}{V_0^2} \qquad \text{Gravity number}$$

$$\zeta = \frac{\rho_2}{\rho_1} \qquad \text{Density ratio}$$

$$m = \frac{\mu_2}{\mu_1} \qquad \text{Viscosity ratio}$$

$$\tau_1 = \frac{T_+}{T_s} \qquad \text{Hot wall temperature ratio}$$

$$\tau_2 = \frac{T_-}{T_s} \qquad \text{Cold wall temperature ratio}$$

$$r = \frac{R}{H_0} \qquad \text{Relative distance between the two walls.}$$

If we specify the fluids as water and steam, then all material parameters are determined and we are left with four independent parameters $\Re_2, r, \tau_2$ and ω and

$$Pr_1 = 0.9825, \qquad Pr_2 = 1.7343,$$
$$\zeta = 1.6502 \times 10^3$$
$$m = 22.47$$
$$\xi = 27.25.$$

The basic flow in dimensionless form is given by

$$U_1(y) = a_1(r - y)^2 + b_1(r - y), \qquad (10e.1)$$

$$U_2(y) = a_2(r - y)^2 + b_2(r - y) + c_2, \qquad (10e.2)$$

$$T_1(y) = \frac{\tau_1}{1 - \tau_2} - \frac{\tau_1 - 1}{1 - \tau_2}\frac{r - y}{r - 1}, \qquad (10e.3)$$

$$T_2(y) = \frac{1}{1 - \tau_2} - (1 - y), \qquad (10e.4)$$

where

$$a_1 = -\frac{(1 + \beta(m-1))(1-\beta)}{(\zeta + \beta(1-\zeta))\beta}, \qquad a_2 = \frac{\zeta}{m}a_1,$$

$$b_1 = \frac{\beta^2(\zeta + m - 2) + 2\beta(1-\zeta) + \zeta}{\beta(\zeta + \beta(1-\zeta))},$$

$$b_2 = \frac{\beta^2(2\zeta m - \zeta - m) + \zeta}{\beta m(\zeta + \beta(1-\zeta))},$$

$$c_2 = \frac{\zeta(m-1) - \beta(2\zeta m - \zeta - m)}{m(1-\beta)(\zeta + \beta(1-\zeta))}.$$

The governing equations become:

$$u_x + v_y = 0, \tag{10e.5}$$

$$u_t + U u_x + v U' = -p_x + \frac{1}{\Re}\nabla^2 u, \tag{10e.6}$$

$$v_t + U v_x = -p_y + \frac{1}{\Re}\nabla^2 v, \tag{10e.7}$$

$$\theta_t + U\theta_x + vT' = \frac{1}{Pe}\nabla^2\theta \tag{10e.8}$$

with boundary conditions:

$$u_1(x, r, t) = v_1(x, r, t) = \theta_1(x, r, t) = 0,$$

and

$$u_2(x, 0, t) = v_2(x, 0, t) = \theta_2(x, 0, t) = 0, \tag{10e.9}$$

and interfacial conditions at $y = 1$:

$$[\![u + U_y h]\!] = 0, \tag{10e.10}$$

$$[\![m(u_y + v_x + U_{yy}h)]\!] = 0, \tag{10e.11}$$

$$2[\![\frac{\zeta}{\Re}v_y]\!] - [\![\zeta p]\!] + Wh_{xx} + (1-\zeta)\tilde{G}h = 0, \tag{10e.12}$$

$$\frac{\Gamma}{\Re_1}[\![\xi\theta_y]\!] + (h_x U_1 + h_t - v_1) = 0. \tag{10e.13}$$

We shall compare results for two different choices of temperature conditions on the interface. If the temperature is saturated in both phases, then

$$\theta_i = \Pi_i p_i + [-T_i' - \Pi_i\tilde{G}]h \tag{10e.14}$$

and the temperature will be discontinuous. The other choice is saturation on the vapor side

$$\theta_1 = \Pi_1 p_1 + [-T_1' - \Pi_1\tilde{G}]h, \tag{10e.15}$$

together with continuity of temperature

$$[\![\theta]\!] + [\![T_y]\!]h = 0. \tag{10e.16}$$

We seek solutions in terms of normal modes

$$\mathbf{v}(x, y, t) = [u(y), iv(y)]\exp i\alpha(x - ct), \quad p(x, y, t) = p(y)\exp i\alpha(x - ct),$$

$$\theta(x, y, t) = \theta(y)\exp i\alpha(x - ct), \quad h(x, y) = h\exp i\alpha(x - ct).$$

Substituting the above expressions into (10e.5) - (10e.16), we get equations for $u(y)$, $v(y)$, $p(y)$, $\theta(y)$ and h:

$$\alpha u + v' = 0 \tag{10e.17}$$

$$\alpha(U - c)u + vU_y = -\alpha p + \frac{i}{\Re}(\alpha^2 u - u'') \tag{10e.18}$$

$$\alpha(U - c)v = p' + \frac{i}{\Re}(\alpha^2 v - v'') \tag{10e.19}$$

$$\alpha(U - c)\theta + vT_y = \frac{i}{Pe}(\alpha^2\theta - \theta''), \tag{10e.20}$$

together with boundary conditions

$$u_1(r) = v_1(r) = \theta_1(r) = 0, \tag{10e.21}$$

$$u_2(0) = v_2(0) = \theta_2(0) = 0. \tag{10e.22}$$

After p and h have been eliminated, the jump conditions at the unperturbed interface $y = 1$ can be expressed as:

$$\alpha[\![u]\!](U_1 - \zeta U_2) + (v_1 - \zeta v_2)[\![U_y]\!] = \alpha c(1 - \zeta)[\![u]\!], \tag{10e.23}$$

$$(U_{1yy} - mU_{2yy})(-u_1 + u_2) + [\![U_y]\!](u_1' - mu_2' + \alpha(mv_2 - v_1)) = 0, \tag{10e.24}$$

$$[\![U_y]\!]\left(\frac{i}{\alpha\Re_1}u_1'' - \frac{i\zeta}{\alpha\Re_2}u_2''\right) - \left((\frac{i\alpha}{\Re_1} - U_1)[\![U_y]\!] + W\alpha^2 + (1 - \zeta)\tilde{G}\right)u_1$$

$$+ \left(\zeta(\frac{i\alpha}{\Re_2} - U_2)[\![U_y]\!] - W\alpha^2 + (1 - \zeta)\tilde{G}\right)u_2$$

$$+2[\![U_y]\!]\left(\frac{i}{\Re_1}v_1' - \frac{\zeta i}{\Re_2}v_2'\right) + (U_{1y}\frac{v_1}{\alpha} - \zeta U_{2y}\frac{v_2}{\alpha})[\![U_y]\!] = (u_1 - \zeta u_2)c[\![U_y]\!], \tag{10e.25}$$

$$\alpha(u_2 - u_1)U_1 + [\![U_y]\!](-v_1 - \frac{i\Gamma}{\Re_1}\theta_1' + \frac{i\Gamma}{\Re_1}\xi\theta_2') = -\alpha(u_1 - u_2)c. \tag{10e.26}$$

The condition (10e.15) that the temperature on the vapor side of the interface is at saturation can be written as

$$\Pi_1(iu_1'' - (i\alpha^2 - \alpha\Re_1 U_1)u_1 + \Re_1 v_1 U_{1y})$$

$$+\alpha\Re_1(\theta_1 - (T_1' + \Pi_1\tilde{G})\frac{[\![u]\!]}{[\![U']\!]}) = \alpha\Pi_1\Re_1 u_1 c. \tag{10e.27}$$

The same equation with subscript 2 in the place of subscript 1 holds for the water side:

$$\Pi_2(iu_2'' - (i\alpha^2 - \alpha\Re_2 U_2)u_2 + \Re_2 v_2 U_{2y})$$

$$+\alpha\Re_2(\theta_2 - (T'_2 + \Pi_2\tilde{G})\frac{[\![u]\!]}{[\![U']\!]}) = \alpha\Pi_2\Re_2 u_2 c. \qquad (10e.28)$$

The continuity of temperature requires that

$$[\![U_y]\!][\![\theta]\!] - [\![T_y]\!][\![u]\!] = 0. \qquad (10e.29)$$

We can not enforce (10e.27) - (10e.29) simultaneously. A choice must be made.

A long-wave asymptotic analysis was done. The zero eigenvalue was traced numerically, increasing α, and it was found that the real part becomes negative, and with a nonzero imaginary part.

IV.10(f) Two Different Interfacial Temperature Conditions

When (10e.27) and (10e.28) are adopted, thermodynamic equilibrium is required at the interface. Consequently, the temperature continuity is not guaranteed. This closes the mathematical formulation of the eigenvalue problem. We shall designate this problem as case I. Alternatively, we can satisfy thermal equilibrium, namely (10e.29), and require that one of the phases be at saturation temperature, say, the vapor phase. We designate this problem as case II.

The two systems of equations differ on one equation, and generally, it should be expected that they will give different results. For the water-vapor case, the differences are insignificant (table 10.1). Hence, only the results of case II will be reported.

Table 10.1. A comparison of the results from different choices of interfacial temperature conditions for typical parameters in the vertical case, $\omega = 90°$. The eigenvalues with the maximum growth rate are displayed in the table.

Parameters	Case I	Case II
$\Re_2=10$	$\tilde{\alpha}=0.0130$	$\tilde{\alpha}=0.0131$
$r=1.5$	$\tilde{\alpha}c=1.040555, 0.152185i$	$\tilde{\alpha}c=1.040536, 0.152197i$
$\tau_2=0.97$		
$\Re_2=1000$	$\tilde{\alpha}=0.250$	$\tilde{\alpha}=0.249$
$r=2.0$	$\tilde{\alpha}c=0.266496, 0.0072672i$	$\tilde{\alpha}c = 0.265408, 0.0072667i$
$\tau_2=0.97$		

IV.10(g) Energy Analysis

The thermal energy equation is coupled to the mechanical energy equation through the mechanism of phase change. In this case, we may get two energy identities, one for mechanical energy designated with a subscript M and another for thermal energy designated with subscript T. The mechanical energy equation is derived as follows. Suppose (u, v, θ) are the components of an eigenvector associated with the maximum growth rate of one of the problems satisfying the equations (10e.17) to (10e.27) and one of the equations of (10e.28) or (10e.29). To get the equation governing the evolution of the mechanical energy of the disturbance, we multiply (10e.18) and (10e.19) with u^* and v^*, the complex conjugates of u and v respectively, then integrate the sum of them over both the liquid and vapor regions.

$$< \alpha(U-c)(|u|^2+|v|^2) > = < \frac{i}{\Re}(\alpha^2(|u|^2+|v|^2)+|u_y|^2+|v_y|^2) > - < vu^*U_y >$$

$$-[\frac{\zeta i}{\Re}(u'u^* + v'v^*)] + [\zeta pv^*] \tag{10g.1}$$

where $<> = \int_1 + \zeta \int_2$, $|u|^2 = uu^*$, $|v|^2 = vv^*$ and we used the boundary and interfacial conditions to evaluate the integrands. ζ is the density ratio in fluid 2, and is 1 in fluid 1. (10g.1) is the energy balance for the small disturbance, every term in it can be interpreted as some kind of energy. The imaginary part of the right hand side represents the growth of the energy of the disturbance and the left side may be split into three parts:

$$\dot{E}_M = I_M - D_M + B_M \tag{10g.2}$$

where

$$\dot{E}_M = \alpha c_i < |u|^2 + |v|^2 >,$$

$$I_M = \text{Im} < vu^*U_y >, \quad \text{energy production from the basic flow}$$

$$D_M = < \frac{1}{\Re}(\alpha^2(|u|^2 + |v|^2) + |u_y|^2 + |v_y|^2) >, \quad \text{viscous dissipation rate}$$

$$B_M = \text{Im}\left(-[\frac{\zeta i}{\Re}(u'u^*+v'v^*)]+[\zeta pv^*]\right), \quad \text{energy production from interface.}$$

We may transform the last term of B_M as follows

$$[\zeta pv^*] = [\zeta p]v_2^* + [v^*]p_1, \tag{10g.3}$$

where $[\zeta p]$ can be evaluated from the jump condition (10e.12) as

$$2[\frac{\zeta i}{\Re}v'] - (-\alpha^2 W + (1 - \zeta)\tilde{G})\frac{[u]}{[U_y]}.$$

Then B_M can be further decomposed into four parts

$$B_M = B_1 + P + B_2 + G \tag{10g.4}$$

where

$$B_1 = \alpha^2 W \, \text{Im}\left(\frac{[u]v_2^*}{[U_y]}\right)$$

can be regarded as the energy supply rate due to surface tension;

$$P = \mathrm{Im}\Big([v^*]p_1\Big)$$

arises from the change of phase;

$$B_2 = -\mathrm{Im}\Big([\frac{\zeta i}{\Re}(u'u^* + v'v^*)] - 2[\frac{\zeta i}{\Re}v']v_2^*\Big)$$

represents interfacial friction due to the viscosity difference; and

$$G = -\tilde{G}(1 - \zeta)\mathrm{Im}\Big(\frac{[u]v_2^*}{[U_y]}\Big)$$

is the gravity term.

To get the equation governing the production of thermal energy we multiply (10e.20) by θ^* and integrate over both vapor and water regions

$$< \alpha(U - c)|\theta|^2 > = < \frac{i}{Pe}(\alpha^2|\theta|^2 + |\theta_y|^2) > - < v\theta^*T_y > + [\frac{i}{Pe}\theta'\theta^*].$$
$$(10g.5)$$

(10g.5) can be written as follows:

$$\dot{E}_T = I_T - D_T + B_T \tag{10g.6}$$

where

$$\dot{E}_T = \alpha c_i < |\theta|^2 >,$$
$$I_T = < \frac{1}{Pe}(\alpha^2|\theta|^2 + |\theta_y|^2) >,$$
$$D_T = \mathrm{Im} < v\theta^*T_y >,$$
$$B_T = -\mathrm{Re}\Big([\frac{1}{Pe}\theta'\theta^*]\Big).$$

IV.10(h) Horizontal Case [Huang and Joseph 1992a]

The horizontal case deserves special treatment because in this case, gravity does not drive the basic flow. The basic state is motionless with a linear temperature profile. Since there is no prescribed velocity, a different unit of velocity scale,

$$\frac{\mu_1\zeta}{\rho_1 H_0 m}$$

is used to make the equations dimensionless. The governing equations then become

$$v^{iv} - 2\alpha^2 v'' + \alpha^4 v = i\alpha c\Re(\alpha^2 v - v''), \tag{10h.1}$$

$$\theta'' - \alpha^2\theta - iPe\, vT' = -i\alpha cPe\,\theta. \tag{10h.2}$$

The general solution of this system can be expressed as

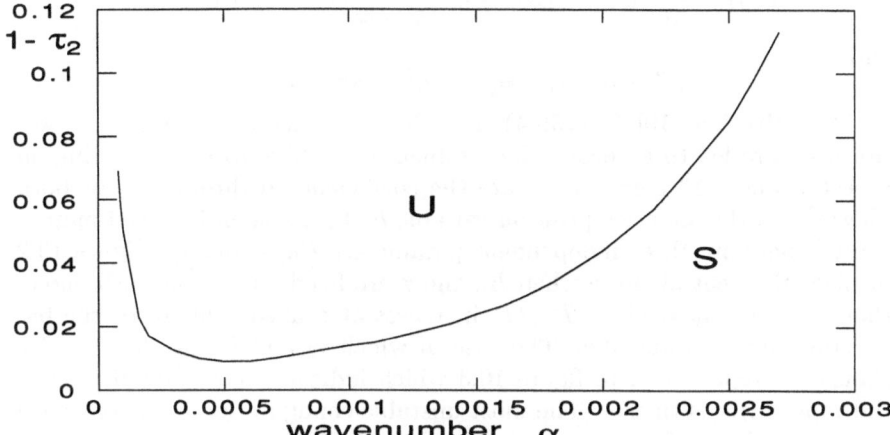

Fig. 10.2. [Huang and Joseph, 1992] Neutral curve when H_0=0.001m, r=1.3, $\tilde{G}$= 0.00434, W =48.4, Γ, Π_1 and Π_2 change with $(1 - \tau_2)$.

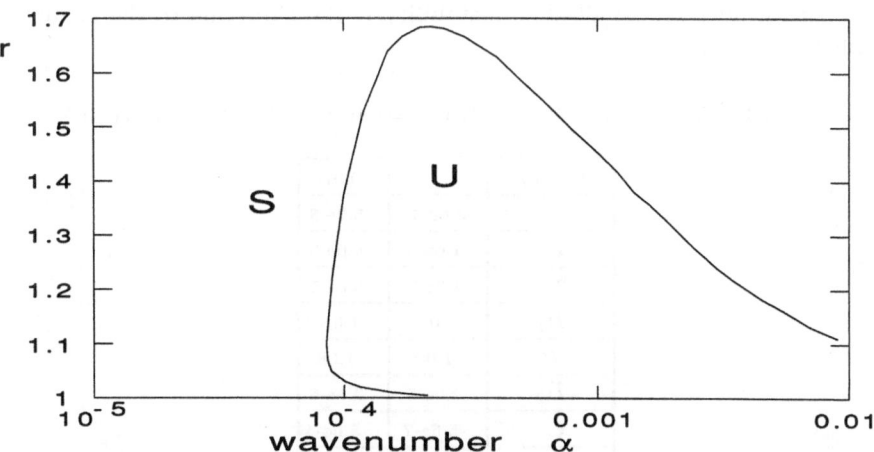

Fig. 10.3. [Huang and Joseph, 1992] Neutral curve when τ_2 (that is, the temperature of the lower plane) is fixed at 0.95, H_0=0.001m, $\tilde{G}$=0.00434, W=48.4, Γ=0.231 e−3, $\Pi_1 = 0.2$e−4, $\Pi_2 = 0.0338$.

$$v(y) = Ae^{\alpha y} + Be^{-\alpha y} + Ce^{\hat{\alpha} y} + De^{-\hat{\alpha} y}, \qquad (10h.3)$$

$$\theta(y) = Ee^{\hat{\gamma} y} + Fe^{-\hat{\gamma} y} + \tilde{\theta}(y), \qquad (10h.4)$$

where

$$\tilde{\theta}(y) = \frac{T'}{\alpha c}[Ae^{\alpha y} + Be^{-\alpha y} + \frac{Pe}{Pe - \Re}\left(Ce^{\hat{\alpha} y} + De^{-\hat{\alpha} y}\right)],$$

and

$$\hat{\alpha}^2 = \alpha^2 - i\alpha c\Re, \qquad \hat{\gamma}^2 = \alpha^2 - i\alpha cPe.$$

Substituting (10h.3)-(10h.4) into the boundary and interface conditions, we are led to an eigenvalue problem for a 13×13 matrix acting on a vector whose 13 components are the coefficients A through F for both phases, and the interface position variable h. This system is solved numerically. There are three independent parameters H_0, r and τ_2. Figure 10.2 displays the neutral curve when H_0 and r are fixed and τ_2 varies. It shows that when $1 - \tau_2 = (T_s - T_-)/T_s$ increases at a fixed value of α, the basic state becomes unstable. The case in which τ_2 and H_0 are fixed and r changes is represented in figure 10.3 which indicates more stability when the vapor layer is thicker. The most unstable configuration is the one with a very thin layer of vapor.

Very thick and very thin vapor layers are stable. This is probably a manifestation of limiting behavior in which layers all of water and all of vapor are stable. The long waves (very small α) are always stable but the system is most unstable to waves of finite but long length. The energy analysis of table 10.2 shows that the phase change term is the dominant destabilizing term, while surface tension and gravity are stabilizing but their magnitudes are small. The instability is therefore due to the change of phase.

Table 10.2. Energy analysis: H_0=0.001m, $\tilde{G}$=0.00434, W=48.4, and τ_1=1.2.

$1 - \tau_2$	0.017	0.070
$\tilde{\alpha}$	3.6e-4	5.2e-3
$\tilde{\alpha}c_i$	1.6e-6	4.0e-3
$\dot{E}_M$	1.2e-3	3.8e-1
B_1	0	-1.0e-8
P	1.00	1.38
B_2	2.0e-5	3.5e-6
G	-5.5e-7	-5.6e-4
$\dot{E}_T$	7.1e-6	1.0e-3
B_T	1.0	1.0
$I_T - D_T$	-9.99e-1	-9.99e-1

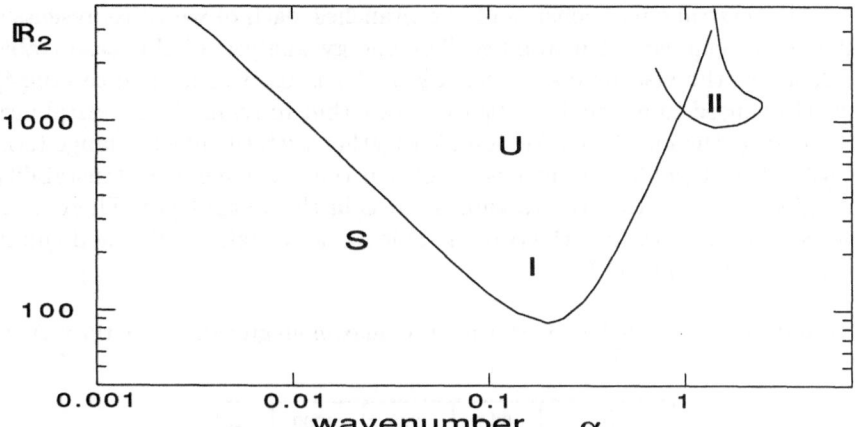

Fig. 10.4. [Huang and Joseph, 1992] Neutral curve when $r=2.0$ and $\tau_2 = 0.97$.

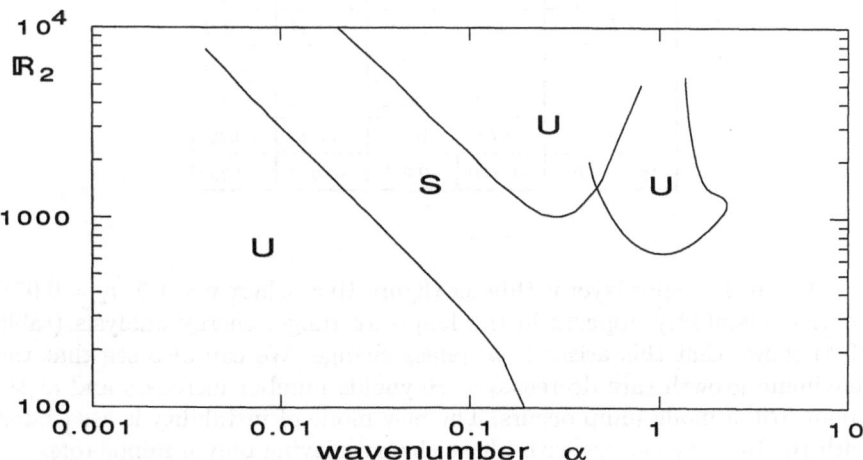

Fig. 10.5. [Huang and Joseph, 1992] Neutral curve for vertical case $\omega = 90°$: $(r, \tau_2) = (1.5, 0.97)$.

IV.10(i) Vertical Case [Huang and Joseph 1992b]

The finite element code of Hu and Joseph [1989a] was adapted to solve the present problem. Figure 10.4 displays neutral curves in the vertical case when $r=2.0$ and $\tau_2 = 0.97$.

The neutral curve consists of two branches, each of which represents a different mechanism of instability. The energy analysis of this case (table 10.3, where the viscous dissipation term D_M is used to normalize (10g.4) and D_T is used to normalize (10g.6)) shows that in region I, the instability is caused by the interfacial friction B_2 together with the phase change term P when the Reynolds number is small. On the other hand, the instability in region II arises from the Reynolds stress in the water layer. There is an overlapping region where the two unstable modes coexist, and a mode jump is observed at $\Re_2 \approx 1130$.

Table 10.3. Energy analysis for the mode of maximum growth rate in the vertical case: $(r, \tau_2) = (2.0, 0.97)$.

$\Re_2$	86.7	900	1200	2000
$\tilde{\alpha}$	0.198	0.243	1.491	1.176
$\tilde{\alpha}c_i$	1.4e-5	7.43e-3	2.07e-2	3.50e-2
$\dot{E}_M$	8.2e-5	1.2e-1	2.16	2.96
$I_M - D_M$	-0.97	-0.98	3.26	3.29
B_1	-2.1e-3	-1.0e-1	-0.873	-0.314
P	0.546	0.072	-0.002	0.002
B_2	0.427	1.130	-0.225	-0.012
$\dot{E}_T$	5.7e-6	3.2e-2	4.2e-2	1.4e-1
B_T	0.72	0.57	0.71	0.72
$I_T - D_T$	-0.720	-0.537	-0.672	-0.583

When the vapor layer is thinner (figure 10.5, where $r = 1.5$, $\tau_2 = 0.97$), another instability appears in the longwave range. energy analysis (table 10.4) shows that this arises from phase change. We can also see that the maximum growth rate decreases as Reynolds number increases and at $\Re_2$ about 670, a mode jump occurs. The new mode of instability is associated with the Reynolds stress with phase change playing only a minor role.

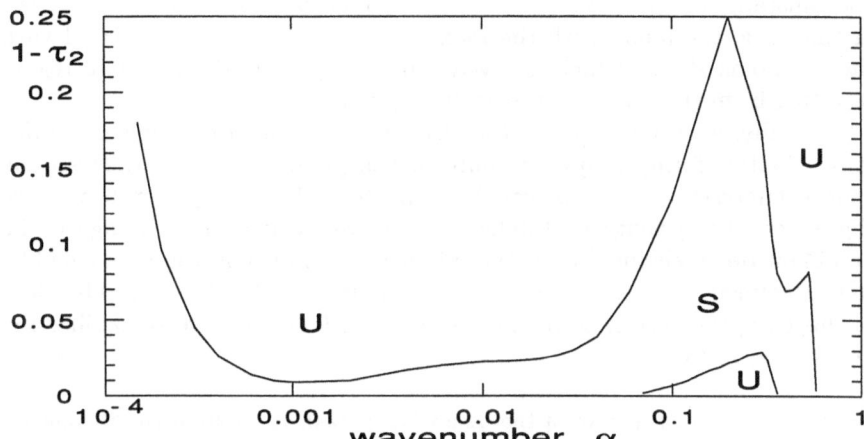

Fig. 10.6. [Huang and Joseph, 1992] Neutral curves when Reynolds number and r are fixed: $\Re_2=1000$, $r=1.5$.

Table 10.4. Energy analysis for the mode of maximum growth rate in the vertical case of figure 10.5: $(r, \tau_2)=(1.5, 0.97)$.

$\Re_2$	100	300	650	700	2000
$\tilde{\alpha}$	0.006	0.0041	0.0031	1.041	1.065
$\tilde{\alpha}c_i$	1.52e-2	5.12e-3	2.28e-3	5.67e-3	8.11e-2
$\dot{E}_M$	6.9e-2	6.9e-2	6.9e-2	6.7e-1	5.61
$I_M - D_M$	-1.0	-0.999	-0.998	1.09	6.06
B_1	0	0	0	-3.1e-1	-4.8e-1
P	1.07	1.06	1.05	0.11	0.07
B_2	3.5e-3	7.9e-3	1.4e-2	-0.21	2.6e-2
$\dot{E}_T$	1.7e-3	1.7e-3	1.7e-3	3.4e-3	0.121
B_T	0.964	0.964	0.964	1.24	1.40
$I_T - D_T$	-0.962	-0.962	-0.962	-1.237	-1.280

Figure 10.6 displays the neutral curves when the Reynolds number and r are fixed ($\Re_2 = 1000$, $r=1.5$) and τ_2 varies. Three different instabilities are identified. For long waves, there is an unstable mode associated with phase change, consistent with the results for the horizontal case. This mode goes unstable when $1 - \tau_2 = (T_s - T_-)/T_s$ increases at a fixed α. However, in this case, the mode with the maximum growth rate is associated with the Reynolds stress, which is always unstable, while the unstable region resulting from the interfacial tension is split into two parts.

Finally, we briefly discuss the effect of the inclination angle ω. In our dimensionless form, ω appears only in the gravity parameter. From the energy analysis, we see, as can be expected when the vapor lies above the water, that gravity is stabilizing. However, when phase change is the dominant destabilizing factor, the influence of gravity is almost negligible. In the inclined cases, there is an unstable mode caused by interfacial friction; in this case, the gravity term is larger but not large enough to stabilize the flow (table 10.5).

Table 10.5. Critical points in the case when $r=2.0$, $\tau_2 =0.97$ with different inclination angles. $\tilde{\Re}_2$ denotes the critical Reynolds number, $\tilde{\alpha}$ denotes the critical wavenumber, and $\mathrm{Re}\,\tilde{\alpha}c$ denotes the corresponding wavespeed.

ω	$90°$	$30°$	$10°$
$\tilde{\Re}_2$	86.7	93.4	106.1
$\tilde{\alpha}$	0.198	0.186	0.169
$\mathrm{Re}\,(\tilde{\alpha}c)$	0.305	0.296	0.272

IV.10(j) Conclusions

The problem of stability of fully developed flow of liquid and its vapor under gravity between hot and cold parallel plates was considered. The water and its vapor were assumed to be incompressible, but vaporization and condensation at the water-vapor interface could occur. In the basic flow, there is no phase change at the flat interface, but the perturbation induces phase change under the condition that the total volume of vapor and the total volume of liquid are conserved. Three kinds of instability can arise: an instability due to the Reynolds stress at higher Reynolds numbers, an instability due to interfacial friction which is associated with the viscosity difference and would arise even in the absence of phase change, and a strong phase change instability at the interface. All these instabilities arise as overstability so that Hopf bifurcation into periodic solutions is expected.

The problem of the influence of phase change on thermal convection has been considered in the literature (see Sotin and Parmentier [1989] for references) on geophysical fluid mechanics, starting with the work of Busse and Schubert [1971]. The assumptions of these authors, small density differences for all authors but Schubert and Strauss [1980] who assume Darcy's law, seem to be compatible with the assumption that the marginal eigenfunctions are steady at criticality (exchange of stability).

The issue of temperature conditions at a phase change interface is an important one. The problem arises whenever a pressure jump across the inteface is allowed even when the phases are solid or liquid. If the pressures on each side of the interface are different, and the temperature is at saturation, then there must be a discontinuity of temperature. Different choices of thermal interface conditions are possible. Although in special case of water and its vapor which we have treated, the stability results do not depend strongly on the choice of conditions of the temperature at the interface, we do not expect weak dependence in general especially when the dependence of the saturation temperature on the pressure is not small. The choice of temperature conditions at the phase change boundary goes beyond continuum thermomechanics and appears to require some form of molecular theory.

References

Numbers [(vm) n] following a reference indicate the volume m and page n on which it is referred.

Acrivos, A. and T. S. Lo, 1978, Deformation and breakup of a single slender drop in an extensional flow, J. Fluid Mech. **86**, 641. [(v1) 143]

Altobelli, S. A., R. C. Givler and E. Fukushima, 1991, Velocity and concentration measurements of suspensions by nuclear magnetic resonance imaging, J. Rheology **35** (5), 721. [(v1) 7, plate I.1.2]

Amarakoon, A. M. D., R. G. Hussey, W. J. Good, and E. G. Grimsal, 1982, Drag measurements for axisymmetric motion of a torus at low Reynolds number, Phys. Fluids **25**(9), 1495. [(v2) 313]

Anderson, P. C., C. J. Veal and V. R. Withers, 1982, Rheology of coal-oil dispersions, Powder Technol. **32**, 45. [(v1) 9]

Anturkar, N. R., T. C. Papanastasiou and J. O. Wilkes, 1990a, Linear stability analysis of multilayer plane Poiseuille flow, Phys. Fluids A **2**, 530. [(v1) 272]

Anturkar, N. R., T. C. Papanastasiou and J. O. Wilkes, 1990b, Stability of multilayer extrusion of viscoelastic liquids, AIChE J. **36**(5), 710. [(v1) 360]

Aoki, K., and C. Cercignani, 1983, Evaporation and condensation on two parallel plates at finite Reynolds numbers, Phys. Fluids **26**, 1163. [(v1) 381]

Arecchi, F. T., P. K. Buah-Bassuah, F. Francini, C. Pérez-Garcia and F. Quercioli, 1989, An experimental investigation of the break-up of a liquid drop falling in a miscible fluid, Europhys. Lett. **9**(4), 333. [(v2) 290-291, 309]

Arney, M., R. Bai, D. D. Joseph and K. Liu, 1992, Friction factor and holdup studies for lubricated pipelines, in preparation. [(v2) 7, 16, 115, 192-202]

Astarita, G. and G. Apuzzo, 1965, Motion of gas bubbles in non-Newtonian liquids, AIChE J. **11**, 815. [(v1) 144]

Atherton, R. W., and G. M. Homsy, 1976, On the derivation of evolution equations for interfacial waves, Chem. Engng. Communications **2**, 57. [(v1) 270, 324, 330; (v2) 266]

Aul, R. W., and W. L. Olbricht, 1989, Stability of a thin annular film in pressure-driven low Reynolds number flow through a capillary, preprint, School of Chem. Engng., Cornell University. [(v2) 8, 13, 50, 66-69, 83]

Babchin, A. J., A. L. Frenkel, B. G. Levich and G. I. Sivashinsky, 1983 a, Flow-induced nonlinear effects in thin liquid film stability, Ann. NY Acad. Sci. **404** 426. [(v1) 333]

Babchin, A. J., A. L. Frenkel, B. G. Levich and G. I. Sivashinsky, 1983 b, Nonlinear saturation of Rayleigh-Taylor instability in thin films, Phys. Fluids **26** (11), 3159 L. [(v1) 333]

Bai, R., K. Chen and D. D. Joseph, 1991, Lubricated pipelining: stability of core-annular flow. Part V: experiments and comparison with theory, submitted to J. Fluid Mech. [(v2) 8, 15, 16, 154-225]

Baird, M. H. I., T. Wairegi and H. J. Loo, 1977, Velocity and momentum of vortex rings in relation to formation parameters, Can. J. Chem. Eng. **55** 19. [(v2) 289]

Barkey, D. P. and P. D. Laporte, 1990, The dynamic diffusion layer in branched growth of a conductive-polymer aggregate in a 2-D electrolysis cell, J. Electrochem. Soc. **137**(5), 1655. [(v2) 343]

Barnett, S.M., A. E. Humphrey and M. Litt, 1966, Bubble motion and mass transfer in non-Newtonian fluids, AIChE J. **12**, 253. [(v1) 144]

Batchelor, G. K., 1970, *An Introduction to Fluid Dynamics*, Cambridge University Press. [(v1) 34; (v2) 155, 294]

Bhattacharji, S., 1967, Mechanics of flow differentiation in ultramafic and mafic sills, J. of Geology **75**, 101. [(v1) 5]

Baumann, N., 1989, *Vortex Rings*, Masters Thesis, U. of Minnesota. [(v2) 288-317]

Baumann, N., D. D. Joseph, P. Mohr and Y. Renardy, 1992, Vortex rings of one fluid in another in free fall, Phys. Fluids A **4**(3), 567. [(v2) 288-323]

Beale, J. T., 1980, The initial value problem for the Navier-Stokes equations with a free surface, Comm. Pure Appl. Math. **34**, 359. [(v1) 183]

Beale, J. T., 1984, Large time regularity of viscous surface waves, Arch. Rational Mech. Anal. **84**, 307. [(v1) 183]

Beavers, G. S., and D. D. Joseph, 1979, Experiments on free surface phenomena, J. Non-Newtonian Fluid Mech. **5**, 323. [(v1) 15]

Beer, A., 1869, *Einleitung in die Mathematische Theorie der Elastizität und Capillarität*, Leipzig: A. Gresen Verlag. [(v1) 67]

Benguria, R. D., and M. C. Depassier, 1987, Oscillatory instabilities in the Rayleigh-Bénard problem with a free surface, Phys. Fluids **30** (6), 1678. [(v1) 171]

Benguria, R. D., and M. C. Depassier, 1989, On the linear stability theory of Bénard-Marangoni convection, Phys. Fluids A **1**(7), 1123. [(v1) 171, 173]

Benney, D. J., 1966, Long waves in liquid film, J. Math. Phys. **45** 150. [(v1) 270; (v2) 261]

Benney, D. J., and A. C. Newell, 1967, The propagation of nonlinear wave envelopes, J. Math. Phys. **46**, 133. [(v1) 324]

Bentwich, M., 1976, Two-phase axial laminar flow in a pipe with naturally curved interface, Chem. Eng. Sci. **31**, 71. [(v2) 4]

Bird, R. B., R. C. Armstrong and O. Hassager, 1977 *Dynamics of Polymeric Liquids*, volumes 1 and 2, Wiley and Sons. [(v1) 143]

Bisch, C., A. Lasek and H. Rodot, 1982, Hydrodynamic behavior of spherical semi-free liquid volumes in simulated weightlessness, J. de Mécanique théorique et appliquée **1**, 165. [(v1) 15]

Blais, P., D. J. Carlsson, T. Suprunchuk and D. M. Wiles, 1971, Bicomponent composites: preparation from incompatible polymers by corona treatment, Text. Res. J. **41**, 485. [(v1) 11]

Blennerhassett, P. J., 1980, On the generation of waves by wind, Phil. Trans. R. Soc. Lond. A **298**, 451. [(v1) 270, 271, 276, 277, 279, 287, 339, 346-350, 356-359; (v2) 51, 227, 237]

Blennerhassett, P. J., and F. T. Smith, 1987, Short-scale waves on wind-driven water ('cat's paws'), Proc. R. Soc. Lond A **410**, 1. [(v1) 270, 302, 306, 319, 351]

Bolton, B., and S. Middleman, 1980, Air entrainment in a roll coating system, Chem. Engng. Sci. **35**, 597. [(v1) 166]

Bond, R. K., 1957, Designing the gelsonite pipeline, Chem. Eng. **64**, 249. [(v1) 7]

Boomkamp, P. A. M., and R. H. M. Miesen, 1991, Nonaxisymmetric waves in core-annular flow with a small viscosity ratio, submitted to Phys. Fluids A. [(v2) 14, 94]

Bowen, R. M., 1971, *Continuum Physics*, Volume II, ed. A. C. Eringen, Academic Press. [(v2) 352]

Brand, R. H., and S. Backer, 1962, Mechanical principles of natural crimp of fiber, Text. Res. J. **32**, 39. [(v1) 11]

Bretherton, C. S., and E. A. Spiegel, 1983, Intermittency through modulational instability, Phys. Letts. **96A**, no.3, 152. [(v2) 228]

Brown, R. A. and L. E. Scriven, 1980, The shape and stability of rotating liquid drops, Phil. Trans. R. Soc. Lond. A **297**, 51. [(v1) 90]

Buckley, R. A., and R. J. Phillips, 1969, The development of bicomponent fibers, Chem. Eng. Prog. **65** (10), 41. [(v1) 11]

Buckmaster, J.D., 1972, Pointed bubbles in slow viscous flow. J. Fluid Mech. **55**, 385. [(v1) 143]

Buckmaster, J.D., 1973, The bursting of pointed drops in slow viscous flow, ASME J. Appl. Mech. **40**, 18. [(v1) 143]

Busse, F. H., 1978, A model of time-periodic mantle flow, Geophys. J. R. Astr. Soc. **52**,1. [(v1) 15, 172]

Busse, F. H., 1981, On the aspect ratio of two-layer mantle convection, Physics of the Earth and Planetary Interiors **24**, 320. [(v1) 15, 172]

Busse, F. H., 1982, Multiple solutions for convection in a two component fluid, Geophysical Research Letters **9** (5), 519. [(v1) 15, 31]

Busse, F. H., 1989, Fundamentals of thermal convection, from *Mantle Convection, Plate Tectonics and Global Dynamics*, W. R. Peltier ed., Gordon and Breach. [(v1) 15]

Busse, F. H., and G. Schubert, 1971, Convection in a fluid with two phases, J. Fluid Mech. **46**(4), 801. [(v1) 378, 399]

Busse, W. F., 1964, Two decades of high-polymer physics: a survey and forecast, Phys. Today **9**. [(v1) 376]

Buzano, E., and M. Golubitsky, 1983, Bifurcation on the hexagonal lattice and the planar Bénard problem, Phil. Trans. R. Soc. London A **308**, 617. [(v1) 176, 263]

Cahn, J. and J. Hilliard, 1954, Free energy of a nonuniform system I: interfacial free energy, J. Chem. Phys. **28**, 258. [(v2) 339]

Calderbank, P.H., 1967, Review Series No. 3 – Gas absorption from bubbles, Trans. Instn. Chem. Engrs. **45**, 209. [(v1) 144]

Calderbank, P.H., D. S. Johnson and J. Loudon, 1970, Mechanics and mass transfer of single bubbles in free rise through some Newtonian and non-Newtonian liquids, Chem. Engng. Sci. **25**, 235. [(v1) 144]

Canuto, C., M. Y. Hussaini, A. Quarteroni and T. A. Zang, 1988, *Spectral Methods in Fluid Dynamics*, Springer Verlag. [(v2) 29]

Carr, J., 1981, *Applications of Center Manifold Theory*, Springer, New York. [(v1) 176]

Carrigan, C. R., and J. C. Eichelberger, 1990, Zoning of magmas by viscosity in volcanic conduits, Nature **343**, No. 6255, 248. [(v1) 6, 15]

Castillo, J. L., and M. G. Velarde, 1982, Buoyancy-thermocapillary instability: the role of interfacial deformation in one- and two- component fluid layers heated from below or above, J. Fluid Mech. **125**, 463. [(v1) 171, 173]

Catton, I., and J. H. Lienhard V, 1984, Thermal stability of two fluid layers separated by a solid interlayer of finite thickness and thermal conductivity, J. Heat Transfer **106**, 605. [(v1) 188]

Cercignani, C., W. Fiszdon and A. Frezzotti, 1985, The paradox of the inverted temperature profiles between an evaporating and a condensing surface, Phys. Fluids **28**, 3237. [(v1) 381]

Cerisier, P., C. Jamond, J. Pantaloni and J. C. Charmet, 1984, Déformation de la surface libre en convection de Bénard-Marangoni, J. Phys. (Paris) **45**, 405. [(v1) 173]

Chandrasekhar, S., 1965, The stability of a rotating liquid drop, Proc. R. Soc. Lond. A **286**, 1. [(v1) 72]

Chandrasekhar, S., 1981 (1st ed. 1961), *Hydrodynamic and Hydromagnetic Stability*, Dover Publications Inc., New York. [(v1) 72, 115, 170, 171, 203, 209; (v2) 23, 37-38, 48, 70-73, 300]

Chang, H.-C., 1986, Traveling waves on fluid linterfaces: Normal form analysis of the Kuramoto-Sivashinsky equation, Phys. Fluids **29**, 3142. [(v1) 337]

Chapman, D. S., and P. R. Critchlow, 1967, Formation of vortex rings from falling drops, J. Fluid Mech. **29**, 177. [(v2) 289]

Charles, M. E., 1963, The pipeline flow of capsules. Part 2: theoretical analysis of the concentric flow of cylindrical forms, Can. J. Chem. Engng. April, 46. [(v2) 4, 6, 176]

Charles, M. E., G. W. Govier and G. W. Hodgson, 1961, The horizontal pipeline flow of equal density oil-water mixtures, Can. J. Chem. Eng. **39**, 17. [(v1) 3, 4, 5, 14; (v2) 2, 6, 12, 16-50, 61-65, 83, 126, 162, 164-165, 174-175, 177, 195-196, 231, 245]

Charles, M. E., and L. U. Lilleleht, 1966, Correlation of pressure gradients for the stratified laminar-turbulent pipeline flow of two immiscible liquids, Can. J. Eng. **44**, 47. [(v1) 4; (v2) 4]

Charles, M. E., and R. J. Redberger, 1962, The reduction of pressure gradients in oil pipelines by the addition of water: numerical analysis of stratified flow, Can. J. Chem. Engng. **40**, 70. [(v1) 4; (v2) 3, 4]

Chen, K., 1990, Lubricated pipelining: stability of core-annular flows, Ph. D. thesis, University of Minnesota. [(v2) 233, 241, 245]

Chen, K., 1991a, Interfacial instability due to elastic stratification in concentric coextrusion of two viscoelastic fluids, J. Non-Newtonian Fluid Mech. **40**, 155. [(v1) 360, 373]

Chen, K., 1991b, Elastic instability of the interface in Couette flow of viscoelastic liquids, J. Non-Newtonian Fluid Mech. **40**, 261. [(v1) 359, 360, 373, 375-378]

Chen, K., 1992, Shortwave instability of core-annular flow, Phys. Fluids A **4**(1), 186. [(v1) 359; (v2) 15]

Chen, K., R. Bai and D. D. Joseph, 1990, Lubricated pipelining III: stability of core-annular flow in vertical pipes, J. Fluid Mech. **214**, 251. [(v2) 16, 114-153, 196]

Chen, K., and D. D. Joseph, 1990, Application of the singular value decomposition to the numerical computation of the coefficients of amplitude equations and normal forms, Applied Num. Math. **6**, 425. [(v2) 243]

Chen, K., and D. D. Joseph, 1991a, Lubricated pipelining: stability of core-annular flows. Part IV: Ginzburg-Landau equations, J. Fluid Mech. **227**, 587. [(v2) 15, 16, 226-260]

Chen, K., and D. D. Joseph, 1991b, Long wave and lubrication theories for core-annular flow, Phys. Fluids A **3**(11), 2672 [(v2) 15, 16, 261-287]

Chen, K., and D. D. Joseph, 1992, Elastic short-wave instability in extrusion flows of viscoelastic liquids, J. Non-Newt. Fluid Mech. in press. [(v1) 359, 361, 375, 377]

Chernikin, V. I., 1956, Combined pumping of petroleum and water in pipes, Trudi. Mock. Neft in-ta **17**, 101. [(v2) 3]

Chow, S.-N., and J. K. Hale, 1982, *Methods of Bifurcation Theory*, Springer. [(v1) 176, 256, 343]

Chung, H. S., and R. Hogg, 1985, Stability criteria for fine-particle dispersions, Colloids and Surfaces **15**, 119. [(v1) 9]

Clark, A. F., and A. Shapiro, 1949, Method of pumping viscous petroleum, U. S. Patent No. 2,533,878. [(v2) 4, 5]

Clift, R., J. R. Grace and M. E. Weber, 1978, *Bubbles, drops and particles*, Academic Press. [(v2) 294, 299]

Clifton, E. G., and L. R. Handley, 1958, Method and apparatus for lubricating pipe lines, U. S. Patent No. 2,821,205. [(v2) 5]

Cohen, B. I., J. A. Krommes, W. M. Tang and M. N. Rosenbluth, 1976, Nonlinear saturation of the dissipative trapped-ion mode by mode coupling, Nuclear Fusion **16**, 971. [(v1) 271, 333-335]

Cohen, Y., and A. B. Metzner, 1985, Apparent slip flow of polymer solutions, J. Rheol. **29** (1), 67. [(v1) 375, 377]

Cotton, F. W., and H. Salwen, 1981, Linear stability of rotating Hagen-Poiseuille flow, J. Fluid Mech. **108**, 101. [(v2) 93]

Craik, A. D. D., 1966, Wind generated waves in thin liquid films, J. Fluid Mech. **26**, 369. [(v1) 338]

Craik, A. D. D., 1983, *Wave interactions and fluid flows*, Cambridge Univ. Press. [(v1) 343, 345; (v2) 241]

Dandy, D. S., and L. G. Leal, 1989, Buoyancy-driven motion of a deformable drop through a quiescent liquid at intermediate Reynolds numbers, J. Fluid Mech. **208**, 161. [(v2) 294, 304, 306]

Dauber, C. A., 1957, Pipeline coal transportation, Coal Age **62**(4),84. [(v1) 7]

Davey, A., L. M. Hocking and K. Stewartson, 1974, On the nonlinear evolution of three-dimensional disturbances in plane Poiseuille flow, J. Fluid Mech. **63**, 529. [(v1) 346-349; (v2) 244]

Davies, A.R., 1988, Reentrant corner singularities in non-Newtonian flow. Part I. Theory, J. Non-Newtonian Fluid Mech. **29**, 269. [(v1) 157]

Davis, H. T., 1988, A theory of tension at a miscible displacement front, in *Numerical Simulation and Oil Recovery*, ed. M. Wheeler, Institute for Mathematics and its Applications **2**. [(v2) 341-342, 374]

Davis, R. H., E. Herbolzheimer and A. Acrivos, 1983, Wave formation and growth during sedimentation in narrow tilted channels, Phys. Fluids **26**, 2055. [(v1) 295]

Davis, S. H., and G. M. Homsy, 1980, Energy stability theory for free-surface problems: buoyancy-thermocapillary layers, J. Fluid Mech. **98** (3), 527. [(v1) 171, 173, 178]

Dean, W.R. and P. E. Montagnon, 1949, On the steady motion of viscous liquid in a corner, Proc. Cambridge Phil. Soc. **45**, 389. [(v1) 151]

Denham, E., D. R. Wall and D. M. Whitehead, 1982, The handling and combustion characteristics of stable coal fuel oil dispersions, Proc. of 4th Int. Symp. on Coal Slurry Combustion, Orlando, Florida. [(v1) 9]

Denn, M. M., 1990, Issues in viscoelastic fluid mechanics, Annu. Rev. Fluid Mech. **23**, 13. [(v1) 375]

Deryagin, B. M., and S. M. Levi, 1964, *Film Coating Theory*, Focal Press, London and New York. [(v1) 166]

Dijkstra, H. A., and P. H. Steen, 1991, Thermocapillary stabilization of the capillary breakup of an annular film of liquid, J. Fluid Mech. **229**, 205. [(v2) 13]

Drazin, P. G., and W. H. Reid, 1982, *Hydrodynamic Stability*, Cambridge University Press, 1st paperback ed. [(v1) 24, 115, 171, 193, 203, 209, 220, 306; (v2) 98, 165]

Drew, D. A., 1983, Continuum modelling of two-phase flows, in *Theory of Dispersed Multiphase Flow*, ed. R. E. Meyer, Academic Press, 173. [(v1) 103, 352]

Drew, D. A., 1983, Mathematical modeling of two-phase flow, A. Rev. Fluid Mech. **15**, 261. [(v1) 103, 352]

Drew, D. A., 1986, Flow structure in the Poiseuille flow of a particle-fluid mixture, SIAM Workshop on Multiphase Flow, June 2-4. [(v1) 6]

Dunn, J. E., 1986, Interstitial working and a nonclassical thermodynamics, in *New Perspectives in Thermodynamics*, ed. J. Serrin, Springer Verlag. [(v2) 346]

Dussan V., E. B., 1979, On the spreading of liquids on solid surfaces: static and dynamic contact lines, *Annual Review of Fluid Mechanics*, Annual Reviews Inc., Palo Alto, **11**, 371. [(v2) 173]

Dussan V., E. B., and S. H. Davis, 1974, On the motion of a fluid-fluid interface along a solid surface, J. Fluid Mech. **65**, 71. [(v1) 105]

Engelman, M. S., 1982, FIDAP – A Fluid Dynamics Analysis Package, Adv. Eng. Software **4**, 163. [(v1) 158]

Engelman, M.S. and R. I. Sani, 1986, Finite element simulation of incompressible fluid flows with a free/moving surface. In: C. Taylor, J. A. Johnson and W. R. Smith (eds.), *Computational Techniques for Fluid Flow*, 47. Pineridge Press, Swansea. [(v1) 158]

Everage, A. E., 1973, Theory of bicomponent flow of polymer melts, I. Equilibrium Newtonian tube flow, Trans. Soc. Rheol. **17**, 629. [(v1) 10, 11, 33, 38, 42]

Falco, R. E., J. C. Klewicki and D. G. Nocera, 1990, A study of flow properties of wet solids using laser-induced photochemical anemometry.

Abstract at the NSF-DOE Workshop on Flow of Particulates and Fluids, Oct. 1-3, NIST, Gaithersburg, MD. [(v1) 6]

Feeny, B.F. and F. C. Moon, 1989, Autocorrelation on symbol dynamics for a chaotic dry friction oscillation, Phys. Letters A **141**(8),9. [(v1) 115, 138, 139]

Fortes, A., D. D. Joseph, and T. S. Lundgren, 1987, Nonlinear mechanics of fluidization of beds of spherical particles, J. Fluid Mech. **177**, 467. [(v1) 127]

Frei, H., and M. Schiffer, 1947, Separation by diffusion in fields of ultrasonic waves, Phys. Rev. **71**, 555. [(v2) 395]

Frenkel, A. L., 1988, Nonlinear saturation of core-annular flow instabilities, Proc. Sixth Symposium on Energy Engineering Sciences, Argonne National Laboratory. [(v1) 333; (v2) 11, 261-271]

Frenkel, A. L., A. J. Babchin, B. G. Levich, T. Shlang and G. I. Shivashinsky, 1987, Annular flows can keep unstable films from breakup: nonlinear saturation of capillary instability, J. Colloid and Interface Sci. **115**, 225. [(v1) 333; (v2) 11, 261, 267-271]

Freundlich, H., 1926, *Colloid and Capillary Chemistry*, Mathuen and Co. Ltd. London. [(v2) 337]

Fujimura, K., 1991a, Methods of centre manifold and multiple scales in the theory of weakly nonlinear stability for fluid motions, Proc. R. Soc. Lond. A **434**, 719. [(v1) 343]

Fujimura, K., 1991b, Nonlinear equilibrium solutions for travelling waves in a free convection between vertical parallel plates, Eur. J. Mech. B/Fluids **10**(2), 25. [(v1) 171]

Funatsu, K., and M. Sato, 1984, in *Advances in Rheology*, 4, eds. B. Mena, A. Garcia-Rejon and C. Rangel-Nafaile, UNAM, Mexico, 465. [(v1) 376]

Galdi, G. P., D. D. Joseph, L. Preziosi and S. Rionero, 1991, Mathematical problems for miscible incompressible fluids with Korteweg stresses, European J. Mech. B/Fluids **10**(3). [(v2) 324, 351]

Galdi, G. P., and B. Straughn, eds., 1988, *Energy Stability and Convection*, Research Notes in Mathematics, Longman. [(v2) 56]

Garcia-Ybarra, P. L., and M. G. Velarde, 1987, Oscillatory Marangoni-Bénard interfacial instability and capillary-gravity waves in single- and two- component liquid layers with or without Soret thermal diffusion, Phys. Fluids **30** (6), 1649. [(v1) 171]

Garik, P., J. Hetrick, B. Orr, D. Barkey and E. Ben-Jacob, 1991, Interfacial cellular mixing and a conjecture on global deposit morphology, Phys. Rev. Lett. **66**(12), 1606. [(v2) 343]

Gebhart, B., 1961, 1971, *Heat Transfer*, second edition, McGraw-Hill Book Company. [(v1) 381]

Gemmell, A. R., and N. Epstein, 1962, Numerical analysis of stratified laminar flow of two immiscible Newtonian liquids in circular pipes, Can. J. Chem. Eng. **40**, 215. [(v1) 4; (v2) 4]

408   References

Ginzburg, V. L., and L. D. Landau, 1950, Zh. Eksper. i. Teor. Fiz. **20**. [(v2) 227]

Gjevik, B., 1970, Occurrence of finite-amplitude surface waves on falling liquid films, Phys. Fluids **13**, 1918. [(v1) 336]

Glass, W., 1961, Water addition aids pumping viscous oils, Chem. Eng. Prog. **57**, 116. [(v2) 6, 8]

Goenaga, A. and B. G. Higgins, 1991, Kinematic and dynamic constraints for flow separation from free surfaces and interfaces, paper no. 111c, AIChE 1991 Annual Meeting, Los Angeles. [(v1) 150]

Golub, G. H. and C. F. Van Loan, 1983, *Matrix computations*, Johns Hopkins University Press, Baltimore. [(v2) 241]

Golubitsky, M., and I. Stewart, 1985, Hopf bifurcation in the presence of symmetry, Arch. Rat. Mech. Anal. **87**, 107. [(v1) 176, 257]

Golubitsky, M., J. W. Swift and E. Knobloch, 1984, Symmetries and pattern selection in Rayleigh-Bénard convection, Physica 10D, 249. [(v1) 175, 176, 263]

Gorodtsov, V. A., and A. I. Leonov, 1967, On a linear instability of a plane parallel Couette flow of viscoelastic fluid, J. Appl. Math. Mech. **31**, 310. [(v1) 364-366]

Gottlieb, D., and S. A. Orszag, 1983, *Numerical analysis of spectral methods: theory and applications*, CBMS-NSF Regional Conference Series in Applied Math., SIAM. [(v1) 191, 373]

Goto, T., and N. Waku, 1985, The preparation of high transition temperature superconducting Pb-Bi-Ge alloy filaments using the method of glass-coated melt spinning, J. Mat. Science **20**, 532. [(v1) 12]

Grace, J.P., 1971, Dispersion phenomena in high viscosity immiscible fluid systems and application of static mixers as dispersion devices in such systems, Engng. Found. 3rd Res. Cong. Mixing, Andover, New Hampshire. [(v1) 143]

Graham, A. L., S. A. Altobelli, E. Fukushima, L. A. Mondy and T. S. Stephens, 1991, Note: NMR imaging of shear-induced diffusion and structure in concentrated suspensions undergoing Couette flow, J. Rheol. **35** (1), 191. [(v1) 7]

Greenspan, H., 1968, *The Theory of Rotating Fluids*, Cambridge University Press. [(v1) 16]

Griffiths, R. W., 1986 a, Thermals in extremely viscous fluids, including the effects of temperature-dependent viscosity, J. Fluid Mech. **166**, 115. [(v2) 326, 328, 332]

Griffiths, R. W., 1986 b, Particle motions induced by spherical convective elements in Stokes flow, J. Fluid Mech. **166**, 139. [(v2) 326]

Grimshaw, R., and A. P. Hooper, 1991, The non-existence of a certain class of travelling wave solutions of the Kuramoto-Sivashinsky equation, Physica D **50**(2), 231. [(v1) 337]

Guckenheimer, J., and P. Holmes, 1983, *Nonlinear Oscillations, Dynamical Systems, and Bifurcation of Vector Fields*, Springer, New York. [(v1) 176]

Guillopé, C., D. Joseph, K. Nguyen and F. Rosso, 1987, Nonlinear stability of rotating flow of two fluids, Journal de Mécanique théorique et appliquée **6** (5), 619. [(v1) 45, 56, 114, 117]

Gumerman, R. J., and G. M. Homsy, 1974, Convective instabilities in concurrent two-phase flow: Part I. Linear stability, AIChE J. **20**, 981; Part II. Global stability, AIChE J. **20**, 1161; Part III. Experiments, AIChE J. **20**, 1167. [(v1) 173, 196, 276, 278]

Han, C. D., 1973, A study of bicomponent coextrusion of molten polymers, J. Appl Polym. Sci. **17**, 1289. [(v1) 11]

Han, C. D., 1975, A study of coextrusion in a circular die, J. Appl. Polym. Sci. **19**, 1875. [(v1) 11]

Hansen, E. B., 1987, Stokes flow down a wall into an infinite pool, J. Fluid Mech. **178**, 243. [(v1) 164]

Happel, J., and H. Brenner, 1983, *Low Reynolds number hydrodynamics*, Martinus Nijhoff Publishers, Boston. [(v2) 155, 294]

Hassager, O., 1985, The motion of viscoelastic fluids around spheres and bubbles. In: A. S. Lodge, M. Renardy and J. A. Nohel (eds.), *Viscoelasticity and Rheology*, Academic Press, Orlando, 1. [(v1) 144]

Hasson, D., U. Mann and A. Nir, 1970, Annular flow of two immiscible liquids, I: Mechanisms, Canadian J. Chem. Eng. **48**, 514. [(v2) 8, 174]

Hasson, D., and A. Nir, 1970, Annular flow of two immiscible liquids, II: Analysis of core-liquid ascent, Canadian J. Chem. Eng. **48**, 521. [(v2) 8]

Hatzikiriankos, S. G., and J. M. Dealy, 1991, Wall slip of molten high density polyethylene. I. sliding plate rheometer studies, J. Rheol. **35** (4), 497. [(v1) 377]

Herbert, T., 1980, Nonlinear stability of parallel flows by higher order amplitude expansions, AIAA J. **18**, 243. [(v2) 240]

Hermanrud, B., 1981, The compound jet: a new method to generate fluid jets for ink jet printing, Report 1/1981, Dept. of Electrical Measurements, Lund Institute of Technology. [(v1) 11]

Hertz, C. H., and B. Hermanrud, 1983, A liquid compound jet, J. Fluid Mech. **131**, 271. [(v1) 11, 12]

Hesla, T. I., F. R. Pranckh and L. Preziosi, 1986, Squire's theorem for two stratified fluids, Phys. Fluids **29**, 2808. [(v1) 276, 277]

Heywood, N. I., 1986, A review of techniques for reducing energy consumption in slurry pipelining, Hydrotransport 10, Tenth International Conference on the Hydraulic Transport of Solids in Pipes, Innsbruck, Austria, Oct. 29-31, 319. [(v1) 8]

Hickox, C. E., 1971, Instability due to viscosity and density stratification in axisymmetric pipe flow, Phys. Fluids **14**, 251. [(v1) 269; (v2) 12, 13, 31, 50, 65, 132]

410   References

Hicks, E. M., J. F. Ryan, R. B. Taylor and R. L. Tichenor, 1960, Reversible crimp in an acrylic fiber, Textile Res. J. **30**, 675. [(v1) 11]

Hicks, E. M., E. A. Tippets, J. V. Hewett and R. H. Brand, 1967, *Man-made fibers*, Vol. 1, H. Mark, S. M. Atlas, and E. Cernia, Eds., Wiley (Interscience), New York. [(v1) 11]

Hiemenz, P. C., 1977, *Principles of Colloid and Surface Chemistry*, Marcel Dekker, Inc., New York. [(v2) 338-339]

Hinch, E. J., 1984, A note on the mechanism of the instability at the interface between two shearing fluids, J. Fluid Mech. **144**, 463. [(v1) 297]

Hinch, E.J. and A. Acrivos, 1979, Steady long slender droplets in two-dimensional straining motion, J. Fluid Mech. **91**, 401. [(v1) 143]

Hocking, L. M., and K. Stewartson, 1972, On the nonlinear response of a marginally unstable plane parallel flow to a two-dimensional disturbance, Proc. R. Soc. Lond. A. **326**, 289. [(v2) 240]

Hoffman, R. L., 1975, A study of the advancing interface. I. Interface shape in liquid-gas systems, J. Colloid Interface Sci. **50**, 228. [(v1) 166]

Holmes, P., 1986, Spatial structure of time - periodic solutions of the Ginzburg - Landau equation, Physica **23**D, 84. [(v2) 240-241]

Homsy, G., 1974, *Lectures in Applied Mathematics*, ed. A. C. Newell, Am. Math. Soc., volume 15, 191. [(v1) 270, 324]

Homsy, G., 1987, Viscous fingering in porous media, Ann. Rev. Fluid Mech. **19**, 271. [(v2) 374, 379]

Hooper, A. P., 1985, Long-wave instability at the interface between two viscous fluids: Thin layer effects, Phys. Fluids **28**, 1613. (Erratum, **28**, 3182) [(v1) 269, 270, 279, 290-296]

Hooper, A. P., 1988, A note on the energy stability equation for Couette flow of two superposed viscous fluids, *Energy Stability and Convection*, eds. G. P. Galdi and B. Straughn, Research Notes in Mathematics, Longman. [(v1) 271; (v2) 56]

Hooper, A. P., and W. G. C. Boyd, 1983, Shear flow instability at the interface between two viscous fluids, J. Fluid Mech. **128**, 507. [(v1) 199, 269, 270, 271, 290, 294, 296-302, 305-306, 320, 349; (v2) 34-35, 56, 106]

Hooper, A. P., and W. G. C. Boyd, 1987, Shear flow instability due to a wall and a viscosity discontinuity at the interface, J. Fluid Mech. **179**, 201. [(v1) 270, 303, 305-319; (v2) 36, 56, 80, 83, 106]

Hooper, A. P., and R. Grimshaw, 1985, Nonlinear instability at the interface between two viscous fluids, Phys. Fluids **28**, 3. [(v1) 270, 287, 324-338; (v2) 262-264]

Hooper, A. P., and R. Grimshaw, 1988, Traveling wave solutions of the Kuramoto-Sivashinsky equation, Wave Motion **10**, 405. [(v1) 337]

Hu, H., and D. D. Joseph, 1989 a, Lubricated pipelining: stability of core-annular flow. Part 2, J. Fluid Mech. **205**, 359. [(v1) 396; (v2) 13, 16, 50-84]

Hu, H., and D. D. Joseph, 1989 b, Stability of core-annular flow in a rotating pipe, Phys. Fluids A **1** (10), 1677. [(v2) 13, 16, 93-94]

Hu, H., and D. D. Joseph, 1992, Miscible displacement in a Hele-Shaw cell, Army High Performance Computing Research Center, University of Minnesota, Preprint 92-007. [(v2) 324]

Hu, H., T. Lundgren and D. D. Joseph, 1990, Stability of core-annular flow with a small viscosity ratio, Phys. Fluids A **2** (11), 1945. [(v2) 14, 16, 94-113]

Huang, A., and D. D. Joseph, 1992a, Instability of the equilibrium of a liquid below its vapor between horizontal heated plates, J. Fluid Mech., to appear. [(v1) 378-399]

Huang, A., and D. D. Joseph, 1992b, Stability of liquid-vapor flow down an inclined channel with phase change, Army High Performance Computing Research Center, University of Minnesota, Preprint 91-98. [(v1) 378-399]

Hurle, D. T. J., and E. Jakeman, 1971, Soret-driven thermosolutal convection, J. Fluid Mech. **47**, 667. [(v1) 171]

Hwang, C. C., and C.-I. Weng, 1987, Finite-amplitude stability analysis of liquid films down a vertical wall with and without interfacial phase change, Int. J. Multiphase Flow **13** (6), 803. [(v1) 378]

Iooss, G., and D. D. Joseph, 1990, *Elementary Stability and Bifurcation Theory*, second edition, Springer-Verlag New York, Inc. [(v1) 60, 61, 204, 241, 256, 343]

Isaacs, J. D., and J. B. Speed, 1904, Method of piping fluids, U. S. Patent No. 759,374. [(v2) 3]

ISF-85, 1985, *Proceedings of the International Symposium on Fiber Science and Technology*, The Society of Fiber Science and Technology, Hakone, Japan, Elsevier Applied Science Publishers. [(v1) 12]

Ishii, M., 1975, *Thermo-Fluid Dynamic Theory of Two-Phase Flow*, Eyrolles. [(v1) 381]

Jeong, J. T., and H. K. Moffatt, 1992, Free surface cusps associated with flow at low Reynolds number, submitted to J. Fluid Mech. [(v1) 140, 141, 143, 153, 154, 162, 168]

Jones, R. S., and O. D. J. Thomas, 1989, The coextrusion of two incompressible elastico-viscous fluids through a rectangular channel, J. Appl. Math. and Phys. (ZAMP) **40**, 425. [(v1) 11]

Joseph, D. D., 1973, Domain perturbations: the higher order theory of infinitesimal water waves, Arch. Rat. Mech. Anal. **51**, 295. [(v1) 184]

Joseph, D. D., 1976, *Stability of fluid motions* , I and II, Springer-Verlag New York, Inc. [(v1) 23, 24, 51, 115, 171, 203, 209, 272; (v2) 93]

Joseph, D. D., 1987, Two fluids heated from below, *Energy Stability and Convection*, eds. G. P. Galdi and B. Straughn, Research Notes in Mathematics, Longman. [(v1) 178; (v2) 57]

Joseph, D. D., 1988, videocassette on coal-oil dispersions. [(v1) 9]

Joseph, D. D., 1990 a, *Fluid Dynamics of Viscoelastic Liquids*, Springer-Verlag New York, Inc. [(v1) 15, 18, 25, 184, 361; (v2) 382]

Joseph, D. D., 1990 b, Fluid dynamics of two miscible liquids with diffusion and gradient stresses, Eur. J. Mech. B/Fluids **9** (6), 565. [(v2) 324-365, 369]

Joseph, D. D., 1990 c, Separation in flowing fluids, Nature **348**, 487. [(v1) 6]

Joseph, D. D., 1992, Understanding cusped interfaces, J. Non-Newt. Fluid Mech., to appear. [(v1) 152, 153]

Joseph, D. D., M. Arney, G. Gillberg, H. Hu, D. Huttman, C. Verdier and H. Vinagre, 1992, A spinning drop extensiotensiometer, J. Rheology. [(v1) 83]

Joseph, D. D., M. Arney and G. Ma, 1992, Upper and lower bounds for interfacial tension using spinning drop devices, J. Colloid and Interface Science **148**(1), 291. [(v1) 81]

Joseph, D. D., and G. S. Beavers, 1977, Free surface problems in rheological fluid mechanics, Rheol. Acta **16**, 169. [(v1) 15]

Joseph, D. D., G. S. Beavers, A. Cers, C. Dewald, A. Hoger and P. T. Than, 1984, Climbing constants for various liquids, J. Rheol. **28** (4), 325. [(v1) 15]

Joseph, D. D., and S. Carmi, 1969, Stability of Poiseuille flow in pipes, annuli and channels, Q. Appl. Math. **XXVI**(4), 576. [(v2) 93]

Joseph, D. D., and H. Hu, 1991, Interfacial tension between miscible liquids, Army High Performance Computing Research Center, University of Minnesota, preprint 91-58; Non-solenoidal effects and Korteweg stresses in simple mixtures of incompressible liquids, preprint 91-03. [(v2) 324-378]

Joseph, D. D., T. S. Lundgren, R. Jackson and D. A. Saville, 1990, Ensemble averaged and mixture theory equations for incompressible fluid-particle suspensions, Int. J. Multiphase Flow **16** (1), 35. [(v2) 352, 354, 356]

Joseph, D. D., J. Nelson, M. Renardy and Y. Renardy, 1991, Two-dimensional cusped interfaces, J. Fluid Mech. **223**, 383. [(v1) 140, 145, 152]

Joseph, D. D., K. Nguyen and G. S. Beavers, 1984, Nonuniqueness and stability of the configuration of flow of immiscible fluids with different viscosities, J. Fluid Mech. **141**, 319. [(v1) 3, 27, 29, 31, 34, 45, 95, 101, 111; (v2) 187]

Joseph, D. D., K. Nguyen and G. S. Beavers, 1986, Rollers, Phys. Fluids **29**, 2771. [(v1) 45]

Joseph, D. D., and L. Preziosi, 1987, Stability of rigid motions and coating films in bicomponent flows of immiscible liquids, J. Fluid Mech. **185**, 323. [(v1) 45, 48, 52, 69, 71, 77, 114]

Joseph, D. D., M. Renardy and Y. Renardy, 1984, Instability of the flow of immiscible liquids with different viscosities in a pipe, J. Fluid Mech.

141, 309; 1983, Mathematics Research Center Technical Summary Report 2503, University of Wisconsin. [(v1) 40, 42; (v2) 13, 16, 18, 29, 30, 59, 130]

Joseph, D. D., M. Renardy, Y. Renardy and K. Nguyen, 1985, Stability of rigid motions and rollers in bicomponent flows of immiscible liquids, J. Fluid Mech. **153**, 151. [(v1) 45, 52, 63, 65, 95]

Joseph, D. D., and D. H. Sattinger, 1972, Bifurcating time periodic solutions and their stability, Arch. Rat. Mech. Anal. **45**, 79. [(v2) 240]

Joseph, D. D., and J.-C. Saut, 1990, Short-wave instabilities and ill-posed initial-value problems, Theoretical and Computational Fluid Dynamics **1**, 191. [(v1) 297; (v2) 13, 382]

Joseph, D. D., P. Singh and K. Chen, 1990, Couette flows, rollers, emulsions, tall Taylor cells, phase separation and inversion, and a chaotic bubble in Taylor-Couette flow of two immiscible liquids, in *Nonlinear Evolution of Spatio-temporal Structures in Dissipative Continuous Systems*, ed. F. H. Busse and L. Kramer, Plenum Press, New York. [(v1) 114]

Kao, T. W., 1965 a, Stability of two-layer viscous stratified flow down an inclined plane, Phys. Fluids **8**, 812. [(v1) 272]

Kao, T. W., 1965 b, Role of the interface in the stability of stratified flow down an inclined plane, Phys. Fluids **8** (12), 2190. [(v1) 272]

Kao, T. W., 1968, Role of viscosity stratification in the stability of two-layer flow down an incline, J. Fluid Mech. **33**, 561. [(v1) 272]

Kao, T. W., and C. Park, 1972, Experimental investigations of the stability of channel flows. Part 2. Two-layered co-current flow in a rectangular channel, J. Fluid Mech. **52**, 401. [(v1) 4, 287; (v2) 4]

Karnis, A., H. L. Goldsmith and S. G. Mason, 1966, The kinetics of flowing dispersions: I. Concentrated suspensions of rigid particles, J. Colloid and Interf. Sci. **22**, 531. [(v1) 6]

Kawahara, T., 1983, Formation of saturated solutions in a nonlinear dispersive system with instability and dissipation, Phys. Rev. Lett. **51**, 381. [(v1) 334-336]

Kelly, R. E., D. A. Goussis, S. P. Lin and F. K. Hsu, 1989, The mechanism for surface wave instability in film flow down an inclined plane, Phys. Fluids A **1**, 819. [(v1) 271]

Kennedy, R. J., 1966, Towards an analysis of plug flow through a pipe, Can. J. Chem. Eng., December 1, 354. [(v1) 7]

Kevorkian, J. and J. D. Cole, 1980, *Perturbation Methods in Applied Mathematics*, Springer, Berlin. [(v2) 99]

Khan, A. A. and C. D. Han, 1976, On the interface deformation in the stratified two-phase flow of viscoelastic fluids, Trans. Soc. Rheol. **20**:4, 595. [(v1) 11]

Khan, A. A. and C. D. Han, 1977, A study on the interfacial instability in the stratified flow of two viscoelastic fluids through a rectangular duct, Trans. Soc. Rheol. **21**, 101. [(v1) 359]

Knappe, W., and E. Krumbock, 1984, in *Advances in Rheology*, eds. B. Mena, A. Garcia-Rejon and C. Rangel-Nafaile, UNAM, Mexico, **4**, 417. [(v1) 376]

Knobloch, E., 1980, Convection in binary fluids, Phys. Fluids **23** (9), 1918. [(v1) 172]

Koh, C. J., and L. G. Leal, 1990, An experimental investigation on the stability of viscous drops translating through a quiescent fluid, Phys. Fluids A **2**(12), 2103. [(v2) 295, 326, 328, 333]

Kohn, R. V., and R. Lipton, 1986, The effective viscosity of a mixture of two Stokes fluids, in *Advances in Multiphase Flow and Related Problems*, ed. G. Papanicolaou, SIAM, 123. [(v1) 42]

Kojima, M., E. J. Hinch and A. Acrivos, 1984, The formation and expansion of a toroidal drop moving in a viscous fluid, Phys. Fluids **27** (1), 19. [(v2) 290, 292, 309, 361-362, 372]

Korteweg, D., 1901, Sur la forme que prennent les equations du mouvement des fluids si l'on tient compte des forces capillaires causees par des variations de densite, Arch. Neerl. Sciences Exactes et Naturelles, Series II **6**, 1. [(v2) 325, 334-337, 363, 369]

Krantz, W. B., and S. L. Goren, 1970, Industrial Engineering Chem. Fundamentals **9**, 107. [(v1) 324]

Krishna, M. V. G., and S. P. Lin, 1977, Nonlinear stability of a viscous film with respect to three-dimensional side-band disturbances, Phys. Fluids **20** (7), 1039. [(v1) 324]

Kuramoto, Y., and T. T. Tsuzuki, 1976, Persistent propagation of concentration waves in dissipative media far from thermal equilibrium, Prog. Theor. Phys. **55**, 356. [(v1) 333]

Lamb, H., 1932, *Hydrodynamics*, Cambridge University Press. [(v1) 142]

Landau, L. D., and E. M. Lifshitz, 1959, *Fluid Mechanics*, Pergamon Press. [(v2) 354, 357-358]

LaQuey, R. E., S. M. Mahajan, P. H. Rutherford and W. M. Tang, 1975, Nonlinear saturation of trapped-ion mode, Phys. Rev. Lett. **34**, 391. [(v1) 270, 333]

Leach, R. W., 1957, Pipeline designed for viscous crude, The Pipeline Engineer, November. [(v2) 5]

Leal, L.G., J. Skoog and A. Acrivos, 1971, On the motion of gas bubbles in a viscoelastic liquid, Can. J. Chem. Engng. **49**, 569. [(v1) 144]

Lebovitz, N. R., 1982, Perturbation expansions on perturbed domains, SIAM Review **24**, 381. [(v1) 184]

Lee, G. W. T., P. Lucas and A. Tyler, 1983, Onset of Rayleigh-Bénard convection in binary liquid mixtures of ^{3}He in ^{4}He, J. Fluid Mech. **135**, 235. [(v1) 171]

Lee, B. L., and J. L. White, 1974, An experimental study of rheological properties of polymer melts in laminar shear flow and of interface deformation and its mechanisms in two-phase stratified flow, Trans. Soc, Rheol. **18:3**, 467. [(v1) 11]

Leighton, D., and A. Acrivos, 1986, Viscous resuspension, Chem. Eng. Sci. **41**, 1377. [(v1) 7]

Leslie, F., 1985, Measurements of rotating bubble shapes in a low-gravity environment, J. Fluid Mech. **161**, 269. [(v1) 76]

Li, C.-H., 1969 a, Stability of two superposed elasticoviscous liquids in plane Couette flow, Phys. Fluids **12**, 531. [(v1) 269, 360]

Li, C.-H., 1969 b, Instability of three-layer viscous stratified flow, Phys. Fluids **12**, 2473. [(v1) 272, 279]

Li, C.-H., 1970, Role of elasticity on the stability of stratified flow of viscoelastic fluids, Phys. Fluids **13**, 1701. [(v1) 269, 272, 359]

Lin, S. P., 1969, Finite amplitude stability of a parallel flow with a free surface, J. Fluid Mech. **36**, 113. [(v1) 324, 336]

Lin, S. P., 1983 a, Effects of surface solidification on the stability of multi-layered liquid films, J. Fluids Engng. **105**, 119. [(v1) 272]

Lin, S. P., 1983 b, Film waves, in *Waves on Fluid Interfaces*, R. E. Meyer ed., Academic Press Inc. [(v1) 272]

Lin, S. P., and E. A. Ibrahim, 1990, Instability of a viscous liquid jet surrounded by a viscous gas in a vertical pipe, J. Fluid Mech. **218**, 641. [(v2) 13]

Lin, S. P., and M. V. G. Krishna, 1977, Stability of a liquid film with respect to initially finite three-dimensional disturbances, Phys. Fluids **20** (12), 2005. [(v1) 324]

Lin, S. P., and C. Y. Wang, 1983, Modeling wavy film flows, in *Encyclopedia of Fluid Mechanics*, chapter 28. [(v1) 272]

Lions, J. L., and E. Magenes, 1972, *Non-Homogeneous Boundary Value Problems and Applications*, Springer-Verlag. [(v1) 58]

Lister, J. R., 1987, Long-wavelength instability of a line plume, J. Fluid Mech. **175**, 413. [(v2) 15]

Loewenherz, D. S., and C. J. Lawrence, 1989, The effect of viscosity stratification on the stability of a free surface flow at low Reynolds number, Phys. Fluids A **1**(10), 1686. [(v1) 16, 272]

Looman, M. D., 1916, Method of conveying oil, U. S. Patent No. 1,192,438. [(v2) 3]

Lundgren, T. S., and N. N. Mansour, 1990, Vortex ring bubbles, J. Fluid Mech. **224**, 177. [(v2) 290]

Lurie K., A. V. Cherkaev and A. V. Fedorov, 1982, Regularization of optimal design problems for bars and plates, J. Opt. Theor. Appl. **37**, 499. [(v1) 42]

Mackrodt, P. A., 1976, Stability of Hagen-Poiseuille flow with a superimposed rigid rotation, J. Fluid Mech. **73**, 153. [(v2) 93]

MacLean, D. L., 1973, A theoretical analysis of bicomponent flow and the problem of interface shape, Trans. Soc. Rheol. **17:3**, 385. [(v1) 11, 33, 38]

Macosko, C. W., M. A. Ocansey and H. H. Winter, 1982, Steady planar extension with lubricated dies, J. Non-Newtonian Fluid Mech. **11**, 301. [(v1) 13]

Manfré, G., G. Servi and C. Ruffino, 1974, Copper microwires spun from the melt, J. Mat. Sci. **9**, 74. [(v1) 12]

Manneville, P., 1981, Statistical properties of chaotic solutions of a one-dimensional model for phase turbulence, Phys. Lett. **84A**, 129. [(v1) 333]

May, S. E. and J. V. Maher, 1991, Capillary wave relaxation for a meniscus between miscible liquids, submitted to Phys. Rev. Lett. [(v2) 343]

Merkle, C. L., and S. Deutsch, 1990, Drag reduction in liquid boundary layers by gas injection, in *Viscous Drag Reduction in Boundary Layers*, ed. D. M. Bushnell and J. M. Hefner **123**, Progress in Astronautics and Aeronautics, published by AIAA, 351. [(v1) 5]

Mhatre, M.V. and R. C. Kintner, 1959, Fall of liquid drops through pseudoplastic liquids, Ind. Engng. Chem. **51**, 865. [(v1) 144]

Michael, H.D., 1958, The separation of a viscous liquid at a straight edge, Mathematika **5**, 82. [(v1) 151]

Miesen, R., G. Beijnon, P. E. M. Duijvestijn, R. V. A. Oliemans and T. Verheggen, 1991, Interfacial waves in core-annular flow, J. Fluid Mech. to appear. [(v2) 14, 94]

Miles, J. W., 1960, The hydrodynamic stability of a thin film, J. Fluid Mech. **8**, 593. [(v1) 338]

Miller, M. C., 1990, Elimination of viscous liquid-fill flight instability by means of lower viscosity, immiscible liquid additives, Proceedings of the 1990 U.S. Army Chemical Research, Development and Engineering Center Scientific Conference on Chemical Defense Research (preprint). [(v1) 16]

Minagawa, N., and J. L. White, 1975, Coextrusion of unfilled and TiO_2-filled polyethylene: influence of viscosity and die cross-section on interface shape, Poly. Eng. and Sci. **15**, 825. [(v1) 11]

Moffatt, H. K., 1977, Behavior of a viscous film on the outer surface of a rotating cylinder, J. de Mécanique **16** (5), 651. [(v1) 31, 48, 49, 66, 85, 86, 87, 88, 92, 110]

Moller, K., and G. G. Duffy, 1978, An equation for predicting transition-regime pipe friction loss, Tappi **61** (1), 63. [(v1) 8]

Moon, H. T., 1982, Transition to chaos in the Ginzburg-Landau equation, Ph. D. thesis, University of Southern California. [(v2) 240]

Moon, H. T., P. Huerre and L. G. Redekopp, 1983, Transition to chaos in the Ginzburg-Landau equation, Physica **7D**, 135. [(v2) 228]

Moynihan, R. H., D. G. Baird and R. R. Ramanathan, 1990, Additional observations on the surface melt fracture behavior of linear low-density polyethylene, J. Non-Newtonian Fluid Mech. **36**, 256. [(v1) 375-376]

Nagata, W., and J. W. Thomas, 1986, Bifurcation in doubly-diffusive systems, Parts I and II, SIAM J. Math. Anal. **17**, 91; Part III, ibid, 289. [(v1) 171, 176, 243]

Nakoryakov, V. Ye., and S. V. Alekseyenko, 1981, Waves on a film flowing down an incline, Fluid Mech. **10** (3), 18. [(v1) 270, 324]

Nataf, H. C., S. Moreno and Ph. Cardin, 1988, What is responsible for thermal coupling in layered convection?, J. Phys. France **49**, 1707. [(v1) 175]

Nelson, J., P. Mohr and D. D. Joseph, 1992, Nonlinear stability of two fluids in a combined Couette-Poiseuille flow, in preparation. [(v1) 320]

Newell, A. C., 1974, Envelope equations, in Lectures in Applied Math. **15**, ed. A. C. Newell, 157. [(v2) 235, 240]

Newell, A. C., 1985, *Solitons in mathematics and physics*, CBMS-NSF Regional Conference Series in Applied Mathematics, SIAM publication. [(v2) 241]

Newell, A. C., T. Passot and M. Souli, 1989, The phase diffusion and mean drift equations for convection at finite Rayleigh numbers in large containers I. private communication. [(v2) 243]

Newell, A. C., and J. A. Whitehead, 1969, Finite bandwidth, finite amplitude convection, J. Fluid Mech. **38**, 279. [(v2) 227]

Norman, B., K. Moller, R. Ek and G. G. Duffy, 1977, Hydrodynamics of papermaking fibers in water suspension, paper presented to the Sixth Fundamental Research Symposium, B. P., B. M. A., Oxford, England, September. [(v1) 8]

Northrup, E., 1912, A photographic study of vortex rings in liquids, Nature **88**, 463. [(v2) 308, 313]

Nunziato, J., S. Passman, C. Givler, D. MacTigue and J. Brady, 1986, Continuum theories for suspensions. In *Advancements in Aerodynamics, Fluid Mechanics and Hydraulics* (Proc. ASCE Special Conf., Minneapolis, Minn.) Ed. R. Arndt, A. Stefan, C. Farrell and S. N. Peterson, 465. [(v2) 352]

O'Brien, V., 1961, Why raindrops break up – vortex instability, J. of Meteorology **18**, 549. [(v2) 290, 292, 308, 313]

O'Brien, V., 1985, On spheroidal gravity-driven vortices, personal communication. [(v2) 309]

Oliemans, R. V. A., 1986, *The Lubricating Film Model for Core-Annular Flow*, Delft University Press. [(v1) 5, 338; (v2) 1, 195-196, 200]

Oliemans, R. V. A., and G. Ooms, 1986, Core-annular flow of oil and water through a pipeline, *Multiphase Science and Technology*, Vol.2, eds. G. F. Hewitt, J. M. Delhaye and N. Zuber, Hemisphere Publishing Corporation. [(v1) 4, 5; (v2) 1, 19]

Oliemans, R. V. A., G. Ooms, H. L. Wu and A. Duÿvestin, 1985, Core-annular oil/water flow: the turbulent-lubricating model and measurements in a 2 in. pipe loop, presented at the Society of

Petroleum Engineers 1985 Middle East Oil Technical Conference and Exhibition, held in Bahrain, March 11 - 14, 1985. [(v2) 7, 10]

Onishi, Y., 1984, 14th International Symposium on Rarefied Gas Dynamics, Tsukuba, Japan. [(v1) 381]

Ooms, G., 1971, Fluid-mechanical studies of core-annular flow, Ph. D. Thesis, Delft University of Technology. [(v2) 12]

Ooms, G., A. Segal, S. Y. Cheung and R. V. A. Oliemans, 1985, Propagation of long waves of finite amplitude at the interface of two viscous fluids, Int. J. Multiphase Flow, 481. [(v1) 5, 271, 338; (v2) 11]

Ooms, G., A. Segal, A. J. Van der Wees, R. Meerhoff and R. V. A. Oliemans, 1984, A theoretical model for core-annular flow of a very viscous oil core and a water annulus through a horizontal pipe, Int. J. Multiphase Flow 10, 41. [(v1) 338; (v2) 10]

Orszag, S. A., 1971, Accurate solutions of the Orr-Sommerfeld stability equation, J. Fluid Mech. 50, 689. [(v1) 191, 303, 342, 350; (v2) 28]

Orszag, S. A., and L. C. Kells, 1980, Transition to turbulence in plane Poiseuille and Couette flow, J. Fluid Mech. 96, 159. [(v2) 28]

Pao, Y. P., 1971, Application of kinetic theory to the problem of evaporation and condensation, Phys. Fluids 14, 306. [(v1) 381]

Papageorgiou, D. T., C. Maldarelli and D. S. Rumschitzki, 1990, Nonlinear interfacial stability of core annular film flows, Phys. Fluids A 2, 340. [(v2) 11, 98, 261, 268-271]

Papageorgiou, D. T., and Y. S. Smyrlis, 1991, The route to chaos for the Kuramoto-Sivashinsky equation, Theoret. Comput. Fluid Dynamics 3, 15. [(v1) 337]

Palmquist, K. E., and S. F. Kistler, 1992, Formation of cusped interface by liquid film plunging into a pool [(v1) 140, 141, 153, 157, 162, 169]

Passman, S., J. W. Nunziato and E. K. Walsh, 1984, A theory for multi-phase mixtures. In *Rational Thermodynamics*. Ed. C. Truesdell, Springer Verlag, New York, 286. [(v2) 352]

Pearlstein, A. J., 1985, On the two-dimensionality of the critical disturbances for stratified viscous plane parallel shear flows, Phys. Fluids 28, 751. [(v1) 276]

Pearlstein, A. J., 1987, American Physical Society Division of Fluid Dynamics Annual Meeting, Paper HC6; Bull. Am. Phys. Soc. 32, 2087. [(v1) 276]

Pedley, T. J., 1968, The toroidal bubble, J. Fluid Mech. 32, 97. [(v2) 290]

Pedley, T. J., 1969, On the instability of viscous flow in a rapidly rotating pipe, J. Fluid Mech. 35, 97. [(v2) 93]

Pekeris, C. L., and B. Shkoller, 1967, Stability of plane Poiseuille flow to periodic disturbances of finite amplitude in the vicinity of the neutral curve, J. Fluid Mech. 29, 31. [(v1) 346]

Pellew, A. and R. V. Southwell, 1940, On maintained convective motion in a fluid heated from below, Proc. Roy. Soc. A 176, 312. [(v1) 171]

Pérez-Garcia, C., J. Pantaloni, R. Occelli and P. Cerisier, 1985, Linear analysis of surface deflection in Bénard-Marangoni instability, J. Phys. (Paris) **46**, 2047. [(v1) 173]

Petrie, C. J. S., and M. M. Denn, 1976, Instabilities in polymer processing, AIChE J. **22**, 209. [(v1) 375]

Philippoff, W., 1937, The viscosity characteristics of rubber solutions, Rubber Chem. Tech. **10**, 76. [(v1) 144]

Pipkin, A., and R. Tanner, 1972, A survey of theory and experiment in viscometric flows of viscoelastic liquids, Mech. Today **1**, 262. [(v1) 15]

Plateau, J. A. F., 1863, Experimental and theoretical researches on the figures of equilibrium of a rotating liquid mass withdrawn from the action of gravity, Annual Report of the Board of Regents of the Smithsonian Institution, Washington, DC, 270. [(v1) 67, 90]

Plateau, J. A. F., 1873, *Statique Expérimentale et Théorique des Liquides*, Gauthier-Villars. [(v1) 27]

Plesset, M. S., and S. A. Zwick, 1954, The growth of vapor bubbles in superheated liquids, J. Applied Phys. **25**(4), 493. [(v1) 381]

Porteous, K. C., and M. M. Denn, 1972, Linear stability of plane Poiseuille flow of viscoelastic liquids, Trans. Soc. Rheol. **16**, 295. [(v1) 361]

Power, H., and M. Villegas, 1990, Viscous-inviscid model for the linear stability of core-annular flow, ZAMP **41**, 1. [(v2) 13]

Pozrikidis, C., 1990, The instability of a moving viscous drop, J. Fluid Mech., **210**, 1. [(v2) 295, 328, 361, 372]

Preziosi, L., 1986, Ph. D. thesis, University of Minnesota. [(v1) 71]

Preziosi, L., K. Chen and D. D. Joseph, 1989, Lubricated pipelining: stability of core-annular flow, J. Fluid Mech. **201**, 323. [(v2) 13, 16-49, 59]

Preziosi, L., and D. D. Joseph, 1988, The run-off condition for coating and rimming flows, J. Fluid Mech. **187**, 99. [(v1) 76, 85, 90, 92]

Preziosi, L., and F. Rosso, 1991, Interfacial stability in a two-layer shearing flow between sliding pipes, Eur. J. Mech., B Fluids **10** (3), 269. [(v2) 14]

Princen, H. M., I. Y. Z. Zia and S. G. Mason, 1967, Measurements of interfacial tension from the shape of a rotating liquid drop, J. Colloid Interface Sci. **23**, 99. [(v1) 78]

Pukhnachov, V. V., 1991, Mathematical model of natural convection under low gravity, University of Minnesota Institute for Mathematics and Its Applications Preprint 796. [(v2) 352]

Pumir, A., P. Manneville and Y. Pomeau, 1983, On solitary waves running down an inclined plane, J. Fluid Mech. **135**, 27. [(v1) 333]

Quincke, G., 1902, Die Oberfächenspannung an der Grenze von Alkohol mit wässerigen Salzlösungen, Ann. Phys. **9**, 1. [(v2) 338]

Raitum, U. E., 1978, The extension of extremal problems connected with a linear elliptic equation, Sov. Math. Dokl. **19**, 1342. [(v1) 42]

Raitum, U. E., 1979, On optimal control problems for linear elliptic equations, Sov. Math. Dokl. **20**, 129. [(v1) 42]

Rallison, J.M. and A. Acrivos, 1978, A numerical study of the deformation and burst of a viscous drop in an extensional flow, J. Fluid Mech. **89**, 191. [(v1) 143]

Ramamurthy, A. V., 1986, Wall slip in viscous fluids and influence of materials of construction, J. Rheol. **30** (2), 337. [(v1) 377]

Ranger, K. B., and A. M. J. Davis, 1979, Steady pressure driven two-phase stratified laminar flow through a pipe, Can. J. Chem. Eng. **57**, 688. [(v2) 4]

Rasenat, S., 1987, Konvektion in zwei Geschichteten Flüssigkeiten, Diplomarbeit, University of Bayreuth. [(v1) 175]

Rayleigh, Lord, 1879, On the instability of jets, Proc. Roy. London Math. Soc., **10**, 4. [(v2) 38, 48]

Rayleigh, Lord, 1914, The equilibrium of revolving liquid under capillary force, Phil Mag. **28**, 161. [(v1) 72]

Rayleigh, Lord, 1916, On convection currents in a horizontal layer of fluid, when the higher temperature is on the under side, Phil. Mag. (6) **32**, 529. [(v1) 171]

Reid, W. H., and D. L. Harris, 1958, Some further results on the Bénard problem, Phys. Fluids **1**, 102. [(v1) 171, 192, 194]

Renardy, M., and D. D. Joseph, 1986, Hopf bifurcation in two-component flows, SIAM J. Math. Anal. **17**, 894. [(v1) 183, 245; (v2) 19, 51]

Renardy, M., and Y. Renardy, 1986, Linear stability of plane Couette flow of an upper convected Maxwell fluid, J. Non-Newtonian Fluid Mech. **22**, 23. [(v1) 363, 364]

Renardy, M., and Y. Renardy, 1988, Bifurcating solutions in a two-layer Bénard problem, Physica D **32**, 227. [(v1) 201-202, 236-267, 269]

Renardy, M., and Y. Renardy, 1992, Pattern selection in the Bénard problem for a viscoelastic fluid, Zeitschrift für Angewandte Mathematik und Physik **43**(1), 154. [(v1) 171]

Renardy, Y., 1985, Instability at the interface between two shearing fluids in a channel, Phys. Fluids **28**, 3441. [(v1) 270, 304, 320; (v2) 80]

Renardy, Y., 1986, Interfacial stability in a two-layer Bénard problem, Phys. Fluids **29**, 356. [(v1) 222-235]

Renardy, Y., 1987 a, The thin-layer effect and interfacial stability in a two-layer Couette flow with similar liquids, Phys. Fluids **30**, 1627. [(v1) 304]

Renardy, Y., 1987 b, Viscosity and density stratification in vertical plane Poiseuille flow, Phys. Fluids **30**, 1638. [(v2) 14, 15, 115-116, 130, 138, 145, 167, 171]

Renardy, Y., 1988 a, Instabilities in steady flows of two fluids, Rocky Mount. J. Math. **18**, Spring, 455. [(v1) 174]

Renardy, Y., 1988 b, Stability of the interface in two-layer Couette flow of upper convected Maxwell liquids, J. Non-Newt. Fluid Mech. **28**, 99. [(v1) 271, 359-375]

Renardy, Y., 1989, Weakly nonlinear behavior of periodic disturbances in two-layer Couette-Poiseuille flow, Phys. Fluids A **1**, 1666. [(v1) 270, 271, 338-359; (v2) 51]

Renardy, Y., and D. D. Joseph, 1985 a, Two-fluid Couette flow between concentric cylinders, J. Fluid Mech. **150**,381. [(v1) 31, 114, 117, 118, 269, 270, 296, 303; (v2) 36]

Renardy, Y., and D. D. Joseph, 1985 b, Oscillatory instability in a two-fluid Bénard problem, Phys. Fluids **28**, 788. [(v1) 173, 174, 269, 296]

Renardy, Y., and M. Renardy, 1985, Perturbation analysis of steady and oscillatory onset in a Bénard problem with two similar liquids, Phys. Fluids **28**,2699. [(v1) 203-235, 269]

Reynolds, W. C., and M. C. Potter, 1967, Finite-amplitude instability of parallel shear flows, J. Fluid Mech. **27**, 465. [(v1) 346-347]

Richardson, S., 1968, Two-dimensional bubbles in slow viscous flows, J. Fluid Mech. **33**, 475. [(v1) 141, 142, 144]

Richter, F. M., and C. E. Johnson, 1974, Stability of a chemically layered mantle, J. Geophys. Res. **79**, 1635. [(v1) 15]

Roberts, M., J. W. Swift and D. H. Wagner, 1986, The Hopf bifurcation on a hexagonal lattice, *Multiparameter Bifurcation Theory*, AMS Series: Contemporary Mathematics, eds. M. Golubitsky and J. M. Guckenheimer, Vol. 56, 283. [(v1) 176, 177, 239, 242, 243]

Robinson, I., 1976, An algorithm for automatic integration using the adaptive Gaussian technique, Aust. Comput. J. **8**, 106. [(v2) 107]

Romanov, V. A., 1973, Stability of plane parallel Couette flow, Func. Anal. and Its Applic. **7**, 137. [(v1) 277]

Rosenthal, D. K., 1962, The shape and stability of a bubble at the axis of a rotating liquid, J. Fluid Mech. **12**, 358. [(v1) 72, 78]

Ross, D. K., 1968, The shape and energy of a revolving liquid mass held together by surface tension, Austral. J. Phys. **21**, 823. [(v1) 72]

Rothman, D. H., and J. M. Keller, 1988, Immiscible cellular-automaton fluids, J. Statistical Physics **52** (3/4), 1119. [(v1) 6]

Rowell, R. L., S. R. Vasconcellos, R. J. Sala and R. S. Farinato, 1981, Surfactant effectiveness on coal-oil mixture stability measured with a sedimentation column device, Ind. Eng. Chem. Process Res. Dev., **20**, 283. [(v1) 9]

Ruelle, D., 1973, Bifurcations in the presence of a symmetry group, Arch. Rat. Mech. Anal. **51**, 136. [(v1) 174, 178]

Rumscheidt, F.D. and S. G. Mason, 1961, Particle motion in sheared suspensions. XII. Deformation and burst of fluid drops in shear and hyperbolic flow, J. Colloid Sci. **16**, 238. [(v1) 143]

Russell, T. W. F., and M. E. Charles, 1959, The effect of the less viscous liquid in the laminar flow of two immiscible liquids, Can. J. Chem. Eng. **39**, 18. [(v1) 4, 6, 33; (v2) 4, 6, 187]

Russell, T. W. F., G. W. Hodgson and G. W. Govier, 1959, Horizontal pipeline flow of mixtures of oil and water, Can. J. Chem. Eng. **37**, 9. [(v1) 3, 4; (v2) 4, 6, 196]

Russo, M. J. and P. H. Steen, 1986, Instability of rotund capillary bridges to general disturbances, experiment and theory, J. Colloid Interface Sci. **113**, 154. [(v1) 90]

Russo, M. J., and P. H. Steen, 1989, Shear stabilization of the capillary breakup of a cylindrical interface, Phys. Fluids A **1**(12), 1926. [(v2) 13]

Saffman, P., and G. I. Taylor, 1958, The penetration of a fluid into a porous medium or a Hele-Shaw cell containing a more viscous liquid, Proc. Roy. Soc., **A245**, 312. [(v1) 2]

Salwen, H., F. W. Cotton and C. E. Grosch, 1980, Linear stability of Poiseuille flow in a circular pipe, J. Fluid Mech. **98**, 273. [(v2) 30]

Salwen, H. and C. E. Grosch, 1972, The stability of Poiseuille flow in a pipe of circular cross-section, J. Fluid Mech. **54**, 93. [(v2) 30]

Saric, W. S., and B. W. Marshall, 1971, An experimental investigation of the stability of a thin liquid layer adjacent to a supersonic stream, A. I. A. A. J. **9**, 1546. [(v1) 338]

Schechter, R. S., M. G. Velarde and J. K. Platten, 1974, The two component Bénard problem, Adv. Chem. Phys. **26**, 265. [(v1) 171]

Schlichting, H., 1960, *Boundary Layer Theory*, Pergamon Press, New York. [(v2) 192]

Schrage, R., 1953, *A Theoretical Study of Interphase Mass Transfer*, Columbia University Press. [(v1) 381]

Schreiber, H. P., and S. H. Storey, 1965, Molecular fractionation in capillary flow of polymer fluids, Polymer Letters **3**, 723. [(v1) 376]

Schreiber, H. P., S. H. Storey and E. B. Bagley, 1966, Molecular fractionation in the flow of polymeric fluids, Transactions of the Soc. of Rheol. **10**:1, 275. [(v1) 376]

Schrenk, W. J., and T. Alfrey, 1978, Coextruded multilayer polymer films and sheets, in *Polymer Blends*, vol. 2, D. R. Paul and N. Seymour, eds., Academic Press, New York, 129. [(v1) 11]

Schubert, G., and J. Strauss, 1980, Gravitational stability of water over steam in vapor-dominated geothermal systems, J. Geophys. Res. **85**, 6505. [(v1) 378, 399]

Schulten, A., D. G. M. Anderson and R. G. Gordon, 1979, An algorithm for the evaluation of the complex Airy function, J. Comp. Phys. **31**, 60. [(v2) 107]

Scriven, L. E., and C. V. Sternling, 1964, On cellular convection driven by surface tension gradients: effects of mean surface tension and surface viscosity, J. Fluid Mech. **19**, 321. [(v1) 171]

Segel, L. A., 1969, Distant side-walls cause slow amplitude modulation of cellular convection, J. Fluid Mech. **38**, 203. [(v2) 227]

Segré, G., and A. Silberberg, 1962, Behavior of macroscopic rigid spheres in Poiseuille flow. Part 1: Determination of local concentration by statistical analysis of particle passages through crossed light beams, J. Fluid Mech. **14**, 115. Part 2: Experimental results and interpretation, J. Fluid Mech. **14**, 136. [(v1) 6]

Segur, J. B., 1953, Physical properties of glycerol and its solutions, in *Glycerol*, eds. C. S. Miner and N. N. Dalton, Reinhold Publishing Corp. [(v2) 347-348, 357, 365-366]

Sen, P. K., and D. Venkateswarlu, 1983, On the stability of plane Poiseuille flow to finite-amplitude disturbances, considering the higher-order Landau coefficients, J. Fluid Mech. **133**, 179. [(v2) 240]

Shankar, P. N., and M. D. Deshpande, 1990, On the temperature distribution in liquid-vapor phase change between plane liquid surfaces, Phys. Fluids A **2**(6), 1030. [(v1) 381]

Shertok, J. T., 1976, Velocity profiles in core-annular flow using a laser-Doppler velocimeter, Ph.D. thesis, Princeton University. [(v2) 8]

Sherwood, J.D., 1981, Spindle shaped drops in a viscous extensional flow, Math. Proc. Cambridge Phil. Soc. **90**, 529. [(v1) 143]

Shirtcliffe, T. G. L., 1969, An experimental investigation of thermosolutal convection at marginal stability, J. Fluid Mech. **35** (4), 677. [(v1) 172]

Shirtcliffe, T. G. L., 1973, Transport and profile measurements of the diffusive interface in double diffusive convection with similar diffusivities, J. Fluid Mech. **57**, 27. [(v1) 171]

Shlang, T., G. I. Sivashinsky, A. J. Babchin and A. L. Frenkel, 1985, Irregular wavy flow due to viscous stratification, Le Journal de Physique **46**, 863. [(v1) 270, 324]

Shlang, T., and G. I. Sivashinsky, 1982, Irregular flow of a liquid film down a vertical column, J. Physique **43**, 459. [(v1) 324]

Sijbrand, J., 1981, *Studies in Non-Linear Stability and Bifurcation Theory*, Ph. D. thesis, Rijksuniversiteit, Utrecht. [(v1) 176]

Simons, G. A., and R. S. Larson, 1974, Formation of vortex rings in a stratified atmosphere, Phys. Fluids **17**, 8. [(v2) 289]

Sinclair, A. R., 1970, Rheology of viscous fracturing fluids, J. Petroleum Technology, June, 711. [(v2) 196, 200]

Singh, P. and D. D. Joseph, 1989, Autoregressive methods for chaos on binary sequences for the Lorenz attractor, Phys. Letters A **135**, 247. [(v1) 133, 135, 138, 139]

Singh, P., P. Mohr and D. D. Joseph, 1992, Application of binary sequences to problems of chaos, Video Journal of Engineering Research, to appear. [(v1) 133]

Sisson, W. A., and F. F. Morehead, 1953, The skin effect in crimped rayon, Text. Res. J. **23**, 152. [(v1) 11]

Sivashinsky, G. I., 1977, Nonlinear analysis of hydrodynamic instability in laminar flames – I. Derivation of basic equations, Acta Astronaut. **4**, 1177. [(v1) 333]

424   References

Sivashinsky, G. I., 1983, Instabilities, pattern formation, and turbulence in planes, Ann. Rev. Fluid Mech. **15**, 179. [(v1) 271, 333, 334]

Sivashinsky, G. I., and D. M. Michelson, 1980, On irregular wavy flow of a liquid film down a vertical plane, Prog. Theor. Phys. **63**, 2112. [(v1) 324]

De Smedt, C., and S. Nam, 1987, The processing benefits of fluoroelastomer application in LLDPE, Plastics and Rubber Process. Appl. **8**, 11. [(v1) 376]

Smith, J. M., and H. C. Van Ness, 1975, *Introduction to Chemical Engineering Thermodynamics*, 3rd ed., McGraw-Hill. [(v2) 354]

Smith, M. K., 1989, The axisymmetric long-wave instability of a concentric two-phase pipe flow, Phys. Fluids A **1**, 494. [(v1) 269, 271; (v2) 15]

Smith, M. K., 1990 a, The mechanism for the long-wave instability in thin liquid films, J. Fluid Mech. **217**, 469. [(v1) 271]

Smith, M. K., 1990 b, The long-wave instability in heated or cooled inclined liquid layers, J. Fluid Mech. **219**, 337. [(v1) 271]

Smith, M. K., and S. H. Davis, 1982, The instability of sheared liquid layers, J. Fluid Mech. **121**, 187. [(v1) 338]

Smith, M. K., and S. H. Davis, 1983, Instabilities of dynamic thermocapillary liquid layers, Part 2. Surface-wave instabilities, J. Fluid Mech. **132**, 145. [(v1) 171]

Smith, P. G., M. Van Den Ven and S. G. Mason, 1981, The transient interfacial tension between two miscible fluids, J. Colloid Interface Sci. **80**(1), 302. [(v2) 338-342]

Smyrlis, Y. S., and D. T. Papageorgiou, 1991, Predicting chaos for infinite dimensional dynamical systems: the Kuramoto-Sivashinsky equation, a case study, Proc. Natl. Acad. Sci. USA Applied Mathematics **88**. [(v1) 337]

Sone, Y., and Y. Onishi, 1978, Kinetic theory of evaporation and condensation - hydrodynamic equation and slip boundary condition, J. Phys. Soc. Jpn. **44**, 1981; **45** 1054. [(v1) 381]

Sotin, C., and E. M. Parmentier, 1989, On the stability of a fluid layer containing a univariant phase transition: application to planetary interiors, Phys. Earth and Planetary Interiors **55**, 10. [(v1) 378, 398]

Southern, J. H., and R. L Ballman, 1973, Stratified bicomponent flow of polymer melts in a tube, Appl. Polymer Symp. **20**, 175. [(v1) 10, 11, 33, 38]

Sparrow E. M., R. B. Husar and R. J. Goldstein, 1970, Observations and other characteristics of thermals, J. Fluid Mech. **41**, 793. [(v2) 363]

Spindler, B., 1982, Linear stability of liquid film with interfacial phase change, Int. J. Heat Mass Transfer **25**, 161. [(v1) 379]

Squire, H. B., 1933, On the stability of three-dimensional disturbances of viscous flow between parallel walls, Proc. Roy. Soc. A **142**, 621. [(v1) 276, 278]

Stein, M. H., 1978, Concentric annular oil-water flow, Ph. D. Thesis, Purdue University. [(v2) 7]

Sternling, C. V., and L. E. Scriven, 1959, Interfacial turbulence: hydrodynamic instability and the Marangoni effect, AIChE J. **5**(4), 514. [(v1) 173]

Stewartson, K., and J. T. Stuart, 1971, A nonlinear instability theory for wave system in plane Poiseuille flow, J. Fluid Mech. **48**, 529. [(v1) 356; (v2) 227, 235]

Stockman, H. W., C. T. Stockman and C. R. Carrigan, 1990, Modelling viscous segregation in immiscible fluids using lattice-gas automata, Nature **348**, 523. [(v1) 6]

Stuart, J. T., and R. DiPrima, 1978, The Eckhaus and Benjamin-Feir resonance mechanisms, Proc. Roy. Soc. Lond. A **362**, 27. [(v2) 240]

Stuke, B., 1954, Zur Bildung von Wirbelringen, Zeitschrift für Physik **137**, 376. [(v2) 308-309]

Sturges, L., and D. D. Joseph, 1980, A normal stress amplifier for the second normal stress difference, J. Non-Newtonian Fluid Mech. **6**, 325. [(v1) 15]

Tartar, L., 1975, Problèmes de contrôle des coefficients dans des équations aux dérivées partielles, Lecture Notes in Economics and Mathematical Systems, Springer **107**, 420. [(v1) 42]

Taylor, G.I., 1931, Effect of variation in density on the stability of superposed streams of fluid, Proc. R. Soc. Lond. A **132**, 499. [(v1) 279]

Taylor, G.I., 1934, The formation of emulsions in definable fields of flow, Proc. Roy. Soc. London A **146**, 501. [(v1) 143]

Taylor, G.I., 1964, Conical free surfaces and fluid interfaces, *Proc. 11th Int. Congr. Appl. Mech.*, Munich. [(v1) 143]

Taylor, T. D., and A. Acrivos, 1964, On the deformation and drag of a falling viscous drop at low Reynolds number, J. Fluid Mech. **18** 466. [(v2) 295]

Temam, R., 1979, *Navier-Stokes Equations*, North-Holland. [(v1) 42, 205]

Than, P. T., L. Preziosi, D. D. Joseph and M. Arney, 1988, Measurement of interfacial tension between immiscible liquids with the spinning rod tensiometer, J. Colloid and Interface Sci. **124**(2), 552. [(v1) 79]

Than, P. T., F. Rosso and D. D. Joseph, 1987, Instability of Poiseuille flow of two immiscible liquids with different viscosities in a channel, Int. J. Eng. Sci. **25**, 189. [(v1) 269; (v2) 15]

Thomson, J. J., and H. F. Newall, 1885, On the formation of vortex rings by drops falling into liquids, and some allied phenomena, Proc. Roy.

426   References

Soc. Lond. **39**, 417. [(v2) 289, 292, 313, 334, 336, 338]

Tipman, E. and G. W. Hodgson, 1956, Sedimentation in emulsions of water in petroleum, J. Petroleum Tech., September, 91. [(v2) 155]

Tlapa, G., and B. Bernstein, 1970, Stability of a relaxation-type viscoelastic fluid with slight elasticity, Phys. Fluids **13**, 565. [(v1) 363]

Topper, J., and T. Kawahara, 1978, Approximate equations for long nonlinear waves on a viscous fluid, J. Phys. Soc. Japan **44**, 663. [(v1) 270]

Tordella, J. P., 1958, An instability in the flow of molten polymers, Rheol. Acta **1**, 216. [(v1) 375]

Truesdell, C. and W. Noll, 1965, *The Non-Linear Field Theories of Mechanics*, Handbuch der Physik, ed. S. Flügge, **3**, Springer-Verlag. [(v2) 345]

Tryggvason, G., and S. O. Unverdi, 1990, Computations of three - dimensional Rayleigh-Taylor instability, Phys. Fluids A **2**(5), 656. [(v1) 272]

Tryggvason, G., and S. O. Unverdi, 1991, Full numerical simulations of multifluid flows, Phys. Fluids A **3**(5), 1455. [(v1) 272; (v2) 294]

Turner, J. S., 1979, *Buoyancy Effects in Fluids*, Cambridge University Press, Great Britain. [(v1) 172; (v2) 289, 326, 332]

Valenzuela, G. R., 1976, The growth of gravity-capillary waves in a coupled shear flow, J. Fluid Mech. **76** (2), 229. [(v1) 270]

Van der Waals, M., 1895, Théorie thermodynamique de la capillarité dans l'hypothèse d'une variation de densité, Arch. Neerl. Sci. Ex. Nat. **28**, 121. [(v2) 339, 344]

Van Dyke, M., 1975, *Perturbation Methods in Fluid Mechanics*, The Parabolic Press, Stanford, California. [(v2) 99, 289]

Van Dyke, M., 1982, *An Album of Fluid Motion*, The Parabolic Press. [(v1) 171]

Veal, C. J., D. R. Wall and A. J. Groszek, 1979, Stable coal/fuel-oil dispersions, Proc. 2nd Int. Symp. on Coal-Oil Mixture Combustion, Danvers, Massachusetts. [(v1) 9]

Wahal, S., and A. Bose, 1988, Rayleigh-Benard and interfacial instabilities in two immiscible liquid layers, Phys. Fluids **31**, 3502. [(v1) 173]

Walmsley, M. R., and G. G. Duffy, 1987, Hydraulic transport of solid particles and capsulized materials in pipelines with friction losses lower than water, J. of Pipelines **6**, 33. [(v1) 8]

Walters, J. K., and J. F. Davidson, 1963, The initial motion of a gas bubble formed in an inviscid liquid, Part 2. The three-dimensional bubble and the toroidal bubble, J. Fluid Mech. **17**, 321. [(v2) 290, 316]

Walters, K., 1984, Some modern developments in non-Newtonian fluid mechanics, *Advances in Rheology*, Proc. IX Intl. Congress on Rheology, Mexico, ed. B. Mena, A. Garcia-Rejon, C. Rangel-Nafaile, 31. [(v1) 13]

Walters, K., 1985, Overview of macroscopic viscoelastic flow, *Viscoelasticity and Rheology*, ed. A. S. Lodge, M. Renardy and J. A. Nohel, Acad. Press. [(v1) 13]

Wang, C. K., J. J. Seaborg and S. P. Lin, 1978, Instability of multi-layered liquid films, Phys. Fluids **21**, 1669. [(v1) 269, 272]

Wang, T. G., R. Tagg, L. Cammack and A. Croonquist, 1981, Non-axisymmetric shapes of a rotating drop in an immiscible system, in *Proc. 2nd Int. Colloq. on Drops and Bubbles* (ed. D. H. LeCroisette), pp. 203 - 213, NASA-JPL. [(v1) 90]

Warshay, M.E., E. Bogusz, M. Johnson, and R. C. Kintner, 1959, Ultimate velocity of drops in stationary liquid media, Can. J. Chem. Engng. **37**, 29. [(v1) 144]

Waters, N. D., 1983, The stability of two stratified "power-law" liquids in Couette flow, J. Non-Newtonian Fluid Mech. **12**, 85. [(v1) 269, 359, 360]

Waters, N. D., and A. M. Keeley, 1987, The stability of two stratified Non-Newtonian liquids in Couette flow, **24**, 161. [(v1) 269, 359, 360]

Wedemeyer, E. H., 1964, The unsteady flow within a spinning cylinder, J. Fluid Mech. **20**, 383. [(v1) 16]

Weinstein, S. J., and M. R. Kurz, 1991, Long-wavelength instabilities in three-layer flow down an incline, Phys. Fluids A **3**(11), 2680. [(v1) 272, 279]

White, J. L., and B.-L. Lee, 1975, Theory of interface distortion in stratified two-phase flow, Trans. Soc. Rheol. **19**, 457. [(v1) 11]

Whitham, G. B., 1974, *Linear and Nonlinear Waves*, John Wiley and Sons, New York. [(v1) 332, 333]

Wilkinson, J. H., 1977, Some recent advances in numerical linear algebra, in *The state of the art in numerical analysis*, ed. D. Jacobs, Academic Press. [(v2) 242]

Williams, M. B., and S. H. Davis, 1982, Nonlinear theory of film rupture, J. Colloid Interface Sci. **90**, 220. [(v2) 264-267]

Williams, P. R. and R. W. Williams, 1985, On the planar extensional viscosity of mobile liquids, J. Non-Newtonian Fluid Mech. **19**, 53. [(v1) 13, 14]

Wollkind, D. J., and J. I. D. Alexander, 1982, Kelvin-Helmholtz instability in a layered Newtonian fluid model of the geological phenomenon of rock folding, SIAM J. Appl. Math. **42** (6), 1276. [(v1) 16]

Wooding, R. A., 1969, Growth of fingers at an unstable diffusing interface in a porous medium or Hele Shaw cell, J. Fluid Mech. **39**, 477. [(v2) 379, 387]

Xu, J.-J., and S. H. Davis, 1985, Instability of capillary jets with thermocapillarity, J. Fluid Mech. **161**, 1. [(v2) 13]

Yakubovich, V. A., and V. M. Starzhinskii, 1975, *Linear Differential Equations with Periodic Coefficients 1*, Halsted, Jerusalem. [(v1) 204, 220]

Yiantsios, S. G., 1988, Ph.D. thesis, University of California, Davis. [(v1) 302]

Yiantsios, S. G., and B. G. Higgins, 1988, Linear stability of plane Poiseuille flow of two superposed fluids, Phys. Fluids. **31**, 3225. (Erratum, Phys. Fluids A **1**, 897) [(v1) 269, 270, 276, 278, 279, 287, 296, 300, 302, 350]

Yiantsios, S. G., and B. G. Higgins, 1989, Rayleigh-Taylor instability in thin viscous films, Phys. Fluids A **1** (9), 1484. [(v1) 272]

Yih, C.-S., 1960, Instability of a rotating liquid film with a free surface, Proc. R. Soc. Lond. A **258**, 63. [(v1) 66]

Yih, C.-S., 1963, Stability of liquid flow down an inclined plane, Phys. Fluids **6**, 321. [(v1) 378]

Yih, C.-S., 1967, Instability due to viscosity stratification, J. Fluid Mech. **26**,337. [(v1) 173, 193, 196, 268, 269, 279-287, 320, 329, 331, 338, 339, 350, 360; (v2) 31, 132]

Yih, C.-S., 1986, Instability resulting from stratification in thermal conductivity, Phys. Fluids **29**, 1769. [(v1) 174]

Yortsos, Y. C., and M. Zeybek, 1988, Dispersion driven instability in miscible displacement in porous media, Phys. Fluids **31**(12), 3511. [(v2) 379]

Yu, H. S., and C. D. Han, 1973, Stratified two-phase flow of molten polymers, J. Appl. Polym. Sci. **17**, 1203. [(v1) 11]

Yu, H. S., and E. M. Sparrow, 1967, Stratified laminar flow in ducts of arbitrary shape, AICHE J. **13**(1),10. [(v2) 3]

Yu, H. S., and E. M. Sparrow, 1969, Experiments on two-component stratified flow in a horizontal duct, Trans. ASME C:J. Heat Transfer, **91**, 51. [(v1) 5, 339]

Zana, E. and L. G. Leal, 1978, The dynamics and dissolution of gas bubbles in a viscoelastic fluid, Int. J. Multiphase Flow **4**, 237. [(v1) 144]

Zeren, R. W., and W. C. Reynolds, 1972, Thermal instabilities in two-fluid horizontal layers, J. Fluid Mech. **53**, 305. [(v1) 173, 192]

Zondek, B., and L. H. Thomas, 1953, Stability of a limiting case of plane Couette flow, Phys. Rev. (2) **90**, 738. [(v1) 294]

Index

The index entry "(vm) n" refers to "volume m, page n".